D1325447

DATE DUE FOR RETURN

CIRCUMSTELLAR MATTER

INTERNATIONAL ASTRONOMICAL UNION

UNION ASTRONOMIQUE INTERNATIONALE

CIRCUMSTELLAR MATTER

PROCEEDINGS OF THE 122ND SYMPOSIUM OF THE
INTERNATIONAL ASTRONOMICAL UNION
HELD IN HEIDELBERG, F.R.G.,
JUNE 23-27, 1986

EDITED BY

I. APPENZELLER

Landessternwarte Heidelberg-Königstuhl, F.R.G.

and

C. JORDAN

*Department of Theoretical Physics,
University of Oxford, U.K.*

D. REIDEL PUBLISHING COMPANY

A MEMBER OF THE KLUWER ACADEMIC PUBLISHERS GROUP

DORDRECHT / BOSTON / LANCASTER / TOKYO

Library of Congress Cataloging in Publication Data

International Astronomical Union. Symposium (122nd: 1986: Heidelberg, Germany)
    Circumstellar matter.

    At head of title: International Astronomical Union = Union astronomique interna-
tionale.
    Includes indexes.
    Circumstellar Matter—Congresses. 2. Stars—Atmospheres—Congresses.
3. Interstellar matter—Congresses. I. Appenzeller, I. (Immo), 1940–
II. Jordan, C. III. Title.
QB792.I58   1986   523.8′6   87–9663
ISBN 90–277–2511–X
ISBN 90–277–2512–8 (pbk.)

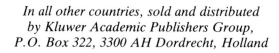

Published on behalf of
the International Astronomical Union
by
D. Reidel Publishing Company, P.O. Box 17, 3300 AA Dordrecht, Holland

Sold and distributed in the U.S.A. and Canada
by Kluwer Academic Publishers,
101 Philip Drive, Assinippi Park, Norwell, MA 02061, U.S.A.

In all other countries, sold and distributed
by Kluwer Academic Publishers Group,
P.O. Box 322, 3300 AH Dordrecht, Holland

Printed in The Netherlands

TABLE OF CONTENTS

CIRCUMSTELLAR SHELLS AND ENVELOPES

CIRCUMSTELLAR DUST AND CHEMISTRY

FUTURE PLANS

CONCLUDING REMARKS

PREFACE

The objective of this meeting was to bring together colleagues from different branches of observational astronomy and
theoretical astrophysics to discuss and analyse the rapid
progress in our knowledge and understanding of the matter
surrounding stars, streaming off stellar surfaces, or falling onto stars. The meeting was sponsored by IAU Commis-
sions 36 (Theory of Stellar Atmospheres), 29 (Stellar Spectra), and 34 (Interstellar Matter). There were two special
reasons for organizing this meeting at Heidelberg in 1986:
During this year the University of Heidelberg celebrated
its 600th anniversary and the IAU symposium joined the many
scientific events accompanying this celebration. Secondly,
the year 1986 also marked the conclusion of a special co-
operative research project ("Sonderforschungsbereich") in
astrophysics at Heidelberg, a major part of which had been
devoted to the physics of circumstellar matter.

The main topics discussed at this meeting were:
(1) circumstellar matter, bipolar flows, and jets from
young stars and protostars; (2) circumstellar envelopes of
evolved stars; (3) stellar coronae; (4) stellar winds from
hot and cool stars; (5) dust formation and circumstellar
chemistry. Many exciting new results were presented in 21
invited or review papers, 26 contributed papers, and 127
poster papers.

This symposium would not have been possible without
the generous financial assistance of the International
Astronomical Union, the German Science Foundation (DFG),
and the State Government of Baden-Württemberg. The practical support of the University of Heidelberg and the Max-
Planck-Society was also very valuable. Finally, the hard
work of the Scientific and Local Organizing Committees and
of many members of the Heidelberg astronomical institutes
contributed greatly to the success of this meeting.

The Editors

SCIENTIFIC ORGANIZING COMMITTEE

K.-H. Böhm (Chairman), B. Baschek, A. Boyarchuk, M. Cohen,
A. G. Hearn, C. Jordan, J. Linsky, A. Omont, D. Reimers,
R. Viotti

LOCAL ORGANIZING COMMITTEE

I. Appenzeller (Chairman), B. Baschek, J. Krautter, R. Mundt,
J. Solf, P. Ulmschneider

LIST OF PARTICIPANTS

B. G. Anandarao	Ahmedabad	India
P. Andre	Gif-Sur-Yvette	France
I. Appenzeller	Heidelberg	FRG
B. Baschek	Heidelberg	FRG
U. Bastian	Heidelberg	FRG
P. Bastien	Montreal	Canada
F. M. Bateson	Greerton/Tauranga	New Zealand
S. Beckwith	Heidelberg	FRG
C. Bertout	Paris	France
M. S. Bessell	Woden	Australia
L. Bianchi	Pino Torinese	Italy
W. Bidelman	Cleveland	USA
H.-J. Blome	Köln	FRG
K. H. Böhm	Seattle	USA
E. Böhm-Vitense	Seattle	USA
J. Bouvier	Paris	France
D. Breitschwerdt	Heidelberg	FRG
A. Brown	Boulder	USA
E. W. Brugel	Boulder	USA
T. Bührke	Heidelberg	FRG
Ch. J. Burrows	Baltimore	USA
S. Cabrit	Paris	France
J.-P. Caillault	Boulder	USA
M. Cameron	London	UK
R. Carballo Fidalgo	La Laguna	Spain
C. Castro Sa	Porto	Portugal
C. Ceccarelli	Frascati	Italy
M. Cerruti-Sola	Firenze	Italy
J. Chapman	Jodrell Bank	UK
C. Chavarria	Mexico, D.F.	Mexico
R. Chini	Bonn	FRG
F. O. Clark	Groningen	Netherlands
M. Cohen	Berkeley	USA
R. J. Cohen	Jodrell Bank	UK
M. Cuntz	Heidelberg	FRG
H. P. Deasy	Dublin	Ireland
M. E. Dollery	Johannesburg	South Africa
R. Dümmler	Münster	FRG
J. E. Dyson	Manchester	UK
C. Eiroa	Madrid	Spain
M. S. El-Nawawy	Cairo	Egypt
H. Elsässer	Heidelberg	FRG

K. Eriksson	Uppsala	Sweden
A. Evans	Keele	UK
N. Evans II	Austin	USA
T. L. Evans	Observatory	South Africa
Y. A. Fadeev	Moscow	USSR
S.A.E.G. Falle	Heidelberg	FRG
U. Finkenzeller	Berkeley	USA
J. Fleming	Tucson	USA
M. Friedjung	Paris	France
H.-P. Gail	Heidelberg	FRG
J. R. Giddings	Heidelberg	FRG
D. Gillet	St.Michel l'Obs.	France
A. E. Glassgold	New York	USA
W. M. Glencross	London	UK
M. Goldsmith	Keele	UK
M. Grewing	Tübingen	FRG
V. Grinin	Nauchny, Crimea	USSR
H. J. Habing	Leiden	Netherlands
R. Hammer	Freiburg	FRG
A. Harpaz	Haifa	Israel
A. G. Hearn	Utrecht	Netherlands
S. R. Heathcote	La Serena	Chile
E. K. Hege	Tucson	USA
A. Heske	Hamburg	FRG
F. V. Hessmann	Heidelberg	FRG
K. H. Hinkle	Tucson	USA
P. Höflich	Heidelberg	FRG
T. E. Holzer	Boulder	USA
P. J. Huggins	New York	USA
G. Hutchinson	Keele	UK
T. Iijima	Asiago	Italy
D. E. Innes	Heidelberg	FRG
C. Jäger	Heidelberg	FRG
C. Jordan	Oxford	UK
U. Ch. Joshi	Ahmedabad	India
P. G. Judge	Oxford	UK
M. Kafatos	Fairfax	USA
F. D. Kahn	Manchester	UK
W. H. Kegel	Frankfurt	FRG
J. Kclemen	Budapest	Ungarn
E. Kirste	Bonn	FRG
G. Klare	Heidelberg	FRG
J. Köppen	Heidelberg	FRG
J. Koornneef	Baltimore	USA
P. Korewaar	Utrecht	Netherlands
J. Krautter	Heidelberg	FRG
G. V. Kurt	Moscow	USSR
M.T.V.T. Lago	Porto	Portugal
H. Lamers	Utrecht	Netherlands
C. Leitherer	Heidelberg	FRG
R. Lenzen	Heidelberg	FRG

J. Lepine	Sao Paulo	Brazil
A. M. Le Squeren	Meudon	France
R. Levreault	Cambridge	USA
J. W. Liebert	Tucson	USA
J. F. Lightfoot	London	UK
W. Liller	Vina del Mar	Chile
J. Linsky	Boulder	USA
D. Lorenzetti	Frascati	Italy
A. M. Magalhaes	Sao Paulo	Brazil
A. Mampasio Recio	La Laguna	Spain
M.-O. Mennessier	Montpellier	France
K. Menten	Bonn	FRG
G. Micela	Palermo	Italy
J. Mikolajewska	Torun	Poland
T. J. Millar	Manchester	UK
D. Muchmore	Heidelberg	FRG
R. Mundt	Heidelberg	FRG
Z. E. Musielak	Cambridge	USA
A. Natta	Firenze	Italy
T. Neckel	Heidelberg	FRG
J. E. Neff	Boulder	USA
C. A. Norman	Baltimore	USA
J. A. Nuth	Greenbelt	USA
A. Omont	Grenoble	France
F. Palla	Firenze	Italy
R. Papoular	Gif-Sur-Yvette	France
M. Perinotto	Firenze	Italy
P. Persi	Frascati	Italy
B. Pettersson	Uppsala	Sweden
F. Praderie	Meudon	France
R. K. Prinja	London	UK
A. Raga	Seattle	USA
T. P. Ray	Castleknock	Ireland
D. Reimers	Hamburg	FRG
B. Reipurth	Santiago	Chile
M. Robberto	Torino	Italy
M. Rosa	Garching	FRG
G. Schatz	Karlsruhe	FRG
J. Schmitt	Garching	FRG
W. Schmutz	Kiel	FRG
M. Scholz	Heidelberg	FRG
U. Schrey	Garching	FRG
K.-P. Schröder	Hamburg	FRG
R. Schulte-Ladbeck	Heidelberg	FRG
A. Schulz	Bonn	FRG
H. E. Schwarz	Dorking, Surrey	UK
S. Sciortino	Palermo	Italy
E. Sedlmayr	Berlin	FRG
S. Serio	Palermo	Italy
F. Shu	Berkeley	USA
B. S. Shylaja	Bangalore	India

G. Silvestro	Torino	Italy
D. Sivagnanam	Meudon	France
Ch. J. Skinner	London	UK
J. Solf	Heidelberg	FRG
L. Spinoglio	Frascati	Italy
S. R. Sreenivasan	Calgary	Canada
O. Stahl	Garching	FRG
J. Staude	Heidelberg	FRG
S. Stefl	Ondrejov	Czechoslovakia
R. E. Stencel	Boulder	USA
F. Strafella	Lecce	Italy
M. Tapia	Mexico, D.F.	Mexico
J.H.F. Telkamp	Seattle	USA
J. Tinbergen	Roden	Netherlands
T. Tsuji	Tokyo	Japan
G. Traving	Heidelberg	FRG
P. Ulmschneider	Heidelberg	FRG
G. Vaiana	Palermo	Italy
K. A. van der Hucht	Utrecht	Netherlands
M. S. Vardya	Bombay	India
R. Viotti	Frascati	Italy
Ch. Waelkens	Heverlee	Belgium
C. M. Walmsley	Bonn	FRG
R. F. Warren-Smith	Durham	UK
P. A. Wayman	Dublin	Ireland
R. Wehrse	Heidelberg	FRG
W. G. Weller	La Serena	Chile
G. Welin	Uppsala	Sweden
V. Werle	Heidelberg	FRG
P. R. Wesselius	Groningen	Netherlands
B. Whitmore	London	UK
P. M. Williams	Edinburgh	UK
M. J. Wilson	Leeds	UK
B. Wolf	Heidelberg	FRG
A. E. Wright	Parkes	Australia
F.-J. Zickgraf	Heidelberg	FRG
H. Zinnecker	Edinburgh	UK
E. Zsoldos	Budapest	Hungary
Ch.-Y. Zhang	Groningen	Netherlands

FIFTEEN YEARS OF "SONDERFORSCHUNGSBEREICH" IN HEIDELBERG:
RESEARCH ON CIRCUMSTELLAR MATTER

B. Baschek
Institute of Theoretical Astrophysics
University of Heidelberg
Im Neuenheimer Feld 561
D-6900 Heidelberg
Federal Republic of Germany

At the end of this year, the Sonderforschungsbereich (Special Collabora-
tive Programme) no. 132 on "Theoretical and Observational Stellar
Astronomy" in Heidelberg will terminate after fifteen years. Although
15 years are only 2 1/2 per cent of the age of the University of Heidelberg, which is celebrating its 600 th anniversary this year, they are
nevertheless a long and important time for astronomical research in
Heidelberg. On the occasion of the termination of the Sonderforschungsbereich, we are now given the opportunity to present an essential part
of its research, namely that on circumstellar matter, at an international
conference, and we are grateful to the International Astronomical Union
that this could be realized, and that we can welcome here so many
participants to this Symposium. As the Speaker of our Sonderforschungsbereich I would like to briefly introduce to you the general concept of
the institution of an SFB and give an overview over its structure and
research activities.
    The concept of support of research in the frame of a Sonder-
forschungsbereich (SFB) was proposed at the end of the 1960s by the
Wissenschaftsrat, our National Science Council. In these new instituties
the federal government and the state governments jointly give funds to
the Deutsche Forschungsgemeinschaft (DFG), the West German Science
Foundation, which in turn financially supports the SFBs in addition to
its various other programs for funding research.
    In 1968, there were 18 SFBs with a total funding of 5 million DM,
this increased to 96 SFBs with 125 million DM in 1972, the year our SFB
was first financed, and at present there are 163 SFBs supported by 317
million DM altogether.
    An SFB is to be approved by the university and by the DFG, and is
equivalent to other institutions of the university, e.g. institutes. The
host university, in establishing an SFB, commits itself to making
available a sufficient base of staff and financial resources ("Grund-
ausstattung") and is applicant and recipient of the DFG support. The
scientists of the SFB decide on the scientific course and the day-to-
day affairs. An important aspect of an SFB is the support by the DFG
for a longer period of time, subject to refereeing in three years'
intervals. The maximum duration of support is 15 years.

1

I. Appenzeller and C. Jordan (eds.), Circumstellar Matter, 1–3.
© *1987 by the IAU.*

Essential for an SFB is joint research by different groups of scientists. Within the SFB 132, four Heidelberg institutes collaborate on problems of theoretical and observational stellar astronomy, the Astronomisches Rechen-Institut, the Institut for Theoretische Astrophysik of the University, the Landessternwarte Königstuhl, and the Max-Planck-Institut für Astronomie.

The SFB is composed of two subprojects:
A. "Objects with Extended Envelopes", and
B. "Astrometry, Stellar Kinematics and Dynamics".

The research topic of project A is essentially equivalent to "Circumstellar Matter", and reaches from stellar atmospheres out into the interstellar medium, hereby excluding conventional (compact) atmospheres as well as the general interstellar medium, or - in other words - deals with the transition and the interaction between stars and the interstellar matter. The "objects with extended envelopes" are very different, in particular, are in different evolutionary states. The basic physical conditions, however, may be similar irrespective of the evolutionary phase so that common observational techniques as well as common theoretical methods are appropriate.

Influenced in part by the original interests of the Heidelberg institutes, the research first concentrated (a) on objects associated with star formation and early evolutionary phases such as molecular clouds, infrared sources, protostars, bipolar nebulae, maser, H II regions, T Tau and YY Ori stars, and Herbig Ae and Be stars, and (b) on objects in later evolutionary phases such as giants, supergiants, miras, planetary nebulae, novae, and again maser and H II regions. In the course of the 15 years, additional topics were added: chromospheres, coronae, bipolar structures and outflows (jets, Herbig Haro objects) and dust formation. These objects have been investigated by combined observational and theoretical approaches with varying degree of detail from surveys to very detailed studies of individual objects.

Regarding the observational side, development of instrumentation for the optical and infrared was carried out at the Landessternwarte and the Max-Planck-Institut. Within the period of the SFB, new instruments became available at the new observatory on the Calar Alto in Spain, in particular the 2.2 m-telescope in 1978, and the 3.5 m-telescope in 1984.

Essential for observing circumstellar matter was also the access to other spectral ranges which bear important information on the surrounding gas/dust envelope. Radio telescopes and satellites such as IUE (launch in 1978), Einstein (1978), and IRAS (1983) have also been used by the scientists of the SFB.

As for the theoretical interpretation, methods had to be developed which are "intermediate" between those for the classical compact, static atmospheres and those for the interstellar medium and equilibrium Strömgren spheres. In particular one has to deal with the occupation of atomic and molecular levels which are far from LTE, radiative transfer in spherical and extended systems, velocity fields comprising turbulence, winds, mass loss and accretion, jets, and mechanical heating (acoustic and magnetohydrodynamic waves), radiative pressure and dust formation.

This overview over a variety of research topics illustrates that the astronomical institutes in Heidelberg have been very fortunate to experience a substantial additional funding in the frame of the SFB for a period of 15 years only through which has research on such a scale been possible. Let me emphasize that we do not take this for granted and that we are deeply indebted for this support. On behalf of the Sonderforschungsbereich 132 I should like to take this opportunity to express our sincere gratitude to the Deutsche Forschungsgemeinschaft, to the State of Baden-Württemberg, and to the University of Heidelberg for their active and efficient support to our research.

BIPOLAR FLOWS, JETS, AND PROTOSTARS

STAR FORMATION AND THE CIRCUMSTELLAR MATTER OF YOUNG STELLAR OBJECTS

Frank H. Shu and Fred C. Adams

Astronomy Department, University of California, Berkeley CA 94720, USA

ABSTRACT. We propose that the formation of low mass stars in molecular clouds takes place in four stages. The first stage is the formation of slowly rotating cloud cores through the slow leakage of magnetic (and turbulent) support by ambipolar diffusion. These cores asymptotically approach quasistatic states resembling singular isothermal spheres, but such end states cannot actually be reached because they are unstable. The second phase begins when a condensing cloud core passes the brink of instability and collapses dynamically from "inside-out," building up a central protostar and nebular disk. The emergent spectral energy distributions of theoretical models in the infall stage are in close agreement with those of recently found infrared sources with steep spectra. As the rotating protostar gains mass, deuterium will eventually ignite in the central regions and drive the star nearly completely convective if its mass is less than about 2 $M_\odot$. This initiates the next step of evolution – the bipolar outflow phase – in which a stellar wind pushes outward and breaks through the infalling envelope. The initial breakout is likely to occur along the rotational poles, leading to collimated jets and bipolar outflows. The intense stellar wind eventually widens to sweep out gas in nearly all 4π steradian, revealing the fourth stage – a T Tauri star with a surrounding remnant nebular disk. Radiation from a disk adds an infrared excess to the expected spectral energy distribution of the revealed source. The detailed shape of this infrared excess depends on whether the disk is largely passive and merely reprocesses stellar photons, or is relatively massive and actively accreting. Both extremes of spectral shapes are observed in T Tauri stars; the amount of circumstellar material in the form of disks around nearly formed stars may be related to the dual issues of the origins of binary-star and planetary systems.

I. INTRODUCTION

Star formation occurs today primarily in large molecular clouds which have masses of 10^4-10^6 $M_\odot$ (Zuckerman and Palmer 1974, Burton 1976, Solomon and Sanders 1985). In contrast, the masses of the stars which result from the process are commonly a solar mass or less. This raises one of the basic questions of star formation – *Why does the interstellar medium, which has ample raw material to produce very massive objects, mostly produce self-gravitating balls of gas that are only marginally capable of thermonuclear fusion?*

7

I. Appenzeller and C. Jordan (eds.), Circumstellar Matter, 7–22.

This question is highlighted by the realization that star formation in molecular clouds is generally an inefficient mechanism; usually, only $\sim 1\%$ of the bulk matter in a molecular cloud is transformed into stars during a generation of star birth (see, *e.g.*, Scoville 1986). This inefficiency results in a lot of material being left over in the immediate vicinity of the formed star; associated with young stellar objects (YSOs) is a wide variety of forms of circumstellar matter: infrared-emitting shells, disks, jets, Herbig-Haro objects, bipolar (molecular) lobes, and cometary (reflection) nebulae. *What determines whether a given parcel of matter becomes incorporated into the newly formed star, or is left over as debris from the star formation process? What accounts for the diversity of forms of circumstellar matter?*

This paper approaches the above questions by considering a comprehensive picture of star formation. The theory schematically breaks down the complete process into four evolutionary phases. The first phase is the formation of molecular cloud cores – centrally condensed regions within larger molecular clouds. Since the existence of such entities is fairly well established observationally (see Myers and Benson 1983), and since the theory of core formation by ambipolar diffusion has been discussed elsewhere (see, *e.g.*, Lizano and Shu 1986), we limit our discussion to the three later stages of evolution. We shall also confine our attention to the kinds of cloud cores which lead to the birth of low mass stars, the extreme conditions needed to produce high mass stars being more controversial (*cf.* Appenzeller and Tscharnuter 1974; Klein, Whitaker, and Sandford 1985; Shu 1986).

The second phase of star formation begins when a molecular cloud core becomes unstable and starts to collapse. This evolutionary phase is characterized by the existence of a central protostar and disk with a surrounding infalling envelope of dust and gas. We discuss the hydrodynamical properties of such objects in §II, their evolution in §III, and the corresponding emergent spectra in §IV.

As a low mass protostar accretes matter, it naturally evolves toward a state with a stellar wind, but at first the ram pressure from material falling directly onto the stellar surface suppresses breakout. Gradually, the "lid" of direct infall will weaken as the incoming material falls preferentially into the disk rather than onto the star. The stellar wind then rushes through the channels of weakest resistance (the rotational poles), and the protostar enters the bipolar flow phase. We discuss this third evolutionary phase in §V.

Eventually, the spreading stellar wind will reverse the infall over almost the entire celestial sphere centered on the object. The object then enters the final phase of protostellar evolution and becomes a newly revealed T Tauri star with a surrounding nebular disk. The theory of the infrared excesses associated with such configurations is outlined in §VI.

There is a fifth phase which is not addressed by us, namely, the final disappearance of the nebular disk matter as it is incorporated into the bodies of planetary or stellar companions, or is dispersed and eroded by the energetic outflow. The topic of "naked T Tauri stars" (Mundt *et al.* 1983, Walter 1986) more properly belongs to the subject of pre-main-sequence evolution than to star formation, although it can also be viewed as logically flowing from the developments of the four prior stages.

II. INSIDE-OUT COLLAPSE OF UNSTABLE CLOUD CORES

As a starting point for the present discussion, we assume that the density distribution ρ of a molecular cloud core which has reached the brink of gravitational instability can be approximated as a singular isothermal sphere, whose self-gravity is exactly but precariously balanced by a thermal pressure gradient (Chandrasekhar 1939):

$$\rho = \frac{a^2}{2\pi G r^2},$$

(1)

where $a = (kT/m)^{1/2}$ is the isothermal sound speed and T and m are the (constant) temperature and mean molecular weight of the gas. (If magnetic fields and turbulence make non-negligible contributions to the mechanical support before collapse, we assume their effects can be mimicked by appropriately redefining an "effective" sound speed a.) Bodenheimer and Sweigart (1968) showed that subsonic evolution of a non-rotating, non-magnetic cloud would always tend to produce a $\rho \propto r^{-2}$ density distribution (see also Larson 1969). Shu (1977) discovered that the singular isothermal sphere (1), long known to be unstable to gravitational contraction, has a self-similar collapse solution, so that the solution at any instant in time looks like the solution at a previous instant except for scaling factors. The form of the similarity solution could be found analytically apart from the integration of some simple ordinary differential equations. Since the central density of the singular isothermal sphere is infinite, the adoption of equation (1) for the initial configuration of the collapse problem has been criticized as being artificial by Whitworth and Summers (1985). However, Nakano (1981) found that including grain coupling in ambipolar diffusion calculations (Elmegreen 1979, 1986; Nakano and Umebayshi 1980), produces very high densities (compared to envelope values) in the central regions of a molecular cloud core before dynamical collapse occurs. As long as the power-law portion of a real core solution spans several decades in density, its gravitational collapse is likely to be well represented by the self-similar solution for the singular isothermal sphere.

Notice that equation (1) asserts that there is *no such thing* as a *typical density* for a molecular cloud core – a power law has no characteristic scale. In particular, equation (1) predicts that the sizes of the maps obtained by radio-line observers will depend on the molecular transition studied; lower density tracers should produce larger maps than higher density tracers, in rough qualitative and quantitative agreement with the actual observations (Myers 1987, Walmsley 1987). Because equation (1) has no characteristic density scale, it has no characteristic Jeans mass – in fact, every radius r contains about one Jeans mass, $2a^2 r/G \propto a^3 \rho^{-1/2}$. In Taurus, where $a \approx 0.2$ km/s, 1 $M_\odot$ is contained roughly within $r = 0.05$ pc where $n_{H_2} \sim 10^4$ cm^{-3} (*cf.* Myers and Benson 1983); in Ophiuchus, where $a \approx 0.35$ km/s, 1 $M_\odot$ is contained roughly within $r = 0.02$ pc where $n_{H_2} \sim 10^5$-10^6 cm^{-3}

(cf. Martin-Pintado et al. 1983; Zeng, Batrla, and Wilson 1984; Wadiak et al. 1985). Going to lower density contours than traced by NH_3 or H_2CO or HCO^+, e.g., as defined by various isotopes of the CO molecule, gives larger maps and larger masses.

In any case, with the adoption of equation (1) to characterize the core before it undergoes gravitational collapse, there is no unique mass scale that can be associated with the resulting star. Instead of a mass, it is the *rate* at which the central object accumulates matter through infall which is well defined; in the self-similar solution, this rate is given by

$$\dot{M} = m_0 a^3 / G, \tag{2}$$

where $m_0 = 0.975$. For $a = 0.2$ km/s, $\dot{M} = 2 \times 10^{-6}$ $M_\odot$/y; for $a = 0.35$ km/s; $\dot{M} = 1 \times 10^{-5}$ $M_\odot$/y.

In the self-similar collapse solution, infall is initiated from "inside-out" by an expansion wave which propagates outward at the speed of sound into the ambient (static) molecular cloud core. The head of the expansion wave reaches the radius $r_h = at$ in time t; and supersonic inflow velocities are generated interior to $r_s \approx 0.4\,at$, inside of which the density begins to approach the free-fall form,

$$\rho = C r^{-3/2}, \tag{3}$$

where $C \equiv \dot{M}/4\pi(2GM)^{1/2}$.

The above discussion applies only to purely spherical collapse. When rotation is included in the problem as a small perturbational effect, the axisymmetric time-dependent collapse (with variations in two spatial dimensions) can still be followed analytically (Terebey, Shu, and Cassen 1984). In the inner parts the rotationally-modified similarity solution asymptotically joins onto free-fall ballistic trajectories, which are parabolae in the limit of a very concentrated mass distribution for the central object (star plus disk). The density distribution (3) is a good approximation outside a centrifugal radius,

$$R_C \equiv G^3 M^3 \Omega^2 / 16 a^8, \tag{4}$$

the position where infalling matter in the equatorial plane encounters a centrifugal barrier if it conserves its initial specific angular momentum. In equation (4), $M = \dot{M}t$ is the infallen mass, and Ω is the assumed (constant) rotation rate of the cloud core before collapse. Inside R_C the isodensity contours become highly flattened and increase less steeply inward as a spherical average than in the nonrotating case. When averaged over all angles, the density increases inward of R_C approximately as $\rho \propto r^{-1/2}$. When R_C is greater than the stellar radius R_*, a nebular disk forms around the protostar whose outer dimension extends at least to R_C (beyond, if the disk has internal mechanisms for appreciable transport of mass and angular momentum in an infall time scale $M/\dot{M}$).

III. PROTOSTELLAR EVOLUTION

When the infalling dust and gas become optically thick to the emergent infrared radiation, the isothermal approximation begins to break down badly. In models of low-mass protostars where the effects of rotation are ignored, this occurs inside a radius r_e of approximately 10^{14} cm (Stahler, Shu, and Taam 1980a,b; 1981; hereafter SST). Fortunately, within the region bounded by the the dust photosphere, r_e, and the stellar radius, $R_* \sim 10^{11.5}$ cm (defined by the accretion shock), the material falls nearly freely toward the mass at the center. Hence, the crossing time for both matter and radiation is short compared to the evolutionary time, and a condition of steady-state flow holds to a high degree of approximation. The inner region of complicated radiative hydrodynamics may thus be solved (using carefully chosen closure relations for the frequency-integrated moment equations of the radiation field) as a set of ordinary differential equations, greatly simplifying the computational demands of the problem. For this inner problem, the similarity solution discussed in the previous section provides outer boundary conditions, namely, inflow at free-fall speeds with mass infall rate (2).

Using an infall rate $\dot{M} = 1 \times 10^{-5}\ M_\odot/\mathrm{y}$, SST found that the protostar accumulated matter (processed through an accretion shock) of ever increasing specific entropy. Thus, the star remained radiative until deuterium ignited near the center when the stellar mass was about 0.3 $M_\odot$. A convection zone then spread outward through the star until it became almost entirely convective at a mass of about 0.5 $M_\odot$. Except for this event, nothing happened to distinguish a particular mass scale for the accreting protostar; in the actual calculations, the infall was artificially terminated after 10^5 y when the star had accumulated 1 $M_\odot$. The surface of the star, which had been kept abnormally hot and luminous by a standing shock, then cooled in less than a day and joined a convective pre-main-sequence track (Hayashi, Hoshi, and Sugimoto 1962). The disappearance of the infall region would first make the star optically visible at this point. If the loci of such points in the Hertzsprung-Russell diagram for different masses are joined, the result is a "birthline" for pre-main-sequence stars of low mass. Stahler (1983) showed the birthline corresponding to spherical mass accretion at the rate $\dot{M} = 1 \times 10^{-5} M_\odot/\mathrm{y}$ gave a remarkable fit to the upper envelope for T Tauri stars in Taurus-Auriga, Orion, NGC 7000/IC 5070, and Ophiuchus (Cohen and Kuhi 1979).

Unfortunately, the agreement between the theoretical birthline and the observed one can be subjected to a number of criticisms (*cf.* Mercer-Smith, Cameron, and Epstein 1984). First, the results described above are limited to purely spherical infall; even a small amount of rotation in the original molecular cloud core would have produced accretion partially through a disk. Second, although the adopted infall rate of $1 \times 10^{-5}\ M_\odot/\mathrm{y}$ is appropriate for Ophiuchus, it is ~ 5 times too large for Taurus. Using a scaling law $R_* \propto \dot{M}^{1/3}$ appropriate to stars which derive all of their energy from accretion (see SST's comparison with Winkler and Newman 1980), the spherical-collapse birthline for Taurus becomes underluminous by a factor of ~ 3.

Shu (1985, see also Shu and Terebey 1984) pointed out that the observed birth-line for T Tauri stars can be understood, not as a special result of a spherically symmetric scenario for star formation, but in terms of an end condition that the star must be able to burn deuterium in order to clear the shroud of placental material. The basic idea is that when the convection driven by deuterium burning in a low mass protostar is coupled with the presence of differential rotation, dynamo action results (Parker 1979), which taps the rotational energy stored in the star. The released energy is speculated to power the intense stellar surface activity observed in young stellar objects. The resulting stellar wind eventually reverses the infall and allows the newly formed star to become visible.

The luminosity of such an object includes sources other than accretion. The primary consideration in this case is that for any given mass, there is a unique radius for which a completely convective star will have a central temperature (Chandrasekhar 1939),

$$T_c = 0.54\,GM_*\mu H/kR_*,\tag{5}$$

high enough for deuterium to burn. In equation (5) we have assumed that the convective star is an ideal gas of constant mean molecular weight $m \equiv \mu H$ (i.e., a polytrope of index 1.5). For cosmic abundances and $0.01\,M_\odot < M_* < 2\,M_\odot$, we can obtain an approximate criterion for deuterium burning by setting $T_c = 1 \times 10^6$ K in equation (5):

$$R_*/R_\odot \approx 0.15 + 7.6\,(M_*/M_\odot),\tag{6}$$

where the small correction term 0.15 on the right-hand side enters to take into account the effects of partial degeneracy at low stellar masses (cf. Nelson, Rappaport, and Joss 1986). Stahler (1986, private communication) points out that the derivation of equation (6) should really include the notion that deuterium burning occurs in steady state; however, the required modification is small because the reaction rate is extremely sensitive to temperatures near 1×10^6 K.

Equation (6) yields a birthline in the Hertzsprung-Russell diagram for pre-main-sequence stars which does not differ much from that drawn originally by Stahler (1983). The main difference is that we have now identified the specific mechanism responsible for the unveiling of low-mass protostars and that this mechanism defines a birthline for pre-main-sequence evolution through physics (thermonuclear fusion and stellar structure) that is relatively insensitive to the detailed manner in which central mass is accumulated (i.e., by direct infall or disk accretion).

The rate of stellar mass accumulation $\dot{M}_*$ affects only the timing of deuterium ignition, i.e., the eventual mass of the star M_* on the birthline. When $\Omega \neq 0$, the rate of stellar mass accumulation,

$$\dot{M}_* = \dot{M}_*(\text{direct infall}) + \dot{M}_*(\text{disk accretion}),\tag{7}$$

does not equal the total infall rate $\dot{M}$ given by equation (2) unless all of the mass which falls into the disk is shortly incorporated into the central star (i.e., unless the parameter η_D to be discussed below equals unity). Higher rates of stellar mass accumulation allow less time for the radiation of the stellar binding energy, leading to larger values of R_* at every M_*, and thus producing higher masses before the onset of deuterium burning. From computed radiation losses of pre-main-sequence stars on Hayashi contraction tracks, we estimate that an average rate of $\dot{M}_* = 10^{-7}$ $M_\odot/y$ will produce a protostar which begins to drive a wind after it accumulates ~ 0.1 $M_\odot$; 10^{-6} $M_\odot/y$, after ~ 0.3 $M_\odot$; 10^{-5} $M_\odot/y$, after ~ 0.5 $M_\odot$; and 10^{-4} $M_\odot/y$, after ~ 2 $M_\odot$. The final stellar masses could be higher than the above listed values if the infall is sufficiently intense to suppress a breakout of the stellar wind despite the convection induced by deuterium burning or if significant amounts of stellar accretion still occurs from a disk during the bipolar flow phase of protostellar evolution.

In any case, for stars which accumulate ~ 2 $M_\odot$ before deuterium ignites, the extra interior luminosity released by deuterium burning can be carried out by radiative diffusion without the inducement of convection. The termination of stellar accumulation for high mass stars must therefore occur by a different process than outlined above (see also the discussion of Cassen, Shu, and Terebey 1985). Since $\dot{M}_*$ cannot on average exceed the total infall rate $\dot{M}$, equation (2) implies that the production of high mass stars requires effective values of a greater than or comparable to 0.7 km/s if $\dot{M}$ is to exceed 10^{-4} $M_\odot/y$. This requires gas temperatures in excess of 100 K in the original molecular cloud core if the primary contribution to a is thermal (but non-negligible magnetic and "turbulent" contributions are likely to be also present). A possible route for obtaining warm molecular cloud cores and high effective values of a is discussed by Shu, Lizano, and Adams (1987), by Shu (1986), and by Lizano and Shu (1986).

IV. INFRARED EMISSION FROM PROTOSTARS

In order to test the fundamental notion that stars form through a process of infall, it is helpful to calculate the expected emergent spectral energy distributions of protostars. Spherically symmetric calculations have been performed by numerous groups (Larson 1969, Yorke and Krugel 1977, Bertout and Yorke 1978, Yorke and Shustov 1981), and Adams and Shu (1985, 1986; hereafter AS) have developed a fast approximate technique applicable to multi-dimensional situations as well.

The AS models are characterized by five parameters: the isothermal sound speed a of the initial cloud core, its angular velocity Ω, the mass $M = M_* + M_D$ that has accumulated to date in the central star and disk, and the efficiencies η_* and η_D with which the star and disk dissipate the energy of differential rotation. Given a deduced value of a, the quantities M, η_*, and η_D combine to produce a single observable – the total protostellar luminosity L. In practice, we fixed η_* and η_D at standard values, and the mass M was varied in order to match the observed luminosity L. This left a single free parameter, the cloud angular velocity Ω, with which to fit the spectrum. The resulting theoretical spectral energy distributions gave good agreement with observations of a class of low-luminosity infrared sources, those which have steep spectra and are found near

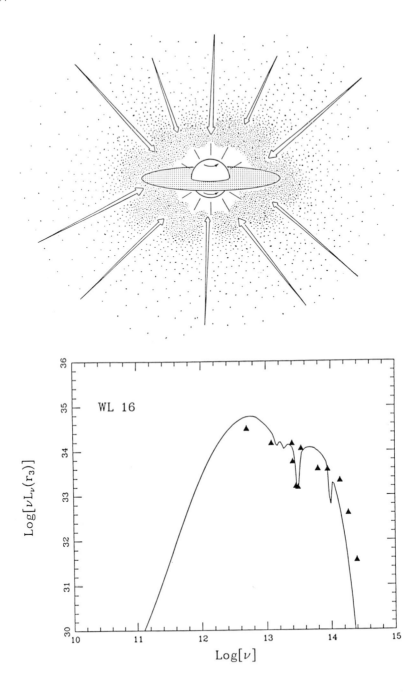

Figure 1. Comparison of theoretical and observed emergent spectral energy distributions of protostellar candidate WL 16 (Ophiuchus): data taken from Wilking and Lada (1983), Lada and Wilking (1984), and Young, Lada, and Wilking (1986); theoretical model assumes $a = 0.35$ km/s, $M = 0.5 M_\odot$ and $\Omega = 5 \times 10^{-13}$ rad/s; (all quantities are given in cgs units).

the centers of dense molecular cloud cores (see Figure 1 for an example). The angular velocities required in order to match the observed emission spectra of seven protostellar candidates (see Adams, Lada, and Shu 1987; hereafter ALS) lie in the range $\Omega = 2 \times 10^{-14}$ s^{-1} to 5×10^{-13} s^{-1}. Such rates are compatible with measured values (*e.g.*, Myers and Benson 1983, Wadiak *et al.* 1985, Goodman 1986), lending support to the entire theoretical development.

The shape of the emergent spectral energy distributions is especially sensitive to the centrifugal radius R_C defined by equation (4). To a good approximation, the total extinction to the central source along a typical line of sight is given by

$$A_V \approx 3.5 \, \kappa_V \, C R_C^{-1/2}, \tag{8}$$

where κ_V is the opacity at visual wavelenths ($\approx$ 200-250 cm^2/g). The column density of matter reprocessing the stellar and disk photons to radiation of longer wavelengths is the primary determinant of both the depth of the 10μm silicate feature and the overall spectral breadth (see ALS).

V. BIPOLAR OUTFLOW PHASE

The rotating infall models described above have the interesting property that they correctly predict the emergent spectra for some infrared sources which are known bipolar outflow sources, *i.e.*, objects that have entered the third phase of protostellar evolution (see Figure 2 for an example). This finding supports the idea that well-collimated sources represent objects in which inflow and outflow are taking place *simultaneously*. Thus, these objects yield the transitional phase of evolution between a purely accreting protostar and a fully revealed pre-main-sequence star (Shu and Terebey 1984; Cassen, Shu, and Terebey 1985).

Observationally, of course, outflows from young stellar objects are known often to take a well collimated form. At least two manifestations of the phenomenon are known: long and narrow optical jets (*e.g.*, Mundt and Fried 1983) and CO bipolar outflows (*e.g.*, Bally and Lada 1983). A complete theory of star formation should explain the origins of both kinds of features, as well as establish if there is a relation between them.

The general notion that stellar outflows from young stars should be directed in two diametrically opposed directions finds a natural explanation in the scenario for star formation posed above. During the infall phase, the swirling inflowing matter acts like the lid of a pressure cooker and suppresses any incipient stellar outflow. As the lid weakens, breakout will occur through the channel of least resistance (the safety valve) at the rotational pole(s) of the accreting protostar where the total column of infalling material is least (see Figure 6 of Terebey, Shu, and Cassen 1984). Since the system is smoothly transforming from a subcritical state (inflow) to a supercritical state (outflow), breakout in the absence of perfect spherical symmetry should first occur at one point (on each hemisphere). Thus,

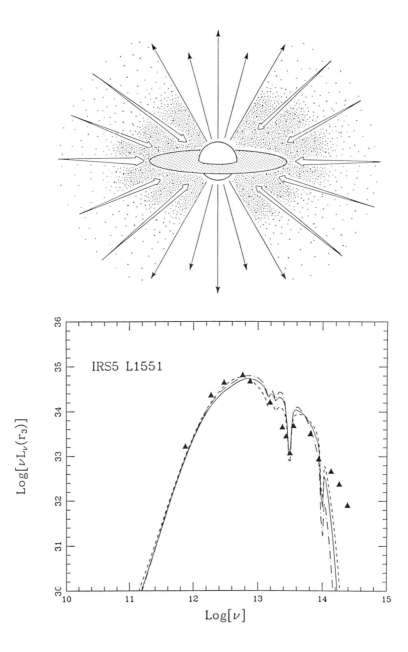

Figure 2. Theoretical and observed spectrum of bipolar outflow source IRS5 L1551 (Taurus): data from Cohen and Schwartz (1983), Cohen *et al.* (1984), and Davidson and Jaffe (1984); the theoretical model corresponding to the dashed curve assumes $a = 0.35$ km/s, $M = 0.5 M_\odot$ and $\Omega = 5 \times 10^{-13}$ rad/s; that of the dashed-dotted curve, to $a = 0.325$ km/s, $M = 0.675 \ M_\odot$, and $\Omega = 1 \times 10^{-13}$ rad/s; and that of the solid curve to $a = 0.35$ km/s, $M = 1.0 M_\odot$, and $\Omega = 1 \times 10^{-13}$ rad/s with $(\eta_*, \eta_D) = (1.0, 0.5)$ instead of $(0.5, 1.0)$.

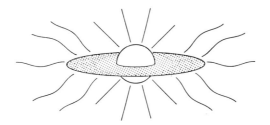

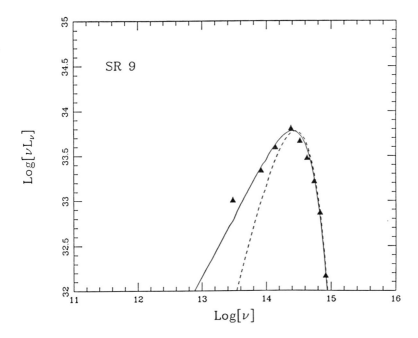

Figure 3.   Theoretical and observed spectra of T Tauri star with infrared excess – SR 9 (Ophiuchus): data from Chini (1981) and Lada and Wilking (1984); theoretical model assumes $L = 3.0L_\odot$, $A_V = 1.0$ mag. and $T_* = 4000$ K. A reddened black body (dashed line) is shown for comparison.

the existence of narrow twin exhaust jets in the youngest outflow objects becomes a natural part of protostellar evolution.

The issue of whether the *observed* optical jets represent this phenomenon is more contentious. At least two alternatives can be envisaged. The first is that an observed YSO outflow starts from within a few stellar radii of the stellar surface already highly collimated, and it later opens up (perhaps because of insufficient radiative cooling as the shocked jets propagate down a declining density gradient in the ambient molecular cloud core) to sweep out the fatter lobes of CO emission observed by millimeter-wave radio astronomers (*e.g.*, Fig. 4 of Shu, Lizano, and Adams 1987). The second is that a YSO outflow starts out from the stellar surface more-or-less as an isotropic stellar wind, but it later becomes focused through interactions with an anisotropic and inhomogeneous ambient medium (*e.g.*, Fig. 2 of Konigl 1982). Klein, Sandford, and Shu (1986) are in the process of evaluating the viability of these alternative scenarios.

VI. T TAURI STARS WITH CIRCUMSTELLAR DISKS

As time proceeds, we expect more and more of the rotating inflowing matter to fall preferentially on the disk rather than on the star. In any reasonable picture of stellar outflows, we expect the opening angle of the wind to widen with time, eventually sweeping outwards over all 4π steradians. This marks the beginning of the fourth evolutionary phase – an optically visible T Tauri star, newly born on a convective (Hayashi) pre-main-sequence track, surrounded by a remnant nebular disk (Elsasser and Staude 1978, Cohen 1983, Beckwith *et al.* 1984, Grasdalen *et al.* 1984, Simon *et al.* 1985), and bubbling with surface activity (Herbig 1962; Kuhi 1964; Calvet, Basri, and Kuhi 1984). The slow rotation rates observed for T Tauri stars (Vogel and Kuhi 1981), which were quite surprising when first discovered, may now be naturally attributed to the large magnetic braking associated with an earlier period of intense mass loss.

This picture provides a natural explanation for the infrared excesses commonly observed in the spectra of T Tauri stars. Even if the circumstellar disk surrounding a newly formed star has no intrinsic luminosity, Adams and Shu (1986) showed that an optically thick (but spatially thin) disk will intercept and reradiate 25% of the stellar luminosity. The resulting star/disk spectra are in close agreement with observed T Tauri stars with infrared excesses (see Figure 3). In these sources, the equilibrium temperature distribution in the disk is given approximately by $T \propto r^{-3/4}$, similar to the classical result for a Keplerian accretion disk (Lynden-Bell and Pringle 1974). The corresponding spectral energy distribution will assume a power-law form at long wavelengths given by $\nu L_\nu \propto \nu^{4/3}$.

However, there is another class of T Tauri stars (which includes T Tauri itself) with spectral energy distriubtions that are more nearly flat at long wavelenths, *i.e.*, $\nu L_\nu \sim$ constant. These sources can be understood if the disk has an intrinsic luminosity comparable to the stellar luminosity and the disk temperature distribution falls off less steeply than the $T \propto r^{-3/4}$ law appropriate for a Keplerian accretion disk (see ALS). Since a flatter temperature distribution implies a flatter rotation curve in a viscously evolving axisymmetric system, these sources probably have extended mass distributions, *i.e.* disks whose masses are a non-negligible fraction of the total. The case for massive disks can be given a more robust argument. First we note that the objects in question must often have intrinsic disk luminosities of several $L_\odot$ in order to explain the observed infrared excesses. Accretion onto a star (either the central source or a companion embedded in the disk) is the most likely energy source and hence must occur at a rate of 10^{-5} to 10^{-6} $M_\odot/\mathrm{y}$. Since many T Tauri stars have such infrared excesses, and their lifetimes are on the order to 10^5 to 10^6 y, the reservoirs of material for accretion in the disks must often be $0.1 - 1.0$ $M_\odot$ (*cf.* Sargent and Beckwith 1986).

It is interesting to speculate on the modes of companion formation in the two extreme cases of T Tauri disks. If a nebular disk turns out to have enough mass to make another star, it may be an excellent candidate for forming a relatively close binary system (with a period shorter than $\sim 10^2$-10^3 y; see Abt 1983). A passive or a nearly passive disk will contain a smaller amount of material, and it may be a likely candidate for forming a planetary system.

REFERENCES

Abt, H. 1983, *Ann. Rev. Astr. Ap.*, **21**, 343.

Adams, F. C., and Shu, F. H. 1985, *Ap. J.*, **296**, 655.

Adams, F. C., and Shu, F. H. 1986, *Ap. J.*, in press (AS).

Adams, F. C., Lada, C. J. and Shu, F. H. 1987, *Ap. J.*, in press (ALS).

Appenzeller, I., and Tscharnuter, W. 1974, *Astr. Ap.*, **30**, 423.

Bally, J., and Lada, C. J. 1983, *Ap. J.*, **265**, 824.

Beckwith, S., Zuckerman, B., Skrutskie, M. F., and Dyck, H. M. 1984, *Ap. J.*, **287**, 793.

Bertout, C., and Yorke, H. W. 1978, in *Protostars and Planets*, ed. T. Gehrels (Tucson: University of Arizona Press), p. 648.

Bodenheimer, P., and Sweigart, A. 1968, *Ap. J.*, **152**, 515.

Bodenheimer, P. 1980, in IAU Symposium No. 93, Fundamental Problems in the Theory of Stellar Evolution, ed. D. Sugimoto, D. Q. Lamb, and D. N. Schramm (Dordrecht: Reidel), p. 5.

Burton, W. B. 1976, *Ann. Rev. Astr. Ap.*, **14**, 275.

Calvet, N., Basri, G., and Kuhi, L. V. 1984, *Ap. J.*, **277**, 725.

Cassen, P., Shu, F. H., and Terebey, S. 1985, in *Protostars and Planets II*, ed. D. C. Black and M. S. Matthews (Tucson: University of Arizona Press), p. 448.

Chandrasekhar, S. 1939, *An Introduction to Stellar Structure* (The University of Chicago Press).

Chini, R. 1981, *Astr. Ap.*, **99**, 346.

Cohen, M. 1983, *Ap. J. (Letters)*, **270**, L69.

Cohen, M. 1984, *Physics Reports*, **116**, no. 4, 173.

Cohen, M., and Kuhi, L. V. 1979, *Ap. J. Suppl.*, **41**, 743.

Cohen, M., Harvey, P. M., Schwartz, R. D., and Wilking, B. A. 1984, *Ap. J.*, **278**, 671.

Cohen, M., and Schwartz, R. D. 1983, *Ap. J.*, **265**, 877.

Davidson and Jaffe, D. T. 1984, *Ap. J. (Letters)*, **277**, L13.

Elmegreen, B.G. 1979, *Ap. J.*, **232**, 729.

Elmegreen, B.G. 1986, preprint.

Elsasser, H., and Staude, H. J. 1978, *Astr. Ap.*, **70**, L3.

Goodman, A. 1986, private communication.

Grasdalen, G. L., Strom, S. E., Strom, K. M., Capps, R. W., Thompson, D., and Castelaz, M. 1984, *Ap. J. (Letters)*, L57.

Hayashi, C., Hoshi, R., and Sugimoto, D. 1962, *Prog. Theor. Phys. Suppl. No. 22*.

Herbig, G. 1962, *Adv. Astr. Ap.* **1**, 47.

Klein, R. I., Sandford, M. T., and Shu, F. H. 1986, in preparation.

Klein, R. I., Whitaker, R. W., and Sandford, M. T. 1985, in *Protostars and Planets II*, ed. D. C. Black and M. S. Matthews (Tucson: University of Arizona Press), p. 340.

Konigl, A. 1982, *Ap. J.*, **261**, 115.

Kuhi, L. V. 1964, *Ap. J.*, **140**, 409.

Lada, C. J. 1985, *Ann. Rev. Astr. Ap.*, **23**, 267.

Lada, C. J., and Wilking, B. A. 1984, *Ap. J.*, **287**, 610.

Larson, R. B. 1969, *M. N. R. A.S.*, **145**, 271.

Larson, R. B. 1969, *M. N. R. A. S.*, **145**, 297.

Lizano, S., and Shu, F. H. 1986, in preparation.

Martin-Pintado, J., Wilson, T. L., Gardner, F. F., and Henkel, C. 1983, *Astr. Ap.*, **117**, 145.

Mercer-Smith, J. A., Cameron, A. G. W., and Epstein, R. I. 1984, *Ap. J.*, **287**, 445.

Mundt, R., and Fried, J. W. 1983, *Ap. J. (Letters)*, **274**, L83.

Mundt, R., Walter, F. M., Feigelson, E. D., Finkenzeller, U., Herbig, G. H., and Odell, A. P. 1983, *Ap. J.*, **269**, 229.

Myers, P. C., and Benson, P. J. 1983, *Ap. J.*, **266**, 309.

Myers, P. C. 1987, in *Star Forming Regions*, ed. M. Peimbert and J. Jugaku (Dordrecht: Reidel).

Nakano, T., and Umebayashi, T. 1980, *Pub. Astr. Soc. Japan*, **32**, 613.

Nakano, T. 1981, *Prog. Theor. Phys. Suppl. No. 70*, 54.

Nelson, L. A., Rappaport, S. A., and Joss, P. C. 1986, *Ap. J.*, in press.

Parker, E. N. 1979, *Cosmical Magnetic Fields* (Oxford University Press).

Sargent, A., and Beckwith, S. 1986, in preparation.

Scoville, N. Z. 1986, in *Star Formation in Galaxies*, ed. G. Neugebauer and N. Z. Scoville, in press.

Shu, F. H. 1977, *Ap. J.*, **214**, 488.

Shu, F. H. 1985, in *The Milky Way*, ed. H. van Woerden, W. B. Burton, R. J. Allen (Dordrecht: Reidel), p. 561.

Shu, F. H. 1986, in *Star Formation in Galaxies*, ed. G. Neugebauer and N. Z. Scoville, in press.

Shu, F. H., and Terebey, S. 1984, in *Cool Stars, Stellar Systems, and the Sun*, ed. S. Baliunas and L. Hartmann (Berlin: Springer-Verlag), p. 78.

Shu, F. H., Lizano, S., and Adams, F. C. 1987, in *Star Forming Regions*, ed. M. Peimbert and J. Jugaku (Dordrecht: Reidel).

Solomon, P. M. and Sanders, D. B., in *Protostars and Planets II*, ed. D. C. Black and M. S. Matthews (Tucson: University of Arizona Press), p. 59.

Stahler, S. W. 1983, *Ap. J.*, **274**, 822.

Stahler, S. W., Shu, F. H., and Taam, R. E. 1980, *Ap. J.*, **241**, 637 (SST).

Stahler, S. W., Shu, F. H., and Taam, R. E. 1980, *Ap. J.*, **242**, 226.

Stahler, S. W., Shu, F. H., and Taam, R. E. 1981, *Ap. J.*, **248**, 727.

Terebey, S., Shu, F. H., and Cassen, P. 1984, *Ap. J.*, **286**, 529.

Vogel, S., and Kuhi, L. 1981, *Ap. J.*, **245**, 960.

Wadiak, E. J., Wilson, T. L., Rood, R. T., and Johnston, K. J. 1985, *Ap. J. (Letters)*, **295**, L43.

Walter, F. W. 1986, *Pub. Astr. Soc. Pac.*, in preparation.

Walmsley, M. 1986, this volume.

Whitworth, A., and Summers, D. 1985, *M. N. R. A. S.*, **214**, 1.

Wilking, B. A., and Lada, C. J. 1983, *Ap. J.*, **274**, 698.

Winkler, K. H., and Newman, M. J. 1980, *Ap. J.*, **236**, 201.

Yorke, H. W., and Krugel, E. 1977, *Astr. Ap.*, **54**, 183.

Yorke, H. W., and Shustov, B. M. 1981, *Astr. Ap.*, **98**, 125.

Young, E. T., Lada, C. J., and Wilking, B. A. 1986, *Ap. J.* (*Letters*), in press.

Zeng, Q., Batrla, W., and Wilson, T. L. 1984, *Astr. Ap.*, **141**, 127.

Zuckerman, B., and Palmer, P. 1974, *Ann. Rev. Astr. Ap.*, **12**, 279.

CIRCUMSTELLAR MATTER OF YOUNG LOW-MASS STARS: OBSERVATIONS VERSUS THEORY

Claude Bertout
Institut d'Astrophysique de Paris
98bis, Boulevard Arago
F 75014 Paris
France

ABSTRACT.   After presenting NGC 7129 as a prototypical star-forming region, I discuss what can be learned from the radio spectra of embedded infrared sources. I then review available observational evidence for disks around young stellar objects, with emphasis on accretion disks around T Tauri stars. Finally, new results on the role of magnetic fields in the circumstellar activity of T Tauri stars are presented.

I. INTRODUCTION

Observations of molecular clouds offer only an incomplete and unresolved view of what is going on in these star-formation regions, and the large amounts of dust in interstellar clouds alter the spectra of young stellar objects non-linearly. Theoretical ideas about stellar formation must overcome these difficulties.

According to the current pre-main-sequence evolution scenario reviewed of Shu and Adams in this volume, the sequence of events leading to the formation of a low-mass star begins when a rotating molecular core is formed by ambipolar diffusion. This core then undergoes dynamical collapse to make a fast-rotating protostar surrounded by a nebular disk. The central temperature of the protostar increases until it reaches the deuterium fusion ignition point; deuterium burning drives convection within the entire protostar, and thereby turn on dynamo processes which then create a stellar wind. The wind breaks out at the points of least resistance which are the stellar poles where the density of accreting material is lowest. The result is collimated jets and molecular outflows. Finally, the outflow angle widens; the infalling gas is swept out, and the central star and surrounding nebular disk become optically visible. This is the beginning of the T Tauri phase, in which the star is still presumably interacting with its environment. As the strength of this interaction diminishes, the star resembles more and more a normal late-type dwarf.

How observed properties of young stellar objects relate to this theoretical framework is the topic of this review. The first section illustrates

I. Appenzeller and C. Jordan (eds.), Circumstellar Matter, 23–38.

the current observational status of star-forming regions, using as an example the molecular cloud NGC 7129, which displays most phenomena usually encountered in star formation. In the second section we focus on the radio properties of young stellar objects. Since the Very Large Array allows resolving structures down to about 0"1, radio emission from embedded young objects is a most promising tool for investigating their circumstellar matter. While the observational database at our disposal is much larger for young stars which have become visible in the optical range -- the T Tauri stars -- than for younger objects, various aspects of their activity remain puzzling. Current ideas on the circumstellar environment of T Tauri stars are reviewed in this paper's third and fourth sections, with emphasis on observational clues for the presence of disks and on the role of magnetic activity in the T Tauri phenomenon.

II. NGC 7129, A STAR-FORMING REGION

Located in Cepheus at a distance of about 1 Kpc (Racine 1968), the molecular cloud NGC 7129 contains a group of seven early-type stars surrounded by bright reflection nebulosity (Pease 1917). Two of these stars, BD+65° 1637 and LkHα 234, were classified as pre-main-sequence stars with approximate spectral types of respectively B5ne and Ae by Herbig (1960).

Every sign of active star formation is present in NGC 7129. Four Herbig-Haro objects (Strom et al. 1974; Guylbudaghian et al. 1978; Hartigan and Lada 1985) align approximately with LkHα 234 in the NE-SW direction and may be associated with the blue and redshifted CO molecular gas detected in the region by Edwards and Snell (1983). Several water masers are also found in the vicinity of LkHα 234 (cf. Sandell and Olofsson 1981) and of a far-infrared source (FIRS; detected by Bechis et al. 1978) located 2' South of LkHα 234. Finally, continuum radio emission from NGC 7129 was detected at 23 GHz by Bertout and Thum (1982) using the Effelsberg 100m single dish, and subsequent interferometric observations at the VLA by Snell and Bally (1986) revealed that the radio source was centered on LkHα 234.

With all this activity going on in an area less than 10'x10', NGC 7129 appears worth a detailed investigation. In particular, the morphology of the CO molecular flows discovered by Edwards and Snell (1983) needs to be clarified. This was the subject of recent observations by Bertout, Cabrit, and Thum (1987) using the IRAM 30m antenna near Granada, Spain. Figures 1 and 2 show preliminary maps of position versus integrated intensity in the $^{12}CO(J=1-0)$ line. The $(0,0)$ reference position is that of LkHα 234, and offsets are in arc seconds. The beam size is 21", and the mapping has been made at 30" intervals in α and δ.

The first map (Figure 1) shows the ambient CO gas, as made by integrating the CO line profile over the velocity range corresponding to the static gas. The most striking feature of Figure 1 is the CO cavity that surrounds the group of hot stars. This hole in the dark cloud is also apparent in optical photographs of the region, where it shows up as an increased density in the number of background stars. The cavity's morphology suggests that it

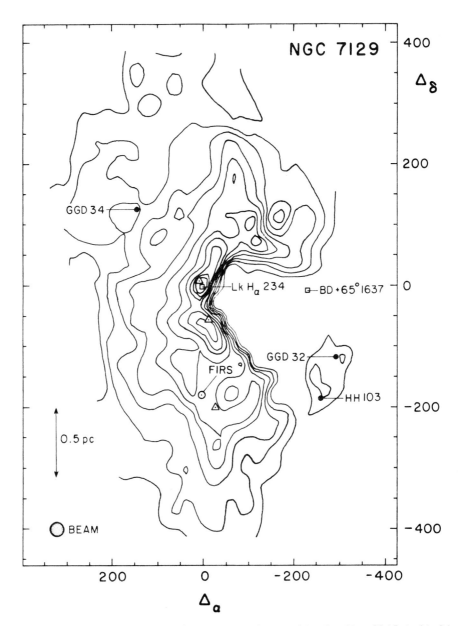

Figure 1 Map of position versus integrated intensity in the CO(J=1-0) line for the static molecular gas in NGC 7129. Open squares denote two Herbig Ae/Be stars, dots mark the positions of Herbig-Haro objects, triangles indicate water masers, and the far-infrared source's position is marked by an open circle.

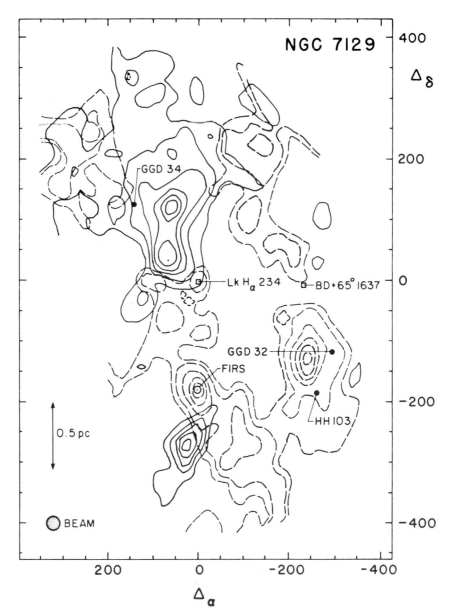

Figure 2 Same as Figure 1, but for the moving CO gas. At least two outflow sources (presumably LkHα 234 and FIRS) are present in NGC 7129.

was created by expansion of a bubble of stellar winds in the direction of least resistance, as suggested by Tenorio-Tagle (1982). Furthermore, both the location of LkHα 234 at the cavity's apex and the alignment of the cavity's symmetry axis with the Herbig-Haro objects mentioned above suggest that the main stellar wind source is LkHα 234, an assumption consistent with its Hα and radio continuum emission.

The complexity of the CO moving gas' map, shown in Figure 2, renders its interpretation difficult. Its main features are an extended, red-displaced emission located NE of LkHα 234 and an extended, weaker blue-displaced emission approximately surrounding the cavity and peaking close to the position of Herbig-Haro objects HH 103 and GGD 32. There is also a nicely symmetric bipolar flow associated with the far-infrared source 2' South of LkHα 234. We show the lowest blue contour at 2 K km s^{-1}, which is close to the noise level, mainly to demonstrate that the blue-displaced emission closely follows the cavity walls.

A tentative interpretation of the extended red and blue CO emission calls for a more or less spherically symmetric wind that originates from LkHα 234 and that is collimated on a large scale by density gradients in NGC 7129 (cf. Königl 1982). This is supported both by the map of ambient gas, which reveals that the excitation temperature peaks in a N-S elongated region surrounding LkHα 234 and by our unpublished CS data. An alternative picture, in which the wind from LkHα 234 is collimated close to the stellar surface by a disk surrounding the star is not supported by the CS interfero-metric observations of Wilking, Mundy, and Schwartz (1986). The lack of an extended blue lobe is presumably caused by the cavity's containing hot ionized gas flowing out of LkHα 234. Full analysis of our ^{12}CO, ^{13}CO, and CS data, now under way, should give us a more precise image of this star-forming region.

It has become clear recently that young embedded stellar objects such as LkHα 234 are often radio sources. Snell and Bally (1986) have shown that many infrared sources associated with high-velocity molecular gas are also radio continuum sources. This property gives us access to the inner regions of the flows, which would otherwise be hidden from us by their surrounding dust.

III. THE RADIO CONTINUUM OF YOUNG STELLAR OBJECTS

Among the best-investigated radio sources associated with pre-main-sequence objects of low-luminosity ($L_{bol} < 100\ L_\odot$) are IRS5 in L1551, T Tauri South--the infrared and radio companion of the T Tauri optical pair--and the radio source between Herbig-Haro objects 1 and 2 discovered by Pravdo et al. (1985). The radio spectra of these 3 objects are shown in Figure 3. Snell and Bally (1986) studied several additional sources associated with young stellar objects of various luminosities and distinguished two classes of objects. The first exhibits typical stellar wind spectra (spectral index ca. 0.6), and the second, which includes L1551 IRS5, has relatively flat spectra which are less easily understood.

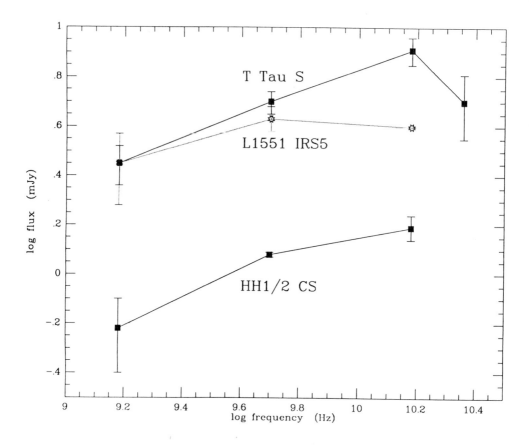

Figure 3 Radio spectra of three embedded low-luminosity sources, as compiled from various sources. The 15 GHz flux of L1551 IRS5 is an upper limit, and the 1.5 GHz data point of T Tauri includes both N and S components.

Interpreting the radio spectra of embedded sources assumes that these spectra are thermal. This assumption is justified by the shape of the spectra and by the frequent partial resolution of the radio emitting regions. In the case of T Tauri South, for which more data are available than for other sources, it is also supported by the apparent lack of strong variability on short and intermediate time-scales. This stability contrasts with the radio emission of T Tauri stars, which often show the fast variations and the spectral shapes typical of non-thermal emission (see Montmerle, André, and Feigelson 1985). I focus here on radio sources associated with embedded low-luminosity objects, and consider several possible models for spectra that show the qualitative behavior seen in Figure 3: more or less flat spectra at high frequencies with the slope apparently increasing at lower frequencies.

III.1. Extended Sources

"Extended" refers here to sources that are partially resolved at 5 GHz by the VLA in A configuration, i.e., to regions with diameter greater than about 0"4, which in turn corresponds to about 60 AU at the distance of the Taurus cloud. After assumption of constant outflow velocity, full ionization and mass conservation in an infinite, spherically symmetric envelope, the emitted radio spectrum varies as $\nu^{-0.1}$ at frequencies where the envelope is optically thin. The flux is given in a good approximation by (cf. Bertout and Thum 1982)

$$S_\nu = ((\pi+\epsilon/3)/4) \ (4\pi N_c^2 R_c^3/D^2) \ B_\nu(T_e) \, \mathcal{X}_\nu , \qquad (1)$$

where D is the distance to the star, $\mathcal{X}_\nu$ the gas opacity, R_c the envelope's inner radius, N_c the electron number density at R_c, and $B_\nu(T_e)$ the Planck function at the local electron temperature that is assumed constant throughout the envelope. The value of ϵ is 1 or 2 depending on whether the central core with radius R_c is opaque or transparent.

The turnover frequency ν_t at which the envelope becomes optically thick is given by

$$\tau_{\nu_t} = 2 \int_{R_c}^{\infty} n_e^2(r) \, \mathcal{X}_{\nu_t} \, dr \approx 1, \qquad (2)$$

Equation 2 then gives the condition on $\mathcal{X}_\nu$ at ν_t of

$$\mathcal{X}_{\nu_t} \approx 1/(2N_c^2 R_c). \qquad (3)$$

Combining Eqs. 1 and 3 leads to the following relationship between the turnover frequency ν_t and the observed flux S_ν at the same frequency:

$$S_{\nu_t} = (\pi/2)(\pi+\epsilon/3)(2k/c^2)(R_c/D)^2 \ \nu_t^2 T_e. \qquad (4)$$

Numbers for a hypothetical "average observed source" can now be inserted into the previous equations. A typical value of the transition frequency ν_t is about 5 GHz, and the flux at that frequency is about 10 mJy. Assuming $T_e = 10^4$K for the purpose of illustration, Eq. 4 is rewritten as a condition on R_c to get $R_c = 2.2 \ 10^{14}$cm $\approx$ 15 AU. Eq. 2 then gives $N_c \approx 8 \ 10^5$ cm^{-3}, which corresponds to a mass-loss rate of about 4 10^{-7} $M_\odot$/yr when assuming a flow velocity of 300 km/s. If the spectrum of a partially resolved source at 5 GHz such as IRS5 in L1551 is formed in a spherical, constant velocity wind, it must therefore originate from a region located at a radius of several AU from the star, because of the requirement of optical thinness at frequencies above about 5 GHz. This condition means that density cannot increase as r^{-2} down to the surface of the infrared source.

Two possibilities then suggest themselves: (i) the gas flowing away from IRS 5 does not become ionized until it is far from the star, or (ii) our basic assumption of spherical symmetry is inadequate. Little is known about the ionization of outflows from young stellar objects. The number of Lyman photons required to photoionize the gas is, in most cases, far higher than provided by a main-sequence star of bolometric luminosity similar to the embedded source (Snell and Bally 1986). But lack of knowledge about the Lyman continuum of pre-main-sequence objects does not allow photoionization to be ruled out. While other mechanisms have been proposed--most notably collisional ionization in the wind (cf. Simon et al. 1983)--, all predict a high degree of ionization close to the central object. The inadequacy of assuming spherical symmetry is therefore a likely possibility, since out-flows are often bipolar on intermediate to large scales. In the case of L1551, there is evidence for anisotropy on scales as small as 70 AU (Mundt, 1985). The possibility of a bipolar wind on even smaller scales should thus also be considered.

The present lack of precise estimates for the flow's degree of collima-tion makes it difficult to predict its detailed emission spectrum when the assumption of spherical symmetry is relaxed (cf. Reynolds 1986). One can check, however, whether the conditions of both small optical depth (neces-sary to reproduce the shape of the spectrum at high frequencies) and observed flux level can be met for any reasonable choice of main parameters. To do this, it suffices to assume that the flow occurs within two cylinders of radius R_d extending from R_c to R_{max} in opposite directions and that the density is constant within the cylinder. The flux in the observer's direc-tion, defined by the view angle i measured from the symmetry axis, is approximately (exactly when i = 90°) given by

$$S_\nu \approx (2\pi/\sin i)(R_d/D)^2 N_0^2 \chi_\nu B_\nu(T_e)(R_{max}-R_c) \tag{5}$$

and as before the condition on optical depth is written as

$$\chi_{\nu_t} \approx \sin i/(2N_0^2 R_d) \tag{6}$$

Combining Eqs. 5 and 6 and assuming that $R_{max} \gg R_c$ and that i = 90°, we get a condition on the product $R_d R_{max}$. With the same choice of parameters as before, we find $R_{max} R_d \approx 10^{29} cm^2$. Since the ionized collimated flow extends at most to about $10^{17} cm$ (Mundt 1985), R_d is at least of the order $10^{12} cm$. Equation 5 then gives $N_0 \approx 10^7 cm$, which corresponds to a mass-loss rate of $2 \, 10^{-10} M_\odot yr^{-1}$. The derived value of N_0 should now be compared to the hydrogen densities in the range 15-130 cm^{-3} that are typically found in optical jets (Mundt 1986). The discrepancy seems to indicate that the out-wards density decrease in the jet is quite strong (optical studies sample primarily the outer parts of the jet), and that the bulk of emission is produced primarily in the innermost, denser ionized regions, where jet collimation is presumably more effective. An outwards decreasing electron density is also necessary in order to account for the low-frequency part of the radio spectrum.

Thus, extended radio emission around young stellar objects probably originates in highly collimated flows (see also Snell et al. 1985). An approximate value of the mass-loss rate derived in the case discussed above is on the order of a few $10^{-10} M_\odot yr^{-1}$, a value apparently much smaller than inferred from the CO flows associated with these sources. There is, however, considerable uncertainty in mass-loss rates derived from CO flows (cf.

Cabrit and Bertout 1986). Alternatively, one can argue either that today's stellar mass-loss is lower than the mass-loss that was once responsible for the large-scale CO flow, or that the ionized component represents only a small fraction of the stellar wind's mass.

III.2. Compact Sources

For unresolved sources, radio spectra with moderate slopes are formed most easily in regions where the density does not fall off too rapidly, e.g. in gravity-driven flows, whether ballistically decelerated outflows or gravitational infall. The gas velocity is $v(r) = V_c(R_c/r)^{1/2}$, where the sign of V_c indicates the flow's direction. Radio observations do not indicate the flow's direction, but only the electron density distribution, which in both cases is given by $n_e(r) = N_c(R_c/r)^{3/2}$ when assuming mass conservation and full ionization. In this case, it is impossible to distinguish between optically thin and thick regimes any more, since the flux varies as $^{-0.1}$ in both cases (Panagia and Felli 1975). Because the flux integration diverges with infinite envelope radius, one cannot derive a simple formula for the emitted flux equivalent to Eq. 4 above. Instead, one must perform a numerical integration up to a maximum envelope radius. The change in spectral slope is then attributed to the finite size of the envelope rather than to optical depth effects.

This model allows easy reproduction of radio spectra like those of Figure 3 (Bertout 1983). Accretion is thus a viable alternative to winds for explaining the unresolved radio emission of young stellar objects. It is also a physically pleasing one since (i) the gas's potential energy provides a reservoir from which energy needed for ionizing the infalling gas can easily be extracted, and (ii) it is consistent with the scenario of Shu and Adams. Because deviations from spherical symmetry in the accreting material occur close to the star in their model, they would go unnoticed in the radio centimetric spectrum, which samples a region located at 10^{14}-10^{15} cm from the stellar center.

There is however next to no other observational evidence for gas accretion in young embedded stars such as T Tauri South or L1551 IRS5, which are both associated with molecular CO outflows. Instead, a spectrogram of IRS 5 obtained by Mundt et al. (1985) reveals a Hα Type-I P Cygni profile, which further confirms that the ionized gas around the star is flowing out. But then, IRS5 is an extended radio source; and it is conceivable that it represents a later stage than T Tau S in Shu's scenario, in which the emission measure of the outflowing gas would largely dominate emission of accreting material. At any rate, results are the same for ballistic outflow and accretion flow. With the same input data as above, one finds the following representative values for the parameters of the radio emitting regions: $R_c \approx$ $2R_\odot$, $R_{max} \approx 5 \ 10^3 R_c$, $N_c \approx 3 \ 10^{10}$ cm^{-3}, which correspond to a mass-flow rate of approximately $6 \ 10^{-9} M_\odot yr^{-1}$ when assuming $V_c = 300$ km s^{-1}.

Both spherical gravity-driven flows and well-collimated, bipolar winds will thus emit radio spectra similar to those observed. Resolution of the source (or lack thereof) provides one means of distinguishing between these possibilities, as do high signal-to-noise measurements of the radio flux at centimetric wavelengths. The two models, which produce similar (but distinguishable, at least in principle) spectra, may correspond to different phases in the scenario described by Shu and Adams, with the unresolved sources still in their accretion-dominated phase while resolved sources have already developed a sizable wind.

IV. EVIDENCE OF DISKS AROUND YOUNG STARS

IV.1. Large-Scale Structures

Young stellar objects are embedded in molecular clouds, and their light is absorbed and reemitted in the far-infrared by the cloud's dust. The performance of available infrared detectors makes the direct detection of disks around these objects a rather difficult task. The work of Cohen et al. (1985) demonstrates both the successes and still present limitations of far-infrared observations using airborne observatories. Observations of molecular lines at radio wavelengths also reveal somewhat flattened structures around young, massive stellar objects. But the relationship, if any, between these large-scale structures and the underlying nebular disks discussed by Adams and Shu is totally unclear at this point.

Observations offer indirect evidence for the presence of a disk around the T Tauri star HL Tauri. Grasdalen et al. (1984) recorded digital images of HL Tauri at 1.6 and 2.2μ and used maximum entropy image reconstruction techniques to show that the star was surrounded by a somewhat flattened structure that they interpret as light scattered by dust in a disk with diameter about 300 AU surrounding HL Tau. Beckwith et al. (1986) used the Owens Valley millimetric interferometer to map the CO(J=1->0) transition in the vicinity of HL Tau, and conclude that the star is embedded in a small-scale (< 900 AU) concentration of molecular gas which appears gravitationally bound to the star. Beckwith et al. suggest that the condensation represents a disk with some 0.01 $M_\odot$.

IV.2. Small-Scale Accretion Disks Around T Tauri Stars

The infrared excess observed in T Tauri stars and related objects has traditionally been attributed to thermal reemission of stellar photons by a dusty envelope (Mendoza 1968; Cohen and Kuhi 1979). And ultraviolet excess emission is usually thought to be of chromospheric origin (cf. Herbig and Goodrich 1986). The physical processes responsible for chromospheric and envelope heating have not yet been determined, but magnetism has been suggested by several authors. Instead, thermal emission from a dusty accretion disk surrounding the young star provides a more attractive and self-consistent explanation of both the near-infrared and ultraviolet excesses observed in T Tauri stars.

The theoretical structure and evolution of steady-state accretion disks was investigated by Lynden-Bell and Pringle (1974), who find that the temperature distribution $T_D(r)$ of an accretion disk surrounding a central star of mass M_* and radius R_* is related to accretion rate M by

$$T_D(r) = ((3GM_*\dot{M}/8\pi\sigma r^3)(1-(R_*/r)^{1/2}))^{1/4} \tag{7}$$

where G is the gravitational constant and σ the Stefan-Boltzmann constant. Provided the star rotates slower than the inner part of the disk, about half of the accretion energy $E_k = GM_*\dot{M}/2R_*$ is radiated away in a boundary layer near the central star, with temperature

$$T_B \approx 6.5 \; 10^4 \; (M_*\dot{M}/10^{-5}M_\odot yr^{-1})^{1/4}(R_*/R_\odot)^{-3/4} \tag{8}$$

The emergent spectrum can then be calculated from these temperature laws. Several investigators have done this for accretion disks in cataclysmic variables and symbiotic stars and, more recently for FU Orionis objects (Hartmann and Kenyon 1985). Because of the large range of system temperatures, the spectrum exhibits both ultraviolet and infrared excesses over the energy distribution of the central star alone. I am investigating the disk hypothesis for T Tauri stars using a database of photometric measurements gathered simultaneously in the optical (UBVRI filters) and in the near-infrared (JHKL filters). Examples of computed spectra are given in Figure 4 along with the observed energy distributions of four T Tauri stars. Besides the view angle, the main free parameters are (i) the disk's maximum temperature T^d_{max}, (ii) the stellar radius, and (iii) the outer disk radius. These preliminary results are summarized in Table 1, where both the stellar effective temperature and the visual extinction are consistent with Cohen and Kuhi's (1979) data, while the four other disk parameters have been adjusted to make a reasonably good fit to the data. The lack of IUE and far-infrared data does not allow us to constrain the model entirely, however. In particular, far-infrared data are needed for testing the disk's temperature distribution more exactly, which is an important issue since one expects different temperature distributions for different disk support mechanisms (cf. Adams, Lada and Shu 1987). The nebular disks advocated by Adams and Shu do not emit strongly in the ultraviolet, since they have no accretion boundary layer (the stellar photosphere is assumed to rotate as rapidly as the inner parts of the disk). While this may be true for embedded sources, we know that the photospheres of T Tauri stars rotate slowly (cf. Bouvier et al. 1986). It is therefore conceivable that an accretion boundary layer is formed only after the central star has lost angular momentum, presumably as a result of its strong wind.

There have been few attempts to find evidence of disks in line spectra of young stellar objects, mainly because it requires high-resolution spectroscopy of rather faint objects. Hartmann and Kenyon (1985) note that some absorption lines of V 1057 Cygni are double, which they attribute to rotation in the accretion disk. Another indirect indication of the presence of disks around T Tauri stars is also the shape of their forbidden lines formed in the moving gaseous envelopes surrounding the stars. Appenzeller et al. (1984) compare high-resolution ($R > 10^4$) profiles of forbidden lines in T Tauri stars with predictions from model computations using several geomet-

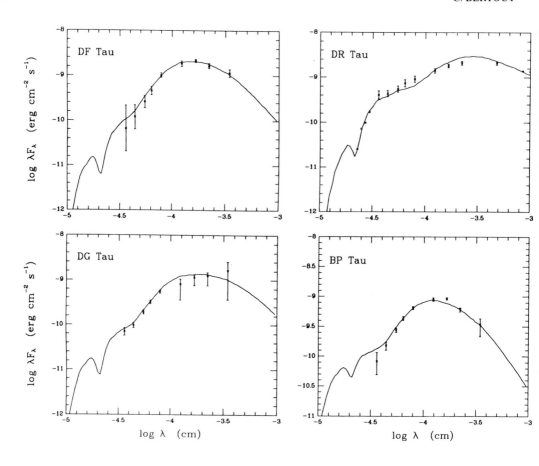

Figure 4 Preliminary comparison of accretion disk spectra with observed energy distributions of T Tauri stars. Error bars on the photometric data do not represent photometric accuracy, but indicate instead the amplitude of observed variations (when available). In the case of DR Tauri, a rather extreme T Tauri star, the bars show typical night-to-night variations.

Table 1 Disk model parameters

Star	R_* (R_0)	R_{disk} (R_0)	T_{dmax} (R_{max})	i (°)	Teff (K)	A_v (mag)
DF Tau	3	6	2700	40	3800	2.0
DR Tau	4	50	2400	0	<2500	1.9
DG Tau	2	50	3500	75	4100	1.4
BP Tau	3	5	2300	75	3800	0.9

ries for the emission region; and they find that the often blue-displaced forbidden lines are formed not in spherically symmetric flows, but rather in cones. Furthermore, the presence of a dusty disk around the star appears necessary to explain occultation of the red side of the line.

V. MAGNETISM AS THE ENGINE OF T TAURI CIRCUMSTELLAR ACTIVITY

More than just a disk is needed to explain the multiciplity of phenomena observed in the circumstellar environment of T Tauri stars (cf. Bertout 1984 for a review). While disks provide a geometry which helps collimate and ionize outflows, it is unlikely that they are actually driving the flows. Bouvier (1986) studied correlations between various emission lines that are known diagnostics of magnetic activity in late-type dwarfs. An unexpected result of this investigation is that the surface fluxes (and total luminosities) in the Hα, CaII K and MgII k lines correlate rather well with one another for a large sample of T Tauri stars of different activity levels. Furthermore, these correlations apparently match the correlations found in main-sequence stars (see the contribution by Bouvier in this volume). These results are not fully understood at this point. It seems unlikely that both K and Hα are formed in exactly the same atmospheric region, since (i) their excitation mechanisms are usually different in late-type dwarfs, and (ii) their profile widths and shapes are usually quite different too. They may, however, be formed in different parts of the same overall region, such as in a stellar wind, while the previous correlation indicates that a common underlying physical process is responsible for both lines. It is nevertheless curious that both Hα and CaII flux should scale in exactly the same way when the strength of the underlying process, presumably the magnetic field, varies. A detailed, quantitative interpretation of these intriguing data is likely to bring valuable insight in the structure of T Tauri envelopes.

Studies of X-ray emission as a function of stellar rotation and of spot properties in T Tauri stars support the view that part of T Tauri activity is magnetically driven. Figure 5 demonstrates that the X-ray flux of T Tauri stars is anti-correlated, albeit weakly, with the rotation period (Bouvier 1986). Such a relationship is also found in other late-type stars, and is usually interpreted as an indication for dynamo-driven coronal heating. Basri, Laurent and Walter (1985) show that the X-ray surface flux of RS CVn stars is not as tightly correlated with rotation period as other activity indicators. A correlation study of UV lines and rotation in T Tauri stars would therefore help confirm the correlation in Fig. 5. Periodic light curves variations--interpreted as rotational modulation by groups of spots on the stellar surface--are found in the whole population of T Tauri stars irrespective of their place in the HRD and of their activity level (Bertout, Bouvier and Bouchet 1987). Computed properties of spots show them to be quite similar to spots found on RS CVn systems. This further confirms that rotational velocity, which is quite similar in T Tauri stars and RS CVn stars, is the main parameter in stellar magnetic activity.

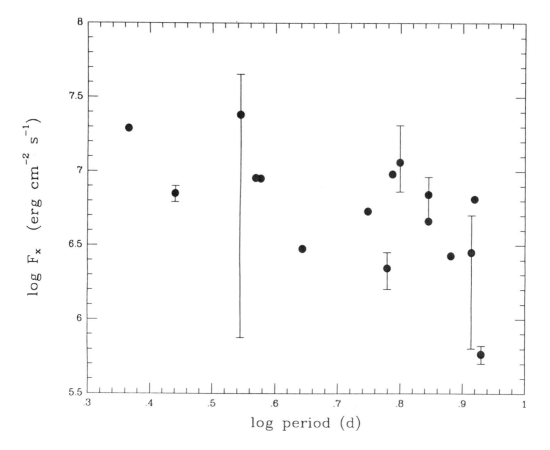

Figure 5 The relationship between X-ray flux and rotation period of T Tauri stars. The observed range of flux variability is indicated by bars when several measurements have been made; in this case, the dots indicate the most often observed flux value.

VI. SUMMARY

The main conclusions of this review are as follows.

i) Compact radio sources associated with young stellar objects may provide evidence for protostellar accretion. High signal-to-noise VLA spectra and direct use of visibility functions would allow testing this possibility further.

ii) More extended radio emission like that from IRS5 L1551 is best interpreted as caused by a bipolar collimated wind, and may represent a somewhat later evolutionary phase than compact sources.

iii) Disks are probably present around many embedded low-luminosity infrared sources and T Tauri stars, and evidence for this is discussed both in Shu and Adams' review and above. While one can argue about the detailed nature of these disks, it is encouraging for future quantitative work that the structure of disks has already been studied in several other astrophysical instances.

iv) Presence of a disk and magnetically-driven atmospheric activity appear to be major ingredients of the T Tauri phenomenon. Their respective roles must be further disentangled.

Many features of young objects --radio emission, bipolar flows, magnetic activity-- were not predicted by theories of star formation. This does not necessarily mean that these theories are at fault, since the appearance of young stars is determined mainly by circumstellar and surface phenomena. Indeed, theorists are doing their best to catch up with observations.

CREDITS It is a pleasure to thank Sylvie Cabrit and Jérôme Bouvier for their contributions to Sections II and V respectively. I also enjoyed valuable discussions with Philippe André and received considerable help from Joli Adams in editing the text.

REFERENCES

Adams, F., Lada, C.J., Shu, F.: 1987, Astrophys. J., in press
Appenzeller, I., Jankovics, I., Oestreicher, R.: 1984, Astron. Astrophys. 141, 108
Basri, G., Laurent, R., Walter, F.M.: 1985, Astrophys. J. 298, 761
Bechis, K.P., Harvey, P.M., Campbell, M.F., Hoffmann, W.F.: 1978, Astrophys. J. 226, 439
Beckwith, S., Sargent, A.I., Scoville, N.Z., Masson, C.R., Zuckerman, B., Phillips, T.G.: 1986, preprint
Bertout, C.: 1983, Astron. Astrophys. 126, L1
Bertout, C.: 1984, Rep. Prog. Phys. 47, 111
Bertout, C., Bouvier, J., Bouchet, P.: 1987, in preparation
Bertout, C., Cabrit, S., Thum, C.: 1987, in preparation
Bertout, C., Thum, C.: 1982, Astron. Astrophys. 107, 368
Bouvier, J. 1986: Doct. Thesis, Paris VII University, Paris, France
Bouvier, J., Bertout, C., Benz, W., Mayor, M.: 1986, Astron. Astrophys., 165, 110
Cabrit, S., Bertout, C.: 1986, Astrophys. J., 307, 313
Cohen, M., Kuhi, L.V.: 1979, Astrophys. J. Suppl. 41, 743
Cohen, M., Harvey, P.M., Schwartz, R.D.: 1985, Astrophys. J. 296, 633
Edwards, S., Snell,R.L.: 1983, Astrophys. J. 270, 605
Grasdalen, G.L., Strom, S.E., Strom, K.M., Capps, R.W., Thompson, D., Castelaz, M.: 1984, Astrophys. J. Letters 283, L57
Guylbudaghian, A.L., Glushkov, Y.J., Denisuyk, E.K.: 1978, Astrophys. J. Letters 224, L127
Hartigan, P., Lada, C.J.: 1985, Astrophys. J. Suppl. 59, 383
Hartmann, L., Kenyon, S.J.: 1985, Astrophys. J. 299, 462
Herbig, G.H.: 1960, Astrophys. J. Suppl. 4, 337
Herbig, G.H., Goodrich, R.W.: 1986, Astrophys. J., in press
Königl, A.: 1982, Astrophys. J. 261, 115

Lynden-Bell, D., Pringle, J.E.: 1974, Mon. Not. R. astr. Soc. **168**, 603
Mendoza, V.E.E.: 1968, Astrophys. J. **151**, 977
Montmerle, T., André, P., Feigelson, E.: 1985 in <u>Nearby Molecular Clouds</u>,
       Ed. G. Serra, Lectures Notes in Physics 237, Springer-Verlag
       Heidelberg
Mundt, R.: 1985 in <u>Nearby Molecular Clouds</u>, Ed. G. Serra, Lectures Notes in
       Physics 237, Springer-Verlag Heidelberg
Mundt, R.: 1986, Canadian Journal of Physics, in press
Mundt, R., Stocke, J., Strom, S.E., Strom, K.M., Anderson, E.R.: 1985,
       Astrophys. J. Letters 297, L41
Panagia, N., Felli, M.: 1975, Astron. Astrophys. 39, 1
Pease, F.G.: 1917, Astrophys. J. 46, 24
Pravdo, S.H., Rodriguez, L.F., Curiel, S., Canto, J., Torrelles, J.M.,
       Becker, R.H., Sellgren K.: 1985, Astrophys. J. Letters 293, L35
Racine, R.: 1968, Astron. J. 73, 233
Reynolds, S.P.: 1986, Astrophys J. 304, 713
Sandell, G., Olofsson, H.: 1981, Astron. Astrophys. 99, 80
Simon, M., Felli, M., Cassar, L., Fischer, J., Massi, M.: 1983, Astrophys.
       J. 266, 623
Snell, R.L., Bally, J.: 1986, Astrophys. J., 303, 683
Snell, R.L., Bally, J., Strom, S.E., Strom, K.M.: 1985, Astrophys. J. 290,
       587
Strom, S.E., Grasdalen, G.L., Strom, K.M.: 1974, Astrophys. J. 191, 111
Tenorio-Tagle, G.: 1982, in <u>Regions of Recent Star Formation</u>, Eds. R.S.
       Roger and P.E. Dewdney (Dordrecht: Reidel), p. 1
Wilking, B.A., Mundy, L.G., Schwartz, R.D.: 1986, Astrophys. J. Letters 303,
       L61

BIPOLAR FLOWS AND JETS FROM STARS OF DIFFERENT SPECTRAL TYPES:
OBSERVATIONS

Martin Cohen
Radio Astronomy Laboratory
University of California
Berkeley, CA 94720
U.S.A.

ABSTRACT. The fact that bipolar flows are widespread among stars of
very different spectral types is emphasized. First, the stars
associated with the phenomenon are divided into broad types: protostars
and pre-main-sequence stars; red giants; symbiotic objects;
protoplanetaries; planetaries; novae and cataclysmic variables; and
peculiar hot stars. Second, the evidence for circumstellar "disks" or
toroids is considered among these different categories of star.
Finally, the possible role of binarity is discussed.

1. INTRODUCTION

There were bipolar flows long before there were molecular radio
astronomers to tell us that they had discovered them. Optical
astronomers knew these already as the class of bipolar nebulae. Defined
strictly morphologically, bipolar nebulae are known to constitute a
heterogeneous body of objects (Cohen 1983). Some are associated with
pre-main-sequence stars or presumed protostars; others with red giants,
both oxygen- and carbon-rich; still others are caught in the brief
transition from red giant to planetary nebula, the so-called
"protoplanetaries"; even some fully-fledged planetaries reveal bipolar
structure rather than spherical.
  We have always felt intuitively that bipolar nebulae have central
stars embedded in equatorial toroids and have undergone recent bipolar
mass loss. In most cases it is hard to prove, in the absence of radio
molecular data, that a bipolar, usually reflection, nebula is actually
driving mass loss through its lobes. Velocity information by optical or
near-infrared techniques is often lacking. A rare exception is GL 2688,
the "Egg" Nebula in whose lobes directly viewed molecular emission bands
(of C2, C3, SiC2) are optically detected (Cohen and Kuhi 1980). The
radial velocities of these bands reveal a clear bipolar flow from this
carbon-rich F supergiant. In general, bipolarity is assumed to extend
from mere morphology to velocity fields if the latter are unknown.
  In this talk I want to focus on three aspects of bipolar nebulae
and flows. First, let me remind you of the very wide range of spectral

I. Appenzeller and C. Jordan (eds.), Circumstellar Matter, 39–50.

types of star with which these flows are allied. Second, we will
examine some of the recent direct evidence in favor of the existence of
circumstellar disks. Finally we will address the possible role of
binarity in the creation of bipolars. To highlight the wide range of
spectral types of central star it is convenient to divide the unevolved
objects into protostellar and pre-main-sequence systems, and the evolved
objects into red giants, symbiotic stars, protoplanetaries, planetaries,
novae and cataclysmic variables, and other peculiar hot stars. It is
important to have a working definition of a "jet" too since this word is
experiencing widespread usage now, and not always with the same
connotations. I shall use "jet" here to signify a highly confined flow
that originates in the immediate vicinity of the responsible star or, if
two-sided, includes the star in its body. This definition eliminates
random filamentary (HH or otherwise) structures the locations of whose
exciting or illuminating stars are not known by any technique, direct or
indirect. Opening angles should be at most a few degrees.

2.   THE SPECTRAL CHARACTER OF THE ASSOCIATED STARS

2.1 Protostars and Pre-Main-Sequence Stars

Since most presumed protostars are optically unseen I am, perhaps,
stretching the title of this talk to include them. However, in a few
cases we can detect and even classify reflected stellar spectra, and in
still rarer circumstances we do directly see the exciting stars. An
obvious subclass of low-luminosity and therefore low-mass protostars is
the group that excites Herbig-Haro (HH) Objects. When visible directly
these are strong emission-line T Tau stars (e.g. AS 353A, HL and DG Tau)
and, indirectly, several HHs scatter the light of stars of this type too
(e.g. HHs 30, 55, 100: Cohen, Dopita and Schwartz 1986a). Not all HHs
are demonstrably jetlike or bipolar at visible wavelengths. Often only
one-sided flows are visible although VLA radio continuum observations
indicate the truly two-sided character of the mass loss (e.g. Bieging,
Cohen and Schwartz 1984; Bieging and Cohen 1985). Usually the red-
shifted jet is lost to optical view because it enters a dark cloud.
Likewise the one-sided fan nebulae associated with some pre-main-
sequence stars, such as PV Cep and R Mon, arise because an extensive
dusty disk overlies one lobe. However, deep CCD frames, especially
those in the I-band, or photographs at fortuitous times, can reveal very
faint traces of a truly bipolar nebula. Examples of this are the faint
southern spike of R Mon recently recorded by Walsh and Malin (1985) and
the very faint southern fan of PV Cep detected by Ray (1986). A rather
unusual example of a jet in which the blue-shifted lobe is extremely
sparse but the red is relatively bright is "DG TauB", discovered by
Mundt and Fried (1983) but not physically related to DG Tau. Spectra by
Jones and Cohen (1986) of this elegant tapered flow show it to be
bipolar and of unprecedentedly low excitation (in one knot the ratio of
[SII] (6717+6731) to H-alpha is almost 12:1!).
     In two cases FU Ori stars are believed to excite bipolar flows,
namely L1551 IRS5 (early K: Mundt et al. 1985) and HH57 (F8III: Cohen,
Dopita and Schwartz 1986b). Optical "jets" sometimes have HH spectra

and may reflect red, presumed stellar, continua though these are usually
too faint for classification. Some even older T Tau stars, not
associated with optical jets or HH objects, lose mass in a bipolar
fashion as demonstrated by the frequent blue shifts of nebular [OI]
lines in their spectra (Appenzeller et al. 1984).

The nebulous early-type stars first studied by Herbig (1960) and
recently reinvestigated by Finkenzeller (1985) also reveal blue-shifted
[OI] lines, indicative of non-spherical flow. In some cases these are
allied with conspicuous bipolar nebulae, like Lk Hα-208 and Lk
Hα-233. Polarimetric images of these clearly mark them as reflecting
the light of a star within the nebulae [from the clear pattern of
centrosymmetric electric vectors: e.g. Shirt, Warren-Smith and Scarrott
(1983); Aspin, McLean and McCaughrean (1985); Aspin, Mclean and Coyne
(1985)].

There are definite indications of alignments between bipolar flows
and the local interstellar magnetic field (e.g. R. J. Cohen, Rowland and
Blair 1984). This global organization is shown, too, by the existence of
close pairs of parallel flows in the R CrA cloud (from R and T CrA:
Ward-Thompson et al. 1985) and in L1551 (from IRS5 and the IRAS object
NE of it: Draper, Warren-Smith and Scarrott 1985).

The impact of these non-spherical flows from both proto- and
pre-main-sequence stars on their environment is registered through the
bipolar patterns of disturbances that extend out into the clouds when
mapped in molecules like CO (e.g. Bally and Lada 1983). Probably all
stars evolve through a phase of bipolar mass loss early in their lives.
Such a conclusion is borne out, also, by the fact that unambiguously
young bipolar systems are allied with stars from B8 (lying in a Bok
globule: Bruck and Godwin 1984) through the typically K-type spectra of
T Tau stars to M3.5 (for HH55's exciting star though the HH is so small
that it has not yet been identified as bipolar) or at least to some
M-type if Parsamyan 13 is included (Cohen et al. 1983).

2.2 Red Giants

Optical spectropolarimetry shows that extreme carbon stars such as
IRC+10216 and IRC+30219 are very highly polarized, up to 30% for GL 2699
(Cohen and Schmidt 1982). For IRC+30219 the abrupt rotation of position
angle and virtual nulling of polarization indicate a definitely bipolar
scattering nebula. Even for IRC+10216, which has not been observed to
show this rotation, there is good agreement between the inferred axis of
scattering (E-vector in p.a. 120 deg.) and the elongation recently
reported in CCD images (p.a. 30 deg.: Crabtree, McLaren and Christian
1986). Velocities as high as 80 km/s have been noted in IRC+30219 in
the form of episodic optical shock spectra (Cohen 1980) although the
morphology of this material flow is unknown.

Among the extreme OH/IR stars the system OH0739-14 seems unique in
the variety of peculiarities displayed. M9 III (or perhaps I?)
starlight is reflected from both lobes of this very red bipolar nebula
as is a curious blue continuum probably due to an otherwise unseen hot
companion. However, along the axis of this nebula are two HH objects,
expanding away from the central star at velocities of ± 150 km/s
(Cohen et al. 1985a)! One even shows greatly enhanced nitrogen

abundance, relating its material to the central evolved star rather than to the ambient medium.

The OH maser peculiarity of this system is the presence of a very broad (100 km/s) plateau of weak satellite-line emission below a sharp, powerful spike Morris and Bowers 1980). This spike is believed to be associated with the line of sight to a hot companion embedded in a substantial disk (Morris, Bowers and Turner 1982); a binary was also postulated by Morris (1981) to account for the morphology of the reflection nebula. Other OH/IR sources may be akin to OH0739-14 because of their maser spectra. These include the bipolar reflection nebula M1-92 (spectral type B0: Cohen and Kuhi 1977) which may have similar maser spectral structure (e.g. Davis, Seaquist and Purton 1979). Roberts 22 (A2I: Allen et al. 1980) may be a member of this small group of objects because its 1612 MHz profile might arguably contain a weak, broad plateau of emission although its satellite lines seem to show the characteristic double-spiked spectra of spherically-shelled late-type OH/IR emitters. There is continuing controversy also about whether to include the OH/IR object IRC+10420, a bright, visible F8Ia star that excites strong 1612 MHz emission (cf. Bowers, Johnston and Spencer 1981) with these other two nebulae. Mutel et al. (1979) saw possible evidence in the OH morphology for a disk, perhaps aligned with the elongated optical nebulosity (Thompson and Boroson 1977), but this is not known to be bipolar. Diamond, Norris and Booth (1983) interpret their MERLIN OH maps in terms of bipolar flow but recent near-infrared speckle results (Ridgway et al. 1986) show no convincing evidence for the asymmetry of its dust shell.

2.3 Symbiotic Stars

In this category we include R Aqr and HM Sge. R Aqr is an M7 giant in a rather complex bipolar nebula whose elegant structure was recently explained by Solf and Ulrich (1985) in terms of two episodes of bipolar outflow. Kafatos, Michalitsianos and Hollis (1986) review the ultraviolet excitation and variability of R Aqr. They discuss the curious optical and radio "jet" that emerges from the star and the history of the "new" knot. Mauron et al. (1985) feel from their UV CCD imagery of R Aqr that the appearance of this latter feature does not connote an episode of outflow but is the response of pre-existing material to sudden ionization by the variable hot source. (This "jet" does not align with the major axis of R Aqr's previous bipolar flows but it does roughly coincide with the brightest part of the inner nebula.)

The red giant/symbiotic system, HM Sge, was also studied by Solf (1984) who found evidence from slit spectra for narrowly confined bipolar, high- velocity (200 km/s) mass loss in the form of small regions of forbidden line emission.

2.4 Protoplanetaries

Most newly-found optically-visible bipolar nebulae belong to this class. They are presumably related to the symbiotic stars, or at least to some red giants, while these are evolving into planetaries. Their central stars are hot; their nebulae are usually small and of high density, and

it is sometimes hard to separate the protoplanetaries from compact
planetaries. Indeed, it may take VLA observations of the latter to
recognize their bipolar morphology (e.g. M1-6: Kwok and Purton 1983).
Calvet and Cohen (1978) and Cohen (1983) list the protoplanetaries and
draw an H-R diagram to show their locations.

2.5 Planetaries

New CCD emission-line images by Balick (1986) indicate that many
planetary nebulae, especially the compact ones or those bright in the
mid- infrared, have bipolar rather than spherical symmetry. Sometimes
the inner nebular structures are quite unlike the outer, as in Abell 30,
for example, where the nucleus is embedded in a pair of orthogonal
bipolar flows of processed material, perhaps 1500 years old, expanding
at roughly 20 km/s from the central star (cf. Reay, Atherton and Taylor
(1983).
        The bipolar planetary NGC 2346 continues to draw attention. The
best picture seems to involve eclipses of a binary system, with unseen
hot companion, by an elongated toroidal dust cloud (recently resolved by
optical speckle work: Meaburn et al. 1985) that rotates around the
stars, orthogonal to which is a bipolar mass flow (cf. Walsh 1983; Roth
et al. 1984; Acker and Jasniewicz 1985).
        The nebula Mz-3, usually thought of as a fully-fledged planetary
(e.g. Lopez and Meaburn 1983), may be interpretable as a protoplanetary
according to Meaburn and Walsh (1985) who have studied the kinematics of
its inner and outer envelopes. The faintest outer shell shows great
line width (up to 450 km/s) compared with the much slower, inner bipolar
structure that is expanding at ± 50 km/s.

2.6 Novae and Cataclysmic Variables

The old nova V603 Aql was studied spectroscopically in great detail.
From these data it is possible to construct a clear picture of the
morphology of its nebular shell and to recognize this as built up from a
series of successive biconical ejections from the nova within its
accretion disk (Weaver 1974). This system was recently reinvestigated
by Haefner and Metz (1985) who give details of the binary and the
accretion disk.
        The novalike cataclysmic variable PG1012-029 shows a non-rotating
component to its line profiles that dominates the character of the
eclipses of the secondary (Honeycutt, Schlegel and Kaitchuk 1986).
These authors attribute this component to a bipolar wind from the
accretion disk.

2.7 Peculiar Hot Stars

If we study the peculiar hot star, MWC 349, we will continue this
bipolar/biconical theme. This star has a vast literature of
peculiarities. I shall confine my remarks here to its VLA maps at 5 GHz
(Cohen et al. 1985b) and 15 GHz (White and Becker 1985). The 5 GHz map
clearly shows an interaction between the winds of MWC 349B, a B0III, and
MWC 349A, the bright central component. Further, despite the good fit

of the visibility function to a spherical mass loss law, this central
component patently shows non-sphericity in the form of what appear to be
biconical projections. At the higher frequency it appears bipolar.

3. EVIDENCE FOR CIRCUMSTELLAR DISKS

3.1 Photometric and Polarimetric Imagery

Usually the fainter lobes of bipolar nebulae are also redder. It is
therefore easy to believe that the fainter lobes are more heavily
extinguished by local obscuration, perhaps due to overlying
circumstellar dust disks. There are even indirect clues to this in the
shapes of some of these lobes whose boundaries, closest to the central
stars, are concave towards the stars (e.g. M1-92). Slightly more direct
evidence comes from the breakdown of the centro- symmetric pattern of
polarization vectors in the vicinities of the stars, or from the
presence of fan-shaped regions of low polarization (due to optically
thick scattering in a toroid by grains aligned parallel to the disk
plane: cf. Aspin et al. 1985a,b). In Lk Hα-208, Shirt, Warren-Smith
and Scarrott (1983) detect an extensive disk partially overlying the
southern nebular lobe by their photometric and polarimetric imaging.
     We might extend the imagery discussed here to include use of the
speckle technique on MWC 349 which Leinert (1986) found to be elongated
in a north-south direction at 2.2 and 3.8 microns, the same axis as
indicated by radio synthesis observations.

3.2 Spatial Variations in Extinction

In particularly fortunate circumstances one can even "see" dust disks by
their obscuration of background stars. This technique works for Mz 3
(Cohen et al. 1978).
     Another method worthy of discussion is valuable in the protostellar
context where the exciting star of an HH chain is not seen optically but
its location can be inferred either from the position of an infrared
source or from the brightness of scattered starlight within different HH
knots. From the Balmer decrements in the HH nebulae one can deduce
approximate extinctions, dominantly foreground since the emission lines
arise in situ. The ratio of infrared to optical luminosities, or the
depth of the 10 μm silicate absorption feature, can be used to
estimate the direct line-of-sight extinction to the central star. Often
it is found that the very large central extinctions drop precipitously
within a very few arc seconds towards the HHs. Examples of this effect
are given in Table 1. They indicate that there is centrally a higher
density of absorbing material than can arise in the foreground
interstellar medium. While it is conceivable to contrive a model in
which this is intracluster (associated with the dark cloud material)
rather than circumstellar (strictly local to the exciting star) this is
only plausible if objects suffer very deep embedding and only one lobe
(the blue-shifted one) of the bipolar flow is visible. It would not
apply to "DG TauB", in which both red- and blue-shifted flows are
clearly visible, nor to DG Tau which star is itself visible, nor to HH57

where, again, both lobes are clearly seen. Therefore one can establish that a relatively thin region of high obscuration exists towards the HH-exciting star that does not extinguish the nearby HH knots appreciably. A dusty disk would account for this.

TABLE I

System	Stellar Extinction	HH Extinction	HH from Star (")
L1551 IRS5	30:	7	< 3
DG Tau	5.4	0.4	8
"DG TauB"	8	0.6	2
HH57 star	5	1:	10

3.3 Near-Infrared Imagery

The highly evolved, almost recombined, planetary nebula Abell 30 was found by Cohen and Barlow (1974) to have a centrally-peaked distribution of mid-infrared-emitting dust grains. Subsequent study by Cohen et al. (1977) showed that, between the Rayleigh-Jeans distribution of the exceptionally hot nucleus and the steeply rising 160K component, there is a population of hot grains whose aperture dependence at 2 and 3 μm seemed closer to $\underline{r}$ than to $\underline{r}$-squared. This could indicate an inclined dusty disk rather than a more spherical distribution of hot grains. Raster maps at 2 and 10 μm (Dinerstein and Lester 1984) have shown that, indeed, such a disk exists. Similarly, for the bipolar nebula NGC 6302, Lester and Dinerstein (1984) near- and mid-infrared rastered images support the conclusion that a dust disk lies around the central star, orthogonal to the bipolar flow axis of this nebula. This infrared work nicely corroborates both the optical polarimetry by King, Scarrott and Shirt (1985) and the VLA observations by Rodriguez et al. (1985), all of which conclude in favor of a similarly oriented disk that must have both dusty and ionized components.

At longer wavelengths (50/100 μm) Cohen and Schwartz (1984) report the existence of a structure, associated with the "Infrared Nebula" in the Cha 1 association, that is spatially-resolved in one dimension (perpendicular to the major axis of the reflection nebula) but unresolved parallel to the "flow". Similarly, for HH-exciting stars, Cohen, Harvey and Schwartz (1985) have discovered 100 micron structures that represent flattened emitting regions, resolved orthogonal to the flows of HH objects but not in the flow directions. One particularly conspicuous flattened zone is associated with the potential FU Ori star that excites HH57. Another surrounds the exciting star of the HH7-11 system. The intrinsic radii deduced for these flattened structures are of order several thousand A.U., with temperatures of the cool dust around 45K.

3.4 Radio Aperture Synthesis Observations

Highly promising results are emerging from aperture synthesis maps of
bipolar nebulae made either at centimeter wavelengths with the VLA
(typically in ammonia lines) or in millimeter wave syntheses in
molecules sensitive to high densities (e.g. CS or HCN). A good example
of the value of these data can be gleaned from new observations of GL
2688. Rieu, Winnberg and Bujarrabal (1986) find a disk in ammonia
surrounding the infrared source that lies between the lobes whereas
other molecules like CO and even HC_7N show emission entirely surrounding
the nebula or at least elongated parallel to the flow·direction,
respectively. He even detects an ammonia "jet", orthogonal to the disk
and moving into the northern lobe. Bieging and Rieu (1986), working
with the 3 mm HCN lines, also note a disk around the waist of GL 2688
with essentially the same diameter as the ammonia toroid. Spectra of
the disk reveal spatial differences with velocity with a pattern
indicative of rotation of the HCN structure.

4. THE ROLE OF BINARITY

4.1 Protostars and Pre-Main-Sequence Stars

At 15 GHz, L1551 IRS5 is resolved into two components that Bieging and
Cohen (1985) interpret as a potential binary. The projection of the
inferred orbital plane is closely orthogonal to the central 5 GHz
contours in the "jet" or extended emission. The jet appears to emanate
from the brighter (the northern) of these two high-frequency sources.
    A similar situation occurs for T Tau. The stronger of the two
radio components lies immediately south of the star and there is an HH
object with perceptible proper motion that is moving to the west,
essentially orthogonal to the line connecting T Tau N and S.
    Mundt and Fried (1983) have suggested that the easternmost blob of
"DG TauB" is elongated roughly perpendicular to the westward line of
knots and that this elongation represents a potential binary. However,
Jones and Cohen (1986) do not confirm the elongation in their direct CCD
image and argue that the exciting star of this system lies between two
bright knots, rather than inside any knot. This is much more plausible,
for if "DG TauB" were such a well-developed, confined flow it must be
very young. If so, nothing should be visible directly in the vicinity
of its exciting star. The binary status of this exciting star should,
therefore, be withdrawn.

4.2 Red Giants and Symbiotic Stars

In at least one model of symbiotics, the red giant is one component in a
binary with a compact hot companion. Such a model was favored by Solf
(1984) for HM Sge and this system exhibits bipolar high-speed mass
outflow. The symbiotic R Aqr shows clear evidence of episodes of
parallel bipolar ejection (Solf and Ulrich 1985), although its "jet" is
not in this same direction. For OH0739-14 the blue continuum reflected
by the nebular lobes speaks for the presence of a blue companion. The

maser model by Morris, Bowers and Turner (1982), that might apply not
only to this OH/IR source but also to M1-92 and Roberts 22, requires
that the strong spike of emission comes from the line of sight through
the equatorial disk toward the hot star. Therefore, it is possible that
all these bipolar OH/IR systems contain binaries.

4.3 Protoplanetaries

So far there is no evidence in favor of binary central stars in any of
the established protoplanetary nebulae.

4.4 Planetaries

The photometric variations of NGC 2346 have been interpreted by many
authors as due to eclipses by a dust cloud orbiting the central star.
The only visible star is too cool to ionize the nebula so there must be
an unseen hot component within the nebula, perhaps within the dust
toroid. 19W32 is a bipolar planetary with a central star that is a
close (<1") double (Kohoutek 1982). The brighter component does not lie
at the nebular center of symmetry and the fainter star is definitely
hot. Perhaps this the faint blue star is the true exciting star and the
bright object just lies in the foreground.

4.5 Novae and Cataclysmic Variables

Implicit in the models for these systems is a binary with an accretion
disk surrounding the active star.

4.6 Other Peculiar Hot Stars

We have already cited the strange high luminosity object, MWC 349. The
VLA map at 5 GHz indicates a region of interaction apparently between
the winds from the two hot stars, MWC 349A and B (BOIII). This
interaction suffices to demonstrate the binary nature of MWC 349 and,
again, the axis of radio bipolarity is clse to orthogonal to the
projected orbital plane of the two stars, in spite of their great
separation.
    HD 44179, the star associated with the "Red Rectangle" nebula is a
very close but visual binary, ADS 4954 (Cohen et al. 1975). Speckle
observations of this star in 1981 (Meaburn and Walsh 1983) successfully
detected the companion, seen last in 1962. Even allowing for the
ambiguity of 180 in position angle the most plausible orbit is not
related to the orientation of the inferred dust disk nor of the nebular
spikes.

5. CONCLUSIONS

In conclusion, it appears that our intuitive picture of bipolar nebulae
as systems incorporating sizable equatorial toroids around the central
star is substantially correct. This toroid may have cool dusty, or
molecular, or ionized components, or any combination of these. Its

orientation is almost invariably perpendicular to the major axis of the nebula.   An appreciable number of stars associated with bipolar nebulae are known to be binaries in which, usually but not always, the inferred orbital plane is orthogonal to the direction of outflow.

Questions still remain about the "disks" and whether they play any significant role in the dynamics of the bipolar flows. For the HH-exciting stars it is, perhaps, more fruitful to think in terms of static confinement of a flow that originates at the protostellar core but must pass through a still-infalling region of gas and dust in order to penetrate to the ambient medium.   Such a picture may well provide confinement, if not collimation, of HH jets (Shu 1987). For the novae like V603 Aql, it is hard to avoid the feeling that activity within the accretion disk is the probable cause of the successive ejections of biconical surfaces.   Cohen et al. (1985a) have built a model for the evolved OH/IR object, OH0739-14, in which the binary creates the highly supersonic HH objects.

In the context of the extreme carbon stars and, for that matter, OH0739-14 too, it is a continuing curiosity (Cohen 1985) that the ancient molecular shells (of CO or OH) indicate a more or less spherical flow while more recent mass loss has somehow generated both an equatorial toroid and a bipolar outflow.   What is the true chronology of the morphology of mass loss in these extremely cool giants?

Certainly the great variety of stellar spectral types allied with bipolar flows indicates that the mechanism, or mechanisms, of non-spherical mass loss are easy to establish and may depend upon common ingredients such as a convective stellar configuration, global mass loss in the presence of an equatorial toroid, a binary system and an accretion disk around the currently active component, stellar rotation, and the circumstellar or interstellar magnetic fields.

6. REFERENCES

Acker, A. and Jasniewicz, G. 1985, A.A., 143, L1.
Allen, D. A., Hyland, A. R. and Caswell, J. L. 1980, M.N.R.A.S., 192, 505.
Appenzeller, I., Jankovics, I. and Ostreicher, J. 1984, A.A., 141, 108.
Aspin, C., McLean, I. S. and Coyne, G. V. 1985, A.A., 149, 158.
Aspin, C., McLean, I. S. and McCaughrean, M. J. 1985, A.A., 144, 220.
Balick, B. 1985, priv. comm.
Bally, J. and Lada, C. J. 1983, Ap.J., 265, 824.
Bieging, J. H. and Cohen, M. 1985, Ap.J. (Letters), 289, L5.
Bieging, J. H., Cohen, M. and Schwartz, P. R. 1984, Ap.J., 282, 699.
Bieging, J. H. and Rieu, N.-Q. 1986, preprint.
Bowers, P. F., Johnston, K. J. and Spencer, J. H. 1981, Nature, 291, 382.
Bruck, M. T. and Godwin, P. J. 1984, M.N.R.A.S., 206, L37.
Calvet, N. and Cohen, M. 1978, M.N.R.A.S., 182, 687.
Cohen, M. 1980, Ap.J. (Letters), 238, L81.
Cohen, M. 1983, proc. IAU Symp. 103, "Planetary Nebulae", ed. D. R. Flower (Dordrecht-Reidel: Holland), p.45.

Cohen, M. 1985, in "Mass Loss from Red Giants", eds. M. Morris and B. Zuckerman (Dordrecht-Reidel: Holland), p.291.

Cohen, M., Aitken, D. K., Roche, P. F. and Williams, P. M. 1983, Ap.J., 273, 624.

Cohen, M. et al. 1975, Ap.J., 196, 179.

Cohen, M. and Barlow, M. J. 1974, Ap.J., 193, 401.

Cohen, M. Bieging, J. H., Dreher, J. W. and Welch, W. J. 1985a, Ap.J., 292, 249.

Cohen, M., Dopita, M. A. and Schwartz, R. D. 1986a, Ap.J. (Letters), Aug. 1.

Cohen, M., Dopita, M. A. and Schwartz, R. D. 1986b, Ap.J. (Letters), 302, L55.

Cohen, M., Dopita, M. A., Schwartz, R. D. and Tielens, A.G.G.M. 1985a, Ap.J., 297, 702.

Cohen, M., Fitzgerald, M. P., Kunkel, W., Lasker, B. and Osmer, P. C. 1978, Ap.J., 221, 151.

Cohen, M., Harvey, P. M. and Schwartz, R. D. 1985, Ap.J., 296, 633.

Cohen, M., Hudson, H. S., O'Dell, S. L. and Stein, W. A. 1977, M.N.R.A.S., 181, 233.

Cohen, M. and Kuhi, L. V. 1977, Ap.J., 213, 79.

Cohen, M. and Kuhi, L. V. 1980, P.A.S.P., 92, 736.

Cohen, M. and Schmidt, G. D. 1982, Ap.J., 259, 693.

Cohen, M. and Schwartz, R. D. 1984, A.J., 89, 277 and 89, 1627.

Cohen, R. J., Rowland, P. R. and Blair, M. M. 1984, M.N.R.A.S., 210, 425.

Crabtree, D. R., McLaren, R. A. and Christian, C. A. 1986, paper presented at the Calgary meeting on "Mass Loss and Evolved Stars", June 1986.

Davis, L. E., Seaquist, E. R. and Purton, C. R. 1979, Ap.J., 230, 434.

Diamond, P. J., Norris, R. P. and Booth, R. S. 1983, A.A., 124, L4.

Dinerstein, H. L. and Lester, D. F. 1984, Ap.J., 281, 702.

Draper, P. W., Warren-Smith, R. F. and Scarrott, S. M. 1985, M.N.R.A.S., 216, 7P.

Finkenzeller, U. 1985, A.A., 151, 340.

Haefner, R. and Metz, K. 1985, A.A., 145, 311.

Herbig, G. H. 1960, Ap.J. Suppl., 4, 337.

Honeycutt, R. K., Schlegel, E. M. and Kaitchuk, R. H. 1986, Ap.J., 302, 388.

Jones, B. F. and Cohen, M. 1986, submitted to A.J.

Kafatos, M., Michalitsianos, A. G. and Hollis, J. M. 1986, Ap.J. Suppl., in press.

King, D. J., Scarrott, S. M. and Shirt, J. V. 1985, M.N.R.A.S., 213, 11P.

Kohoutek, L. 1982, A.A., 115, 420.

Kwok, S. and Purton, C. R. 1983, A.A., 122, 346.

Leinert, C. 1986, A.A., 155, L6.

Lester, D. F. and Dinerstein, H. L. 1984, Ap.J. (Letters), 281, L67.

Lopez, J. A. and Meaburn, J. 1983, M.N.R.A.S., 204, 203.

Mauron, N., Nieto, J. L., Picat, J. P., Lelievre, G. and Sol, H. 1985, A.A., 142, L13.

Meaburn, J. and Walsh, J. R. 1983, M.N.R.A.S., 205, 53P.

Meaburn, J. and Walsh, J. R. 1985, M.N.R.A.S., 215, 761.

Meaburn, J., Walsh, J. R., Morgan, B. C., Hebden, J. C., Vine, H. and
      Stanley, C. 1985, M.N.R.A.S., 213, 35P.
Morris, M. 1981, Ap.J., 249, 572.
Morris, M. and Bowers, P. F. 1980, A.J., 85, 724.
Morris, M., Bowers, P. F. and Turner, B. E. 1982, Ap.J., 259, 625.
Mundt, R. and Fried, J. W. 1983, Ap.J. (Letters), 274, L83.
Mundt, R., Stocke, J., Strom, S. E., Strom, K. M. and Anderson, E. R.
      1985, Ap. J. (Letters), 297, L41.
Mutel, R. L., Fix, J. D., Benson, J. M. and Webber, J. C. 1979, Ap.J.,
      228,771.
Ray, T. 1986, submitted to A.A.
Reay, N. K., Atherton, P. D. and Taylor, K. 1983, M.N.R.A.S., 203,
      1079.
Ridgway, S. T., Joyce, R. R., Connors, D., Pipher, J. L. and Dainty, C.
      1986, Ap.J., 302, 662.
Rieu, N.-Q., Winnberg, A. and Bujarrabal, V. 1986, A.A., in press.
Rodriguez, L. F., Garcia-Barreto, J. A., Canto, J., Moreno, M. A.,
      Torres-Peimbert, S., Costero, R., Serrano, A., Moran, J. M. and
      Garay, G. 1985, M.N.R.A.S., 215, 353.
Roth, M., Echevarria, J., Tapia, M., Carrasco, L., Costero, R. and
      Rodriguez, L. F. 1984, A.A., 137, L9.
Shirt, J. V., Warren-Smith, R. F. and Scarrott, S. M. 1983, M.N.R.A.S.,
      204, 1257.
Shu, F. H. 1987, proc. IAU Symp. 115, "Star Forming Regions", eds. J.
      Jugaku and M. Peimbert (Dordrecht-Reidel: Holland).
Solf, J. 1984, A.A., 139, 296.
Solf, J. and Ulrich, H. 1985, A.A., 148, 274.
Thompson, R. I. and Boroson, T. A. 1977, Ap.J. (Letters), 216, L75.
Walsh, J. R. 1983, M.N.R.A.S., 202, 203.
Walsh, J. R. and Malin, D. 1985, M.N.R.A.S., 217, 31.
Ward-Thompson, D., Warren-Smith, R. F., Scarrott, S. M. and
      Wolstencroft, R. D. 1985, M.N.R.A.S., 215, 537.
Weaver, H. 1974, Highlights of Astronomy, 3, 509.
White, R. L. and Becker, R. H. 1985, Ap.J., 297, 677.

THEORY OF BIPOLAR FLOWS AND JETS FROM YOUNG STARS

Colin A. Norman
Department of Physics and Astronomy, Johns Hopkins University
and
Space Telescope Science Institute

ABSTRACT. Current thinking on the origin, evolution and stability of stellar jets and bipolar flows is reviewed. Particular emphasis is given to the driving mechanisms of bipolar molecular and ionised gas outflows from young stellar objects. General constraints on both hydrodynamic and magnetohydrodynamic flows are presented. The interrelationship between the protostellar outflows and others such as those associated with powerful, highly collimated outflows in active galactic nuclei are discussed. Intimately connected phenomena such as maser sources and Herbig-Haro objects are briefly treated and finally, areas of potentially interesting future research are indicated.

I. INTRODUCTION

Observations of bipolar flows and jets have been made in an extraordinary variety of stellar and protostellar objects (c.f. Cohen, these proceedings) and, when discussed in the broader context of extragalactic radio jets, the range of luminosities, scales and densities attributable to this phenomenon are one of the more remarkable aspects of current astrophysics. While focussing here on flows from young stars I will draw considerably on work done over a longer time concerned with the extragalactic problem. My task is to review theories of these young stellar objects and to bring out the salient physical merits and demerits of various models to try to, at least, indicate where each model may be applicable and what can be learned from it. No single model explains the whole range of observations but I have selected to review those that are generally useful for reasonably large subsets.

The relatively close (10^2-10^3 pc) protostellar flows and jets are particularly fascinating to jet theorists because of the wealth of observational detail over many wave bands with consequently good physical constraints on the relevant parameters such as density, pressure, momentum and energy flow. In fact with the very high resolution and high astrometric accuracy data available with, say, the Hubble Space Telescope or the Very Long Baseline Array, it may be possible to make real time hydrodynamics observations of jet formation

51

I. Appenzeller and C. Jordan (eds.), Circumstellar Matter, 51–59.

in protostellar objects and evolution over, say, a twenty year timescale
for flows in the nearby Taurus-Aurigae association.

An additional, very general aspect of the models is the frequently
invoked presence of an accretion disk. Although there is, as yet, very
little direct evidence for these in protostellar systems it is obvious
they must be there from angular momentum considerations and consequently
may well strongly influence collimated flows. In fact, the evolution of
bipolar flows and protostellar disks are inextricably related over the
whole pre-main sequence phase from massive molecular outflows and
protostellar disks to T-Tauri outflows and disks to the protoplanetary
and planetary systems. When considering the Hertzprung Russell diagram
for pre-main sequence evolutionary tracks both these additional effects
of disks and bipolar flows should be incorporated in the track
calculations.

I shall first discuss generally properties of the omni-present jets
and bipolar flows (§II) and then review in more detail the facts
concerning bipolar flows from young stellar objects. The origin,
evolution and stability for both magnetised and unmagnetised flows are
analyzed in the next two sections (§III and §IV). On small scales much
can be learned by studying structures in the flow that give both OH and
H_2O maser emission and emission-line objects such as Herbig-Haro objects
(§V). The final section (§VI) is devoted to a summary and sketch of
future work.

II. BIPOLAR FLOWS AND JETS

A quick mental run through of properties of the ubiquitous jet and
bipolar flow phenomena shows an enormous dynamic range for the physical
environment in which they are found. Large scale extragalactic radio
jets have scales of ~100 kpc and luminosities of ~10^{44} erg s^{-1},
intermediate scales are of order kiloparsecs with total powers of order
10^{43} erg s^{-1}, and compact jets resolved by VLBI techniques on scales of
~10 pc can have energy outputs of order 10^{45} erg s^{-1}. Closer to home in
the Galaxy, sources such as SS433 and Sco X-1 seem very similar to the
extragalactic sources. The Galactic Centre itself shows signs of a
mildly collimated outflow possibly due to a starburst (Sofue 1984) and
other recently discovered starburst driven extragalactic bipolar
outflows have been found by deep CCD imaging in M82, Arp 220 and other
IRAS selected sources (Heckman 1986, McCarthy et al. 1986). At the
smallest end of the scale are the bipolar outflows from young stellar
objects with scales of at most parsecs, velocities of order 10^2 km s^{-1}
and central object mass of 1-10 $M_\odot$. This is a contrast of 10^6 in
length, 10^4 in velocity and 10^8 in mass of central object between the
powerful jets in active galactic nuclei and those in young stellar
objects. The similarities in these outflows have been documented
recently by Konigl (1986) to be high collimation and two sidedness,
length extending over several decades, association with emission-line
knots, origination in compact objects, termination in extended lobes,
association with surface instabilities and entrainment, a possibly
significant effect played by magnetic fields and quite generally an

invocation of an origin associated with disks.

The specifically relevant physical parameters for molecular outflows and young stellar objects are; the high momentum input with $\dot{M}_w V_w \sim 10^2 - 10^3$ (L/c) (Lada 1985) where $\dot{M}_w$, V_w are the mass flow and velocity in the jet and L is the luminosity of the central object; the realization that the ionized central jets have insufficient momentum to drive the molecular outflow; characteristic velocities of order 10-100 km s^{-1}; Herbig-Haro and H_2O maser velocities of order 200-400 km s^{-1}; and other such details given in relevant reviews in this volume.

III.   HYDRODYNAMIC FLOWS

The original Blandford and Rees (1974) application of the de Laval nozzle concept to jet collimation is applicable here if the dense molecular jets do not cool before they are nozzled (Konigl 1986). Nozzling occurs when the flow speed equals the sound speed in the flow and application of the formula for the ratio of cooling length, L, to characteristic flow scale, R_s,

$$\frac{l_{cool}}{R_s} = 10^{-5} \left[\frac{V_w}{10^2 \text{ km s}^{-1}}\right]^7 \left[\frac{L_{flow}}{L_\odot}\right]^{-1} \left[\frac{R_s}{10^{14} \text{ cm}}\right] \gtrsim 1$$

gives a required velocity in the case of L1551 of

$$V_w > 10^{2.5} \left(\frac{R_s}{10^{14} \text{ cm}}\right)^{-0.14} \text{ km s}^{-1}.$$

Therefore one infers that typically the highest velocity jets could indeed feel this effect. For such flows, as indicated by Smith et al. (1983) bubbles, jets and clouds may form in the flow, depending on the mechanical luminosity in the jet and the sound speed on the cavity walls. Bubbles will form when the Kelvin-Helmhotz instability disrupts the surface of the cavity in which the jet is focussing. This effect is stabilised at higher flow velocities when the flow time across the cavity beats the Kelvin-Helmholtz growth rate. At higher luminosities the high pressure, low density jet will induce a Rayleigh-Taylor instability at the boundary with the cooler higher density external cavity medium. Such instabilities could possibly form structures that will eventually cool and form bullets and knots further along in the flow.

Bipolar flows are not highly collimated. Explosions in an inhomogeneous atmosphere were ruled out as a collimating mechanism for narrow extragalactic jets but in this protostellar context they may well be relevant. Calculations were made by Sanders (1976) and Mollenhof (1976) who concluded that for the exponential atmosphere, an initially point explosion will break out of the atmosphere in a few characteristic exponential scale heights, forming a mildly focussed cone with a break out angle of order the inverse Mach Number in the flow at the break out

point. Focussing was found to give cone or bipolar flow opening angles
of 25°-40°. In a similar spirit calculations of wind-driven flows in
inhomogeneous atmospheres have been made by Sakashita and Hanami (1986),
Konigl (1982), Okuda and Ikeuchi (1986) and Dyson (1984) for both the
energy driven and momentum driven cases. Generally the thin-shell
Laumbach-Probstein method, or variations thereon, was used. Here again
after a few scale heights a mildly collimated outflow was formed. A
careful analysis of existing data led Dyson to conclude that the energy
driven calculations give the best agreement. Once again the absence of
cooling is required, which can really only be satisfied for high jet
velocities.

A completely different but powerful approach to the disk wind
solution has been made by Bardeen and Berger (1978). They sought the
stationary state for a massive self-gravitating disk with a wind by
assuming the flow solutions would be self-similar. Such a viewpoint is
potentially very powerful particularly for assessing focussing at large
distances. Their self-similar solution needs internal heat sources in
the flow to maintain the self-similarity but a modified version of this
could be very relevant for molecular flows.

Expanding collimated flows can undergo reconfinement and even
renozzling (Saunders 1983, Konigl 1982, Smith 1982). If the flow is in
a region of low external pressure relative to that in the jet it will
expand freely with an opening angle of order its internal Mach number.
If it then propagates into a relatively higher pressure region (i.e., it
hits a cloud or intercloud fragment) then pressure waves and shocks will
travel back into the jet and it will reconfine, adjusting to the
increased external pressure. Generally speaking if the external
pressure falls as $p_{ext} \propto r^{-n}$, for n > 2 the jet will be free, for n < 2
there will be reconfinement and for n = 2 the jet will propagate in an
unchanged manner. As discussed by Shu (these proceedings), self-similar
collapse of isothermal clouds has an n = 2 solution. Both the
reconfinement process and over-expansion process during the transition
to confined or free jets can generate a train of Mach disks. In cooling
flows, it has been suggested by a number of authors that the observed
quasi-regularly spaced knots in jets may be a consequence of such a
physical process (Mundt, this volume).

There are now quite rigorous calculations of essentially Kelvin-
Helmholtz type instabilities in shearing, cylindrical jets with and
without magnetic fields. The general form is that of a radially
dependent function with a phase term $\sim e^{i(kz+n\theta-wt)}$. The standard
nomenclature is that n = 0 is a pinching mode, n = 1 is a helical or
kink mode, n $\geqslant$ 2 are classified as flute modes. Ordinary modes exist
even without the cylindrical boundary but reflection modes require
one. It is now clear that these instabilities are hard to stablize.
They can be slowed by increasing the density contrast between jet and
environment for either light or heavy jets; velocity profiles with shear
scale lengths, h, at, for example, the jet-environment boundary layer
can stabilize short wavelength instabilities for wave numbers kh $\gtrsim$ 1.
Smooth profiles can show the growth for wavenumbers ka $\lesssim$ a/h which can
be very significant for reflection modes; jet expansion corresponds to
an expanding, comoving cylindrical radius a = a (t) and allows secular

rather than exponential growth to occur; magnetic fields can stabilize these modes if the Alfven speed greatly exceeds the sound speed, pinch stability is realized if $M \lesssim 2v_A/c_s$ and all modes stabilize if $M < 1$.

Observable consequences of these instabilities have been proposed to be that internal reflection modes shock and produce knots, helical modes produce twists and bends, and the general overall Kelvin-Helmholtz instability can produce flaring of jets. Saturation of this instability occurs when the perturbed flow trajectory diverges from the cylindrical one by an angle more than the inverse Mach number M^{-1}. In this case a shock will form. The final result of the Kelvin-Helmholtz saturated instability is probably to form an inner stable jet system with a much wider and turbulent, shearing, cocoon possibly found by the nor-linear evolution of cats eyes and shocks.

IV. HYDROMAGNETIC FLOWS

That magnetic fields are highly important in cloud structure, star forming regions and collimated outflows has long been suspected by theorists and is now becoming increasingly obvious from careful observational studies. Across the Taurus dark cloud complex the linear polarization measurements of Monetti et al. (1984) show a clear large scale alignment of clouds, subclouds and the overall magnetic field. The polarization data of Hodapp (1984) again indicates a strong correlation between polarization angle and flow direction in a number of sources including DG Tau, NGC 2071 and GL 961. Strom and Strom (1985) illustrate the remarkable alignment of four collimated structures in Orion across a large part of the entire complex which are also aligned with at least two other such systems in another region. While only circumstantial evidence for a general overall organizing mechanism, the cloud's magnetic field could obviously consistently explain this. A more quantitative study of the relationship between the projected directions of bipolar outflows and the interstellar magnetic field has been presented for 10 well studied bipolar sources (Cohen et al. 1984) and there is a clear alignment effect within of order $\sim 30°$. Therefore we feel confident the study of magnetized cloud structures is a very significant undertaking.

Anisotropic gravitational collapse naturally results in a pancake structure (Lin, Mestel and Shu 1985, see however Goodman and Binney 1983). The necessary small initial pressure anisotropy can easily be provided by the magnetic field when the cloud exceeds its magnetic Jeans' mass. A global field structure could, in this way, induce large scale pancake alignment in a cloud complex. More generally, collapse is expected to occur along field lines in a magnetically dominated system. When considering a wind or explosion in a medium dominated by a magnetic field the strongly anisotropic pressure will give rise to stationary flows preferentially elongated along the magnetic field.

The first paper to seriously suggest that hydromagnetic effects gave rise to bipolar flows was that of Draine (1983) who envisaged that one way or another a rapidly spinning, highly magnetized object was formed in the strongly non-equilibrium situation associated with the

protostar formation process. Extremely strong toroidal fields, $B\phi$, are
built up and there is strong pressure gradient $\nabla(B_\phi^2/8\pi)$ upwards and
toward the axis that produces the collimated outflow. This mechanism
certainly works but the credibility of these highly transient, non-
equilibrium initial conditions has been questioned and will be discussed
later.

In a seminal paper Blandford and Payne (1982) discovered and
analysed the self-similar centrifugally driven outflow from a disk. The
similarity assumption is that all velocities scale with disk rotation
curves such that $v_A(r)\alpha\ c_S(r)\alpha\ v_{rot}$ (r). The case studied by these
authors is the Keplerian one, but the flat rotation curve case seems
also to be of relevance for massive self-gravitating protostellar
disks. The wind is generated by particles heated at the surface of the
disk in a corona that can move along corotating field lines on which the
outward directed centrifugally driven 'sling shot' force exceeds the
gravitational binding force at the disk. A smooth hydromagnetic flow
develops and above the Alfvenic surface, as inertial forces become
important, the field lines are swept back and strong toroidal fields
develop with associated inwardly directed pinching forces, or hoop
stresses, that tend to self confine the flow along its symmetry axis.

Uchida and Shibata (1985) have made remarkable axisymmetric
magnetohydrodynamic calculations of a magnetically driven outflow where
the principal driver is the strongly sheared and wound up toroidal
magnetic field whose $\nabla(B_\phi^2/k\pi)$ force focusses the flow toward the axis.
This is a fundamentally non-equilibrium, system and should be understood
as a natural development of Uchida's work on the development and
evolution of coronal transients. Its philosophical basis is quite
similar to that of Draine (1983).

Pudritz and Norman (1983, 1986a, b) have considered in great detail
the physics of a centrifugally driven wind from molecular disks
surrounding the central protostar. The disk component is essential here
since it is the huge angular momentum of the disk that is utilised in
order to provide the basic driver of the outflow. The focussing is
again that due to hoop stress. Typically, there is a central ionized
gas flow at ~300 km s^{-1} with mass loss rate ~10^{-6} $M_\odot$ yr^{-1}, a massive
molecular wind $\dot{M}_{molec}$ ~ 10^{-4} $M_\odot$ yr^{-1} with characteristic velocity ~50 km
s^{-1}. The flow will corotate with rotation velocity roughly of order ~1
km s^{-1} up to the Alfven surface as has been found in LISS1 by Kaifu
et al. (1984). At the termination surface of the flow there is a
magnetohydrodynamic shock that could give characteristic shocked
molecular lines such as H_2. In the flow itself there are Herbig-Hero
objects and H_2O masses. The model can successfully account for the
momentum and energy of the bipolar flow phenomenon and is a natural
consequence of the standard magnetised collapse scenario for the star
formation process. The magnetically driven bipolar flow transports
angular momentum away from the disk thus acting as our effective source
of viscosity. This in turn drives disk evolution further which drives
the wind completing the feedback cycle. Clearly positive feedback
driven outbursts can be generated in the feedback loop here. Corotation
will occur out to the Alfven surface R_A ~ ($\Phi^2/\Omega\dot{M}_w$), where Φ is the
magnetic flux in the disk. Gradual acceleration of the flow will take

place from the surface of the disk to the Alfven surface. The flow pattern is similar to that of a cone with some density profile increasing outwards i.e. a somewhat hollow cone. It is to be expected that once Uchida and Shibata run their code for long enough their initially transient state will with appropriate boundary conditions settle down to a steady state wind model as described in the Pudritz-Norman scenario. This is a model in which I have <u>every confidence</u> but because of the nature of this review it is only fair to give equal time to other alternatives. A major point here is that the theory does require substantial protostellar disks of which more should be observed in the near future.

In a most interesting paper Sakurai (1985) has solved numerically (not simulated) the difficult transfield equation for stationary axisymmetric magnetohydrodynamic flow. His boundary conditions were the simplest possible, that of a magnetic monopole. He finds that interior to the Alfven surface the toroidal field $\nabla(B_\phi^2/k\pi)$ initially focusses the flow and outside the Alfven surface hoop stress acts to focus this further. He is currently planning to calculate the complete wind-disk problem using this method.

There are a number of indications that axisymmetric magnetohydrodynamic flows may focus to the axis under quite a wide range of boundary conditions. Norman and Heyvaerts (1986, in preparation) have shown that; field line curvature is an invariant for open field lines above the Alfven surface for cold flows and consequently any slightly focussed field line at this inner region will remain so at infinity; a perturbation theory developed about the axis shows that focussing will always occur if the density fall off is $p \sim r^{-n}$ where n is less than or equal to 2; and finally all qualitative indications and numerical calculations lead to the conjecture that many, if not all, axisymmetric magnetohydrodynamic flows with only open field lines (i.e., no dead zone) will tend to focus along the axis of symmetry. With merely a conjecture this may give an underlying physical basis for the ubiquity of the jet phenomenon.

Another good way to approach our understanding of magnetic confinement of jets is to study sef-similar solution of cylindrical jets. Achterberg, Blandford and Goldreich (1983) adopted the scaling laws $B\phi \sim 1/r$, $v_A \sim$ constant, $B_r \sim r$, $v_\phi \sim r$, and $\partial p/\partial r \sim r$ and found a solution that for the jet transverse velocity that could be expressed as motion in an effective potential dependent only on the transverse radius. Quasi-periodic oscillations in the transverse cross section of the cylinder were found. The focussing was due to a balance of toroidal magnetic pinching, magnetic pressure, ram pressure, and internal thermal pressure. The solution matched roughly some extragalactic jet observational data and could be pursued further in the content of young stellar objects.

V. H_2O MASERS, HERBIG-HARO OBJECTS, AND OTHER RELEVANT OBSERVATIONS

Herbig-Haro objects are clearly associated with bipolar flows as are the high-velocity H_2O masers. This topic is discussed extensively elsewhere

in this volume by Raga and analysed assuming that they are bullets
(Norman and Silk 1979). Consequently all the beautiful recent data on
this topic could well lead to a far better understanding of cooling
processes, thermal instabilities, densities, temperature and processes
in the flows themselves (Raga and Bohn 1986, Raga 1986, Raga, Bohm and
Solf 1986).

H_2O masers are probably pumped in shock associated with bullets.
These are C-shocks with broad magnetic precursors. The only viable
models that pumps the strong masers as well as the lower luminosity ones
is that in which hot electrons pump the maser up by collisions and cold
neutrals pump it down by collisions. This two temperature system avoids
the saturation problem (Kylafis and Norman 1986a, b). Here one can
infer more accurately physical conditions in masers and their associated
bipolar flow environment. H_2O masers can be used as excellent tracers
of the proper motion of parts of the flow hydrodynamics particularly
with the future VLBA.

Other simple questions concerning observations of bipolar flows
include; the crucial question of whether all the flows corotate as
indicated for L1551 by Kaifu et al. (1984); are bipolar flows associated
with binaries, in particular those with S-shaped mirror symmetry; what
are the further details of the correlations of these objects with
magnetic fields; what are the physical parameters that can be determined
by studying the molecular shocks, cavity boundaries and shells; is there
an unambiquious association with molecular disks; what are statistical
studies able to tell us--are these flows steady or integrated versions
of a number of transient outbursts; can the Hubble Space Telescope
really do real time hydrodynamics or will these be marred by phase
effects where it is not in fact physical knots that seem to move.

VI. SUMMARY

I have only been able to briefly touch on some of the remarkably
beautiful work that is being undertaken in this field. In assimilating
our view of theories of the origin, evolution and stability of bipolar
flows from young stellar objects and their relationship to other jets it
seems obvious how highly interactive this field is between many fields
including active galactic nuclei, and starbursts, star formation,
numerical simulations and analytic hydrodynamics and
magnetohydrodynamics. At this meeting the cross interaction has been
excellent and I thank the organizers for their efforts in making this
possible.

REFERENCES
Achterberg, A., Blandford, R. D. and Goldreich, P. 1983, Nature, 304,
    607.
Bardeen, J. M. and Berger, B. K. 1978, Ap. J., 221, 105.
Blandford, R. D. and Payne, D. G. 1982, MNRAS, 199, 883.
Blandford, R. D. and Rees, M. J. 1974, MNRAS, 169, 395.
Cohen, R. J., Rowland, P. R. and Blair, M. M. 1984, MNRAS, 210, 425.

Draine, B. T. 1983, Ap. J., 270, 519.
Dyson, J. E. 1984, Astrophys. Space Sci., 106, 181.
Goodman, J. and Binney, J. 1983, MNRAS, 203, 265.
Heckman, T. N., Armus, L., McCarthy, P., van Breugel, W. Miley, G. K. 1986, preprint.
Hodapp, K. W. 1984, Astron. Astrophys., 141, 255.
Kaifu, N., Suzuki, S., Hasegawa, T., Morimoto, M., Inatani, J., Nagane, K., Miyazawa, K., Chikada, Y., Kanzawa, T., and Akabane, K. 1984, Astron. Astrophys., 134, 7.
Konigl, A. 1982, Ap. J., 261, 115.
Konigl, A. 1986 in Jets from Stars and Galaxies, ed. R. N. Henriksen, Canadian Institute for Theoretical Astrophysics, Canadian Journal of Physics, 64, 351.
Kylafis, N., and Norman, C. 1986a, Ap. J. (Letters), 300, L73.
Kylafis, N., and Norman, C. 1986b, preprint.
Lin, C. C., Mestel, L. and Shu, F. 1965, Ap. J., 142, 1431.
McCarthy, P., Heckman, T. M. and van Breugel, W. J. M. 1986, A. J., in press.
Möllenhoff, C. 1976, Astron. Astrophys., 50, 105.
Monetti, A., Pipher, J. L., Helfer, H. L., McMillan, R. S., and Perry, M. L. 1984, Ap. J., 282, 508.
Norman, C. and Silk, J. 1979, Ap. J., 228, 197.
Okuda, T. and Ikeuchi, S. 1976, PASJ in press.
Pudritz, R. E. and Norman, C. A. 1983, Ap. J., 274, 677.
Pudritz, R. E. and Norman, C. A. 1986a Ap. J., 301, 571.
Pudritz, R. E. and Norman, C. A. 1986b in Jets from Stars and Galaxies, Proceedings of the CITA meeting ed. R. N. Henriksen, Canadian Journal of Physic 64, 351.
Raga, A. and Bohm, K. H. 1985, Ap. J. Suppl., 58, 201.
Raga, A. C., and Bohm, K. H. 1986a, preprint.
Raga, A. C., 1986, preprint.
Raga, A. C., Bohm, K. H. and Solf, J. 1986, preprint.
Sakashita, S. and Hanami, H. 1986, PASJ, in press.
Sakurai, T. 1985, Astron. Astrophys., 151, 121.
Sanders, R. H., 1983, Ap. J., 266, 73.
Smith, M. D. 1982, Ap. J., 259, 522.
Smith, M. D., Smarr, L., Norman, M. L., and Wilson, J. R. 1973, Ap. J., 264, 432.
Sofue, Y. 1984, PASJ, 36, 539.
Strom, S. E. and Strom, K. E. 1985, Comments Astrophy., 10, 179.
Uchida, Y. and Shibata, K. 1985, PASJ, 37, 515.

RADIO EMISSION FROM YOUNG STELLAR OBJECTS IN THE RHO OPHIUCHI CLOUD

Ph. André, T. Montmerle
Service d' Astrophysique, C.E.N. Saclay
91 191 Gif-sur-Yvette Cedex, France, and
E.D. Feigelson, P. Stine
Penn State University, University Park, PA 16802, USA

1. **Observations.** Following an Einstein X-ray survey of the rho Oph molecular cloud, uncovering several dozen young stellar objects (YSO) (Montmerle et al. 1983), an extensive radio continuum VLA(C) survey was conducted in 1983 to look for counterparts to these sources at 1.4 and 5 GHz (André et al. 1986) To shed light on emission mechanisms, we performed follow-up VLA(A/B) observations of identified radio stars in March 1985 at 1.4, 5 and 15 GHz.

2. **Comparison with known populations of YSO in the cloud.** The X-ray sources make up a population of late-type PMS stars (Bouvier and Appenzeller in prep.) undergoing solar-type flares. In the VLA(C) survey, out of 93 detected radio sources, we found 8 to be definitely stellar. Only 3 of them were previously known. Few X-ray sources (< 10%) are detected, but most (> 60%) stellar radio sources have X-ray counterparts. This can be understood by analogy with the radio vs X-ray properties of RS CVn systems (well-studied flaring stars) since few of them would be seen at the distance of the cloud (160 pc). However, our stellar radio sources are probably not all flaring objects (see 3). They are distinct from optical T Tauri stars (TTS) since we detected none of 10 bona fide TTS in our entire 1983 radio survey. This emphasizes the finding by Bieging et al. (1984) that TTS are not the strongest radio emitters among YSO. Also, among the infrared sources listed by Lada and Wilking (1984), we detect selectively 'class III' (~reddened blackbodies, see Lada 1987) sources: none of the 5 protostars ('class I') and none of the 21 embedded TTS ('class II') are detected, while 4 out of 6 'class III' objects show radio emission. Along with their spatial distribution (loosely clustered around the dense molecular core of the cloud), this suggests that our stellar radio sources make up a new population of relatively evolved YSO.

3. **Nature of emitting objects.** The variety of radio spectra observed in 1983 and 1985 (see fig.) suggests that a single emission mechanism cannot account for them all. A previous interpretation (Falgarone and Gilmore 1981) for all sources in terms of thermal emission from compact HII regions around main-sequence B stars is ruled out (see Montmerle et

I. Appenzeller and C. Jordan (eds.), Circumstellar Matter, 61–62.
© *1987 by the IAU.*

al. 1987). Spherical ionized stellar winds are dismissed too, for they lead to spectra (e.g. Wright and Barlow 1975) steeper than observed (see fig.). Collimated winds (see Reynolds 1986) are more likely, leading to flatter power-law spectra. We best interpret the steady radio emission from a source like VSSG 14 as arising from an ionized envelope such that Ne(r)~r(-3/2) (accretion or ballistic expansion, see Bertout 1986). Non-thermal emission from flares (which produces rather steep and bent spectra, see Klein and Trottet 1984) is certainly the explanation for two highly variable sources like ROX 8 (Feigelson and Montmerle 1985) and ROX 31. But clearly, more information is needed to distinguish non-thermal from thermal models in the case of a source like WL 5 (see fig.), and additional observations are planned.

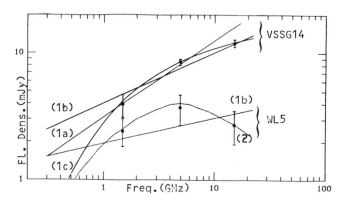

Figure: Possible interpretations of the spectra of VSSG 14 and WL 5. (1) Parameters of thermal emission models: gas temperature, 10(4) K. Wind models: term. velocity: ~ 200 km/s; mass-loss rates : (a) ~ 10(-7) M☉/yr (spherical wind), (b) ~ 10(-8) M☉/yr (collimated winds); widths of the collimated winds proportional to (r/r0)(&) with & = 3/4 and 2/3 for VSSG 14 and WL 5 respectively. (c) Accretion model: M* ~ 1 M☉, Ṁ ~ 10(-8) M☉/yr; size of the ionized flow: rmax ~ 10(15) cm. (2) Parameters of the non-thermal flare model for WL 5: energetic electron spectrum ~ E(-1.5) between 0.5 and 50 MeV in B = 100 G; density of energetic electrons: 10 cm-3; free-free absorption in the flare plasma of density 10(8) cm-3 and temperature 10(6) K; flare size ~ 10(12) cm.

References
André, P., Montmerle, T. and Feigelson, E. D.: 1986, A. J. submitted.
Bertout, C.:1986, in 'Jets and Accretion in Astrophysics'. Ed. J.P.Lafon.
Bieging, J. H., Cohen, M. and Schwartz, P. R.: 1984, Ap. J. 282, 699.
Falgarone, E., and Gilmore, W.: 1981. Astr. Ap. 95, 32.
Feigelson, E. D. and Montmerle, T.: 1985, Ap. J. Lett. 289, L19.
Klein, K. L. and Trottet, G.: 1984, Astr. Ap. 141, 67.
Lada, C. J.: 1987, in Proc. IAU symp. 115.
Lada, C. J. and Wilking, B. A.: 1984, Ap. J. 287, 610.
Montmerle, T. et al.: 1983, Ap. J. 269, 182.
Montmerle, T. et al.: 1987, in Proc. IAU symp. 115.
Reynolds, S. P.: 1986, Ap. J. 304, 713.
Wright, A. E. and Barlow, M. J.: 1975, M.N.R.A.S. 170, 41.

CIRCUMSTELLAR RADIO EMISSION FROM PRE-MAIN SEQUENCE STARS

Alexander Brown
Joint Institute for Laboratory Astrophysics, University
of Colorado and National Bureau of Standards, Boulder,
Colorado 80309-0440, USA

ABSTRACT. VLA radio continuum observations have been obtained for a
number of pre-main sequence (PMS) stars in Corona Australis, Lupus,
Scorpius and Taurus. A variety of PMS sources were detected and for
other stars upper limits to the ionized mass loss rates were determined. A strong double source, showing two radio jets, was found
associated with an embedded infra-red source in the R CrA molecular
cloud. Some of the PMS stars show extended radio emission associated
with ionized circumstellar envelopes, even though photoionization by
EUV photons is not sufficient to produce the ionized regions.

Five fields in Corona Australis, Scorpius and Lupus, centered on
R CrA, TY CrA, CrA X-1, HR 5999 and RU Lup, were observed using the
VLA* in C and C/D array during 1985 September and October. Eleven
6 cm radio continuum sources were found in the CrA molecular cloud;
the only sources with optical counterparts were TY CrA (1.2 mJy) and
the Herbig Haro object HH 101 (0.24 mJy). For the other sources near
R CrA there was exceptionally good correlation between the detected
6 cm sources and the positions of 2 μ infrared sources from the mapping of Taylor and Storey (1984). The strongest source in this field,
first detected by Brown and Zuckerman (1975), is a double source coincident with the 2 μ source IRS7. The double source has two extended
radio "jets" and an integrated flux density of 21.7 mJy. The radio
jets show internal structure and appear to be inclined to the plane
of the sky. IRS7 is the most deeply embedded IR source of Taylor and
Storey with A_V = 20-25 magnitudes. Table 1 presents data on six stars
which were not detected but for which interesting upper limits result.
The case of RU Lupi is particularly interesting because the upper

---

*The Very Large Array (VLA) is part of the National Radio Astronomy
 Observatory operated by Associated Universities, Inc., under contract
 with the National Science Foundation.

I. Appenzeller and C. Jordan (eds.), Circumstellar Matter, 63–64.
© *1987 by the IAU.*

TABLE 1.   Some PMS stars for which only 6 cm upper limits were
           obtained.

Star	Spectral Type	Distance (pc)	3σ Upper Limit (mJy)	Ionized $\dot{M}$ $(M_\odot/yr)^*$	Comment
R CrA	F0e	130	0.18	$\leq 1 \times 10^{-8}$	PMS F star
T CrA	F0e	130	0.18	$< 1 \times 10^{-8}$	PMS F star
CrA X1	K0	130	0.16	$< 1 \times 10^{-8}$	Naked T Tauri star
RU Lup	---	150	0.13	$< 1 \times 10^{-8}$	Extreme T Tauri star
HR 5999	A7 IIIe	270	0.15	$< 3 \times 10^{-8}$	Herbig Ae star
HR 6000	B6 Vp	270	0.15	---	Young Bp star

*Assuming wind velocity of 200 km/s. For HR 5999 $v_{wind} \sim 100$ km/s is
a better estimate giving $\dot{M} < 1.5 \times 10^{-8}$ $M_\odot/yr$.

limit to the ionized mass loss is an order of magnitude smaller than
that suggested by the models of Lago (1984), but consistent with that
of Kuin (1986). Both these models are based on Alfvén wave-driven
wind theory.

VLA observations of HL and XZ Tau have shown evidence for extended radio emission. These stars show spectral indices of 1.0±0.3
and 0.9±0.3, respectively, based on 6 cm and 2 cm data (Brown, Drake,
and Mundt, 1986). These values suggest that the emission is free-free
emission from their stellar winds.

This work was supported by NASA grant NGL-06-003-057 through the
University of Colorado.

REFERENCES

Brown, A., Drake, S. A. and Mundt, R. 1986, in Cool Stars, Stellar
    Systems, and the Sun, eds. M. Zeilik and D. M. Gibson, Lecture
    Notes in Physics No. 254 (Springer-Verlag: Berlin), p. 451.
Brown, R. L. and Zuckerman, B. 1975, Ap. J., 202, L125.
Kuin, N. P. M. 1986, in Cool Stars, Stellar Systems, and the Sun,
    eds. M. Zeilik and D. M. Gibson, Lecture Notes in Physics No. 254
    (Springer-Verlag: Berlin), p. 466.
Lago, M. T. V. T. 1984, Mon. Not. R. astr. Soc., 210, 323.
Taylor, K. N. R. and Storey, J. W. V. 1984, Mon. Not. R. astr. Soc.,
    209, 5p.

CIRCUMSTELLAR MATTER AROUND THE CANDIDATE PROTOSTAR EL29

H. Zinnecker, Royal Observatory, Edinburgh
C. Perrier, Observatoire de Lyon, St. Genis-Lavel
A. Chelli, Instituto de Astronomia, Mexico City

Infrared speckle interferometric observations at 3.6 μm carried out at the ESO 3.6m telescope have indicated that EL29, a deeply embedded infrared source in the Ophiuchus dark cloud (Elias 1978), appears to have a core-halo structure (Zinnecker, Chelli and Perrier 1985). A fit to the data gave angular scales of order of 0.05 arcsec for the core and 3.5 arcsec for the halo (Gaussian FWHM) corresponding to linear diameters of 8 AU and 620 AU, respectively (for a distance to the cloud of 160 pc). New infrared speckle observations confirm that the core is resolved at the 2 sigma level, and carries about 90% of the total 3.6 μm flux (Fig. 1). Furthermore, there is evidence from a recent IRTF lunar occultation experiment in the K-band (2.2 μm) for an additional inner 6 milli-arcsec ($\sim$ 1 AU) core which carries about 75-85% of the total K-flux from this source (Simon et al 1986). Perhaps this inner core is the dust shell inside of which grains are destroyed by the radiation from a central object.

The high resolution speckle data suggest that <u>EL29 could be a low-mass protostar still in the process of accretion,</u> since the substantial infrared emission associated with the outer core is indicative of substantial amounts of hot dust very close to the central object from where the luminosity originates (cf. Yorke 1980). Because radiation pressure is unimportant for objects of low luminosity, the circumstellar material must either be falling in or rotating in a disk. The disk interpretation seems to be more likely in view of the fact that one can see the source at NIR wavelengths and in view of the slight asymmetry apparent from the comparison of the speckle visibility functions in the two orthogonal directions (Fig. 1). This conclusion is supported by the work of Adams, Lada and Shu (1986) who have modelled the emergent spectral energy distribution of EL29 theoretically and could fit the observed spectrum (including IRAS data) satisfactorily with a model of an accreting protostar of instantaneous mass M = 1 $M_\odot$ formed from a rotating parent molecular cloud core of angular velocity Ω = 2.10^{-13} rad/s and temperature T = 35K.

Also we briefly report speckle results of a second candidate protostar in the same cloud, called WL16 (Wilking and Lada 1983). This source is another low-luminosity source for which IRAS data (Young, Lada and Wilking 1986) and infrared spectroscopy (Zinnecker, Webster and Geballe 1985, Thompson 1985) exist. We find that it shows a similar core-halo structure at 3.6 μm as EL29.

I. Appenzeller and C. Jordan (eds.), Circumstellar Matter, 65–66.

Its core with a FWHM angular scale ~ 0.07 arcsec is resolved at the 1 sigma level while the size of the halo is ~ 3.5 arcsec.   Again 2 orthogonal position angles were measured but in this case no spatial asymmetry can be discerned within the errors.

Further speckle data for both sources (EL29 and WL16) in the K and M bands were also taken and will be discussed elsewhere.

References

Adams, Lada and Shu (1986).   Ap.J. (in press).
Elias (1978).   Ap.J. 224, 453.
Simon et al. (1986).   preprint.
Thompson (1985).   Ap.J. 299, L41.
Wilking and Lada (1983).   Ap.J. 274, 698.
Yorke (1980).   Astron. Astrophys. 85, 215.
Young, Lada and Wilking (1986).   Ap.J. 304, L45.
Zinnecker, Webster and Geballe (1985).   in Nearby Molecular Clouds, Springer-
     Verlag, ed. G. Serra, p.81.
Zinnecker, Chelli and Perrier (1987).   in Star Forming Regions, IAU-Symp. 115,
     eds. Jugaku & Peimbert (Dordrecht: Reidel).

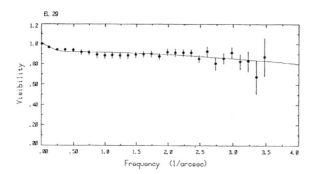

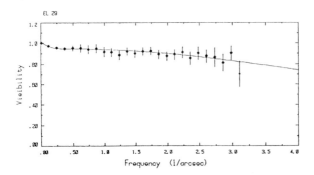

Fig. 1:   Visibility functions of the protostellar object EL29 in the L' band.
          Upper graph: PA = 0°;   lower graph: PA = 90°.

CIRCUMSTELLAR DUST AROUND FU ORIONIS STARS

Gunnar Welin
Astronomical Observatory
Box 515
S-751 20 Uppsala
Sweden

It is immediately apparent from IRAS and other far infrared data (figure
1) that all known FU Orionis stars, or fuors, are associated with substantial amounts of cool circumstellar dust.

The pre-outburst magnitudes for V1057 Cyg indicate that this star
was immersed in circumstellar dust already before its flare-up in 1969-
1970, and hence that this dust did not - as has been claimed - appear
only after the outburst; regrettably there exist no far infrared data
from that time. The colours also exclude the possibility that the star
then was of the alleged spectral type K0, but point to it being of a
rather early type and obscured by both interstellar, local, and circumstellar matter. For details, see Welin 1985.

A number of suspected pre-fuors also show considerable cool dust
emission at far infrared wavelengths. These stars have been selected
by the following criteria, based on what is known for the fuors:
    a) advanced T Tau-type emission line spectra superposed on
      (nearly) featureless continua (for V380 Ori an early type);
    b) moderate light variations (amplitudes less than about 1 m.);
    c) the presence of arcuate reflection nebulae;
    d) position in local dark nebulae.

Exact comparisons of the relative contributions to the total radiation from the underlying stars and from the circumstellar dust are unfortunately impossible until better knowledge has been obtained of truly
interstellar reddenings and extinctions, and of the influence by local
dust clouds (possibly with R_v values well above 3, as seems likely in
the local cloud around V1057 Cyg, judging from data for the nearby star
LkHα 192 given by Haro 1972). Good polarimetric, photometric, and spectroscopic data also for neighbouring stars would be valuable in order to
deduce interstellar reddenings, and might furthermore aid in determining
local cloud properties.

If however, as an example, the flux from the dust associated with
Bretz 4 is extrapolated beyond 100 μ by assuming blackbody radiation
peaking at 100 μ, this object radiates about 35 times more energy at
wavelengths longer than 100 μ than in optical-near infrared wavelengths
(cf. Cohen and Kuhi 1979) - and to this must be added the still stronger
radiation at shorter far infrared wavelengths. Since the circumstellar

67

I. Appenzeller and C. Jordan (eds.), Circumstellar Matter, 67–68.
© *1987 by the IAU.*

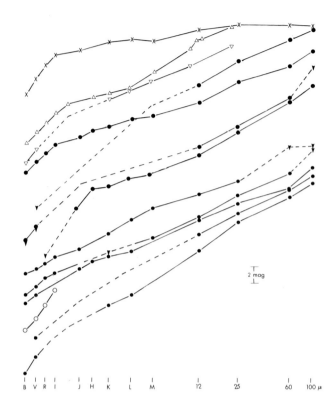

Figure 1. The distributions of optical and infrared magnitudes (from IRAS and various other sources) for: X α Ori (for comparison); V1057 Cyg Δ about 1 yr after its outburst, ∇ 11 yrs after, o before; ● other fuors, from top right downwards HH57IRS8, FU Ori, V1515 Cyg, V1735 Cyg; suggested pre-fuors V380 Ori, V1331 Cyg, V1121 Oph, RNO 33, Bretz 4. ▼ denotes uncertain magnitudes or IRAS upper limits.

dust presumably is heated by the underlying star, the total luminosity of the star must be quite high – as the distance is unknown, so is the luminosity, but Cohen's and Kuhi's data for the neighbouring star Mon R2-5a suggest a distance of about 2 kpc. This would give an optical-near infrared luminosity for Bretz 4 of 15 $L_\odot$, and a total luminosity of the order of $10^3\ L_\odot$. The partial dispersal of circumstellar dust obscuring the star may thus give an increase in the optical brightness of the star of the same order as that observed in the known fuors.

Similarly, the existence of dust around pre-outburst V1057 Cyg, the amount of cool dust observed now, and inferences drawn from the proposed pre-fuors indicate that the luminosity of V1057 Cyg before the flare-up is likely to have been considerably higher than the often quoted 8 $L_\odot$ derived from optical data alone, and that changes in the circumstellar dust shell may have played a part in the 1969-1970 outburst.

REFERENCES

Cohen, M. and Kuhi, L.V. 1979, *Astrophys. Journal Suppl.* **41**, 743
Haro, G. 1972, *Inf. Bull. Variable Stars* No. 714
Welin, G. 1985, *Astrofizika* **23**, 437

THE EXTENDED INFRARED RADIATION FROM THE L1551 BIPOLAR FLOW, $L > 19\ L_\odot$

F.O. Clark[1,2], R.J. Laureijs[1], G. Chlewicki[1], and C.Y. Zhang[1]

[1]Lab. for Space Research
P.O.B. 800
9700 AV Groningen
The Netherlands

[2]Dept. of Phys. and Astronomy
University of Kentucky
Lexington, KY 40506
U.S.A.

ABSTRACT. The infrared bolometric luminosity of the extended infrared emission from the L1551 flow is estimated as 19 (−4 +10) $L_\odot$. Ultraviolet radiation from the shock associated with the flow appears to heat the surrounding dust. The extended infrared emission raises the total energy requirement for the flow over a 10^4 year lifetime to 10^{46-47} ergs. If gravitational in origin, this energy likely originates from a region $<10^{13}$ cm. Infrared radiation offers a new probe for interstellar shocks by sampling the ultraviolet halo surrounding the shock.

1.1. INFRARED LUMINOSITY OF THE L1551 FLOW

Bipolar flows from very young stars have heretofore been detected primarily by means of broad wings on spectral lines. We analyze the energetics of the infrared emission detected from the bipolar flow in L1551 (Clark and Laureijs 1986), and show that infrared emission from dust surrounding bipolar flows is an effective tool for studying such flows. To estimate the extended flux and dust temperature, IRAS data were analyzed and background levels determined over a 16.5° by 33° field. We have estimated the total extended flux by assuming a constant flux from the inner region near IRS5 as 90 Jy (−13 +12) at 60 μm and 870 Jy (−170 +440) at 100 μm. The extended flux above represents an observed infrared luminosity of 10 (−2 +4) $L_\odot$. Correcting for that portion of the Planck curve not seen by IRAS, we estimate the infrared bolometric luminosity as 19 (−4 +10) $L_\odot$. This estimate of the infrared luminosity is some 50% of the bolometric luminosity of the central star.

The conversion of mechanical energy in the shock to infrared radiation is unlikely to be 100% efficient. We estimated the mechanical luminosity of the shock by modeling the ultraviolet heating of the dust to the observed spatially resolved dust temperature using the numerical coefficients from Hollenbach and McKee, resulting in estimates of the mechanical luminosity of 40-140 $L_\odot$, corresponding to a range of infrared dust emissivities which vary with wavelength$^{(-1\ \mathrm{to}\ -1.5)}$.

I. Appenzeller and C. Jordan (eds.), Circumstellar Matter, 69–70.

1.2. THE DUST HEATING MECHANISM

The dust exhibits a nearly constant temperature in the outer parts of the flow of 22-26 K. We exclude large scale radiative heating from IRS5 as the dust temperature is so uniform. We have made calculations of collisional and ultraviolet heating from the shock and find ultraviolet heating generally more efficient. The path length of ultraviolet photons from the shock which heat of the dust offers a simple explanation of the infrared morphology which is twice the CO length and width.

1.4. WHAT DRIVES THE L1551 FLOW?

Emerson et al. (1984) estimate the bolometric luminosity of IRS5 as 38 L_o, which indicates that the observed infrared luminosity is some 50% of the bolometric luminosity of IRS5, and IRS5 may not be energetically capable of driving the L1551 flow with radiation pressure (Draine 1983). A luminosity range of 19 L_o over a flow lifetime of 10^4 years implies an energy of ~10^46 ergs. If IRS5 is estimated as 2 M_o, then the gravitational potential energy from ~ 10^13 cm away from IRS5 is ~10^47 ergs, and could provide the necessary energy with 10% conversion efficiency. Several times 10^48 ergs are released in the collapse of a solar type star to the main sequence. The model of Draine (1983) provides energies of this order and is consistent with the shock parameters derived here.

1.5. SUMMARY

Infrared emission offers a new technique for probing bipolar flows. The infrared luminosity of the L1551 flow is estimated as 19 (-4 +10) L_o, 50% of the estimated bolometric luminosity of IRS5. The dust appears to be heated by ultraviolet radiation from the shock; the excess infrared size being well modeled by the surface of UV opacity one around the shock. Over a 10^4 year lifetime 19 L_o requires an energy of 10^46, two orders of magnitude larger than previous estimates of the energy in the flow based on CO measurements (Snell and Schloerb 1985). The energetic requirements suggest that the phenomenon driving the flow, if gravitational in origin, likely originates from a region of order 10^13 cm. Draine's magnetic bubble model for the bipolar flow is capable of supplying energies of this magnitude. We acknowledge stimulating discussions with E.E. Becklin and F. Shu.

1.6. References

Clark, F.O. and R.J. Laureijs, 1986 A.&A. 154, L26.
Draine, B.T. 1983 Ap.J. 270, 519.
Emerson, J.P., S. Harris, R.E. Jennings, C.A. Beichman, B. Baud, D.A.
    Beintema, P.L. Marsden, and P.R. Wesselius 1984 Ap.J. Letters 278, L49.
Hollenbach, D. and C.F. McKee 1979 Ap.J. Suppl. 41, 555.
Snell, R. 1981 Ap.J. Suppl. 45, 121.
Snell, R. and F.P. Schloerb 1985 Ap.J. 295, 490.

CO OBSERVATIONS OF GAS SURROUNDING B335 (IR) and L1551-IRS5

C.M. Walmsley, K.M. Menten
Max-Planck-Institut für Radioastronomie,
Auf dem Hügel 69,
5300 Bonn 1, F.R.G.

ABSTRACT. Recent submillimeter observations of infrared sources embedded in dark clouds have uncovered the existence of a few highly compact objects. Two examples of this phenomenon are the infrared sources observed towards B335 and L1551 (Keene et al. (1983), Davidson and Jaffe (1984)). One can estimate for these regions gas masses of approximately one solar mass, densities of at least 10^6 cm^{-3}, and diameters less than 10^{17} cm. Both infrared sources are associated with bipolar outflows observed in CO with mass loss rates of order 10^{-6} solar masses per year. It is reasonable in both cases to suspect the existence of matter in some form of disc around the infrared source.

One might also expect to observe molecular counterparts to the submillimeter compact objects. In an effort to do this, we have mapped the $C^{18}O$ $J = 1 \to 0$ transition in a small area surrounding the infrared sources in B335 and L1551 using the IRAM 30-m telescope. The beamwidth was 22 arc sec. which corresponds to $5 \cdot 10^{16}$ cm at the distance of L1551 and $9 \cdot 10^{16}$ cm at the distance of B335. In each case, we do find a compact structure which we believe to be the molecular counterparts of the compact submillimeter sources. We estimate the abundance ratio CO/H_2 to be of order 10^{-4} in L1551 and at least 10^{-5} in B335. Further details of this work are contained in an article submitted to Astronomy and Astrophysics.

REFERENCES

Davidson, J.A., Jaffe, D.T. (1984) Astrophys. J. 277, L13
Keene, J. et al. (1983) Astrophys. J. 274, L43

I. Appenzeller and C. Jordan (eds.), Circumstellar Matter, 71.

HYPERSONIC JETS FROM YOUNG STARS IN MOLECULAR CLOUDS

Hans-Joachim Blome, Inst. f. Theoretische Physik (SFB 301),
D-5 Köln 41
Wolfgang Kundt, Inst. f. Astrophysik, D-53 Bonn 1

ABSTRACT. We argue that the narrow jets which are sometimes seen to escape from YSOs into an ambient molecular cloud consist of $e^{\pm}$-plasma which can be created in stellar magnetospheric discharges and subsequently centrifugally post-accelerated. This high-pressure pair plasma is squeezed into two jets which ram cocoons into the molecular cloud, observed in the form of molecular lobes (outflows).

1. JET SPEED

Twin-jets from stars and galactic nuclei are often thought to be caused by a similar mechanism (Königl 1982, Kundt 1984). Arguments have been given that jets from AGN consist of relativistic pair plasma (Kundt & Gopal-Krishna 1980). Here we estimate the temperature of the medium that fills the high-pressure cavities (lobes, outflows - which we interpret as cocoons). The estimates are based on the pressure needed to blow the cavities, the high required sound speed, the usual invisibility of the flow heads and the condition that a cocoon is filled through its jet. We thus arrive at almost relativistic jet velocities:

Observed fact	Inferred lobe temperature/ jet speed
lobe pressure $= \varrho_a v^2 \lesssim 10^{-8}$ dyn cm^{-2}, no H II region	$T \gg 10^{4}$K
piston subsonic (consisting of shocked jet)	$T > 10^{7}$K
power liberated by piston is unseen (at X-rays)	$T \gtrsim 10^{8}$K
flow heads are only seen in some 5% of all BFs	$T \gg 10^{8}$K
particle number conservation: lobes are filled through jets	$\beta_{jet} \approx 1$
jet near obj. 50 has been seen 'illuminated'	$\beta_{jet} \approx 1$

A young star is unlikely to blow off its envelope at almost relativistic speeds. Instead, we expect relativistic charges to be created in localized discharges inside the corotating magnetosphere. These charges will produce relativistic $e^{\pm}$-pairs on collision with stellar photons if their energy exceeds the necessary thresholds.

2. GENERATION OF PAIR PLASMA NEAR A YSO.

We assume that a star forms at the center of a massive disk. The (proto-) star's convective layer is likely to generate equipartition magnetic fields of order (v_t = turbulent velocity):

I. Appenzeller and C. Jordan (eds.), Circumstellar Matter, 73–74.

$$B \leq (8\pi \varrho v_t^2)^{1/2} \approx 10^{6.5} v_6 \text{ Gauss.}$$

Magnetospheric discharges can accelerate charges up to an energy

$$E = e \int (\vec{B} \times \vec{B}) \cdot d\vec{x} \approx eB \ GM/c^2 = B_4 \ (M/M_\odot) \text{ erg}$$

near the speed-of-light cylinder. Electrons and positrons are created during collision of these charges with stellar photons when their Lorentz factors exceed $2m_e c^2/h\nu$, i.e. when B exceeds 10 kG. The injected number rate of pair plasma can be estimated both from the pressure near the end of the jet, of cross-sectional area A, and from the loss of stellar rotational energy:

$$\dot{N}_{e\pm} = nAc = \dot{E}_{rot}/\gamma m_e c^2 \approx 10^{42} \ s^{-1}/\gamma_1 \ ,$$

where the average Lorentz factor γ has been inserted in units of 10.

3. JET AND COCOON

The ramming of a jet through the ambient molecular cloud (of mass density ϱa) can be described as a succession of explosions taking place along the jet path. For a hypersonic beam of width 2b propagating along the z-axis, the shape of the bowshock can be derived from momentum conservation in z-direction <u>and</u> energy conservation (Sedov-Taylor-wave) transverse to it: $r/a = \sqrt{z/a}$ with $a \approx b \sqrt[4]{(\varrho_j/\varrho_a)}$. This law describes both the bowshocks of young bipolar flows and of certain extragalactic radio sources (3C 33 S).

In the cocoon, the shocked relativistic pair plasma emits both synchrotron radiation at low radio frequencies and inverse Compton radiation at UV frequencies which heat the ionized and neutral component of the ambient molecular cloud, cf. Clark & Laureijs. The jet itself consists of light relativistic pair plasma streaming inside heavy 'walls' of thermal plasma which emit Coulomb-Bremsstrahlung (Snell, Bally & Strom).

The (invisible) power of the jet is related to the rotational energy of the YSO:

$$\int L_{jet} \ dt \approx E_{rot} \approx E_{grav}$$

$$\approx 10^{48} \text{ erg } (\frac{M}{M_\odot})^2 (\frac{R_\odot}{R}).$$

It is worth mentioning that the boundary of the cocoon is Rayleigh-Taylor stable as long as its pressure decreases with time.

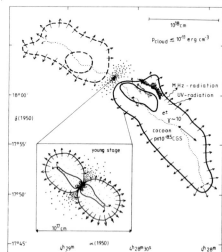

Fig. 1 displays our interpretation of the bipolar flow L 1551/IRS 5!

REFERENCES

Clark, F.O., Laureijs, R.J.: Astron. Astrophys. 154, L 26 (1986)
Königl, A.: Astrophys. J. 261, 115 (1982)
Kundt, W., Gopal-Krishna: Nature 288, 149 (1980)
Kundt, W.: Astrophys. Sp. Sci. 98, 275 (1984)
Snell, R.L., Bally, J., Strom, S.E. & K.M.: Ap.J. 290, 587 (1985)

A PRECESSING JET IN L1551

J. F. Lightfoot
Department of Physics and Astronomy
University College London
Gower Street
London WC1E 6BT

ABSTRACT. An analysis of published observations of the L1551 optical jet leads to the conclusion that the driving region is very small, possibly little larger than the star itself. Further the data suggest that the jet is precessing.

1. THE OPTICAL JET

The jet from L1551 IRS5 is composed of largely neutral gas travelling at about $300 kms^{-1}$ (Snell et al., 1985). It is highly collimated within 100AU of IRS5 (Mundt & Fried, 1983) and is twisted, as if the source is precessing. The absence of shock emission along the leading edge of the twist implies that the jet is moving through very rarefied material.

The jet radius is too small to be observed directly, but the regular spacing of the emission knots along it may provide an indirect measure if this spacing reflects the responsivity of the jet while seeking equilibrium. Thus, if the knots are internal shocks caused by changes in the external pressure (Sanders, 1983), their spacing would be $\sim Mxr$, where M is the Mach number of the jet and r its radius. For a cool, neutral jet ($M\sim 50$) this implies a radius of 1.5×10^{14} cm. Alternatively, if the jet is magnetically collimated, the characteristic wavelength of internal oscillations is $\sim 3M_A xr$, where M_A is the Alfven Mach number of the jet (Chan & Henriksen, 1980). Since the jet is probably highly super-Alfvenic, this would again imply a radius of a few $\times 10^{14}$ cm.

Since all the jet driving mechanisms proposed so far, be they magnetic (e.g. Blandford & Payne, 1982) or thermal (Torbett, 1986), involve a large sideways expansion by a factor of 10-100 before the (magnetic?) collimating force becomes effective, the width of the visible jet is almost certainly much greater than that of its place of origin. This suggests that the driving region has a radius of several $\times 10^{12}$ cm, little larger than the young star itself.

2. THE HERBIG-HARO OBJECTS

Once the jet has cooled and faded, it will speed onward through the ISM

75

I. Appenzeller and C. Jordan (eds.), Circumstellar Matter, 75-76.

as a dark filament, only becoming visible again when it hits dense gas.
HH29 appears to be the brightest section of a large helical emission
structure and we suggest that this is the bow-shock formed along the
length of the cold jet as it advances through chaotic gas remnants
inside the L1551 molecular shell. The shape of the kinked optical jet
near IRS5 and the helix at HH29 can both be quite well explained if the
jet axis has been precessing with opening angle $8°$ and period 100yr
around an axis which itself precesses with opening angle $40°$ and period
2000yr (figure 1).

     If this is correct, then the L1551 jet is behaving very like that
of SS433, save with precession periods longer by a factor 4000. SS433 is
itself poorly understood, but in one model (Katz et al., 1982) the
precessional motion results from the gravitational torque exerted by a
binary companion on the accretion disc around the compact object driving
the jet. If it is valid to extend this model to L1551, it would imply
that IRS5 is a binary composed of $1M_\odot$ stars separated by 10^{15} cm. Radio
observations resolve IRS5 into two point sources at just this spacing
(Bieging & Cohen, 1985).

     Any model of this type again requires that the jet originate in the
central part of the accretion disc, or at the young star itself.

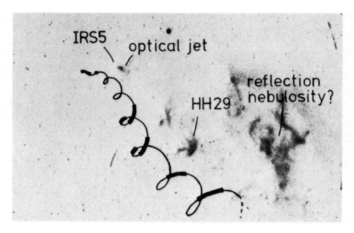

Figure 1. This is an Hα image of L1551 obtained by Snell et al. (1985).
The precessing jet discussed in the text is superimposed as a solid
line, displaced eastward for clarity.

REFERENCES

Bieging, J.H. & Cohen, M. 1985, Astrophys.J.Lett., 289, L5
Blandford, R.D. & Payne, D.G. 1982, Mon.Not.R.Astron.Soc., 199, 833
Chan, K.L. & Henriksen, R.N. 1980, Astrophys.J., 241, 534
Katz, J.I. et al. 1982, Astrophys.J., 260, 780
Mundt, R. & Fried, J.W. 1983, Astrophys.J.Lett., 274, L83
Sanders, R.H. 1983, Astrophys.J., 266, 73
Snell, R.L. et al. 1985, Astrophys.J., 290, 587
Torbett, M.V. 1986, Can.J.Phys., 64, 514

EVIDENCE FOR HELICAL VELOCITY FIELD IN MOLECULAR BIPOLAR FLOWS -- SUPPORT FOR MAGNETODYNAMIC MODEL

Y. Uchida: Tokyo Astronomical Observatory,
             University of Tokyo
K. Shibata: Department of Earth Science,
             Aichi University of Education

ABSTRACT: A search for the helical velocity field that had been predicted in a magnetodynamic theory of Uchida and Shibata was made in the bipolar flows L1551 by using ^{12}CO 115 GHz line, and evidence was obtained for it in the low velocity maps as the skew inter-invasion of the root part of the blue- and red shifted lobes into different sides of the opposite lobes. Theoretical implications of this and other findings are discussed, and the advantage of models with magnetic field is stressed.

Summary of the Poster Paper Presented

The model for the bipolar flows proposed by Uchida and Shibata (1985a, b, Shibata and Uchida 1986) is based on a global picture of the star-formation in which the *magnetized* cloud condenses into a star-disk system. The angular momentum component perpendicular to the large scale magnetic field may damp in the early phase of contraction of the cloud due to the excitation of Alfven waves. In the later part of evolution, the disk thus has a tendency to rotate perpendicularly to the magnetic field, twisting up the part of the field lines brought into the rotating disk in the process of contraction. It is shown by the 2.5-dimensional simulation by Uchida and Shibata that the magnetic twist relaxes out along the large scale field as a nonlinear torsional wave, driving the mass by the $j \times B$ force, and causes a hollow cylindrical jet having a helical velocity field in it. Angular momentum loss due to the production of the spinning jet as well as to the emission of the torsional Alfven waves allows the continued accretion of the disk to the star.
  A series of observations was made in order to check whether such a velocity field actually exists in the bipolar flows. The bipolar flows L1551 were taken as a typical case and observations were made in ^{12}CO J = 1 → 0 115 GHz line at the Nobeyama Radio Observatory in 1985 and 1986 (Uchida *et al.* 1987). The results are presented at the Symposium in the form of the detailed maps at various velocity offsets and the position-velocity diagrams for various base-lines. Among several other points of interest (see Uchida *et al.* 1987 for the details), special attention is attracted to the maps with small velocity offsets (+1.5 km/s and -1.5 km/s from the LSR velocity, 7.6 km/s) which show skew inter-invasion of a part of the blue- and redshifted contours into different sides of the opposite lobes.
  These features, together with others, are shown to be well

I. Appenzeller and C. Jordan (eds.), Circumstellar Matter, 77–78.
© *1987 by the IAU.*

reproduced by the following simple model synthesized from the observed characteristics: outflows are assumed to stream on a hollow cone $z = ar^\gamma$ (a $\sim$ 1, γ $\sim$ 3, r and z   in the cylindrical coordinates with the z-axis taken along the axis of the lobes) with a long range acceleration up to a   point   corresponding   to   0.15 pc and a gradual deceleration thereafter.   The longitudinal velocity reaches $\sim$ 50 km/s in the model, and a small spinning velocity around the axis is assumed to have $\sim$ 1.5km/s in the same direction as the rotation of the disk-like object found by Kaifu et al.   (1984).   It is shown that the iso-velocity contours and PV-diagrams computed from the simple model reproduce (i) the inter-invasion of the part of the blue- and redshifted lobes into different sides of the opposite lobes for the inclination of the axis of the flows, $10°\sim15°$, (ii) the receding ridges in the intensity in the maps for larger velocity offsets,   (iii) the seemingly linear feature   in   the   PV-diagram taken along the axis, and (iv) the semi-elliptic feature having a void in the PV-diagram taken along the strips across the lobes.

Another point of interest in the observed results is the velocity field around the blobs found in the flow.   It is seen that the main body of the blob is visible at lowest velocities, and the contours at higher velocities appear to shift downflow.   This suggests that the main part of the blob is initially at rest, and the mass at the surface of it is being peeled off by being given momenta of the flow.   This is reproduced in the simulation, and the result is shown in the poster (Shibata and Uchida 1986, in preparation).   If this is the case, the estimate of the mass of the molecular flows may go down by a considerable factor to ease the requirement for the theories.

It is pointed out that the characteristics described above, especially those related to the spinning of the lobes and the long-ranged acceleration of the flows, favor Uchida-Shibata model, or those models (Blandford and Payne 1982, Pudritz and Norman 1983 with centrifugal effect due to magnetic lever arms) which take into account of the magnetic field which transfers angular momentum to the lobes from a large enough reservoir, the rotating accretion disk rather than the central star (see details in Uchida et al. 1987, submitted to PASJ).

REFERENCES

Blandford, R.D., and Payne, D.G., 1982, Monthly Notices Roy. Astron. Soc., 199, 883.

Kaifu, N., Suzuki, S., Hasegawa, T., Morimoto, M., Inatani, J., Nagane, K., Miyazawa, K., Chikada, Y., Kanzawa, T., and Akabane, K., 1974, Astron. Astrophys., 134, 7.

Pudritz, R.E., and Norman, C.A., 1983, Astrophys. J., 274, 677.

Shibata, K., and Uchida, Y., 1986, Publ. Astron. Soc. Japan, 38, 631.

Uchida, Y., and Shibata, K., 1985a, in The Origin of Non-Radiative Heating/Momentum in Hot Stars, eds. Underhill, A. and Michalitsianos, A.G., NASA CP 2358 (NASA Printing), p 167.

Uchida, Y., and Shibata, K., 1985b, Publ. Astron. Soc. Japan, 37,515.

Uchida, Y., Kaifu, N., Shibata, K., Hayashi, S.S., and Hasegawa, T., 1987, in Star-forming Regions, eds., Peimbert, M., and Jugaku, J. (D. Reidel).

MOLECULAR OUTFLOWS AND MASS LOSS IN PRE-MAIN-SEQUENCE STARS

Russell M. Levreault
Center for Astrophysics
60 Garden Street
Cambridge, MA   02138 USA

ABSTRACT. We have conducted a survey for molecular outflows toward
71 pre-main-sequence stars using the $J = 2 \rightarrow 1$ transition of CO.
Outflows were detected and mapped toward 20 of these objects and in
an additional six background sources not included in the original
survey. The outflow sources range in mass from 0.5 to 30 M_Θ, in
luminosity from 4 to 1.1×10^5 L_Θ, and in age from $< 10^4$ to $\sim 10^6$
years. In the H-R diagram, the outflow sources form a distinct band
running across the top of the diagram.
    Roughly half of the observed outflows are bipolar at some level;
the rest show a rich variety of morphologies. In addition to the
CO ($J = 2 \rightarrow 1$) data, we have obtained CO ($J = 1 \rightarrow 0$) and ^{13}CO ($J = 1 \rightarrow 0$)
observations that enable us to estimate the excitation temperature and
optical depth in each outflow, leading to the derivation of outflow
masses. The observed outflows range in mass from 0.01 to 56 M_Θ, in
size from < 0.07 to ~ 5 pc, and in age from $\sim 10^3$ to 5×10^5 years old.
    Simple momentum conservation arguments are then used to derive
the mass loss rates necessary to produce the observed outflows. The
results range from $\sim 10^{-8}$ to $\sim 10^{-3}$ M_Θ yr^{-1}. Correlation of these
mass loss rates with other parameters shows that pre-main-sequence
mass loss goes as the square root of the bolometric luminosity and as
the 1.7 power of the stellar mass. Implications for these findings
on the nature of the pre-main-sequence mass loss mechanism, for self-
regulated low-mass star formation, and for formation of the terrestrial
planets are discussed.

I. Appenzeller and C. Jordan (eds.), Circumstellar Matter, 79.

13
INTERFEROMETRIC IMAGES OF CO EMISSION TOWARD HL TAU

Steven Beckwith and Anneila Sargent
MPI fur Astronomie and Owens Valley Radio Observatory

ABSTRACT. Millimeter wave interferometric images of HL Tau show a
clear disk morphology with a nearly keplerian velocity curve. A few
a few tenths of a solar mass are in orbit around the star.

With the angular resolution provided by millimeter wave radio
interferometers, one can directly image the gas in the circumstellar
regions near young stars. HL Tau and R Mon are two pre-main sequence
stars that appear to be surrounded by disks of gas and dust (Cohen
1983; Beckwith et al. 1984). We recently observed CO emission toward
these stars and found it to be spatially (< 5") and spectrally (< 3
km/s) unresolved in both cases (Beckwith, Sargent et al. 1986); the
gas probably resides in the disks. Uncertainties about the amount and
distribution of the gas motivated further observations with better
sensitivity and spectral resolution, a logical choice being the
isotope 13CO. This talk is a brief synopsis of the 13CO observations
of HL Tau.
    We observed the J=1-0 13CO line (110 GHz) with the Owens Valley
Millimeter Interferometer in 1985. Five interferometer configurations
produced a synthesized beam of about 6" x 10" (R.A. x Dec) in each of
thirty two spectral channels (0.13 km/s resolution) and a broad
continuum channel. A description of the instrument and the observing
procedures is given in our previous paper.
    Figure 1 shows maps of the velocity integrated 13CO emission (1a)
and the continuum emission (1b). A cross marks the stellar position
in each figure. The continuum emission is completely unresolved and
coincident with the star; its distribution reproduces the synthesized
beam almost exactly. Most of the 13CO intensity comes from a bright
core at the stellar position with the more extended emission looking
very much like an edge on disk. The disk is oriented along position
angle 147 degrees similar to the angle of 146 degress of the polarized
light near HL Tau (Hodapp 1984) and nearly orthogonal to the nearby
jet (Mundt and Fried 1983), and the outflow directions from several
other nearby stars: L1551 IRS 5 (Snell, Loren, and Plambeck 1980), DG
Tau and HH 30 (Mundt and Fried 1983).
    Figure 2 show maps of the 13CO emission versus velocity.   The

I. Appenzeller and C. Jordan (eds.), Circumstellar Matter, 81–83.

rotation curve derived from positions of maxima is consistent with keplerian velocities. The gas at the stellar velocity (2c, 6.60 km/s) peaks up at the stellar position, gas which is at low red (2b, 7.12 km/s) and blueshifted (2d, 6.08 km/s) velocities peaks at about 10" from the star on opposite sides, and the highest red (2a, 7.64 km/s) and blueshifted (5.9 km/s, not shown) velocities are seen once again close to the star. This general pattern is characteristic of velocities in a central potential.

Assuming the distance to HL Tau is 160 pc and it is a 1 solar mass star, the shift of +/-0.5 km/s at positions +/-10" from the star corresponds to the orbital velocity for material at 1600 A.U. radius. Uncertainties in the stellar mass and peak position make it impossible to show this gas is in precise keplerian motion, but the distribution and magnitude of the velocities are persuasive evidence that the gas is bound to the star, and stellar gravity dominates the potential.

In the central pixel, the 13CO has almost the same brightness temperature as the CO - T(13CO) = 5.2 K and T(CO) = 5.4 K - yielding an optical depth of about 3 for 13CO. Using a previous estimate for the gas excitation temperature of 50 K (Beckwith, Sargent et al 1986), the unknown filling factor reconciling the excitation and brightness temperatures, the total gas hydrogen mass in the central beam is 0.3 solar masses. While the result is uncertain (factor of three, say), it is clear that the disk contains a substantial fraction of a stellar mass.

The continuum radiation is thermal emission from solid particles in the disk. Following the arguments in our earlier paper and using an emissivity coefficient falling as frequency squared (e.g. Hildebrand 1983), the equivalent molecular hydrogen mass implied by the continuum radiation is 0.3 solar masses, in good agreement with that implied by the CO emission.

The disk structure and large mass indicated by the 13CO observations guarantee the stability of the gas against disruption by stellar winds or radiation pressure (Beckwith et al. 1984). It certainly contains enough mass to form a planets similar to those around the sun. With nearly as much mass in the disk as in the star, it is even possible the disk is self-gravitating and influences further accretion of cloud material. It should be possible to directly map the gas distribution in the outer parts of the disk with millimeter wave interferometers, and we expect future observations to play a role in deciding how these disks form and evolve. A more detailed analysis of these data will be presented in a longer article this year (Sargent and Beckwith 1986).

REFERENCES
Beckwith, S., Sargent, A.I., Scoville, N.Z., Masson, C.R., Zuckerman, B., and Phillips, T.G. 1986, Ap. J., in press.
Beckwith, S., Zuckerman, B., Skrutskie, M.F., and Dyck, H.M. 1984, Ap. J., 287, 793.
Cohen, M. 1983, Ap.J.(Letters), 270, L69.
Hildebrand, R.H. 1983, Quart. Journ. R.A.S., 24, 267.
Hodapp, K.W. 1984, Astron. Ap., 141, 255.

Mundt, R. and Fried, J.W. 1983, Ap.J.(Letters), 274, L83.
Sargent, A.I. and Beckwith, S. 1986, in preparation.
Snell, R., Loren, R., and Plambeck, R. 1980, Ap.J.(Letters), 239, L17.

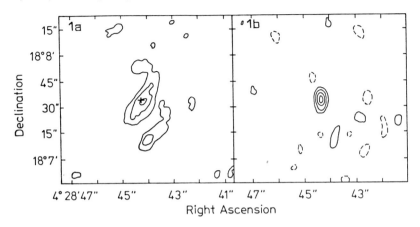

Figure 1: Maps of integrated 13CO (1a) and continuum emission (1b) towa
HL Tau. The continuum emission is unresolved and maps the beam.

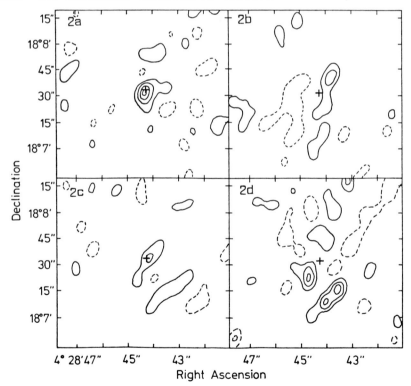

Figure 2: Maps of 13CO emission in different velocity channels.
2a - 7.64 km/s; 2b - 7.12 km/s; 2c - 6.60 km/s; 2d - 6.08 km/s.

V645 Cyg - ON THE STRUCTURE OF A YSO

A. Schulz [1], J.H. Black [2], C.J. Lada [2]
Max-Planck-Institut für Radioastronomie
Auf dem Hügel 69, 5300 Bonn, FRG
Steward Observatory, Univ. of Arizona
Tucson 85721 AZ, USA

ABSTRACT. In the CO(J=3-2) line we observe a bipolar outflow in V645 Cyg. The physical properties of the neutral flow gas and its geometry put new constraints on the structure of this YSO.

INTRODUCTION

V645 Cyg (AFGL2789) is one of the few young stellar objects (YSO's) which can be studied at many wavelengths. The visible object shows a star-like knot (NO) and a few nebulae filaments (N1, N2..). Its spectrum is like a Herbig-Ae star, with displaced absorption at velocities up to -340 km/s (LSR) and strong emission lines some of which show P-Cyg profiles; it rises steeply longward of 1 µm, H I IR emission lines and the 10 µm silicate absorption are observed. H_2O maser emission arises exactly from the position of NO. Radio continuum measurements gave only upper limits at 5 and 15 GHz. The distance of V645 Cyg, the nature of the central star, and the physical conditions in its environment have remained controversial. Associated with V645 Cyg is a molecular cloud (Harvey & Lada 1980). To investigate the distribution and kinematics of the molecular gas closely surrounding V645 Cyg, we performed CO(J=3-2) sub mm line observations. We briefly present here some main implications of our work (detailed discussion and figures in Schulz et al. 1987).

OBSERVATIONS

The map of 25 CO(J=3-2) spectra (2'x2'; 20" spacing, beam 26") was obtained during 17 - 24 April 1984 at the Multiple Mirror Telescope on Mt. Hopkins, Arizona, which was used as a phased sub mm array. The antenna temperatures are accurate to ± 10 %. The spectra show a line core component of gaussian shape at a velocity of -44.3 ± 0.1 km/s (LSR) and high-velocity wings ranging over ± 15 km/s and extending over an area of 1'x1' centered on V645 Cyg. Red-shifted and blue-shifted emission are spatially displaced by 20" at a position angle of -25° ± 15° which is about perpendicular to the optical polarization vectors of NO and N1 (Cohen 1977).

I. Appenzeller and C. Jordan (eds.), Circumstellar Matter, 85–86.
© *1987 by the IAU.*

DISCUSSION

Derived properties of V645 Cyg depend on the distance. The kinematic distance is not reliable because of deviations from circular motions in the
Perseus arm of our galaxy. Chavarria et al. (1986) have determined the
distances of many H II regions in this area and related them to their
velocities; from this analysis, at l^{II} = 94.6^{o} and v_{LSR} = -45 km/s we
obtain a distance of 3.0 $\pm$ 0.5 kpc for V645 Cyg which we adopt hereafter.
     The visible spectrum indicates a strong wind, with no evidence for a
hot O-type photosphere. The absence of noticeable radio continuum emission implies that the central star has not developed a photoionized nebula. In the (H-K)/(K-L) two-colour diagram V645 Cyg is located in the
region of the extreme Herbig-Ae/Be stars (Coodrich 1986). The IR emission
line spectrum (see McGregor et al. 1984) is similar to those of many
other objects regarded as pre-main-sequence stars. All this gives evidence for V645 Cyg being a very young object. Its total luminosity is
4.5 x 10^{4} $L_{\odot}$ (3 kpc dist.).
The 10 μm silicate absorption implies an A_V (star) of at least 10 mag towards the observer, much more than determined for NO and the other filaments. The optical spectra and the polarisation suggest that all visible
features must be regarded as reflection nebulae and, according to very
recent data by Goodrich (1986) and Solf (priv. comm.) have embedded knots
of shock-ionized line emission.
The flow is only poorly collimated. From our CO(3-2) map, we can derive
parameters for the outflow gas assuming 1.) thermalized CO emission, 2.)
T_{ex} between 20 and 100 K, 3.) the outflow region being a cube of 1' on
each side, 4.) CO abundance of 10^{-4}. Then, the total mass of the outflow
gas is m = 7 $m_{\odot}$; its mechanical luminosity (3$L_{\odot}$) is small; the corresponding momentum transfer rate, nevertheless, is close to the value derived for the stellar wind (Kwok 1981) using the expansion velocity of
330 km/s of Humphreys et al. (1980); hence, the ionized stellar wind may
in fact be capable of driving the outflow and we do not have to assume
an additional component to the wind.
We are not yet able to derive a unique model for V645 Cyg: The outflow
seems to define a projected preferential axis; from the A_V values for
the central star and for the nebulae, one finds an anisotropic dust distribution. But the variety of features with different high and low velocity components leave some confusion about the true geometry. Particularly, the fact that blue-shifted visual emission is located on top of
red-shifted molecular gas still needs a comprehensive explanation.

REFERENCES
Chavarria-K., C., Hasse, I., Moreno, M.A., Pismis, P., 1986 (in prep.)
Cohen, M., 1977, Ap. J. 215, 533
Goodrich, R.W., 1986 (in press)
Harvey, P.M., Lada, C.J., 1980 Ap. J. 237, 61
Humphreys, R.M., Merrill, K.M., Black, J.H., 1980, Ap. J. Lett. 237, L17
Kwok, S., 1981, P.A.S.P. 93, 361
McGregor, P.J., Persson, S.E., Cohen, J.G., 1984, Ap. J. 286, 609
Schulz, A., Black, J.H., Lada, C.J., Ulich, B.L. Martin, R.N., 1987
     Ap. J. (in prep.)

ON THE LINE PROFILES OF SHELL-SHAPED BIPOLAR OUTFLOWS

G. Silvestro and M. Robberto
Istituto di Fisica Generale dell'Università
C. M. D'Azeglio, 46
I-10125 Torino
Italy

1. INTRODUCTION

High velocity molecular outflows with bipolar morphology are detected in association with young stellar objects within dense interstellar clouds. Recent observations suggest that the flow could be "confined to a relatively thin, swept-up shell surrounding an evacuated wind cavity" (1). A shell structure characteristic of the wind-cloud interaction had been predicted in earlier theoretical works (see for instance (2)). More recently, models with different (not shell-shaped) geometries were presented, e.g. (3).
   An evaluation of line profiles of the high velocity molecular gas at different angular resolutions could allow comparison with observation, and help choosing between different models. We present preliminary data with reference to the shell model suggested by Barral and Cantò (4), which gives an explicit description of the velocity field along the shell and allows a simple evaluation of the intensity distribution along the shocked surface.
   The model contains several simplifying assumptions: the system is considered to be in a steady state, and "pressure driven"; we are presently studying the time evolution of a thin shell in more general terms.

2. NUMERICAL CODE AND RESULTS

Using the Barral and Cantò model we compute the shock surface configuration for arbitrary angles of the polar axis with the line of sight. The code makes use of a (90x180) matrix with steps of 1 degree in latitude and 2 degrees in azimuth. The gas emissivity is estimated by the equation

$$L = (\text{constant}) \times L_{shock} \times \phi_{beam} \times j_{max} \tag{1}$$

where: $L_{shock} = \rho v^3$, $\phi_{beam} = (1/2)^{(r/B)^2}$ (antenna beam pattern), $j_{max} = \rho c$ (mass mixing, (5)). The code computes the line structure at each point of the array by using Eq. (1) for the intensity, evaluates the li̲

87

I. Appenzeller and C. Jordan (eds.), Circumstellar Matter, 87–88.
© 1987 by the IAU.

ne-of-sight velocity component for the line center, and estimates the line width resulting from turbulence in the region of mixing.

The turbulent velocity width Δv_{turb} is the main source of uncertainty in the computation. We consider two cases: (a)$\Delta v_{turb} = v_s/8$, (b)$\Delta v_{turb} = v_s/4$ (v_s is the mean square post-shock velocity). All line profiles are normalized over the number of points considered. The minimum beam amplitude contains more than 200 points. Our code cannot be used for estimating the luminosity very close to the star, where the relation for L_{shock} would give an unrealistically high emissivity. Our unit of length is the scale factor R_0 (the shock radius at $\theta=0$); we assume a terminal wind velocity $v_* = 100$ km/sec. Some results are presented in Fig. 1.

One can see a double peak is present for the lower value of the turbulent velocity (of order 15 km/sec). The structure is in reasonable agreement with observations of Mon R2 (see (1)). A detailed comparison with observational data is in course.

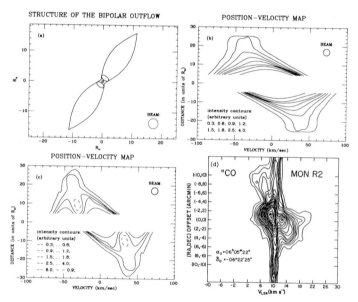

Fig. 1. (a) A typical bipolar outflow having polar axis along the bisectrix of the first octant. (b) Position-velocity map for $\Delta v_{turb} = v_s/4$. (c) The same, for $\Delta v_{turb} = v_s/8$. (d) Observational position-velocity map of the outflow source Mon R2.

LITERATURE

(1) C.J. Lada, Ann. Rev. Astron. Astrophys. 23, 267 (1985).
(2) R. Weaver, R. McCray, J. Castor, P. Shapiro and R. Moore, Astrophys. J., 218, 377 (1977).
(3) R.E. Pudritz and C.A. Norman, Astrophys. J., 301, 571 (1986).
(4) J.F. Barral and J. Cantò, Rev. Mexicana Astron. Astrofis. 5, 101 (1981).
(5) F.D. Kahn, Astron. Astrophys. 83, 303 (1980).

A MODEL OF MOLECULAR EMISSION IN BIPOLAR FLOWS

Sylvie Cabrit and Claude Bertout
Institut d'Astrophysique de Paris
98bis Bd Arago
75014 PARIS
FRANCE

ABSTRACT. We present a model of CO J=1→0 line formation in accelerated bipolar outflows. We find that the integrated intensity maps can be divided into four categories, depending on the view angle and the flow opening angle. We then show that no accurate values of the flow parameters can be derived from the CO observations without careful modelling of the outflow properties.

1. INTRODUCTION

Recent millimeter and optical studies of star-forming regions have provided strong evidence that many young stellar objects are undergoing a phase of energetic, collimated outflow activity (cf. Lada 1985). CO line observations are especially important since they offer a powerful probe of the physical properties, large scale structure, and kinematics of the high-velocity molecular gas.

In order to exploit the wealth of information contained in CO observations, we have developed a code which computes the excitation and the rotational line profiles of CO in axial symetry in the presence of supersonic accelerated velocity fields. Our first investigation was emission from a steady-state radial outflow expanding in two biconical lobes. This short contribution reviews our main assumptions and implications for high-velocity CO excitation and outflow geometry. A complete analysis of our results can be found in Cabrit and Bertout (1986).

2. MAIN ASSUMPTIONS

2.1. Flow geometry

Although many molecular flows have an obvious bipolar geometry, a major uncertainty is whether the observed lobes are filled with high-velocity gas or trace the surface of an expanding hollow shell. While the first case would yield wide, single-peaked profiles similar to the observed ones, the second should produce narrower, double-peaked profiles when the line of sight crosses two distinct thin layers, unless the turbulent velocity within the shell is of the order of magnitude of the outflow velocity, or the shell is made of unresolved clumps moving at different velocities. In at least one flow, L1551-IRS5 (Moriarty-Schieven et al.1986), a bright- rim structure in the high-

I. Appenzeller and C. Jordan (eds.), Circumstellar Matter, 89–92.

velocity gas is seen which is suggestive of the shell geometry. At this time, we present the results for the filled cone model. We assume that the high-velocity gas fills a bicone of opening angle $2\ \theta_{max}$, inner radius r_{min} and outer radius r_{max}, and that the velocity and density laws are exponential in $(1/r)$ with exponents α and δ. Mass conservation within the flow required that $\delta = 2 - \alpha$. Two cases were investigated: a constant velocity outflow with $\alpha = 0$ and a linearly accelerated outflow with $\alpha = -1$.

2.2. CO excitation

In conditions typical of molecular clouds, the CO rotational lines are excited mainly by radiative transitions and collisions with hydrogen molecules. The sources of radiation external to the high-velocity gas are the thermal emission of dust grains, the line emission of the static ambient cloud, the photospheric emission from the central star, and the cosmic background emission at 2.7 K. Computations of the ratio of dust grains to molecular gas emissivity show that the effect of dust-to-gas radiative coupling on the transfer of the first two rotational lines of CO is always negligible, but that it may be important for higher transitions. The static cloud line radiation, being Doppler-shifted in the high-velocity gas rest frame, cannot contribute to the CO excitation in the flow and did not appear in our computations. For moderate high-velocity CO optical depths, we found that heating by stellar radiation was noticeable only at distances $r < r_* \ (T_*/10,000 \ \text{K})^{1/2}$. Since the CO observations suggest that high-velocity emission originates at distances greater than 10^{-2} pc (Goldsmith et al. 1984, Moriarty-Schieven et al. 1986), we also neglected the stellar contribution and considered that the high-velocity gas was heated only by collisions and absorption of local or cosmic background photons.

Because the outflow velocities inferred from molecular observations ($v \geq 10$ km s^{-1}) are much larger than the CO thermal velocity ($v_{th} = 0.02 \ \sqrt{T_k}$ km s^{-1}), we adopted the fast-flow approximation to compute NLTE level populations, but did not use it for the emergent profiles, which were integrated exactly. This hybrid method has been shown to give excellent results for fast flows (Bastian et al. 1980). The CO excitation conditions at a given point in the flow are then uniquely determined by the values of the following independent parameters: T_k, n_{H_2}, α, and the local optical depth $k(r) = 8\pi^3/3h \ \mu^2 \ n_{CO} \ r/v(r)$.

3. PHYSICAL IMPLICATIONS

3.1. CO excitation

In agreement with theoretical expectations, we found that the $J = 1 \rightarrow 0$ transition was thermalized only for:

$$(n_{H_2})^2 > 530 \ v \ (\text{km s}^{-1}) / (\ A_{CO} \ r(\text{pc}) \ ) \ \text{cm}^{-6}.$$

In a typical molecular flow, where $A_{CO} \sim 10^{-4}$, $r \sim 0.1$ pc, $v \sim 10$ km s^{-1}, this condition becomes :

$$n_{H_2} > 2.5 \ 10^4 \ \text{cm}^{-3}.$$

The absence of detectable high-velocity CS emission in the observed CO lobes seems to indicate that in molecular flows the mean density lies below this critical value. NLTE calculations of the CO excitation are thus required if most of the high-velocity emission is formed in an extended region of moderate density.

3.2. Flow geometry

For a radial biconical outflow, four types of configurations can be distinguished, depending on the values of θ_{max} and i, the angle between the flow axis and the line of sight.

Case 1: $i < \theta_{max}$ and $i < \pi/2 - \theta_{max}$, is illustrated in Fig.1a ($i = 10°$, $\theta_{max} = 30°$). The two cones are seen nearly face-on, so that the contours of blue- and red-shifted integrated intensity are circular and overlapping.

Case 2: $i > \theta_{max}$ and $i < \pi/2 - \theta_{max}$, occurs only if $\theta_{max} < 45°$ and is illustrated in Fig.1b ($i = 50°$, $\theta_{max} = 30°$). The blue and red-shifted lobes are now spatially separated and the bipolar structure of the flow is apparent.

Case 3: $i > \theta_{max}$ and $i \geq \pi/2 - \theta_{max}$, is presented in Fig.1c ($i = 80°$, $\theta_{max} = 30°$). The two cones again appear as two distinct lobes but each one is now partly blue-, partly red-shifted.

Case 4: $i < \theta_{max}$ and $i \geq \pi/2 - \theta_{max}$, occurs only for $\theta_{max} > 45°$ and is illustrated in Fig.1d ($i = 50°$, $\theta_{max} = 60°$). It is simply a Case 3 where the two lobes are still overlapping due to their larger opening angle.

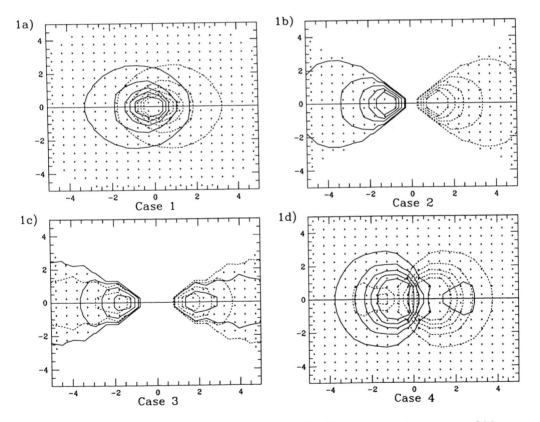

Figure 1a to 1d (left to right and top to bottom). Synthetic contour maps of blue-shifted (solid lines) and red-shifted (dashed lines) CO J=1→0 integrated intensity computed for n_{H_2} (r_{min}) = 10^4 cm^{-3}, T_k = 10 K, k(r_{min}) = 100, α = -1, r_{max} = 10×r_{min}, illustrating the four flow configurations defined above (see text for details).

Although Cases 1 and 2 reproduce quite well the main aspects of the observed flows (see Lada 1985), Cases 3 and 4 have never been reported, even though they are expected whenever $\theta_{max} > 30°$. There are three possible explanations for this lack of detection: First, emission at low radial velocity is undiscernable from the ambient cloud emission, so that Case 3 (resp. Case 4) flows might be classified as Case 2 (resp. Case 1) flows after subtraction of the central low-velocity emission. Second, observations that are attributed to a pair of outflows seen face-on could in fact represent a single Case 3 flow. Finally, most bipolar flows might have $\theta_{max} < 30°$. These conclusions remain valid for a radial outflow confined to a conical shell.

Our calculations have shown that the CO emission is also strongly dependent on the velocity field (Cabrit and Bertout 1986), which makes our model a powerful tool for the study of molecular flows properties.

4. OBSERVATIONAL UNCERTAINTIES IN THE FLOW PARAMETERS

In order to test the method commonly used by observers for deriving flow properties, we applied it to our synthetic maps and compared the resulting estimates with the actual parameters of our model.

We found that the assumption of small optical depth underestimates the CO mass and momentum by a factor of 25 at worst, except in flows with low optical depth and high density gradients where they can be overestimated by up to a factor of 2. The dynamical time scale, obtained by dividing the apparent size of the flow by the maximum observed velocity, is always too small because the flow length is underestimated, especially at small view angles. And if $\alpha = -1$, it can be in error by up to a factor of 20, because in accelerated flows the velocity at the base of the flow is overestimated. In that case, the rate at which momentum is transferred to the CO flow is overestimated by similar factors, but it can also be underestimated by up to a factor 10 when $\alpha = 0$ and the optical depth is large.

The values of the rates of momentum and energy input in the flow derived from the CO observations must therefore be considered with extreme caution.

5. CONCLUSION

This model gives a better understanding of what can be expected and deduced from the geometry and emission characteristics of molecular outflows. Our code is now being improved to handle more complex geometries (thin shells) and velocity fields (decelerated flows), which will be investigated in our next papers. We will also attempt detailed comparisons of our computations with recent high-resolution CO observations. Only after such a study will we understand the kinematics of bipolar flows and the constraints they put upon mass-loss mechanisms in pre-main sequence stars.

REFERENCES

Bastian, U., Bertout, C., Stenholm, L., Wehrse, R. 1980, Astr. Ap. **86**, 105
Cabrit, S., Bertout, C. 1986, Ap. J. **307**, 313
Goldsmith, P.F., Snell, R.L., Hemeon-Heyer, M., Langer, W.D. 1984, Ap. J. **286**, 599
Lada, C.J. 1985, Ann. Rev. Astr. Ap. **23**, 267
Moriarty-Schieven, G., Snell, R.L., Strom, S.E., Schloerb, F.P., Strom, K.M. 1986, Submitted to Ap. J.

STAR FORMATION IN THE SOUTHERN COMPLEX REGION NGC 3576

P. Persi, M. Ferrari-Toniolo, L. Spinoglio
Istituto di Astrofisica Spaziale, CNR,
P.O. Box 67
00044 Frascati
Italy

NGC 3576 (RCW 57) is a very bright optically visible nebula ($\sim$10' in size) located near the galactic plane at a distance of 3.6kpc. The region is associated with the HII region G291.28-0.71 (Retallack & Goss), and with a giant molecular cloud of $M \simeq 9 \times 10^4$ $M_\odot$ (Cheung et al.1980) The source is reported in the IRAS Small Scale Structure Catalog. From the observed far-IR fluxes we derive a total luminosity of $L(FIR) \simeq 7 \times 10^5$ $L_\odot$, suggesting that at least seven possible 07-09 stars could ionize the entire observed region. A cluster of five IR sources and extended 10μm emission were found by Frogel & Persson (1974). IRS 1 is a very compact object ($<3'$), while IRS 2-5 show diffuse infrared emission. In addition IRS 2, very close to IRS 1, is coincident with the radio emission peak. An H_2O maser source and NH_3 emission are present in the complex region.

We present multidiaphragm photometry from 1 to 20μm of IRS 1, IRS 3 and IRS 4 and CVF observations between 2-4μm and 8-13μm of IRS 1, collected at the 1m and 3.6m ESO telescopes.

Results

The observed energy distribution of IRS 1 between 1 and 20μm is reported in Fig.1. Between 1 and 2.2μm the spectrum is very flat. This could be explained by the contamination of the nearby source IRS 2. IRS 1 shows absorption features at 9.7μm and 3.08μm due to "silicate" and "ice" bands. The derived optical depths are respectively: $\tau(9.7\mu m) \simeq 4.6$ and $\tau(3.1\mu m) \simeq 0.73$. These values are very similar to those of massive "protostellar" objects observed by Willner et al.(1982). Adopting the relationship $A_V/\tau(9.7\mu m)=16$ (Rieke & Lebofsky 1985) we derive for IRS 1 $A_V=74$. The observed luminosity between 1-20μm is $L \simeq 6 \times 10^4$ $L_\odot$. This luminosity and the observed features in IRS 1 are consistent with the presence of a very young object, probably a deeply embedded 07-8 (ZAMS) star. In the direction of IRS 1 we observed the Brγ and Brα lines with an intensity ratio of $I(Br\alpha)/I(Br\gamma)=6.75\pm0.63$. Comparing the observed ratio with the nebular ratio for $T_e=10^4$ K and Menzel's case B ($Br\alpha/Br\gamma=2.83$) and assuming $A_\gamma=0.126$ A_V and $A_\alpha=0.04$ A_V, we derive a visual extinction of $A_V=11$. This could imply that the Brackett lines ori-

93

I. Appenzeller and C. Jordan (eds.), Circumstellar Matter, 93–94.

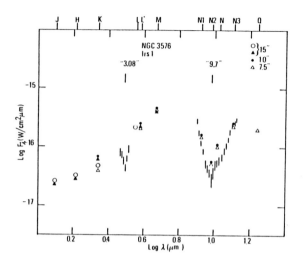

Fig.1: IR energy distribution of IRS 1

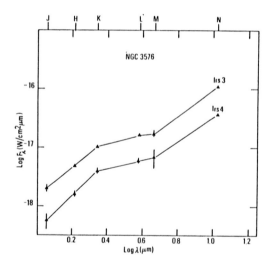

Fig.2: IR energy distributions of
        IRS 3 and IRS 4

ginate in the HII region
(observed from the radio
continuum) located in front
of the molecular cloud.
The sources IRS 3 and IRS 4
show infrared energy distributions very similar (Fig.
2). The derived 1-10μm luminosity for both the sour-
ces is $6.8 \times 10^3 L_\odot$ and $1.8 \times 10^3 L_\odot$, consistent with the
presence of a B0.5 and a B2
(ZAMS) star respectively.
Assuming that no IR excess
is present in the near-IR
(1-2.2μm), we derive A_V=15-
20 for the sources. This i-
mplies that IRS 3 and IRS 4
are associated with the HII
region.

Conclusions

        The complex region NGC
3576 could represent an example of sequential star
formation as described by
Elmegreen & Lada (1977), in
which the optical nebula
lies just outside the cloud.
A cluster of at least seven
OB (ZAMS) stars could be
responsible for the excitation of the nebula and the
younger HII region, located
probably at the edge of the
molecular cloud. Deeply embedded into the cloud, a
very young object (IRS 1)
is present, inferring that
continuum star formation
occurs in NGC 3576.

References

Cheung et al.1980:Astrophys.J.,240,74
Elmegreen & Lada,1977:Astrophys.J.,214,725
Frogel & Persson,1974:Astrophys.J.,192,351
Retallack & Goss,1980:Mon.Not.R.astr.Soc.,193,261
Rieke & Lebofsky,1985:Astrophys.J.,288,618
Willner et al.1982:Astrophys.J.,253,174

STAR FORMATION IN CHA T1

P.R. Wesselius, R. Assendorp, H. Roede
Space Research,
P.O. Box 800,
9700 AV Groningen,
The Netherlands.

ABSTRACT. In a region of 0.4 × 0.5 pc around HD 97300 the amount of mass in gas and dust is < 16 $M_\odot$, while the stellar mass is 6 - 9 $M_\odot$. Thus in this core of a collapsing cloud (Cha T1) several stars have been formed in a very efficient way.

1. OBSERVATIONS AND CALIBRATION

Centered on HD 97300 special high-resolution IRAS observations were executed: 4 with the smallest detectors at the edges of the survey instrument (EDGE), and 6 with CPC. Standard data reduction has resulted in calibrated EDGE maps (see Young et al. 1985) and CPC maps (see Wesselius et al. 1985). We have combined the maps, for EDGE and CPC separately (see Fig. 1). These observations have larger position errors than the IRAS Point Source Catalog (IPSC). Using the optical positions of stars detected at 12 and 25 μm the errors have been reduced to 5".

Our ID	RA m s	DEC ' "	Other Name	12 μm	25 μm	60 μm	100 μm
			Table 1 IRAS sources around HD 97300				
A	07 51	07 00		0.6 P	0.6 P	0.5 P	nd
B	07 56	17 39	HJM:C1-6,C1-3	0.8 D	8 D	13 D	
C	08 16	20 17	HD 97300	11.2 D	12 D	104 D	285 C
D	08 22	19 08	WW CHA, HM23	14.0 D	34 D	60 D	
E	08 26	16 12	HJM:C1-2	0.02 D	con	con	
F	08 33	13 35	JH16	0.5 D	0.7 P	nd	nd
G	10 53	27 59	HJM:E2-4	2.0 P	2.8 P	nd	nd
H	10 49	20 50	HJM:E1-9a	0.5 P	0.4 P	nd	nd

All objects are at 11^h, $-76°$, epoch 1950; C, E, P: flux determined from respectively CPC, EDGE, IPSC; no: not observed; nd: not detected; con: confused; HJM: Hyland, Jones, and Mitchell (1982); HM: Henize & Mendoza (1973); JH: Jones & Hyland (1986).

I. Appenzeller and C. Jordan (eds.), Circumstellar Matter, 95–96.

In the confused region just around HD 97300 gaussians have been fitted
to the apparent sources and the fluxes estimated are accurate to about
50 %; other source fluxes have about 10 to 20 % uncertainty. In Table 1
data on the sources found are summarized.

2. GAS AND STELLAR MASS

The formulae and constants presented by Hildebrand (1983) have been
used, for a wavelength dependence of grain emissivity of $\lambda^{-1.5}$. The
mass of gas and dust, derived using Hildebrand's eq. 10, is extremely
dependent on the value of the temperature; the ratio of the fluxes at 60
and 100 μm indicates T = 50 K, leading to 0.1 $M_\odot$. At 0.5 pc from
HD 97300 the dust temperature is of order 18 K; thus an upper limit to
the dust+gas mass is 16 $M_\odot$.
    For the (proto)stars B, D, E - confused with HD 97300 - the total
luminosity is equal to the IR luminosity (including J, H, K) because of
the heavy obscuration. For HD 97300 we use the mass estimate of Thé et
al (1986). Total luminosity can be converted into spectral type and
spectral type into mass using standard calibration tables for main-
sequence stars. However, these pre-main-sequence stars may be superlumi-
nous (if by a factor 2.5, the mass estimate is 40 % too high). We arrive at a total mass of the stars B, C, D, E of 9 $M_\odot$.

REFERENCES

Henize, K.G., Mendoza, E.E. 1973, Astrophys. J. **180**, 115.
Hildebrand, R.H. 1983, Q. Jl. R. astr. Soc. **24**, 267.
Hyland, A.R., Jones, T.J., Mitchell, R.M. 1982, MNRAS **201**, 1095.
Jones, T.J. et al. 1985, Astron. J. **90**, 1191.
Thé, P.S., et al. 1986, Astron. Astrophys. **155**, 347.
Young, E.T. et al. 1985, PRE-008N, Internal Report IPAC, JPL, USA.
Wesselius, P.R. et al. 1985, Internal Report ROG, Groningen.

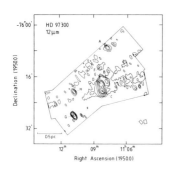
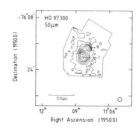
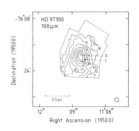

Maps at 12 μm(EDGE) and at 50 and 100 μm(CPC) are shown. 12 μm contour-
levels at 1,2,4,10,20,40,100 $10^{-7} Wm^{-2} Sr^{-1}$; 50 and 100 μm levels at
.14,.28,.56,1.4,2.8,5.6,14,28,56 and at 0.4,0.8,1.6,4,8,16,40 MJy Sr^{-1}.

AN OBSERVATIONAL STUDY OF THE HERBIG Ae STAR VV SERPENTIS, AND OF R-STARS ASSOCIATED WITH ITS DARK CLOUD

C. Chavarría-K.[1,3], J. Ocegueda[2,3], E. de Lara[3],
U. Finkenzeller[4], E. Mendoza[3]
[1] Landessternwarte Königstuhl, 6900 Heidelberg 1,
F.R.G.
[2] Universidad de Michoacán, Morelia Michoacán,
México
[3] Instituto de Astronomía, UNAM, Ap. Postal 70-264,
México, D.F. 04510
[4] Astronomy Department, University of California,
Berkeley CA, USA

In this work we give a revised distance (250 ± 20 pc) to
the star VV Serpentis and its associated dark cloud complex,
redetermine its spectral type (= A2e, see also Herbig, 1960,
ApJ Suppl. 4, 337), and report on observations of previously unknown associated emission line stars, and of stars with
associated reflection nebulosity (R-stars) belonging to the
same region.

The low resolution spectroscopic data of VV Serpentis
We have obtained uvby-Hβ and JHKLM photometry of VV
Serpentis, and of 12 selected R-stars with the 1.5 and 2.1m
telescopes of the Sierra San Pedro Mártir National Observatory (México), respectively. Most of the stars were also observed at intermediate resolution (R ≃ 4000) with the Waltz
72 cm reflector of the LSW, and/or with the 2.2 m at ESO,
La Silla (Chile). VV Serpentis has also been observed at
both high and low resolutions (600 < R < 40.000) by using
the IDS and the Hamilton echelle spectrograph of the Lick
Observatory (USA). For the present study α(16), Λ(9) photometric colours of VV Serpentis were also available. The summarized observational data and results are listed in table 1.
    The low resolution spectroscopic data of VV Serpentis
indicate a spectral type A2, based on the lower Balmer
lines, and the presence of lines of neutral and once ionized
metals, mainly Ca II, Fe I and II, Ti I and II, Cr II (cf.
Figure 1). The observed Balmer discontinuity is accordingly.
Likewise two high resolution spectrograms clearly show Fe II
(42) λλ4923 and 5018 in absorption. These lines are typical
of early A type stars. Both Hα and Na I-D lines are highly
variable. The presence of He I λ5876 A (W_λ = 0.86 A, asymmetric to the red) should be regarded as indicative of stellar activity and/or of an extended atmosphere rather than

97

I. Appenzeller and C. Jordan (eds.), Circumstellar Matter, 97–98.
© 1987 by the IAU.

traces of an early spectral type (see Finkenzeller and Mundt, 1984, Astron. Astrophys. Suppl. 55, 109). The large $\Lambda(9)$ colour index observed for the star gives support to this.

The combined data yielded $A_V/E(B-V) = 4.0 \pm 0.1$ and $d = 250 \pm 20$ pc for the dark cloud complex. Hence the luminosity of VV Serpentis is $\simeq 36\ L_O$, with $L_{ir}/L_{opt} > 5$, and $M \approx 2.5\ M_O$. Furthermore, the star is a photometric variable in the optical and infrared ($\Delta m \leq 0^m.4$), in time scales of 1^d or less. Its location in the H-R diagram is shared with the stars T Ori, HK Ori, BD +46O3471, all being evolved low mass Herbig Ae stars.

In order to check the R-stars for possible H_α emission, the plates taken by Iriarte and Chavira (1956, Bol. Tonantzintla Tacubaya 14, 31) were reinspected by us, and 7 new H_α emission stars in the field were found. Our final results will be published elsewhere.

This work has been partially supported by the Deutsche Forschungsgemeinschaft (SFB 132 A), the Dirección General de Apoyo al Personal Académico de la UNAM, and the Alexander von Humboldt Stiftung (FLF-VB2).

TABLE I.   R-stars associated with VV Serpentis' dark cloud

Nr.	other	coordinates 1950 h m s o ' "		V	K	sp.t.	n	Av	R	dist pc
1	BD-2O4607	18 18 57	-2 4 2	10.36	-	A2V:	1	1.92		259
3	VV Ser	18 26 15	+0 6 34	12.54V	5.55V	A2e	1	4.6	6.6	Pmss
5		18 27 0	+1 6 13	-	9.8	<K2	1			
6		18 27 1	+1 6 20	-	8.3	AOV	1			
7		18 27 24	+1 1 48	12.25	-	<AOV:	1	6.3		
9		18 27 35	+1 1 7	11.74	-	<K2	1	3.5		
10	BD+1O3694	18 27 52	+1 11 26	9.87	7.5	A1V	1	2.53	3.8	185
12	SAO123590	18 28 17	+1 21 22	8.46	6.40	B4V	1	2.97	4.1	239
13	SAO123595	18 28 37	+1 25 14	8.54	6.90	B3V	1	2.62	4.1	319
14		18 28 40	-2 22.9	13.9 V	-	Be	1	5.4		Pmss
15	SAO123661	18 32 35	+0 0 3	8.05	-	B3V	1	2.7		259
16		18 34 24	+0 17 36	10.65	-	<AOV	1	3.7		

Remarks to table 1
spectral type based on spectral data
luminosity class from spectrum or H_β index

Figure 1. VV Serpentis taken the 30. Oct. 1985 with the Lick Observatory 40" and IDS.

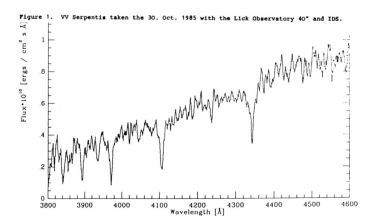

MULTIPLE DUST SHELLS AROUND HERBIG Ae/Be STARS

P.S. Thé   and   D.N. Dawanas
Astronomical Institute   Bandung Institute of
"Anton Pannekoek"   Technology
Roetersstraat 15   Jalan Ganesha 10
1018 WB Amsterdam   Bandung
The Netherlands   Indonesia

Intermediate mass $(2 < M/M_\odot < 9)$ pre-main sequence objects, also named Herbig Ae/Be stars, are known to have excess radiation in the near-infrared. From IRAS o bservations it turns out without doubt (quality 3, high S/N radio), that these objects are very strong far-infrared emitters at 12, 25, 60 and often also at 100 μm. The spectral energy distribution, depicted in Fig. 1 for intermediate mass pre-main sequence stars, show clearly this large excess. From the difference curves it is apparent that this excess radiation is most probably caused by several dust shells. Using very simplified methods it is possible to derive the average temperature of the dust shells (see Thé, Wesselius, Tjin A Djie and Steenman, 1986). If the chemical composition of the mixture of the dust grains and their average size are assumed it is also possible to estimate other characteristics like the distance from the central star and the mass of the dust shells (see Thé, Hageman, Westerlund, Tjin A Djie, 1985).

REFERENCES:

Thé P.S., Hageman, T., Westerlund, B.E., Tjin A Djie, H.R.E.: 1985, Astronomy and Astrophysics, **151**, 391.
Thé, P.S., Wesselius, P.R., Tjin A Djie, H.R.E., Steenman, H.: 1986, Astronomy and Astrophysics, **155**, 347.

I. Appenzeller and C. Jordan (eds.), Circumstellar Matter, 99–100.

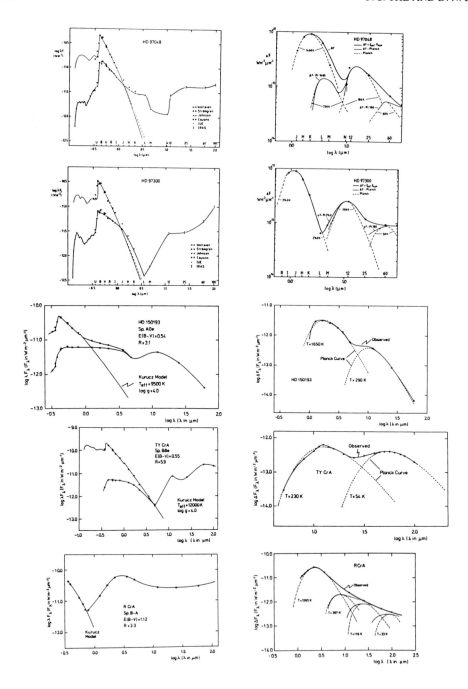

Fig. 1. The spectral energy distribution and difference curve of 5
       intermediate mass pre-main sequence stars.

NEUTRAL OXYGEN IN HERBIG Ae STARS

F. Praderie[1], C. Catala[2], J. Czarny[1], P. Felenbok[1]
(1) Observatoire de Paris, Section de Meudon
    92195 MEUDON (France)
(2) High Altitude Observatory
    PO Box 3000
    BOULDER, Colorado, 80307 (USA)

EXTENDED ABSTRACT. In stars with dense envelopes such as classical Be and T Tauri stars, the near infrared lines of OI have been observed : λ 8446 Å in emission, λ 7772 Å in absorption or in emission (e.g. Polidan and Peters, 1976 ; Herbig and Soderblom, 1980). In Herbig stars, only low resolution (2-7 Å) material is available on a few objects (Shanin et al. 1975 ; Andrillat and Swings, 1976).

We observed a sample of Herbig, intermediate mass, pre-main sequence stars in the near IR OI lines in view of studying the formation mechanism of these lines and of comparing them to the UV resonance triplet at λ 1302 Å.

The observations consist in high resolution ($\Delta\lambda \sim 0.2$ Å) spectra of the regions λ 7772 and λ 8446 Å in 7 Herbig Ae stars : BD + 61°154, AB Aur, HD 250550, HR 5999, HD 150193, HD 163296 and BD + 46°3471. These spectra were obtained at the Canada-France-Hawaii Telescope, Hawaii, and at the Coudé Auxiliary Telescope, ESO, Chile. We also obtained short wave, high resolution (0.1 Å) images from the International Ultraviolet Explorer for the star AB Aur only.

The remarkable characteristics of the lines are as follows :
1. Near infrared OI lines in 7 Herbig Ae stars. Three groups can be recognized among the stars observed.
   Group I.   Both the lines λ 8446 Å and λ 7772 Å are in emission (4 stars)
   Group II. Both sets of lines are in absorption (2 stars)
   Group III. λ 7772 Å is in absorption, λ 8446 Å in emission (1 star).
Group I contains the stars with the largest emission in other lines. The near IR OI line profiles are very different in shape from star to star, with more pronounced differences than observed in other spectral features (Catala et al., 1986).

2. The space UV OI lines in AB Aur. In the resonance multiplet (UV2), the λ 1302.17 Å line presents a wind profile without P Cyg emission.

I. Appenzeller and C. Jordan (eds.), Circumstellar Matter, 101–102.
© *1987 by the IAU.*

The blue edge velocity is very similar to that observed in C IV and Mg II at the same date. We infer from this that OI resonance line blue wing traces depths of the wind where V_{max} has been reached. On 4 coadded spectra, we searched for the presence of the intercombination multiplet UV1 : the lines λ 1355.60 and λ 1358.51 Å are absent, whether in absorption or in emission.

An interpretation of the observed behaviour of the near IR and resonance OI lines is presented on the example of AB Aur. The ionization equilibrium of OI is studied in the expanding chromosphere of the star, for which a model has been constructed (Catala et al, 1984). The similarity of the blue edge velocity of the resonance lines of OI, C IV and Mg II favours a formation zone for the OI lines in the upper chromosphere, where OI is radiatively ionized but still abundant enough.

The formation of the near IR lines can follow a scheme where λ 8444 Å results from the fluorescent excitation of $3d\ ^3D°$ by Ly β from $2p^4\ ^3P$ (the wavelength coincidence is to $\Delta\lambda = 0.05$ Å). λ 7772 Å in emission can be formed in the same way, first by excitation of $3d\ ^5D°$ by Ly β from $2p^4\ ^3P$ but this time $\Delta\lambda = 0.766$ Å, or $\Delta v = 216$ km s^{-1}. A preliminary computation by one of us (C.C.) of the mean intensity of Ly β, centered at the wavelength of each of the $2p^4 - 3d$ OI lines, expressed in the observer's frame, for the wind model of AB Aur, shows that the whole mechanism is possible. Full computations will be published as soon as the lacking transition probabilities in OI are available. Also, one can understand the lack of P Cyg emission in λ 1302 Å and the absence of λ 1355 Å by the consideration of the optical depth and of the major populating and depopulating processes in each of these lines.

The still partially qualitative mechanism invoked above to explain both λ 8444 and λ 7772 Å in emission involves the broadening of Ly β by the velocity field and is able to give account of cases where λ 7772 Å is in emission and of cases where it is in absorption, depending on the velocity law v(r) in these stars and on the possibility to fluorescently pump the $3d\ ^5D°$ level in OI.

REFERENCES

1. Andrillat, Y., Swings, J.P., Ap.J. 204, L 123.
2. Catala, C., Kunasz, P.B., Praderie, F., 1984, Astron. Astrophys. 134, 402.
3. Catala, C., Czarny, J., Felenbok, P., Praderie, F., 1986, Astron. Astrophys. 154, 103.
4. Herbig, G.H., Soderblom, D.R., 1980, Ap.J. 242, 628
5. Polidan R.S., Peters G.J., 1976, IAU Symp. 70, p.59
6. Shanin, G.I., Shevchenko, V.S., Shcherbakov, A.G., 1975, IAU Symp. 67, p. 117.

THE CIRCUMSTELLAR ENVIRONMENT OF CHROMOSPHERICALLY ACTIVE T TAURI STARS

U. Finkenzeller and G. Basri
Astronomy Department
University of California
Berkeley, CA 94720

ABSTRACT. We discuss new spectroscopic material on 7 T Tauri stars of low to intermediate activity level which have envelopes of low optical thickness and small circumstellar/interstellar extinction. We show that difference plots between the target star and appropriate standards are a powerful tool to probe the stellar envelope structure. In our sample we find 1 object with a P Cyg type, 3 with inverse P Cyg type, and 3 with symmetrical Balmer line profiles. We conclude that the physical processes in these T Tauri stars do not differ qualitatively from the ones found in extremely active ones. In particular, the inverse P Cyg type profiles are not restricted to stars with very opaque envelopes and are possibly a much more common attribute of young stellar objects.

1. DATA BASE

The work discussed here is part of a comprehensive study of T Tauri stars with low to intermediate activity levels which were selected from the low resolution catalogue of Appenzeller *et al.* (1983). Our observational material is innovative since it is at high resolution (R=12000), is flux calibrated, has a wide wavelength span (3900 to 8700 Å), and was obtained quasi-simultaneously. The program intentionally focuses on objects showing low to intermediate activity as referenced by weak or moderate emission line strength, i.e. the least complex manifestation of the T Tauri phenomenon. For more details, see Finkenzeller and Basri (1985).

2. DIFFERENCE PLOTS AS A DIAGNOSTIC TOOL

The arithmetic difference between a target star and an appropriate photospheric spectral standard turned out to be a powerful tool to analyze high S/N spectrograms. For this purpose, the spectrogram of the standard has to be shifted to the rest frame of the T Tauri star and convolved with a rotational function to match the $v*sini$ of the target star.

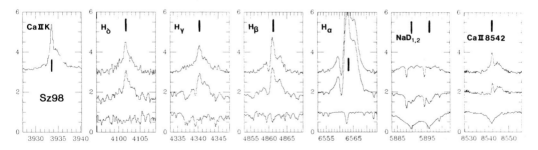

Fig. 1 For the T Tauri star Sz65 all major emission lines are given as difference and original data (top and middle tracing). The lower tracing is the processed standard star. A vertical tick indicates the systemic velocity, and each frame is ± 500 km/s wide.

I. Appenzeller and C. Jordan (eds.), Circumstellar Matter, 103–104.

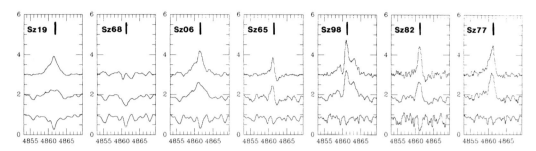

Fig. 2 H_β for all the T Tauri stars, same representation as in Fig. 1. From left to right, the H_β surface flux $(ergs/cm^2 s)$ and $A_v (mags)$ are 4.2e7, 2.5; n.a., 1.7; 2.6e7, 1.9; n.a., 0.9; 1.3e7, 0.8; 7.1e6, 0.9.

Applied to the lines of H_α, H_β, H_γ, and H_δ one can show that all line profiles closely resemble each other (Fig. 1), indicating formation in a region with common physical properties. CaII H, CaII K, and the IR triplet lines of Calcium share this property as well.

The appearance of upper photospheric and chromospheric lines when using this differential comparison method is already described by Finkenzeller and Basri (1985) and will not be discussed here.

3. LINE PROFILE TYPES

The selected T Tauri stars represent the least active species of their class, but display in the same variety of line profiles as their violently active counterparts. In particular, we were surprised to find 3 objects (Sz65, Sz82, and Sz77) whose line profiles have emission wings more steep on the red than on the blue side. A weak redshifted absorption component, vaguely present in the original data, is more distinct in the difference representation (Fig. 2) and reminiscent of the so-called YY Ori profiles (interpreted by some authors as indicative of mass infall). Sz98 shows a Beals type III P Cyg profile, whereas Sz06, Sz19, and Sz68 have symmetrical line profiles.

Until now, the occurence of inverse P Cyg profiles with T Tauri stars was known to occur only with the most active (or extreme) T Tauri stars, generally thought to be the youngest in their class. The 3 low activity objects found here are only marginally reddened ($A_v = 0.8 - 0.9 mag$) yet show this phenomenon. The formation of inverse P Cyg profiles might not directly be related to the presence of infalling material with high column densities. The combination of radiation transport effects with non-spherical geometries (see, e.g. Ulrich and Knapp 1985 or Bertout 1979) may be a more acceptable scenario.

REFERENCES:

Appenzeller, I., Jankovics, I., Krautter, J. 1983, Astron.Astrophys.Suppl., **53**, 191
Bertout, C. 1979, Astron.Astrophys., **80**, 139
Finkenzeller, U., Basri, G. 1985, The ESO Messenger, **42**, 20
Ulrich, R., Knapp, G. 1985, preprint

STELLAR WIND FLOWS IN T TAURI STARS

C. Sá[1] , M.T.V.T. Lago[1] , M.V. Penston[2]
1. Grupo de Matemática Aplicada, Universidade do Porto,
   Rua das Taipas, 135, 4000 Porto, Portugal
2. Royal Greenwich Observatory, Herstmonceux Castle,
   Hailsham, East Sussex BN27 1RP, UK

ABSTRACT. Following the successful modelling of the wind from RU Lupi using data at moderate and high dispersion we report on similar observations of other T Tauri stars where the general pattern of the wind, as deduced from the widths of the emission lines of the various species, seems to be similar.

T Tauri stars form a class of young (pre-main sequence) variable stars showing strong stellar wind outflows (Kuhi 1964, Penston & Lago 1983) and rather high mass loss rates. One of the most attractive and most completely explored models to provide an explanation for the large proportion of the photospheric energy in these stars being carried by a mechanical rather than a radiative flux invokes driving by hydromagnetic waves (Lago 1979, De Campli 1981, Hartmann et al., 1982). In the extreme T Tauri star RU Lupi, the observational constraints were satisfied by an Alfvén wave driven wind if the (mean) surface magnetic field is $\approx$ 600 G (Lago 1979, 1982, 1984 and Lago & Penston 1982). The magnetically-driven wind accelerates rapidly reaching its maximum speed very near the stellar surface. Further out and before the flow reaches the escape velocity, the wind deccelerates due to gravitational forces.

Being aware that models of "average" T Tauri stars may not be meaningful we are extending Lago's work to several of the brightest T Tauri stars of high emission class. In our spectroscopic observations of RW Aur, DF Tau and V380 Ori a similar pattern to the one found in RU Lupi is suggested but in DI Cep the wind is more solar-like with a continuous acceleration out to the hotest regions (Sá et al.,1986).

In the present paper we report another group of recent observations. The spectra were taken in October 1985 using the 500mm camera of the IDS and IPCS (slit width set to 0.5 arc seconds and dispersion of 17Å mm^{-1}) on the Isaac Newton Telescope at the Observatorio del Roque de los Muchachos on the Island of La Palma. The resolution ($\approx$ 0.5 Å) is adequate to study the overall line profiles and, in particular, to measure good values for the line widths. The data were calibrated and sky subtracted using standard SPICA programmes on the Starlink Vax 11/780 at RGO.

The spectra of the stars LkHα 120, LkHα 264 and BP Tau in the blue region are shown in Fig. 1. The spectrum of Lk Hα 120 is dominated by FeI and FeII emission lines; no proper identification of HeI can be done above the threshold limit of noise and of resolution. In the Lk Hα 264 the absence of strong FeI emission lines is outstanding but the Na D lines and HeI λ5876 are strongly in emission. BP Tau has no "visible" Fe lines and the outstanding emission lines of the optical spectrum of this star are H, He, SiI, NaID and Ca H and K lines.

I. Appenzeller and C. Jordan (eds.), Circumstellar Matter, 105–106.

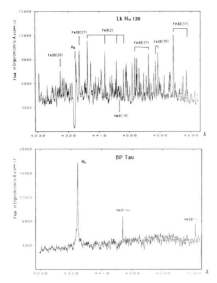

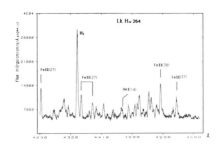

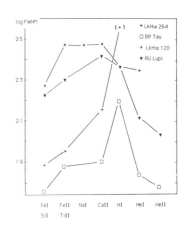

Fig.1 - The bue spectra of Lk Hα 120,
Lk Hα 264 and BP Tau.

For the purpose of this paper the FWHM of the identified emission lines are interpreted as due to a velocity field and the mean widths of each species are used.The data are summarized in Fig. 2.

Fig.2 - The stars involved are late type stars and we believe that ordering the species by increasing excitation energy of the levels corresponds to an increasing distance from the photosphere. Interpreting the plot as wind velocity against a measure of height above the atmosphere the general trend is similar to the one presented by RU Lupi (shown for comparision): an inicial acceleration in the inner chromosphere followed by decceleration in a balistic zone far from the star.

Although only a very preliminar analysis has been done so far there seems to be some indication of a common behaviour on the runs of the velocity of RU Lupi (the only detailed model so far) and the more recently observed T Tauri stars.

Acknowledgments. This work was partially supported by Reitoria da Universidade do Porto through Proj. de Investigação 25/85. MTVTL acknowledges the support of the Henri Chretien Award (1985) administrated by the American Astronomical Society. CS acknowledges the hospitality of the Royal Greenwich Observatory.

References
De Campli, W.M., 1981,Ap.J.244,124.
Hartmann,L.,Edwards,S.,Avrett,E.,1982,
Ap.J.261,279.
Kuhi, L., 1964, Ap. J.140,1409.
Lago, M.T.V.T.,1979,D.Phil. thesis, Univ. Sussex

Lago,M.T.V.T.,Penston,M.V.,1982, M.N.R.a.S.198,429.
Lago, M.T.V.T., 1982, M.N.R.a.S.198,445.
Lago,M.T.V.T., 1984, M.N.R.a.S.210,323.
Penston,M.V.,Lago,M.T.V.T.,1983, M.N.R.a.S,202,77.
Sá,C.,Penston M.V.,Lago M.T.V.T.,1986,
M.N.R.a.S.in press

THE NAKED T TAURI STARS: THE LOW MASS PRE-MAIN SEQUENCE UNVEILED

Frederick M. Walter
Center for Astrophysics and Space Astronomy
University of Colorado, Campus Box 391
Boulder, CO 80309-0391 (USA)

ABSTRACT. I discuss a survey of X-ray sources in regions of star formation. The survey has revealed at least 30 low mass PMS, naked T Tauri stars (NTTS) in Tau-Aur, and a comparable number in Oph. I summarize the properties of these stars, and argue that the spectra of the classical T Tauri stars are due to the interaction of an underlying NTTS with a dominant circumstellar environment. I discuss the impact the NTTS are likely to have on our understanding of the PMS evolution of low mass stars.

The classical T Tauri stars (CTTS) are often considered synonymous with the low mass pre-main sequence (LMPMS) stars. Despite their extreme properties, all we know of LMPMS stars is based on obervations and interpretations of the CTTS. However, there do exist LMPMS stars which lack the complications afflicting most of the CTTS, and which may lead to a better understanding of the LMPMS stars and their evolution. Here I summarize the properties and implications of recently discovered LMPMS stars, the naked T Tauri stars (NTTS). See Walter (1986b) for a complete discussion.

A method of finding the less active LMPMS stars is through X-ray surveys. Walter (1986a) and Walter et al (1986) discuss the X-ray discovered LMPMS stars found to date. Expanding on this work, I observed more optical counterparts of X-ray sources in regions of star formation, confirming 30 NTTS in Tau, with another 6 possible. The identification criteria include strong Li 6707 absorption, a radial velocity consistent with that of the Tau T association, and CaII emission. Less complete studies in Oph and CrA have revealed 33 more NTTS.

The properties of the NTTS are unremarkable, as predicted by Herbig (1978). $H\alpha$ profiles are symmetric and are in absorption in stars hotter than about K5. The colors are normal. Many NTTS are low amplitude variables; none are catalogued variable stars. Only one NTTS is coincident with an IRAS source. Overall, the NTTS look like normal, but active, cool stars. They certainly cannot be mistaken for the typical CTTS. The NTTS comprise roughly half the LMPMS stars now known in Tau-Aur, and a larger fraction in Oph. They are distributed throughout regions of star formation, with no tendency to concentrate near the dark clouds like the CTTS. In Tau and Oph, prominent clumpings

I. Appenzeller and C. Jordan (eds.), Circumstellar Matter, 107–108.

of NTTS lie 10 to 20 pc from the nearest dark clouds. Ages range from about 10^6 yrs (comparable to the CTTS) to about 2×10^7 yrs. The youngest stars lie closest to the clouds, and the oldest stars tend to lie furthest away.

I argue that NTTS and the underlying stars in the CTTS are the same, as established by the identical radial and rotational velocity distributions. These similarities cease once we consider those characteristics which define the CTTS. But these (the IR excesses and the $H\alpha$ and forbidden line emission) are likely not photospheric or chromospheric in origin. The CTTS emission line profiles imply extended atmospheres with significant mass motions, and there is strong evidence for disks around some stars. The properties that make a LMPMS star a CTTS are more likely symptomatic of the circumstellar (CS) environment than of the underlying star. Without the CS material (CSM), one would have an unveiled LMPMS star - a star with no IR or UV excesses, and no line emission beyond that generated by a solar-like atmosphere - i.e., a naked T Tauri star.

The NTTS afford an opportunity to study the PMS evolution of low mass stars and not the of the CSM. They will allow tests of the evolutionary tracks, because the luminosities and the temperatures are known. NTTS counts will improve estimates of the star formation efficiency. The NTTS can probe the relation between activity, rotation and age for $T < 10^7$ years. Now that we can see the underlying stars, perhaps we will be able to understand the CTTS and how a young star interacts with its CS environment. The NTTS are more representative of the true LMPMS stars than are the CTTS because in the latter we observe primarily the interaction of a star with a dominant CS environment. This may explain the lack of correlation of CTTS activity with either age or rotation, since the observed activity has little contribution from a solar-like chromosphere. The extreme CTTS are stars where the circumstellar material dominates the spectrum; in the NTTS the stellar contribution dominates at all wavelengths.

One would expect that the LMPMS stars follow an evolutionary sequence from extreme to less extreme CTTS to NTTS, as the CSM dissipates. This cannot be strictly an age sequence. It is likely that the rate of dissipation of the CSM depends on initial conditions. Stars which form in less dense regions of clouds, or which move out of the dense regions, might dissipate their CSM before those stars which remain in the dense regions. Indeed, the CTTS tend to be associated with dark clouds or nebulosity on the sky, while the NTTS are found in unobscured regions. The age of an LMPMS star is not a good predictor of its evolutionary state.

This work is supported by a NASA grant to the University of Colorado. Coworkers include A. Brown, E. Feigelson, R. Mathieu, P. Myers, and F. Vrba.

REFERENCES

Herbig, G.H. 1978, in "Problems of Physic and Evolution of the Universe", (Yervan: Academy of Sciences of the Armenian SSR), pg. 171.

Walter, F.M. 1986a. *Ap. J.*, **306**, 573.

Walter, F.M. 1986b. *Pub.A.S.P.* (in press).

Walter, F.M., et al. 1986, *Ap. J.* (in press).

THE CIRCUMSTELLAR ENVIRONMENTS AND VARIABILITY OF RY AND RU LUP

M. G. Hutchinson[1], A. Evans[1], J. Davies[2], M. Bode[3],
D. Whittet[3], D. Kilkenny[4]

1. Department of Physics, University of Keele.
2. Department of Space Research, University of Birmingham.
3. Department of Astronomy, Lancashire Polytechnic.
4. South African Astronomical Observatory.

ABSTRACT.   The variability of the two T Tauri stars RY and RU Lup has
been monitored between 1980 and 1984 at both optical and infrared
wavelengths.   We present here a preliminary analysis of the data and
suggest possible mechanisms for the observed variability.

1. OBSERVATIONS

Our data were taken at the South African Astronomical Observatory
between 1980 and 1984 on the 0.5 and 0.75 m telescopes.   We have
broadband photometry from U (0.36 μm) to N (10 μm), together with Hβ
photometry.   The complete set of observations is given in
Kilkenny et al. (1985).

2. LIGHT AND COLOUR VARIATIONS

Fig. 1 shows the observed variation of B-V with V for both stars. For
RY Lup there is clearly more scatter, and a change in slope, for
V > 11.4,   otherwise   the   slopes   for   the   two   stars   are
$R = \Delta V/\Delta(B-V) \simeq 3.2$,   approximately   that   expected   for   variable
extinction by interstellar-like grains (R = 3.1). The visual lightcurve
of RY Lup has several distinct sharp minima (see Fig. 2), which give
rise to the change in R for V $\gtrsim$ 11.4; at each of these minima the Hβ
line goes into nett emission.
In the case of RY Lup two distinct mechanisms could account for the
observed deep minima. Large, cool (with respect to T_*) starspot groups
on the stellar surface could effectively smother the observable
photosphere at minimum light, with the effect of enhancing the
chromospheric contribution (in the form of the line emission) to the
total light output of the star. Alternatively, occultation of the
observed stellar disc by optically thick circumstellar dust aggregations
would also give rise to enhanced chromospheric emission by reducing the
effective   photospheric   contribution   to   the   light   output
(cf. Evans et al. 1982).

I. Appenzeller and C. Jordan (eds.), Circumstellar Matter, 109–112.

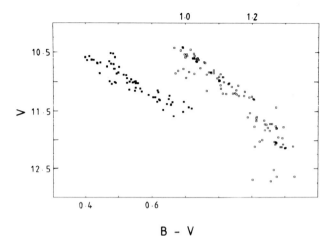

Fig. 1: [V, (B-V)] diagram for RU Lup
(filled squares) and RY Lup (open
squares). Upper (B-V) scale refers to RY,
lower scale to RU.

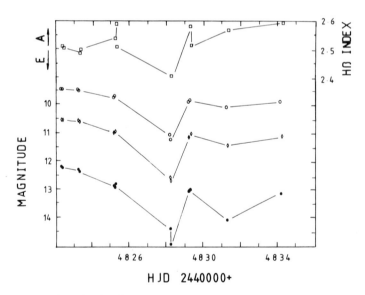

Fig. 2: U (filled circles), V (diamonds),
I (open circles) light curves, and Hβ
light curve (open squares), for RY Lup.
For Hβ, E denotes nett emission, A nett
absorption.

We suggest that smaller starspots and/or flaring are responsible for the small amplitude variations outside deep minima. To investigate the possibility that starspots cause the small amplitude light variations, we attempted to fit sine curves to the lightcurves (cf. Vrba et al. 1986); attempts to fit more complex functions are currently under way. We assumed that the rotation period of RY Lup was 3.7 days (P. Bastien, private communication), which corresponds with a peak in the fourier power spectrum algorithm of Deeming (1975) for each light curve. The rotation period of RU Lup was taken to be 2.9 days, although none of the peaks in our power spectra were particularly dominant. The irregularity of the light variations suggests that large flares may also contribute to the observed flux: in the case of RU Lup this would certainly be consistent with its ultraviolet excess.

Although the lightcurves were irregular, we had some measure of success in fitting sections of the light curve with sine curves having various phases. Such changes in phase would be due to the continual appearance and disappearance of several spot groups as the star rotates. We found that the change of amplitude with wavelength can best be approximated by (Bopp and Noah 1980)

$$\Delta m = 2.5 \log\{1 - f[1 - (B_\lambda(T_*)/B_\lambda(T'))]\}$$

for a single spot of characteristic effective temperature T' covering a maximum fraction f of the observed disc. From a non-linear least squares fit we find that $T_* = 6000$ K, f = 0.9 and T' = 10000 K for RY Lup, and $T_* = 4400$K, f = 0.6 and T' = 5150K, for RU Lup. The fit for RY Lup is clearly not physically realistic and casts doubt on the possibility that all the observed minima are due to variable starspot coverage.

3. STELLAR AND DUST PARAMETERS

From our estimated bolometric luminosity and T_* for RY Lup, we find that $R_* \simeq 2 R_\odot$. Using the measured rotation velocity v sin i =25 ± 4 km s^{-1} (Bouvier et al. 1986), and a rotation period of 3.7 days, we see that the star's rotation axis is close to the plane of the sky.

We can extend our wavelength coverage to 100 μm using the IRAS PSC data (Beichmann et al. 1985). We deredden our data assuming that our (V-I) and (V-R) colours are least affected by reddening, and use the colour excess ratios given by Mendoza (1968). To the resultant flux distributions we have fitted blackbody functions. We find that both stars have infrared excesses which can be modelled by a hot (≃ 1600 K), a tepid (≃ 250 K) and a cool (≃ 50 K) dust shell. Thus the hypothesis that minima in the lightcurve of RY Lup could be caused by optically thick circumstellar "clumps" is tenable, and it remains possible that both stars may have associated protoplanetry discs and expanding shells of dust in which cometry formation may occur. Many pre-main sequence stars (including RY Lup) are also variable polarimetrically (e.g. Bastien 1985); models to account for the polarimetric variability, involving scattering by circumstellar dust clumps, are currently being developed by P. Bastien (private communication).

4. CONCLUSIONS

We find evidence for irregular modulation of the lightcurves of both RY and RU Lup.   For RU Lup the modulation is possibly due to starspot activity,   and   also   to   flaring   activity.   A   simple   starspot interpretation   for   RY Lup gives unrealistic results, and it   is more likely   that   other processes contribute to the light modulation (e.g. flares   and   circumstellar   extinction).   Both   stars may be prime candidates to investigate planetary and cometary formation.   Further co-ordinated photometric, polarimetric and spectroscopic studies of these stars are highly desirable.

ACKNOWLEDGEMENTS

MGH and MFB are supported by the SERC.

REFERENCES

Bastien, P., 1985, Ap. J. Supp. October 1986.
Beichmann, C. A., Neugebauer, G., Habing, H. J., Clegg, P. E.
     and Chester, T. J., 1985, IRAS Explanatory Supplement.
Bouvier,C., Bertout, C., Benz, W. and Mayor, M., 1986, Astron.
     Astrophys., 165, 110.
Bopp, B. W. and Noah, P. V., 1980, Pub. Astron. Soc. Pacific,
     92, 717.
Deeming, T. J., 1975, Astrophys. Space Sci., 36, 137.
Evans, A., Bode, M. F., Whittet, D. C. B., Davies, J. K.,
     Kilkenny, D. and Baines, D. W. T., 1982, MNRAS, 199, 37P.
Kilkenny, D., Whittet, D. C. B., Davies, J. K., Evans, A.,
     Bode, M. F., Robson, E. I. and Banfield, R. M., 1985,
     SAAO Circulars, No. 9, 55.
Mendoza, E. E., 1968, Ap. J. 151, 977.
Vrba, F. J., Rydgren, A. E., Chugainov, P. F., Shakovskaya, N. I.
     and Zak, D. S., 1986 Ap. J., 306, 199.

NEAR-IR OBSERVATIONS OF THE SSV 13, SSV 9, SSV 5 SOURCES IN NGC 1333

M. Busso (1), P. Persi (2), M. Robberto (3), F. Scaltriti (1),
G. Silvestro (3)
(1) Osservatorio Astronomico di Pino Torinese, Torino, Italy
(2) Istituto di Astrofisica Spaziale del C.N.R., Frascati,
Italy
(3) Istituto di Fisica Generale dell'Università, Torino, Italy

ABSTRACT. The sources (1) SSV 13 (identified with the IRAS (2) source
03259+3105), SSV 9 (IRAS 03256+3107), SSV 5 (IRAS 03262+3108) were observed photometrically in the JHKL bands with the In-Sb photometer at
the Italian IR Telescope (TIRGO, Gornergrat, Switzerland). In addition
CVF spectrophotometry between 2.0 and 2.4 μm of SSV 13 was carried out.
We present a study of the energy distribution between 1 and 100 μm, including both our photometry and IRAS data.

SSV 13

This is the exciting star (probably a T Tauri star) of the H-H 7-11
chain, well studied by Cohen and Schwartz (C-S) (3). The total energy
distribution between 1 and 60 μm is reported in Fig. 1. A luminosity
$L_{IR} \sim 87$ $L_{\odot}$ has been inferred for the source by integrating the whole e-
nergy distribution. This luminosity is larger than the bolometric luminosity found by C-S of about 29.8 $L_{\odot}$.
We derive a visual extinction A_V = 43 from the optical depth at
9.7 μm ($\tau_{9.7} \sim 2.7$) obtained from narrow-band photometry (3), using the
relationship $A_V / \tau_{9.7}$ = 16. No Brγ line emission at λ=2.167 μm is observed from the 2.0-2.4 μm spectrum, while the H_2 (v = 1 - 0) S(3) line is
present (Fig. 2), in agreement with the extended emission of molecular
hydrogen observed in the H-H 7-11 chain. The J and H fluxes observed
with large diaphragm (D=27") appear to be greater than those observed
with the 17" diaphragm. This is probably due to contamination by a near
by optical star.

SSV 9

The source is associated with a visible and probably variable star.
The star has been classified as K 7 (3) and is surrounded by an optical
nebulosity. By comparing our photometry with that previously obtained
(1), we derive a variation of about 1 mag in both H and K bands for
this source. The near-IR colour cannot be explained simply by a reddened K 7 star: the observed colour index K-L= 2.31 implies the presence

I. Appenzeller and C. Jordan (eds.), Circumstellar Matter, 113–114.

of dust at a temperature T_C (K-L) ~ 950 K. The derived IR luminosity $L_{1-25} \sim 4 \, L_\odot$ agrees with the classification given for this star.

SSV 5

The source is associated with the Herbig-Haro object H-H 17. Our photo-metry gives a K band magnitude lower by 3-4 mag than the data of Strom et al. (1), suggesting the source might be variable. On the other hand, our photometry agrees with the IRAS data. We estimate L_{1-25} of order $1 \, L_\odot$; the source could be associated with a low-mass protostellar ob-ject in the region.

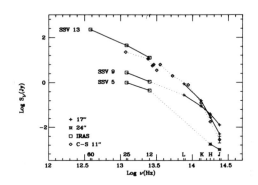

Fig. 1. Flux distribution of SSV 13, SSV 9, SSV 5. Our data are presen-ted with those of IRAS and of Cohen and Schwartz (C-S).

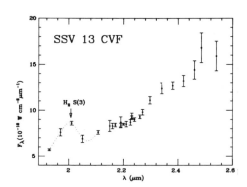

Fig. 2. CVF spectrophotometry in the K band region.

LITERATURE

(1) S.E. Strom, F.J. Vrba, K.M. Strom, Astron. J. 81, 314 (1976).
(2) A.G. Beichman, G. Neugebauer, H.J. Habing, P.E. Clegg, T.J. Chester, IRAS Catalogues and Atlases, Explanatory Suppl. (1985).
(3) M. Cohen and R.D. Schwartz, Astrophys. J., 265, 877 (1983).

SERPENS - SVS 20: A NEW INFRARED DOUBLE SOURCE

C. Eiroa[1], R. Lenzen[2], K. Leinert[2], K. Hodapp[2,3]
1: Observatorio Astronomico Nacional, Madrid
2: Max-Planck-Institut für Astronomie, Heidelberg
3: Institut For Astronomy, University of Hawaii

ABSTRACT.   0.9 μm CCD images and 2.2 μm slit scan observations reveal the double character of the infrared source Serpens - SVS 20; the separation, position angle and brightness ratio are 1.''6, 10°, and Δm $\approx$ 1.6 mag respectively. The southern source is the brightest at both spectral ranges. At 0.9 μm the observed polarization of each component is p = 9.0 %, Θ = 155° and p = 10.4 %, Θ = 135° for SVS 20-South and SVS 20-North respectively. These results combined with 2.2 μm polarimetry and JHKLM photometry suggest that both SVS 20 sources are in fact low mass, probably PMS, stars.

1.   INTRODUCTION

The Serpens cloud is an active star formation region in which a high number of young objects, IR-sources with H_2O masers, molecular outflow, etc., is found. We are now carrying out an extensive observational study of this region, by means of CCD and near infrared techniques. In this work we report about our results in one of the most interesting Serpens objects: the IR-source SVS 20 (Strom et al., 1976, Churchwell and Koornneef, 1986), which is thought to be a low mass PMS star.

2.   OBSERVATIONS AND MAIN RESULTS

The 3.5 m and 2.2 m telescopes of the Calar Alto Observatory, Spain, were used to realize the following observations: 0.9 μm CCD direct images and polarimetry; 2.2 μm slit scan and polarimetry; JHKLM photometry and 2-4 μm CVF spectralphotometry. The main results of these observations are:
a) Both CCD images and slit scan observations clearly show that Serpens - SVS 20 is a double source. An isocontour plot of the CCD image and the 2.2 μm visibility data are shown in Fig. 1 and 2. At both wavelengths the brightest source is the southern one (SVS 20-South); the separation and the brightness ratio are 1.''6, NS direction, and Δm $\approx$ 1.6 mag respectively. A position angle of 10° is seen in the CCD frame.

I. Appenzeller and C. Jordan (eds.), Circumstellar Matter, 115–116.
© *1987 by the IAU.*

b) 0.9 μm polarization values of each source are: p = 9.0 %, Θ = 155o (SVS 20-South), and p = 10.4 %, Θ = 135° (SVS 20-North). The integrated SVS 20 polarization (6" beam) at 2.2 μm is p = 2.6 %, Θ = 158°.
c) The SVS 20 near infrared fluxes (6" beam) can be reasonably fitted by a black body of 960 K. This is also the colour temperature of each individual SVS 20 source.
d) The 2-4 μm CVF spectrum (6" beam) of SVS 20 shows the 3.1 μm ice absorption feature, $\tau \approx 1$.

3. DISCUSSION

The observed separation between both SVS 20 sources corresponds to a projected distance of 400 AU, assuming a distance of 250 pc for the Serpens cloud (Chavarria et al., 1987). One interesting point concerns the nature of the sources. Three possibilities are discussed: a) both sources are stars, b) both are knots of nebulosity, and c) one of the sources is a reflection nebula illuminated by the other. Taking into account all the results reported above, the first possibility seems most probable. Lower and upper limits of the luminosity of each star can be deduced from our observations and from the FIR measurements of Harvey et al. (1984). We obtain L (SVS 20-South) $\gtrsim$ 5.5 L_O, L (SVS 20-North) $\gtrsim$ 1.3 L_O, and L (SVS 20-South + SVS 20-North) $\lesssim$ 45 L_O.

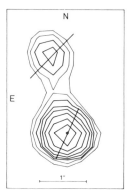

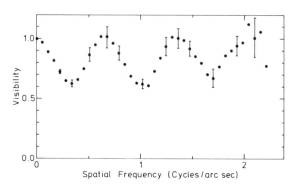

Figure 1: Isocontour plot of Serpens-SVS 20 (= 0.9 μm) logarithmic spacing by $\sqrt{2}$. The lines represent the 0.9 μm polarization. North is up, and East to the left

Figure 2. Visibility data of Serpens-SVS 20 at 2.2 μm Error bars are 1σ.

REFERENCES

Chavarria, C., Ocegueda, J., de Lara, E., Finkenzeller, U., Mendoza, E.: 1987, This Symposium
Churchwell, E., Koornneef, J.: 1986, Ap. J. 300, 729
Harvey, P.M., Wilking, B.A., Joy, M.: 1984, Ap. J. 278, 156
Strom, S.E., Vrba, F.J., Strom, K.M.: 1986, A. J. 81, 638

GSS 31: Another T Tauri star with an infrared companion[*]

Hans Zinnecker, Royal Observatory, Edinburgh.
Alain Chelli, Instituto de Astronomia, Mexico City.
Luis Carrasco, Instituto de Astronomia, Mexico City.
Irene Cruz-Gonzales, Instituto de Astronomia, Mexico City.
Christian Perrier, Observatoire de Lyon, Saint Genis-Laval.

Using a rapid slit scan technique in the infrared (JHKL), the source GSS 31 (alias EL 22 or Do-Ar 24E), a T Tauri star in the Ophiuchus dark cloud, was discovered to be double at a projected separation of $1\overset{''}{.}95\pm0\overset{''}{.}10$ (i.e. 320 AU). The position angle is almost exactly north-south. This is the second discovery of its kind, and the first in the southern hemisphere, the only previous case known being T Tauri itself (Dyck *et al.* 1982, Schwartz *et al.* 1984). The important point is that we were able to secure infrared photometry separately for both components while for T Tau the separation into two components was model-dependent due to their small separation (0."6). IR-photometry for the joint system GSS 31 had previously been obtained by Grasdalen, Strom & Strom (1973) and Elias (1978), while Chini (1981) obtained UBVRI photometry.

We find that the infrared companion of GSS 31 becomes brighter as wavelength increases from 1.6 μm to 2.2 μm, and dominates the optical component in flux at 3.6 μm (see Fig. 1). By integrating the dereddened energy distribution, we estimate a bolometric luminosity of 3.3 L_0 for the optical star and 0.9 L_0 for the infrared companion, using $A_V = 4$.

We have similar data on the binary PMS star Cham I (Glass 1979). This object is also double at a projected separation of 2."7, position angle roughly east-west.

Assuming both components formed at the same time from the same cloud, we can estimate individual masses and a mass ratio for these components in the following way: given the derived bolometric luminosity and an observed spectral type (KO for GSS 31; Bouvier, priv. commun.) we can place the optical component in an HR-diagram with PMS tracks and isochrones (see Cohen & Kuhi 1979). The infrared companion can also be placed into the HR-diagram according to its derived bolometric luminosity and the condition that it must lie on the same isochrone as the optical component (under the above assumption of identical age). Thus the masses of both components and therefore the mass ratio can be determined, provided the PMS tracks used are correct (the principle still applies even when better tracks become available). For GSS 31 we find $M(\text{optical}) \approx 1.5 M_0$ and $M(\text{infrared}) \approx 1.0 M_0$ from Fig. 7 in Cohen & Kuhi (1979).

[*]Based on observations obtained at the ESO 3.6m telescope, La Silla, Chile.

I. Appenzeller and C. Jordan (eds.), Circumstellar Matter, 117–118.

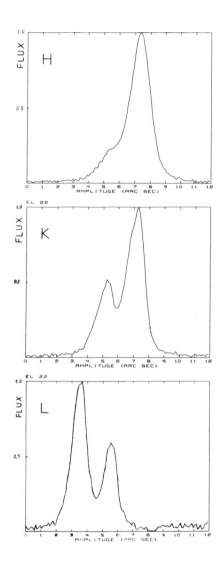

Fig. 1: Near infrared slit scans of the PMS-star GSS 31 (= EL 22) showing
    the binary nature of the object (projected separation ∼ 2 arcsec).

References

Chini (1981). *Astron.Astrophys.* 99, 346.
Cohen & Kuhi (1979). *Ap.J. Suppl.* 41, 731.
Dyck *et al.* (1982). *Ap.J.* 255, L103.
Elias (1978). *Ap.J.* 224, 453.
Glass (1979). *MNRAS* 187, 305.
Grasdalen, Strom & Strom (1973). *Ap.J.* 184, L53 (GSS)
Schwartz *et al.* (1984). *Ap.J.* 280, L23.

THE EINSTEIN SURVEY OF THE YOUNG STARS IN THE ORION NEBULA

Jean-Pierre Caillault[1] and Saied Zoonematkermani[2]
[1]Joint Institute for Laboratory Astrophysics, University
of Colorado and National Bureau of Standards, Boulder,
Colorado 80309-0440
[2]Columbia Astrophysics Laboratory, 538 W. 120th St.,
New York, New York 10027

ABSTRACT. We report here on the complete EINSTEIN survey of Orion
within the central 2° × 2° region centered on the Trapezium. We present an X-ray mosaic of the Nebula and a complete X-ray catalog (200
sources) for this very young cluster. In addition, we discuss in detail variability, early-type stars, solar-type stars, and K-M stars.

Variability. Of the 200 X-ray sources detected, at least 69 of them
have exhibited variability in the sense that they were not detected in
at least one field with a longer exposure time than a field in which
they were detected. Of the 76 stellar candidates with color and/or
spectral types, at least 27 are variable including 19 in the K-M
spectral range. Most of the observed X-rays in Orion may arise as a
consequence of continual "flaring," as suggested with regard to the
pre-main sequence stars in the similarly youthful ρ Oph cloud.[1]

Early-Type Stars. The fourteen O-B5 stars detected (out of 22 observed) all exhibit X-ray luminosities and L_x/L_{bol} ratios similar to
previously observed O and early B stars, possibly slightly more active. However, it is the later B stars that provide the newest and
most interesting information. Previously, $\lesssim 3$ main sequence stars
between spectral types B6-B9 were known to be X-ray emitters (F.
Walter, private communication). With this investigation of the Orion
Nebula, we have more than tripled the known sample (11 detected out of
24 observed). However, whether or not the X-ray emission can be properly attributed to these hot stars or, rather, to possible late-type
companions remains to be determined; studies of radial-velocity variations of these stars and of the late-type B stars not detected should
help us to answer that question.

Solar-Type Stars. These stars in Orion have yet to complete their
evolution to the ZAMS; however, it is still instructive to compare the
mean X-ray luminosity of these stars to that of the solar-type stars
in older clusters such as the Pleiades[2] and the Hyades[3] and, also,

119

I. Appenzeller and C. Jordan (eds.), Circumstellar Matter, 119–120.

with the solar-type field stars.[4] A $t^{-3/2}$ decay relationship[4] for
these solar-type stars is inadequate in describing their behavior over
more than three decades in log t.

In addition, using the reprocessed EINSTEIN data for these F7-G8
stars, we show that the weak correlation of X-ray activity with rotational velocity obtained for these solar-type stars by Smith, Pravdo,
and Ku (1983)[5] is now even more doubtful. Hence, their conclusions
concerning the roles that rapid rotation, youth, and circumstellar
disks play in determining the X-ray luminosity of these stars are now
questionable; in particular, the inverse relation between X-ray emission and circumstellar disk strength (based on absorption and emission
signatures in the Na D lines[6]) is no longer substantiated by these
data.

K-M Type Stars. A large fraction of these stars (39 of the 176 have
already been identified as K-M stars) will almost certainly prove to be
T-Tauri or "Naked" T-Tauri stars,[7] since their mean X-ray luminosity
(log L_x = 30.96 ergs s^{-1} for just the detected ones) and mean L_x/L_{bol}
ratio (=-2.86) are values consistent with those exhibited by those
types of stars. In addition, as mentioned above, the fact that many
(19) of these stars are variable in X-ray emission also supports the
likelihood of their being pre-main sequence objects. With this large
sample added to the large number already collected by Walter[7] and
collaborators from the Tau-Aur and Oph regions, problems such as the
population characteristics and X-ray mechanism of these stars should
prove solvable.

REFERENCES

1. Montmerle, T. et al. 1983, Ap. J., **269**, 182.
2. Caillault, J.-P. and Helfand, D. J. 1985, Ap. J., **289**, 279.
3. Stern, R. A. et al. 1981, Ap. J., **249**, 647.
4. Maggio, A. et al. 1986, Ap. J., submitted.
5. Smith, M. A., Pravdo, S. H., and Ku, W. H.-M. 1983, Ap. J., **272**, 163.
6. Smith, M. A., Beckers, J. M. and Barden, S. C. 1983, Ap. J., **271**, 237.
7. Walter, F. M. et al. 1986, Ap. J., in press.

CONSISTENT SPHERICAL NLTE-MODELS FOR BN-LIKE OBJECTS

Peter Höflich
Institut für Theoretische Astrophysik
Im Neuenheimer Feld 561,D-6900 Heidelberg

ABSTRACT. Spherical NLTE-model photospheres surrounded by envelopes are calculated in order to interpret BN-like objects. Both the existence and the emitted spectra of the HII-region can be understood by such models.

1. INTRODUCTION

In recent years a number of objects similar to the Becklin-Neugebauer-object (BN) in Orion have been found. These are thought to be hot protostars or young stars which are surrounded by compact high density HII-regions and neutral dust shells. The luminosities are those of main sequence B stars (Wynn-Williams, 1982).

2. THE MODEL CONSTRUCTION

The models can be characterised by the following:

i) Spherical geometry with a run of density that is given by a power law r^{-n} or by the hydrostatic equation in the photosphere of an underlaying main sequence B star. All parts of the configuration (e.g. the radiation field and the run of temperature in the photosphere and envelope) are calculated consistently.

ii) For hydrogen eight levels are allowed to deviate from LTE. All transitions are included both in the statistical equation and radiative transfer. Since it is not clear which parts of the observed velocity fields are due to outflow, rotation or turbulence, we mimic the velocity by a (micro-)turbulent velocity.

iii) Continuous opacities are taken into account for the whole wavelength range from the EUV to the cm region.

iv) Radiative equilibrium is assumed for the whole object.

3. COMPARISON WITH OBSERVATIONS

The observational indicators for the free parameters are shown in table II. If suitable observations are not available, we took v_{turb} =100 km/s as a standard value.

I. Appenzeller and C. Jordan (eds.), Circumstellar Matter, 121–123.

TABLE I. Parameters which are not changed in almost all calculations

Model parameter	assumptions
Luminosity class	V
Density distribution	r^{-2}
Condition for the outer radii	ionisation bounded
$v_{turb}(r)$	constant
Chemical composition	solar

TABLE II. Free model parameters and their indicators

Model parameter	Indicator
Effective temperature T_{eff}	Integrated continuum fluxes and distances; Linie ratios
Footpoint density N_o	Flux of $Br\alpha$
v_{turb}	Mean half widths of lines
Extinction A_V	Ratio $F_{Br\gamma}/F_{Br\alpha}$

TABLE III. Models of BN-like objects. The effective temperature T_{eff}, the stellar radius $r_{\ast}$, the footpoint density N_o at the inner boundary of the nebulae, the turbulent velocity v_{turb}, the calculated radius of the HII-region r_{HII}, the extinction A_V as determined by the line ratio of $F_{Br\alpha}/F_{Br\gamma}$ and the mass loss rate $\dot{M}$ (indicated by the models if the observed line widths are entirely due to an outflow) are given.

Object	T_{eff} [K]	$r_{\ast}$ [$R_{\odot}$]	N_o [cm^{-3}]	v_{turb} [km/s]	r_{HII} [AU]	A_V [mag]	$\dot{M}$ [$M_{\odot}$/year]	Remarks
S106 IRS4	28000.	5.2	1.10^{12}	100.	15.0	16.8	$5 \cdot 10^{-7}$	
BN	28000.	5.2	3.10^{12}	20.	6.7	25.6	$3 \cdot 10^{-7}$	
M17-IRS1	25400.	4.3	1.10^{13}	100.	2.5	20.7	$5 \cdot 10^{-6}$	
S140-IRS1	25400.	4.3	1.10^{13}	100.	–	–	–	outer radius is 43 $R_{\odot}$ and $N(r) \sim r^{-3}$
CRL961	22400.	4.1	1.10^{13}	100.	1.7	29.8	$4 \cdot 10^{-6}$	
MonR2-IRS3	22400.	4.1	5.10^{12}	100.	–	>42	$2 \cdot 10^{-6}$	model of CRL961 but scaled N_o
CLR490	17500.	3.8	1.10^{12}	130.	0.8	12.6	$6 \cdot 10^{-7}$	

The given fluxes (see Table IV) are extinction corrected by the van de Hulst curve No. 15. They are normalized to $r_{\ast}$. For the given mass loss, we assume that all parts of the observed velocity fields are connected to an outflow. Therefore $\dot{M}$ should be regarded as an upper limit.

TABLE IV. Comparison of predicted with observed fluxes (see Hall et al., 1978; Thompson und Tokunaga, 1979; Simon et al., 1981; Simon et al., 1983), corrected for extinction (in brackets).

Objects	S106-IRS4	BN	M17-IRS1	S140-IRS1	CLR961	CLR490
$lg(F_{Br\alpha})$	8.60(8.7)	8.77(8.7)	9.54(9.6)	7.53(7.6)	9.44(9.5)	8.54(8.6)
$F_{Br\alpha}/F_{Br\beta}$	1.75(-)	1.86(-)	1.18(-)	0.45(-)	1.20(-)	1.75(-)
$F_{Br\alpha}/F_{Br\gamma}$	2.24(2.3)	2.40(2.4)	1.26(1.3)	0.23(-)	1.38(1.4)	1.90(1.9)
$F_{Br\alpha}/F_{Pf}$	6.17(-)	5.24(-)	4.85(-)	2.18(-)	5.50(-)	4.80(-)
$F_{Br\alpha}/F_{Pf}$	7.08(6.7)	4.50(4.6)	3.55(3.7)	1.09(1.0)	4.25(4.2)	4.30(2-3)
$v_{HW\ Br\alpha}$(km/s)	110(110)	28.(30)	140.(-)	120.(-)	125.(-)	145.(-)
$v_{HW\ Br\gamma}$(km/s)	95.(110)	27.(30)	130.(135)	120.(-)	125.(-)	145.(150)
$lg(F_{1.3cm})$+12	8.0(8.1)	7.9(-)	7.25(<8.7)	–	6.7(<8.)	6.0 (8.2)
$lg(F_{6.0cm})$+12	7.2(7.6)	6.8(<7.2)	5.92(9.1)	–	5.3(-)	4.7(<7.7)

4. CONCLUSION

i) The existence of ionised envelopes around B stars can be understood by such models. The main mechanisms are ionisation by Balmer continuum photons, strong bound-bound collisions, small radiative net rates, and small dilution factors in contrast to classical HII-regions.

ii) There is a complicated interplay between densities, temperatures and occupation numbers. The occupation numbers for higher levels (main quantum number greater than 6) are almost completely determined by the collisional rates.

iii) The backwarming of the outer photospheric layers caused by the envelope may amount to several thousand degrees.

iv) The lines of BN-like objects can be fitted by models which in most cases have a run of density $\sim r^{-2}$. The line profiles can be explained by turbulence of the order of 100 km/s and by Stark broadening.

v) The models are compatible with the radio observations.

vi) For several BN-like objects for which suitable observational data are available, reliable luminosities can be determined from the models.

vii) The intrinsic reddening was determined.

For more detailed information see (Höflich,1986; Höflich, Hartmann and Wehrse, 1986).

REFERENCES

Hall,D.N.B.;Kleinmann,S.G.; Ridgway,S.T.; Gillet,F.C. 1978, Astrophys.J.Letters 223, L47

Höflich,P. 1986, Doktorarbeit, Heidelberg

Höflich,P.;Hartmann,L.; Wehrse,R. 1986, submitted to Astron.Astrophys.

Simon,M.;Righini-Cohen,G.;Fischer,J.;Cassar,L. 1981, Astrophys.J. 251, 552

Simon,M.; Felli,M.,Cassar,L.,Fischer,J.; Massi,M.; Astrophs.J. 266, 623

Thompson,A.T.; Tokunaga,R.I. 1979, Astrophys.J.229, 583

Wynn-Williams,C.G. 1982, Ann.Rev.Astr.Ap. 20, 587

INFRARED OBSERVATIONS OF GGD OBJECTS

R. Carballo[1], C. Eiroa[2] and A. Mampaso[1]

[1] Instituto de Astrofísica de Canarias, 38071 La Laguna. Tenerife, Spain.
[2] Observatorio Astronómico Nacional. Alfonso XII 3. 28014 Madrid, Spain

We present accurate positions and near infrared photometry (Table I) of 11 point-like objects in the neighbourhood of GGD objects obtained on the 1.55 m and on the 1.23 m in Teide Obs. and Calar Alto Obs. respectively, in Spain. Several of the near infrared sources are directly associated with the GGD nebulae and/or are candidate for their excitation. In addition some of them seem to be the near infrared counterparts of IRAS sources. We believe, on the basis of their infrared excess, far infrared emission (IRAS), association with nebulosity, coincidence with H_2O masers or the fact that in most cases the observed luminosities are higher than those expected for main sequence stars, that most of them (9/12) are young stars embedded in the dark clouds which contain the GGD objects. The loci of the detected sources in an (H-K,K-L) infrared two-colour diagram is the same as that obtained for known pre-main sequence stars, such as T Tauris and Herbig Ae-Be stars, indicating the presence of dust shells with temperatures in the range 800-1500 K. The observed range in luminosity, 10-4600 L_o, added to other different characteristics found between them, such as the presence, or absence, of H_2O masers, indicates the interest for a detailed study of the infrared sources and related GGD nebulae.

For GGD1, GGD2-3, GGD9, GGD19, GGD29 and GGD36 no sources were found, to a limiting magnitude of K=10. The searched area were approximately 60x60 arcsec^2 around the nominal GGD position.

I. Appenzeller and C. Jordan (eds.), Circumstellar Matter, 125–126.

TABLE I . Near infrared positions (1950.0) and photometry of GGD/ IR-sources.

Infrared source		Source position		Photometric magnitudes				
		α	δ	J	H	K	L	M
GGD4	IRS	$05^h37^m21.7^s$	$+23°49'23''$	11.46±0.09	9.88±0.02	7.56±0.02	5.33±0.02	4.14±0.04
GGD7	IRS1	05 38 24.2	−08 06 03	10.49±0.03	9.53±0.03	9.27±0.04	>8.7	
	IRS2	05 38 23.9	−08 06 54	10.79±0.04	9.53±0.02	8.92±0.02	7.40±0.11	>6.0
	IRS3	05 38 25.9	−08 07 25	10.12±0.02	9.34±0.02	8.99±0.03	8.03±0.17	>6.0
	IRS4	05 38 24.4	−08 08 39			EXTENDED		
	IRS5	05 38 21.9	−08 09 00	>13.37	11.50±0.13	10.77±0.08	>8.7	
GGD8	IRS1	05 48 16.3	+03 07 13	12.47±0.20	10.70±0.11	9.59±0.05	8.31±0.10	>5.7
	IRS2	05 48 16.3	+03 06 43	9.27±0.01	8.21±0.02	7.62±0.14	6.68±0.05	>6.0
GGD10	IRS1	05 59 53.8	−09 06 31	>13.37	11.78±0.12	10.20±0.09	7.7±0.1	6.0±0.2
	IRS2	05 59 54.4	−09 06 02	12.52±0.10	10.73±0.04	9.98±0.03	>8.6	
GGD16−17	IRS1	06 10 23.0	−06 12 55			10.50±0.15	7.43±0.20	
	IRS2	06 10 24.0	−06 12 20	11.70±0.08	10.77±0.02	10.02±0.03	>7.7	

Error positions are ±2 arcsecs. Magnitude errors are 1σ ; limiting magnitudes are 3σ .

DUST DISTRIBUTION NEAR YOUNG STARS OF BIPOLAR FLOWS DEDUCED FROM CCD POLARIMETRY AT 1 μm

R. Lenzen
Max-Planck-Institut für Astronomie, Königstuhl,
D-6900 Heidelberg, FRG

ABSTRACT. Deep direct 1 μm images and polarization maps have been taken for several objects of bipolar molecular outflow, using a CCD camera at the prime-focus of the 3.5 m telescope on Calar Alto/Spain. Compact reflection nebulae are found near GL2591, GL2884 = S140/IRS1, GL2789 = V645 Cyg, L1551/IRS5 and within the red reflection nebula in Serp. The position angles of polarization vectors coincide in all cases with that of the 2.2 μm polarization of the central source itself, and it is perpendicular to the direction of mass outflow (if it is well defined). The visual extinction in front of the central source, is at least 20 mag higher than that which is affecting the scattered light. This is consistent with a model of bipolar outflow, where the active source is surrounded by a dust torus or ring.
As an example, the results for GL2591 are given in Fig 1 and 2.

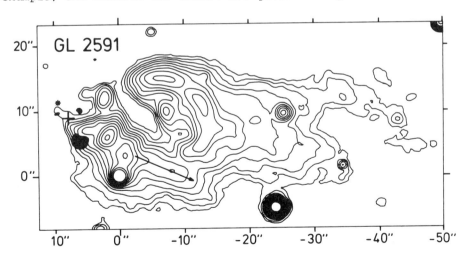

Fig. 1 GL2591
Isophotal contour map (logarithmic scaling by a factor of $\sqrt{2}$).
Coordinates are given relative to the star at the position:
$20^h27^m35^s.13$, $40°01'05".0$ (1950)

I. Appenzeller and C. Jordan (eds.), Circumstellar Matter, 127–128.

+   Position of infrared source (Lada et al. 1984)
·   5 GHz - VLA HII-region (Campbell 1985)
*   Position of the H₂O-maser source (White et al. 1975)
A jet-like feature is marked by an arrow, which gives at the same
time the direction of the blue CO wing (Lada et al. 1984)

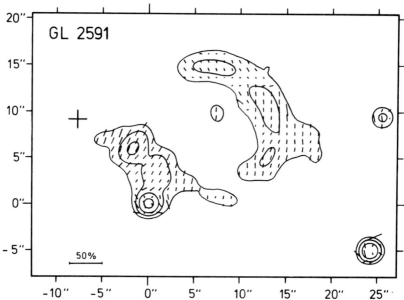

Fig. 2
     The corresponding polarization map (electrical vectors).
     +   Position of infrared source

References

Campbell B., 1984, Ap.J. 287, 334
Lada, Ch.J., Thronson, H.A., Jr., Smith, H.A., Schwartz, P.R., Glaccum,
W., 1984, Ap.J. 286, 302
White, G.J., Little, L.T., Parker, E.A., Nicholson, P.S., McDonald,
G.H., Bale, F., 1975, MNRAS, 170, 37p

EVIDENCE FROM OPTICAL POLARIMETRY FOR SPIRAL STRUCTURE IN THE MAGNETIC FIELD AND CLOUD DENSITY AROUND NEWLY-FORMED STARS

R.F. Warren-Smith, P.W. Draper & S.M. Scarrott
Physics Department
University of Durham
South Road
Durham DH1 3LE
England

ABSTRACT. Deep CCD imaging of the Serpens bipolar nebula shows it to be surrounded by molecular cloud material having spiral density structure. Polarization mapping indicates that the magnetic field in this material also exhibits spiral structure and we interpret this as the remains of the magnetically-braked collapse of a protostellar cloud. A binary star system has formed in the cloud core.

1. INTRODUCTION

We have performed deep CCD imaging and polarimetry of the Serpens bipolar nebula (RA=18:27:24, Dec=1°12'40") in an investigation of remnant protostellar cloud material and magnetic field structure around newly-formed stars. Our purpose is to clarify the detailed geometry of protostellar collapse and to investigate the role which magnetic fields may play in this process. We also aim to assess the suitability of the resulting circumstellar density and magnetic field structure for the acceleration and collimation of bipolar outflows.

The Serpens nebula is located in an active star-forming molecular cloud, and it is the site of a pre-main-sequence star IRS2 (age $\leq 10^5$ yr), which illuminates the surrounding cloud optically through cavities produced by a bipolar outflow. Density and field structure in the accreting cloud which existed during the formation of IRS2 should still be observable in regions as yet undisturbed by this bipolar outflow.

2. OBSERVATIONS

Our CCD images (Plate 1) show that the nebula is partly encircled by a radially extensive dark spiral filament, which provides clear evidence of spiral structure in the accreting material. High linear polarization is also seen in the scattered light, forming the expected approximately circular pattern around the illuminator IRS2. However (as in similar objects elsewhere), substantial deviations from

I. Appenzeller and C. Jordan (eds.), Circumstellar Matter, 129–130.
© 1987 by the IAU.

circularity are also present which can be attributed to selective extinction by magnetically aligned foreground dust in the circum-nebular medium. Details of the surrounding field structure may be deduced from these deviations from circularity in the polarization pattern.

3. INTERPRETATION

Substantial deviations which form a band of parallel polarization around the centre of the nebula can be attributed to an approximately toroidal magnetic field in a circumstellar disk. A small but significant spiral distortion is also seen throughout the polarization map, appearing to mimic the spiral structure in the underlying cloud density. We interpret this as due to the outer field structure of a non-axisymmetric magnetically-braked collapse in a cloud whose rotation was initially inclined to the magnetic field (Fig 1).

The centre of the polarization pattern, which is dominated by magnetically aligned grains (rather than scattering), lies to the NE of IRS2, implying that IRS2 does not lie at the centre of the magnetic structure of the nebula. A nearby faint red star appears to have formed as a companion to the NE of IRS2 due to fragmentation of the cloud core, with the resulting magnetic structure enveloping both stars.

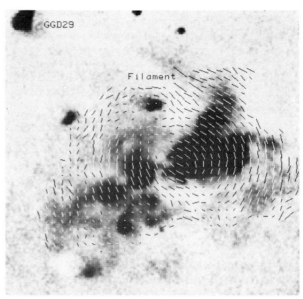

Plate 1. A 780nm wavelength linear polarization map of the Serpens nebula with a 670nm CCD image.

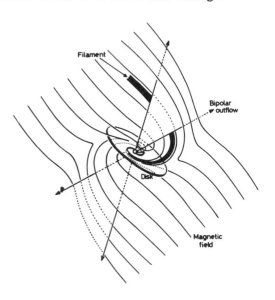

Figure 1. The proposed field structure around the nebula, resulting from magnetically-braked star formation.

A fuller account of these observations will be submitted to MNRAS.

CIRCULAR POLARIZATION IN T TAURI STARS

P. Bastien and R. Nadeau
Département de Physique and Observatoire du mont Mégantic
Université de Montréal
B.P. 6128, Succ. A, Montréal, Québec   H3C 3J7   Canada

ABSTRACT. We report the detection of circular polarization in three T
Tauri stars with known intrinsic linear polarization. A circumstellar
origin is required.

T Tauri stars are known to be surrounded by dust grains from their
linear polarization (Bastien and Landstreet 1979) and their infrared
excess (Cohen and Kuhi 1979). The linear polarization is variable in
at least 60% of the stars with sufficient data (Ménard 1986; Ménard and
Bastien, in preparation). Circular polarization has been reported
previously in only one young star, an FU Orionis star, V1057 Cyg
(Wolstencroft and Simon 1975), although with a signal/noise ratio of
$\simeq 3$.

In a search for circular polarization in young stars, we detected
a significant polarization in the T Tauri stars RY Tau, T Tau, and SU
Aur, with a signal/noise ratio of 5 to 7. Three sigma upper limits can
be given for two other stars, DG Tau and FU Ori. These measurements
and previous circular polarization measurements are presented in
Table I.

Two mechanisms are likely for producing circular polarization in
T Tauri stars: multiple scattering by grains in a circumstellar
envelope, or scattering by aligned, non-spherical grains. Woltencroft
and Simon interpreted the variable circular polarization in V1057 Cyg
as due to changes in the alignment of elongated particles distributed
in a disk- like configuration which we are looking at from the pole.
However, a careful analysis of circular polarization data presented
here and published linear polarization data shows that both mechanisms
occur in different stars.

I. Appenzeller and C. Jordan (eds.), Circumstellar Matter, 131–132.

TABLE I

Star	JD– 244 000.0	λ	V/I		Ref
V1057 Cyg	1984.5	6250	− 0.9 ±	2.4	1
		4500	− 14.0	6.0	
	2012.5	6250	− 1.9	2.6	
		4500	− 1.3	1.7	
	2193.5	6250	+ 1.2	2.4	
		4500	− 3.7	1.4	
	2247.5	6250	− 4.0	3.6	
		4500	+ 17.5	4.7	
RY Tau	3746.89	7543	+ 4.1	3.9	2
RY Tau	6388.654	7925	+ 6.4	1.0	3
T Tau	6385.613	7925	− 5.5	1.0	3
	6482.578	7925	− 6.0	2.5	3
SU Aur	6385.731	7925	+ 4.9	1.0	3
DG Tau	6385.865	7675	+ 0.6	1.1	3
FU Ori	6390.859	7925	+ 0.8	1.3	3

1: Wolstencroft and Simon (1975)
2: Bastien (1982)
3: Nadeau and Bastien (1986)

REFERENCES

Bastien, P.: 1982, Astron. and Astrophys. Suppl. 48, 153, and 48, 513.
Bastien, P., and Landstreet, J.D.: 1979, Astrophys. J. (Letters) 229, L137.
Cohen, M., and Kuhi, L.V.: 1979, Astrophys. J. Suppl. 41, 743.
Ménard, F.: 1986, M. Sc. Thesis, Univ. of Montréal.
Nadeau, R., and Bastien, P.: 1986, Astrophys. J. (Letters) 307, in press.
Wolstencroft, R.D., and Simon, T.: 1975, Astrophys. J. (Letters) 199, L169.

A POLARIZATION OUTBURST IN THE T TAURI STAR UY AURIGAE

F. Ménard and P. Bastien
Département de Physique and Observatoire du mont Mégantic
Université de Montréal
B.P. 6128, Succ. A, Montréal, Québec H3C 3J7 Canada

During the course of a monitoring programme of the linear polarization
of various T Tauri stars, UY Aur was observed to undergo a strong polarization burst between 1984 October and 1985 January. The linear polarization rised from about 1.5% early in 1984 to a maximum of 7.6% in
1984 October and declined back to the earlier value after 3 to 4
months. It is the first time that such a large increase in linear
polarization is reported in a T Tauri star. However the interpretation
is complicated by the fact that UY Aur is a visual binary with a separation of 0.8 and both components were included in the measurements.

The linear polarization in T Tauri stars is due mostly to scattering of light from the stars and their immediate line emitting regions
by grains outside these inner regions (Bastien and Landstreet 1979).
This is supported in the case of UY Aur by a strong infrared excess
(e.g. V-L = 6.50, Rydgren and Vrba (1983)) and also by its IRAS detection at 12, 25, 60 and 100 μm. Assuming single scattering, one can
compute the wavelength dependence of polarization $P(\lambda)$ to be expected
from a distribution of various grain particles. No single grain composition and grain radius could fit the observed $P(\lambda)$, see Figure 1, near
the end of the polarization burst. A good fit can be obtained, to
within a scale factor, with of mixture of spherical amorphous carbon
grains of 0.16 and 0.23 μm in radius. The details of the fit can be
found in Ménard (1986).

REFERENCES

Bastien, P., and Landstreet, J.D.: 1979, Astrophys. J. (Letters) 229,
    L137.
Ménard, F.: 1986, M. Sc. Thesis, Univ. of Montréal.
Rydgren, A.E., and Vrba, F.J.: 1983, Astron. J. 88(7), 1017.

I. Appenzeller and C. Jordan (eds.), Circumstellar Matter, 133–134.

UY AURIGAE

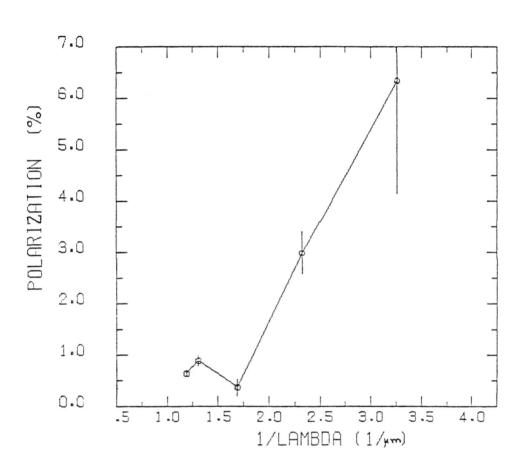

<u>FIGURE 1</u>   Observed polarization curve of UY Aurigae. The measurements were taken on 1985 January 31st.

POLARIZATION MEASUREMENTS OF SOME T TAURI STARS

U. C. Joshi, M. R. Deshpande and A. K. Kulshrestha
Physical Research Laboratory
Navrangpura
Ahmedabad-380009
India

T Tauri stars show linear polarization typically between 1-3%. A two band linear polarization survey of some T Tauri star was reported earlier by Bastien (1982). Most of the stars show pronounced time variability in polarization and position angle (Bastien, 1980; 1982). Wavelength dependence of polarization is important in determining the specific mechanism(s) producing polarization. For a systematic study of polarization in T Tauri stars, we have taken up an observing programme to measure linear polarization of some stars in Taurus-Auriga region. Polarization measurements of 9 T Tauri stars are reported here. Observations were made on January 8-11, 1984 with MINIPOL (Frecker and Serkowski, 1976) on 61" telescope of University of Arizona.

Figure 1 shows the wavelength dependence of polarization and position angle. A few important results are discussed here. HL Tau shows larger degree of polarization increasing towards I band, P_I being $\sim$ 15%. DG Tau also shows high degree of polarization which is found to be variable in a time scale of a day. Polarization vector in DG Tau is perpendicular to the "jet" detected by Mundt and Fried (1984). Position angle changes significantly with wavelength in all stars discussed here except the RY Tau. RY Tau shows large λ - independent polarization ($P \sim 3.4\%$); position angle is also λ - independent. Stars DR Tau, SU Aur and RW Aur show small degree of polarization ($\sim 0.4\%$) in all bands but the position angle show strong λ -dependence, especially in RW Aur θ change by 65 degrees from U to I band. Percent polarization slowly increases from U to I band in XZ Tau (1.2 to 3.0%) and DG Tau (2.7 to 6%) whereas θ is almost λ - independent. Stars VY Tau and BP Tau show λ - dependence for both - P and θ ; BP Tau shows very peculiar behaviour of λ - dependence of position angle.

135

I. Appenzeller and C. Jordan (eds.), Circumstellar Matter, 135–136.

References

Bastien, P. : 1980, Astron. Astrophys. 94, 294.
Bastien, P. : 1982, Astron. Astrophys. Suppl. 48, 153.
Frecker, J. E. and Serkowski, K. : 1976, Appl. Opt. 15, 605.
Mundt, R. and Fried, J. W. : 1983, Astron. J. 274, L83.

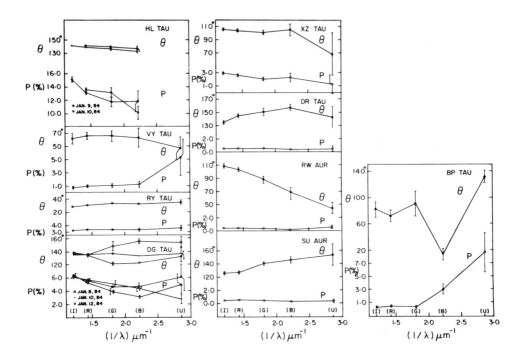

Figure 1.   Wavelength dependence of percent polarization (P)
            and Position angle (θ) for 9 T Tauri stars.   Error
            bars (± 1 sigma) are marked.

POLARIZATION MEASUREMENTS IN B5, L134 AND HEILES CLOUD 2 AND EVIDENCE OF NEWLY BORN STARS

U.C. Joshi, P.V. Kulkarni, M.R. Deshpande, A. Sen and
A.K. Kulshrestha
Physical Research Laboratory
Navrangpura
Ahmedabad-380009
India

Polarization measurement of the background stars in the region of dark globules is important to study the magnetic field geometry and grain's characteristics in the globule. These parameters are important for the formation and evolution of dark globules. We made polarimetric observations of stars in three nearby dark clouds - B5, L134 and Heiles Cloud 2. Polarization measurements of stars in the region of B5 were made with 'MINIPOL' (Frecker and Serkowski, 1976) on 61" telescope of University of Arizona. Observations for the stars in the region of L134 and Heiles Cloud 2 were made using PRL polarimeter (Deshpande et al. 1985) on 1 meter telescope of Indian Institute of Astrophysics, Bangalore. Results are presented and discussed here. Figure 1 shows the polarization vectors projected on the sky plane for the above globules.

We assume Davis-Greenstein mechanism for grain alignment. In this situation polarization vectors are the direction of projected magnetic field. In B5 polarization vectors are generally aligned in the NE and SW direction, except for the NW region where the vectors are more or less perpendicular to the polarization vectors in other region. Detailed discussion on B5 is given elsewhere (Joshi et al. 1984). Following are the main inferences -

a) Strong stellar wind from embedded IR sources (Beichmann et al. 1984) especially from IRS1 has retarded the collapse along NE-SW direction, which is also the direction of rotation axis.

b) the central object (IRS1?) form with a torus around it; the orientation of torus being perpendicular to the rotation axis. The torus collimates the mass loss from the central object, which has further restricted the collapse. The magnetic field is constricted by the collapse and follows the torus. The polarization vector for stars 13, 16 and 17 may be thus explained.

137

I. Appenzeller and C. Jordan (eds.), Circumstellar Matter, 137–139.
© *1987 by the IAU.*

c) Some stars (e. g. star no. 6 and 16) show λ - dependence in position angle. This is possible when light passes through inhomogeneous regions, with various orientation of dust grains (Coyne, 1974). The anisotropic stellar wind can produce inhomogeneity in the nearby regions of embeded IR sources. The mass loss from newly born stars have perhaps disturbed the otherwise homogeneous distribution of grains significantly.

Polarization vectors in dark clouds L134 and Heiles Cloud 2 show high degree of alignment. This probably indicates either the star formation has not yet set in or has not reached to a stage where vigorous mass loss is expected. Degree of polarization in Heiles Cloud 2 is quite high ($P \sim$ 6%) which means the degree of alignment is also high. This is possible only if the magnetic field in the surrounding region is high or the disturbing factors are at minimum.

References

Beichmann, C.A., Jennings, R.E., Emerson, J.P., Baud, B., Harris, S., Rowan-Robinson, M., Aumann, H.H., Gautier, T.N., Gillet, F.C., and Young, E. : 1984, Astrophys. J. 278, L45
Coyne, G.V. : 1976, Astron. J. 79, 565
Deshpande, M.R., Joshi, U.C., Kulshrestha, A.K., Banshidhar., Vadher, N.M., Mazumdar, H.S., Pradhan, S.N., and Shah, C.R.: 1985, Bull. Astron. Soc. India 13, 157
Frecker, J.E. and Serkowski, K.: 1976, Appl. Opts., 15, 605
Joshi, U.C., Kulkarni, P.V., Bhatt, H.C., Kulshrestha, A.K. and Deshpande, M.R. : 1985, Mon. Not. R. Astron. Soc. 215, 275

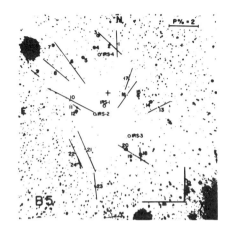

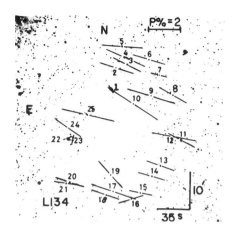

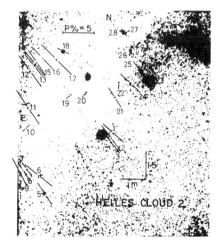

Figure 1. Polarization map for
globules B5, L134 and Heiles
cloud 2. Polarization vectors are
superimposed on the star fields,
taken from Palomar Sky Survey
prints.

ON THE NATURE OF PROTOSTELLAR H_2O MASERS

G. M. Rudnitskij
Sternberg Astronomical Institute
Moscow State University
Moscow V-234, 119899
USSR

Most sources of H_2O maser radio emission at 1.35 cm, associated with star formation regions, show strong variability with, sometimes, rapid bursts of emission (see, e.g., Liljeström 1984, Rowland and Cohen 1986, and references therein). A preliminary conclusion on the possible cyclicity of H_2O maser variability can be drawn (Lekht et al. 1982, 1983), with a quasiperiod of several years. The "quiet" state of a maser source, with moderate, slowly varying values of the line flux density, turns to the "active" phase with H_2O line bursts (Lekht et al. 1983). The H_2O maser generation region is probably located in a rotating gas-and-dust disc (torus) around a protostar (or young star). This is pointed to by VLBI observations showing in some sources maser features arranged in an ellipsoidal structure around a common centre (presumably, the protostellar object - see Downes et al. 1979), as well as by symmetrical character of H_2O line profiles of many masers (Lekht et al. 1982). As an excitation mechanism for H_2O, collisional pumping in two-temperature medium behind a shock front (with hot heavy particles and cold free electrons or vice versa) is widely accepted (Bolgova et al. 1982, Kylafis and Norman 1986).

I suggest two models explaining variability of protostellar H_2O masers, including their cyclic activity. In Model 1 H_2O variability is connected with variable luminosity of the central object, due to nonstationary accretion of matter onto it. As it was first shown by Yorke and Krügel (1977), the accretion onto a protostar can be unstable and can oscillate with a period of several years. At each luminosity rise, the star creates a shock in the encompassing gas-and-dust torus. Behind the shock front, the conditions are favourable for the H_2O maser pumping. At larger distances, the shock expires, and the maser intensity fades. In Model 2 the star itself is surrounded by a small circumstellar gas-and-dust ("protoplanetary")

141

I. Appenzeller and C. Jordan (eds.), Circumstellar Matter, 141–142.

disc, with the axis tilted to that of the external torus.
The small disc would precess in the gravitational field of
the torus. The stellar wind blowing from the poles of the
disc would impact the surface of the torus' internal
cavity, creating there a shock wave. Time- and space-
variable shadowing of the torus' surface from the stellar
wind by the small disc may also cause maser variability.
This model was earlier applied by Rudnitskij (1987) to the
explanation of optical variability of cometary reflection
nebulae in star formation regions.

Both models can be tested observationally. In Model 1
correlated variations of several emission features in the
H_2O line profile should be observed (that was really seen
in, e.g., W49N by Liljeström (1984)), connected with the
central object's variability (which can be observed in the
infrared). In Model 2 the drift of the bright spot of
maser emission along the circumstellar torus, due to the
small disc's precession, must be observable.

REFERENCES

Bolgova, G.T., Strel'nitskij, V.S., Umanskij, S.Ya. 1982.
    Sci. Inf. Astr. Council Acad. Sci. USSR, 50, 22.

Downes, D., Genzel, R., Moran, J.M., Johnston, K.J.,
    Matveyenko, L.I., Kogan, L.R., Kostenko, V.I.,
    Rönnäng, B. 1979. Astr. Ap., 79, 233.

Kylafis, N.D., Norman, C. 1986. Ap. J. (Letters), 300, L73.

Lekht, E.E., Pashchenko, M.I., Rudnitskij, G.M.,
    Sorochenko, R.L. 1982. Soviet Astr., 26, 168.

Lekht, E.E., Pashchenko, M.I., Sorochenko, R.L. 1983.
    In: 15th All-Union Conference on Galactic and
    Extragalactic Radio Astronomy, Held in Kharkov
    11-13 October 1983. Abstracts of Papers, p. 214.

Liljeström, T. 1984. Rept Radio Lab. Helsinki Univ.
    Technol., No. 162, 40.

Rowland, P.R., Cohen, R.J. 1986. M.N.R.A.S., 220, 233.

Rudnitskij, G.M. 1987. In: Star Formation Regions. Proc.
    IAU Symp. 115. Eds. J. Jugaku and M. Peimbert.
    D. Reidel Publ. Co., Dordrecht.

Yorke, H.W., Krügel, E. 1977. Astr. Ap., 54, 183.

MAGNETIC FLUX DISSIPATION DURING THE CONTRACTION OF
A MAGNETIC PROTOSTELLAR GAS CLOUD

M.S.El-Nawawy,A.Z.Aiad and M.A.El-shalaby
Cairo University
Faculty of Science,Astronomy department
Cairo
Egypt

The role of the magnetic field and the possibility of its leakage during the contraction of an interstellar cloud need more studies both quantitatively and qualitatively.

Because the physics of the interstellar cloud collapse is relatively complicated,we studied the hydromagnetics of a self-gravitating magnetic cloud neglecting both the chemistry and rotation. An adiabatic relation is used for the temperature variation. A hydromagnetic computer code has been constructed in order to follow numerically the collapse of an axially symmetric (2D) magnetized cloud. The method of explicit donor-cell was applied to the hydromagnetic equations with a moving grid of cylindrical coordinates. This ensures mass and momentum conservation and allows the grid to follow the collapse of the fluid near the center of the grid. The cloud is assumed to be axisymmetric and initially uniform. The contraction has been studied for densities in the range of 10 cm$^{-3} \lesssim n \lesssim 10^{13}$ cm^{-3}. The ion density has been calculated by the relation $x_i \propto n^{-1}$.

We have carried out nine numerical models. The first four models were developed assuming that the magnetic field is frozen in the cloud. The rest of the models were devoted to study magnetic flux dissipation.

In case of frozen in magnetic field,the central magnetic field increases according to the relation $B \sim \rho^{k}$,$k \lesssim 1/2$,in agreement with Scott and Black(1980). The contraction was followed up to density increase by about five orders of magnitude. All the models predict flattening.

Comparison of the magnetic flux to mass ratio of different objects reveals that there are about three orders of magnitude difference between a molecular cloud and a magnetic star,El-Nawawy and Aiad(1986) and Nakano(1983). The magnetic flux dissipation through Joule heating was studied in five models (5-9),in order to explain the difference in the mass flux ratio between a molecular cloud and the sun.

The results indicate that the magnetic flux dissipation

143

I. Appenzeller and C. Jordan (eds.), Circumstellar Matter, 143–144.

by Joule heating starts at densities $n \gtrsim 5\times10^{10}$ cm^{-3}.   The
magnetic flux dissipation is sensitive to both the ion den-
sity and the magnetic field strength and its gradient.   At
densities $n \gtrsim 10^{13}$ cm^{-3}, the central magnetic flux has decreased by more than three orders of magnitude; thus explaining
how the magnetic flux can decrease to stellar values.   The
resulting magnetic flux dissipation in our study represents
only the contribution of ions. Therefore our results represent the maximum values for the magnetic flux dissipation.
At densities $n \gtrsim 5\times10^{10}$ cm^{-3}, the electrons in addition   to
ions increase the magnetic flux dissipation. At lower dens-
ities ($n \le 5\times10^{10}$ cm^{-3}), $n_i/n_e \le 100$ as given by Nakano(1984).
In this case, the electron conductivity opposes the magnetic
flux dissipation. Hence the minimum density for the begining
of magnetic flux dissipation   cannot   become   earlier   than
5×10^{10} cm^{-3}.
Due to magnetic flux dissipation, the magnetic   field
tends to increase isotropically especially at higher densities. In addition, the pressure increases with temperature
according to an adiabatic relation. Therefore as long as the
magnetic dissipation becomes effective, and the pressure increases, the core of the evolved protostellar cloud tends   to
be spherical and magnetically isotropic, especially in clouds
of initially small magnetic to gravitational energy ratio.
This can explain why the observed new-born stars are approximately spherical while the dense dark clouds are   mostly
flattened.
Up to densities 5×10^{10} cm^{-3}, one   may expect   that the
magnetic field   effectively transfers the angular momentum
from a contracting cloud.
For more details see El-Nawawy(1985).

References

El-Nawawy, M.S. 1985, Ph.D. Thesis Cairo University, Faculty of
                Science.
El-Nawawy, M.S. and Aiad, A.Z. 1986, in preprint.
Nakano, T. 1983, Pub. Astron. Soc. Japan, 35, 87.
--------- 1984, Fundamentals of cosmic Physics, Vol.9, 139.
Scott, E.H. and Black, D.C. 1980, Ap. J., 239, 166.

HERBIG-HARO OBJECTS

RECENT OBSERVATIONS OF HERBIG-HARO OBJECTS, OPTICAL JETS, AND THEIR SOURCES

Reinhard Mundt
Max-Planck-Institut für Astronomie
Königstuhl
D-6900 Heidelberg 1
Federal Republic of Germany

ABSTRACT. Recent observations of Herbig-Haro objects and optical jets are reviewed, including observations of the stellar objects responsible for these and related outflow phenomena. The review discusses observations obtained in the following wavelength bands: radio, far-infrared, infrared, optical, and ultraviolet.

1. INTRODUCTION

The occurence of energetic mass outflows from all types of young stars and their possible impact on pre-main sequence evolution has only recently become fully appreciated (e.g. Lada 1985). In the case of low-mass stars (M $\lesssim$ 3 M$_\odot$) these outflows are mostly studied through CO line observations of the so called "high-velocity" molecular gas accelerated by these stars or through optical observations of Herbig-Haro (HH) objects and "optical" jets.

HH objects and "optical" jets are tracing the outflowing matter with the highest velocity and highest degree of collimation (e.g. Schwartz 1983, Mundt 1985a, 1986). Their radial velocities reach values of about 400 km/s and several jets have length-to-diameter ratios of about 20 (Mundt, Brugel, and Bührke 1986). In contrast, the "high-velocity" molecular flows associated with low-mass stars have typical velocities of 5-30 km/s and their length-to-diameter ratio is about 2 to 3 (e.g. Lada 1985). HH objects and "optical" jets (hereafter called jets only) are intimately related phenomena. They show the same emission-line spectrum, which is very probably formed behind shock waves with velocities of 40-100 km/s (e.g. Schwartz 1983; Mundt, Brugel and Bührke 1986). Furthermore, several (often long known) HH-objects form the brightest knots of a jet. On the basis of such observations it has been suggested (e.g. Mundt 1985a,b) that many HH objects simply represent the locations of the most brightest radiative shocks in a jet from a young star. These shocks have to be rather oblique, since the typical shock velocities are considerably lower than the flow velocities of

147

I. Appenzeller and C. Jordan (eds.), Circumstellar Matter, 147–158.

about 200-400 km/s (Mundt 1986).

A large and rapidly growing number of papers have been published within the past 5 years on HH objects and other outflow phenomena associated with young stars. There have also been several conferences at which these topics have been discussed. For conference proceedings the reader is referred to Roger and Dewdney (1981), Canto and Mendoza (1983), Black and Matthews (1985), Serra (1985), and Henrikson (1986). For reviews on HH objects, jets, and molecular flows the reader is referred to Schwartz (1983), Mundt (1985a,b), and Lada (1985), respectively. In order to avoid too much overlap with these other reviews, I will concentrate here on the observational results obtained during the past three years on HH objects, optical jets, and on the stars responsible for these outflow phenomena. These observations have been obtained through a variety of methods over a very broad wavelength range (20 cm - 10^{-5} cm). The observations in the various wavelength bands will be discussed in order of decreasing wavelength.

2.   VLA RADIO CONTINUUM OBSERVATIONS

2.1.   HH1, HH2 and their "central" star

HH1 and HH2 are the only HH objects detected so far with the VLA (Pravdo et al. 1985). In all other cases only the outflow source has been detected. In some of these latter cases, however, part of the radio emission may originate in compact HH-like or jet-like nebulae near the source (see §2.2.).

At 6 cm HH1 and HH2 were detected by Pravdo et al. at a flux level of 0.55±0.04 mJy and 1.22±0.04 mJy, respectively. The observed spectral indices are -0.2±0.3 (HH1) and -0.2±0.1 (HH2) leading Pravdo et al. to conclude that one is observering free-free emission from an optically thin region. The Hα flux measured for these two objects is consistent with this interpretation.

Pravdo et al. (1985) also detected the "central" star of HH1 and HH2 at a flux level of 1.2±0.04 mJy (λ = 6 cm). Its spectral index of +0.4±0.2 is significantly different from HH 1/2 and is consistent with free-free emission from an ionized stellar wind. These VLA observations together with recent CCD observations (Strom et al. 1985, Scarott et al. 1986) showed that this VLA source is the source of the bipolar outflow traced by HH1 and HH2 and not the Cohen-Schwartz star.

2.2. Outflow Sources

The existing VLA data on sources of HH objects and jets are summarized in Table 1. In general these sources are rather weak with fluxes often of the order of 0.1 mJy, even for sources with distances of only 150 pc. It has been pointed out by Bieging, Cohen, and Schwartz (1984) that the liklihood of detecting an outflow source with the VLA is much higher than in the case of a "normal" T Tauri star (TTS). This suggests that the former objects have stronger winds on average. However, the observed spectral indices suggest that only in some objects is the wind

Table 1: Radio Fluxes and Spectral Indices of Jet and HH Object Sources

Source	Peak Flux at 6 cm (mJy)	Spectral Index α	Ref.	Source	Peak Flux at 6 cm (mJy)	Spectral Index α	Ref.
SSV13 (HH7-11)	0.27[a]	–	1	VLA1-HLTau	0.13	–	2
				HH34-IRS	0.08[a]	–	2
Haro 6-5B (=FS TauB)	0.15	0.1	2	VLA1-HH1/2	1.2	0.4±0.2	4
T Tau N	0.7	1	3	HH43-IRS1	0.2[a]	–	1
T Tau S	5	0.44	3	HH24-IRS	0.6	–	1
DG Tau	0.51	0.55±0.3	1	HH26-IRS	0.2[a]	–	1
Haro 6-10	0.8	–	1	R Mon	0.4[a]	–	5
L1551-IRS5	1.7	0.05±0.06	1	AS 353A	0.5[a]	–	6
HH30-star	0.08	–	2	1548C27	0.1[a]	–	2

a= 2δ upper limit; 1 = Bieging, Cohen, and Schwartz 1984; 2= Brown, Drake, and Mundt 1985, 1986, 3 = Schwartz, Simon, and Campbell 1986; 4 = Pravdo et al. 1985; 5 = Cohen, Bieging, and Schwartz 1982; 6 = Snell and Bally 1986.

observed directly (i.e.$\alpha \approx 0.6$). Instead an optically thin ($\alpha \approx 0$) and relatively compact - but spatially resolved - radio emission region is observed around the star. For example, the radio emission region around L1551-IRS5 and DG Tau is clearly resolved along the jet axis (Bieging, Cohen, and Schwarz 1984) and radio emission regions with sizes of the order of 100 AU have been found around XZ Tau, HL Tau, and Haro 6-5B (Brown, Drake, and Mundt 1985, 1986). These "extended" radio emission regions are probably ionized by shocks in the outflowing matter.

3. CO LINE OBSERVATIONS OF HIGH VELOCITY MOLECULAR GAS NEAR HH OBJECTS

The reader is refered to Edwards and Snell (1984) and references there-in for a detailed discussion of this subject. These authors showed that "high-velocity" molecular flows (v=5-30 km/s) are relatively common in molecular clouds with associated HH objects. In total they found 17 anisotropic or bipolar molecular flows in the vicinity of 58 HH objects which are characterized by ^{12}CO full velocity widths $\geq$10 km/s. 75% of the 58 HH objects lie within 10' of the outflow center. However, only 25% of these 58 lie within 1' of the outflow center and are those objects which are more likely to be directly related to the molecular outflow, since "optical" flows have typical length of 0.1 pc (2.5' at 150 pc; Mundt, Brugel, and Bührke 1986).

However, all these numbers are not very meaningful, since for many HH object (e.g. for about 70% in Herbig's 1974 catalogue) the source is not known. In addition, many HH object and jet sources are low-lumino-

sity objects (L ≈ 1 $L_\odot$) and are expected to have weaker and slower outflows requiring sensitive, high-spatial resolution observations in
order to be detected. Indeed, about 80% of the molecular flow sources
in the list of Edwards and Snell (1984) with known luminosities have
L ≲ 20 $L_\odot$ and about 30% have L = 200-10^4 $L_\odot$. Nevertheless, it is quite
an important task to investigate the relationship between "optical"
flow phenomena and molecular ones. To find out, for example, whether
the molecular flows might be driven by the high-velocity gas traced by
HH objects and jets. This requires definite source identifications and
in many cases sensitive molecular line observations at high spatial
resolution.

   Mundt, Brugel, and Bührke (1986) have estimated the mass fluxes of
16 jets and other highly collimated flows emanating from low luminosity stars (0.1-100 $L_\odot$) and investigated whether these jets could drive
the molecular flows of these young stellar objects. They concluded that
momentum driven molecular flows are very unlikely, since the momentum
fluxes of the jets are typically 100 times smaller than those of the
molecular flows associated with low-luminosity stars. Energy driven
molecular flows, however, would be consistent with the currently available data.

4. IR AND FIR OBSERVATIONS OF JET AND HH OBJECT SOURCES

A relatively large number of IR (1-20 μ) and in particular FIR (20-
160 μ) observations have been carried out in recent years. In the
latter wavelength region very little data predate 1983 (for details see
Schwartz 1983). Relevant recent publications in the IR and FIR are:
Cohen and Schwartz (1983); Cohen, Harvey, and Schwartz (1985, and
references therein); Vrba, Rydgren, and Zak (1985); Harvey et al.
(1986). The FIR observations discussed by these authors have all been
carried out with the help of NASA's Kuiper Airborne Observatory. So far
relatively few FIR observations of these objects based on IRAS data
have been published (Emerson et al. 1984; Clark and Laureijs 1986).
Polarimetric observations, at IR and optical wavelengths, of these and
related objects have been discussed by Hodapp (1984). All of these new
observations together with the data obtained before 1983 (see Schwartz
1983) are important for the identification of sources, for measuring
their spectral energy distribution and bolometric luminosities, and
for determining the spatial distribution of their circumstellar dust.

   From the existing data it is evident that these sources are in
general much more obscured than "normal" TTS. For the latter stars A_V
is typically 0.5-3 (Cohen and Kuhi 1979), while A_V values of 10-20
are not unusual for the jet and HH object sources. This strongly
indicates that the outflow sources are younger than "normal" TTS.
Nevertheless, most sources have optical counterparts on deep CCD images
taken in the 0.6-1 μ region. The CCD images also show that many sources
are associated with (cometary) reflection nebulae. Their spectral
energy distributions are also quite different compared with those of
"normal" TTS. For the sources with known K-L colors (≈ 20) the average
K-L value is 2.0, while the corresponding value for the TTS in the

Taurus-Auriga dark cloud is 0.8 (Cohen and Kuhi 1979). On average, about 70% of the source energy is radiated at wavelength larger than 20 μ. The corresponding value for the TTS in Taurus-Auriga, which are not outflow sources (like DG Tau or T Tau), is 10-20% (Rucinski 1985). The FIR energy distribution of the   sources is typical of cool dust grains with T = 40-60 K.

In about 30 cases, the probable source of flow (traced by HH objects and jets) has been identified by various means. For about 20 sources FIR measurements have been carried out. However, only for about two-thirds of these there is no obvious confusion problem with other sources in the beam. The bolometric luminosities derived from the FIR data for 22 objects (the confusion problem is neglected here) have the following distribution: 50% have $\lesssim$ 20 $L_\odot$, 30% have 20-100 $L_\odot$ and 20% have about 10^3-10^4 $L_\odot$. The latter sources are R Mon, LKHα 234, GGD 37/Cep A (e.g. Lenzen, Hodapp, and Solf 1984), and IRC2-M42 (Jones and Walker 1985; Taylor et al. 1986). In the latter 3 cases there are seveve source confusion problems. Nevertheless, it obvious that about 80% of the known sources are low-luminosity (low-mass) stars with L $\lesssim$ 100 $L_\odot$.

5. H_2 LINE OBSERVATIONS

Since the first extensive observations of HH objects in the IR H_2 lines by Elias (1980) many more HH objects have been observed and several flows have been mapped. Table 2 gives an overview of the objects which have been detected and which have been mapped. For the latter cases, a beam size is given. These maps have been in general obtained in the ν = 1—— 0 S(1) line at 2.12 μm.

Table 2: HH Objects detected in the IR H_2 lines

Object	beam size of map (arcsec)	Ref.	Object	beam size of map (arcsec)	Ref.
HH 12	15	1	HH 19	-	5
HH 7-11	12, 19.6	2, 3	HH 46	-	6
HH 6	15	1, 3	HH 52	5	8
HH 5	-	3	HH 53	5	6, 8
T Tau/ Burnham's Neb.)	-	4, 5	HH 54	5	6, 8
HH 40	-	6	HH 101	-	9
HH 1/2	7.5	6, 7	HH 32	5	8, 10

References: 1 = Lane and Bally 1986; 2 = Zealey, Williams, and Sandell 1984; 3 = Lightfoot and Glencross 1986; 4 = Beckwith et al. 1978; 5 = Zinnecker et al. 1985; 6 = Elias 1980; 7 = Harvey et al. 1986; 8 = Zealey et al. 1984; 9 = Brown et al. 1983; 10 = Zealey et al. 1985

A comparison of the H_2 maps with the optical images shows that the H_2 emission occurs in the same general area as the optical emission and its intensity is correlated with the optical intensity. This is consistent with the detection of HH19 and HH12, but not of their associated jets. In a few cases, there are significant differences. One example is HH2 where the peak of the H_2 emission is 7" NW of the brightest optical knot (Harvey et al. 1986) However, a detailed comparison is often hampered by the large beam sizes used for the H_2 observations.

From the existing data very little is known on the structure of the shocks generating the H_2 emission or the dynamics of the pre- and post-shock gas. For example, are we observing molecular gas entrained in high-velocity jets? Alternatively, is the emission produced by shocks, formed as a result of the high velocity flow ploughing into the ambient molecular matter (e.g. at the jet's working surface)? To answer these questions radial velocity measurements and mapping at higher spatial resolution are needed. Such observations would be very interesting for those HH objects where a significant fraction of the optical emission is apparently formed in a bow shock(see § 6.2.2. and Mundt, Brugel, and Bührke 1986). In these cases the H_2 emission should be generated in the wings of the bow shock (where the normal shock velocity is low) and the optical emission near the apex.

6. OPTICAL OBSERVATIONS

6.1. Proper Motion Measurements

Since the discovery of large proper motions for HH28 and HH29 by Cudworth and Herbig (1979) the proper motions of 66 HH knots have been measured (see Schwartz, Jones, and Sirk 1984, and references therein). These 66 knots are associated with about 25 outflows. A histogram showing the frequency distribution of tangential velocities of 63 HH knots has been published by Schwartz (1986). In this histogram most HH knots have small proper motions and the number of HH knots in each velocity bin is approximately decreasing monotonically with increasing tangential velocities. The highest measured tangential velocity (350 km/s for HH1, Herbig and Jones 1981) is nearly as high as the highest radial velocity reported for an HH object (410 km/s for HH32, Hartigan, Mundt, and Stocke 1986). It has to be emphasized, however, that the tangential velocities with nominal values $\lesssim 100$ km/s are very unreliable. About 65% of these values have errors larger than about 50%.

For most of the relatively faint knots in the jets recently discovered through CCD imaging no proper motion data are available. The existing data on the bright knots in these jets show that not only the knots at the jet's end are moving (like HH47A), but also some of the knots along the jet (e.g. HH11). This suggest, that at least some of these flows are not in steady state conditions, where only the knots at the jet's end (the working surface) should be moving.

6.2. CCD Imaging

CCD detectors have been frequently used in recent years to study high velocity outflows through their shock induced optical line emission. These studies have led to the discovery of jets (e.g. Mundt and Fried 1983; Strom, Strom, and Stocke 1983; Mundt et al. 1984; Mundt, Brugel, and Bührke 1986; Krautter 1986; Ray 1986) and new HH objects and related nebulae (e.g. Strom et al. 1986). Furthermore, imaging in HH emission lines and in the continuum have helped to distinguish between scattered light and in situ formed shock emission in complex nebulae. These latter studies have been supplemented in several cases by CCD imaging polarimetry (e.g. Lenzen 1986, Scarott et al. 1986)

6.2.1 Optical Jets. More than 20 jets are known today. Their typical observational properties are summarized in Table 3. For a recent detailed discussion of the existing observational data the reader is refered to Mundt, Brugel, and Bührke 1986 (see also Mundt 1985a, 1985b, 1986). Morphologically, the jets are seen as radially projected single or bipolar elongations with an observed aspect ratio (length to width) of typically 10-20. In high dynamic range images all jets show knots. In four cases a series of 4-6 roughly equidistant knots is observed.

Table 3: Typical Observational Properties of the Jets

projected length:	0.02 - 0.5 pc
length-to-width ratio:	10 - 20
opening angle:	$3 - 10^{\circ}$
radial velocity:	$\lesssim 400$ km/s
tangential motion of knots:	$\lesssim 300$ km/s
collimation length:	3×10^{-3} pc
electron density:	$500 - 2000$ cm^{-3}
spectrum:	shock-excited emission line spectrum with $v_{shock} = 40-100$ km/s
sources:	T Tauri-like stars with $L \approx 1-10$ $L_{\odot}$

The brightest knots in these jets are often known as HH objects and have been discovered many years before their associated jet. In about 60% of all jets, a bright knot, or knotty region, is observed at the (presumable) end of the jet. These bright knots or knotty regions have been interpreted as the working surface of the jet (Mundt 1985b, Dyson 1987), i.e. as that region, where the jet is colliding with the ambient medium. This interpretation is strongly supported by the radial velocity decrease observed in these knots (see § 6.3.) and the bow shock structures associated with them (see § 6.2.2.). The knots along the jet have been interpreted as internal shocks, which - in principal - can be excited through various mechanisms. One mechanism, pressure gradients in the external gas, seems to be relatively important in this respect (Mundt, Brugel, and Bührke 1986; Falle, Innes, and Wilson 1986).

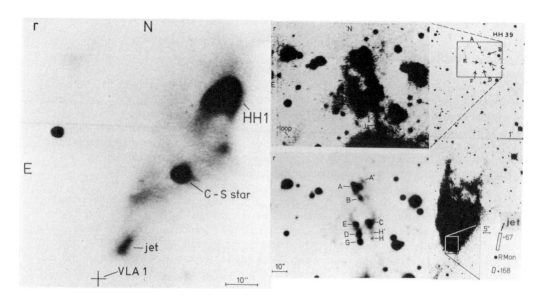

Fig. 1: CCD images of HH1 and HH39 illustrating their bow shock like
        structures.

      As mentioned above, many jet sources are associated with cometary
(or cone-like) reflection nebulae. In all these cases the jet axis is
lying approximately parallel to the nebular axis (e.g. Mundt, Brugel,
Bührke 1986). It is not yet clear how this very interesting morphological association is to be interpreted. In several cases, spectroscopy of
these reflection nebula has allowed a study of the often highly
obscured jet source. For example, it was shown by this method that
L1551-IRS5 probably belongs to the rare class of FU Orionis objects
(Mundt et al. 1985).

6.2.2. Bow Shock Structures. Recent CCD imaging has shown that HH1,
HH39, and in particular HH34 show bow shock like structures (Reipurth
et al. 1986; Mundt, Brugel, and Bührke 1986). CCD images of HH1 and
HH39 are reproduced in Fig. 1. For an Hα image of HH34 see paper by
Bührke and Mundt in this volume. All three objects have the following
properties in common:
1. The convex side of the bow shock structure points away from the
jet source.
2. They are internally highly structured (i.e. knotty and patchy). This
suggest that the knots in these HH objects are not isolated entities
but part of one large-scale flow pattern.
3. A relatively short jet is pointing from the source towards all three
HH objects. This jet can't be traced all the way to the HH object.
      The observed structures have been interpreted as a bow shock,
created by the rapid propagation (v ≈ 100-200 km/s) of the jet's working
surface through the ambient gas (Mundt, Brugel, Bührke 1986). This
interpretation is consistent with the large proper motions of HH1 and
HH39 and the radial velocities of HH34. The large spatial extend of the

bow shocks can probably be explained by the partial ionization of the pre-shock gas (Raga 1986).

6.3. Long-Slit Spectroscopy

HH objects and jets show in general a complex spatial structure in velocity, velocity dispersion, electron density and line excitation. Often strong gradients are observed, which in several cases can't be properly resolved by ground based observations. Long-slit spectroscopy is therefore one of the most powerful methods to study these flows (in particular when one deals with one-dimensional structures). High quality long-slit spectra with a velocity resolution of 15-100 km/s have recently been obtained of the following flows: GGD37 (Lenzen, Hodapp, and Solf 1986), HH7-11 (Böhm and Solf 1986), T Tau/HH1555 (Bührke, Brugel, and Mundt 1986), L1551-IRS5, HH28/29 (Sarcander, Neckel, and Elsässer 1985; Stocke et al. 1986), HH34 (Reipurth et al. 1986; Bührke and Mundt 1987), HH1/2 (Böhm and Solf 1985), HH24 (Solf 1987), R Mon/HH39 (Brugel, Mundt, and Bührke 1984; Walsh and Malin 1985), HH46/47 (Graham and Elias 1983; Meaburn and Dyson 1986), Th-28 (Krautter 1986), HH32 (Hartigan, Mundt, and Stocke 1986; Solf, Böhm, and Raga 1986), Haro 6-5B, DG Tau, DG TauB, HH30, VLA1-HL Tau, HH33/40, HH19, 1548C27 (Mundt, Brugel, and Bührke 1986).
     In the case of jets and other highly collimated flows traced by HH objects, the following correlations and trends have been derived from the currently available data by Mundt, Brugel, and Bührke (1986):
1. In 9 out of 12 flows with reasonable [SII] $\lambda\lambda$ 6716, 6731 data the electron density is decreasing with increasing distance from the source (or at least more distant regions have a smaller average electron density). This correletation has been interpretet by a flow of approximately constant mass flux of which the cross section is increasing with increasing distance from the source.
2. All jets which show a bright knot at their end, and which have sufficiently high radial velocities ($\gtrsim$100 km/s) to do detailed kinematical studies, show a decrease in radial velocity within (or near) the knot. These knots are interpreted as the jet's working surface. The observed radial velocity decrease is due to momentum exchange with the external gas and due to dissipation of part of the jet's kinetic energy in the shocks of the working surface.
3. There is no indication that the radial velocities are in general decreasing along the flow channel, even if several knots (i.e. internal shocks) are present. This suggests that the kinetic energy is in general transported relatively efficiently along the jets. This is also expected on theoretical grounds, if the flow velocities are higher than about 200 km/s.
     The long slit data of HH1, HH32, and HH34 have been discussed extensively in the context of bow shock models of HH objects (e.g. Solf, Böhm, and Raga 1986). As discussed in § 6.2.2. only for HH1 and HH34 CCD imaging suggests that at least part of the emission originates in a bow shock. The long-slit spectra of HH1 show that the velocity dispersion is highest near the apex of the presumed bow shock structure and decreases towards the central source. This correlation is in

(qualitative) agreement with simple bow shock models (e.g. Raga 1986 and references therein).

7. IUE OBSERVATIONS

HH objects have been observed extensively in the UV spectral range with the IUE satellite, which went into operation more than eight years ago. IUE observations showed that HH objects have a surprisingly strong continuum in the UV. Two-photon emission of hydrogen has been suggested as a likely source of that continuum. Furthermore, the studied HH objects can be divided into low- and high-excitation objects. The former group shows only H_2 fluorescence lines in the short wavelength range IUE spectra, while in the latter ones these lines are not observed but instead emission from CIV, CIII, SiIV or OIII is seen. For more details, the reader is refered to the reviews of Böhm (1983) and Schwartz (1983).

   Strong variations in the CIV 1550 and CIII 1909 line of HH1 have recently been reported by Brugel et al. (1985). The IUE data indicate a monotonic decrease in the fluxes of these lines by a factor of at least 4-6 between 1979 and 1983. Surprisingly, no indications of drastic changes in the optical range (specifically in the [OIII] 5007 line) have been found.

   Using 12 and 14 hour exposures in the short wavelength range of IUE Böhm et al. (1986) have obtained new spectra of HH1 and HH2 with a relatively high signal-to-noise ratio. The continua of both objects peak at about 1575 Å and not near 1410 Å, where the two-photon continuum of hydrogen has its maximum. The wavelength dependence of that continuum suggests that significant parts of it are due to H_2 continuum emission. The data obtained by these authors also showed that the contribution of the "Orion Reflection Nebulosity" to the continuum emission of these objects in the 1300-1900 Å wavelength range is not higher than 20-30%. (see also Mundt and Witt 1983).

The authors thanks Drs. F. Hessmann and T. Ray for valuable comments and critically reading the manuscript.

References

Black, D.C., and Matthews, M.S. 1985, "Protostars and Planet II", The
     University of Arizona Press, Tucson
Beckwith, S., Gatley, I., Matthews, K., Neugebauer, G. 1978, Ap.J.
     (Letters) 233, L41
Bieging, J.H., Cohen, M., and Schwartz, P.R. 1984, Ap.J. 282, 699
Böhm, K.H. 1983, Rev.Mex.Astr.Astrofis. 7, 55
Böhm, K.H., and Solf, J. 1985, Ap.J. 294, 533
Böhm, K.H., and Solf, J. 1986, in preparation
Böhm, K.H., Bührke, T., Raga, A.C., Brugel, E.W., Witt, A.N., and
     Mundt, R. 1986, Ap.J., submitted

Brown, A., Drake, S.A., and Mundt, R. 1985, in "Radio Stars", eds.
    R.M. Hellming and D.M. Gibson (Reidel, Dordrecht), p. 105
Brown, A., Drake, S.A., and Mundt, R. 1986, in prep.
Brown, A., Millar, T.J., Williams, P.M., and Zealey, W.J. 1983,
    M.N.R.A.S. 203, 785
Brugel, E.W., Böhm, K.H., Shull, J.M., and Böhm-Vitense, E. 1985,
    Ap.J.(Letters) 292, L75
Brugel, E.W., Mundt, R., and Bührke, T. 1984, Ap.J. (Letters) 287, L93
Bührke, T., Brugel, E.W., and Mundt, R.: 1986, Astr.Ap., 163, 83
Bührke, T., and Mundt, R. 1987, this volume
Cantó, J., and Mendoza, E.E. 1983, Rev.Mex.Astr.Astrofis. 7
Clark, F.O., and Laureijs, R.J. 1986, Astr.Ap. 154, L26
Cohen, M., Harvey, P.M., and Schwartz, R.D. 1985, Ap.J. 296, 633
Cohen, M., and Kuhi, L.V. 1979, Ap.J. Suppl. 41, 743
Cohen, M., and Schwartz, R.D. 1983, Ap.J. 265, 877
Cudworth, K.M., and Herbig, G.H. 1979, A.J. 84, 548
Dyson, J.E. 1987, this volume
Edwards, S., and Snell, R.L. 1984, Ap.J. 281, 237
Elias, J.H. 1980, Ap.J. 241, 728
Emerson, J.P. et al. 1984, Ap.J. (Letters) 278, L49
Falle, S.A.E.G., Innes, D., and Wilson, M.J. 1986, M.N.R.A.S.,
    submitted
Graham, J.A., and Elias, J.H. 1983, Ap.J. 272, 615
Hartigan, P., Mundt, R., and Stocke, J. 1986, A.J. 91, 1357
Harvey, P.M., Joy, M., Lester, D.F., and Wilking, B.A. 1986,
    Ap.J. 301, 341
Henrikson, R.N. 1986, "Jet from Stars and Galaxies", Can.J.Phys. 64,
    p. 351-535
Herbig, G.H. 1974, "Draft Catalogue of Herbig-Haro Objects", Lick Obs.
    Bulletin, No. 658
Herbig, G.H., and Jones, B.F. 1981, A.J. 86, 1232
Hodapp, K.-W. 1984, Astr.Ap. 141, 255
Jones, B.F., and Walker, M. 1985, A.J. 90, 1320
Krautter, J. 1986, Astr. Ap. 161, 195
Lada, C.J. 1985, Ann. Rev. Astron. Astrophys. 23, 267
Lane, A.P., and Bally, J. 1986, Ap.J., in press
Lenzen, R. 1986, Astr.Ap., in press
Lenzen, R., Hodapp, K.-W., and Solf, J. 1984, Astr.Ap. 137, 202
Lightfoot, J.F., and Glencross, W.M. 1986, M.N.R.A.S. 221, 993
Meaburn, J. and Dyson, J.E. 1986, preprint
Mundt, R. 1985a in "Protostars and Planets II", eds. D. Black, and M.
    Matthews, University of Arizona Press, Tucson, p. 414
Mundt, R. 1985b in "Nearby Molecular Clouds", ed. G. Serra, Lecture
    Notes in Physics, 217, p. 160, Springer Verlag, Heidelberg
Mundt, R. 1986, in "Jets from Stars and Galaxies", Can.J.Phys. 64,
    407
Mundt, R., Brugel, E.W., Bührke, T. 1986, Ap.J., submitted
Mundt, R., Bührke, T., Fried, J.W., Neckel, T., Sarcander M., and
    Stocke, J. 1984, Astr. Ap. 140, 17
Mundt, R., and Fried, J.W. 1983, Ap.J. (Letters) 274, L83

Mundt, R., Stocke, J., Strom, S.E., Strom, K.M., and Anderson, E.R.
      1985, Ap.J. (Letters) **297**, L41
Mundt, R., and Witt, A.N. 1983, Ap.J. (Letters) **270**, L59
Pravdo, S.H., Rodriguez, L.F., Curiel, S., Cantó, J., Torrelles,
      M.,Becker, R.H., and Sellgren, K. 1985, Ap.J. (Letters) **293**, L35
Raga, A.C. 1986, A.J., **92,** 637
Ray, T. 1986, Ast. Ap., in press
Reipurth, B., Bally, J., Graham, J.A., Lane, A., and Zealey, W.J. 1986,
      Astr.Ap., **164,** 51
Roger, S., and Dewdney, P.E. 1981, "Regions of Recent Star Formation",
      Reidel Dordrecht
Rucinski, S.M. 1985, A.J. **90,** 2321
Sarcander, M., Neckel, T., and Elsässer, H. 1985, Ap.J.(Letters)
      288, L51
Scarott, S.M., Warren-Smith, R.F., Droper, P.W. and Gledhill, T.M. 1986
      in "Jets from Stars and Galaxies", Can.J.Phys. **64,** 426
Schwartz, R.D. 1983, Ann. Rev. Astron. Astrophys. **21,** 209
Schwartz, R.D. 1986, in "Jets from Stars and Galaxies", Can. J. Phys.
      64, 414
Schwartz, R.D., Jones, B.F., and Sirk, M. 1985, A.J. **89,** 1735
Schwartz, P.R., Simon, T., and Campbell, R. 1986, Ap.J. 303, 233
Serra, G. 1985, "Nearby Molecular Clouds", Lecture Notes in Physics
      237, Springer Verlag, Heidelberg
Snell, R.L., and Bally, J. 1986, Ap.J. **303,** 683
Solf, J., 1987, this volume
Solf, J., Böhm, K.H., and Raga, A.C. 1986, Ap.J. **305,** 795
Stocke, J., Strom, S.E., Strom, K.M., and Hartigan, P. 1986, in prep.
Strom, K.M., Strom, S.E., and Stocke, J. 1983, Ap.J. (Letters) **271,**
      L23
Strom, S.E., Strom, K.M., Grasdalen, G.L., Sellgren, K., Wolff, S.,
      Morgan, J., Stocke, J. and Mundt, R. 1985, A.J. **90,** 2281
Strom, K.M., Strom, S.E., Wolff, S.C., Morgan, J., and Wenz,
      M. 1986, Ap.J. Suppl. **62,** 39
Taylor, K., Dyson, J.E., Axon, D.J., and Hughes, S. 1986, M.N.R.A.S.,
      in press
Vrba, F.J., Rydgren, A.E., and Zak, D.S. 1985, A.J. **90,** 2074
Walsh, J.R., and Malin, D.F. 1985, M.N.R.A.S. **217,** 31
Zealey, W.J., Williams, P.M., and Sandell, G. 1984, Astr.Ap. **140,** L31
Zealey, W.J., Williams, P.M., Storey, J., Taylor, K., and Sandell, G.
      1984, in "Edinburgh Star Formation Workshop", p. 109, ed. R.D.
      Wolstencroft
Zealey, W.J., Williams, P.M., Taylor, K., Storey, J., and Sandell, G.
      1985, preprint
Zinnecker, H., Mundt, R., Williams, P.M., Zealey, W.J. 1985,
      Mittl.Astr.Ges. **63,** 234

THEORETICAL MODELS OF HERBIG-HARO OBJECTS

J. E. Dyson
Department of Astronomy
University of Manchester
Manchester M13 9PL
England

ABSTRACT. A brief overview of the observational characteristics of HH objects is given. Current models for their production by the interaction of stellar winds and jets with interstellar gas are critically discussed. Models for two specific systems of HH objects, namely, the Orion HH objects and the HH46-47 system are described with reference to the general production mechanisms.

1.   INTRODUCTION

It would be hard to imagine more deceptively uninteresting objects than the inconspicuous semi-stellar knots of nebulosity seen against the dark clouds of NGC 1999 first brought to the attention of the astronomical world independently by Herbig (1951) and Haro (1952). Herbig (1951) realized immediately that their bright [OI] line emission set them apart from the relatively well understood photoionized HII regions, and, with considerable prescience, suggested that their excitation involved some mechanical process which involved stellar participation. Many years later, these seemingly unremarkable objects are the subject of extensive observational and theoretical investigation, and considerable controversy surrounds their interpretation. To some extent, the controversy is artificial, specifically in regard to mechanisms for physically producing these objects, since there has been a marked tendency to look for a unique model to describe what is most probably a collection of objects produced in a variety of ways. This is not to say that these objects do not have features in common, in particular, there seems little doubt that the emission from HH objects is due to the mechanism of shock excitation (though see Section 5.2 for a possible caveat to this statement).
  The astrophysical significance of HH objects can hardly be overstressed - at least not in this meeting! Their existence is bound up with the structure and stellar (or proto-stellar) content of dark molecular clouds. Not all that long ago, it would have almost certainly provoked cries of outrage (not least from the author) to suggest that dark clouds are much more interesting than the observationally far more spectacular HII regions. However, the richness of dynamical, physical and chemical phenomena

159

I. Appenzeller and C. Jordan (eds.), Circumstellar Matter, 159–172.
© 1987 by the IAU.

occurring in them revealed by radio, infra-red and mm-wavelength investigations over the past few years strongly support this viewpoint.

The study of HH objects has unearthed a number of largely unresolved problems in theoretical astrophysics: for example, the structure of cooling flows behind complex shock structures, the interaction of various forms of stellar mass loss with their environment and, arguably most important of all, the production and collimation of remarkably energetic stellar mass loss from relatively low luminosity stars. This review deals with a restricted sub-set of these problems, namely the gas dynamical interactions which can - possibly - lead to the formation of HH objects. It is not, however, possible - or even sensible - to attempt to discuss these interactions without at least some passing references to the other problems, and these will be made as appropriate.

2. OBSERVATIONAL CHARACTERISTICS OF HERBIG-HARO OBJECTS

Extensive discussion of HH characteristics are given by Schwartz (1983), Mundt (these Proceedings) and in the recent Symposium edited by Canto and Mendoza (1983), and only a few salient details will be reviewed here.

Optical spectra imply that a wide range of excitation conditions exist from one HH object to another, and equally importantly, within a given object. Böhm (1983) has compared the characteristic spectrum of a high-excitation HH object (HH2H) with that of a low-excitation object (HH7). Striking differences are apparent; for example strong [OIII]5007Å emission in the former but not in the latter, extremely strong [SII]6724Å emission in the latter, much weaker in the former. Both classes of object show [OI]6300, 6363Å emission, but the emission from this low ionization state ion is much stronger in the latter. Any model of any particular HH object should model its spectrum as well as its kinematics, but there has been a strong tendency to concentrate on this second aspect.

A few HH objects have been detected in the UV, although their close association with the dusty dark clouds clearly militates against this, HH1, 2 and 32 (all classed as high excitation optically) show lines from very high excitation ions such as C^{+3} and O^{+3}. The presence of these ions in conjunction with that of, for example, O°, has important implications for the structure of HH objects. Two low excitation objects, (HH43, 47), show UV Lyman band lines of H_2, but do not show the high excitation ionic emission seen in the other objects.

Near infra-red observations have also indicated the association of H_2 and HH objects. In some cases the molecular emission appears to envelop the object.

A strong blue continuum emission has been observed in some HH objects. Its origin is the subject of debate. It may be two-photon emission from hydrogen, in which case there are very important implications for the structure of shocks in HH objects (Dopita, Binette and Schwarts, 1982). Table I lists various important physical characteristics of HH objects which have been derived from their spectra. In the main they have been taken from Bohm (1983).

TABLE I

Parameter	Characteristic Values
Scale size (AU)	300 - 2000
Electron temperature (K)	7500 - 12000
Electron density (cm^{-3})	$2 \times 10^3 - 6 \times 10^4$
Fractional ionization	0.07 - 0.8
Mass (earth masses)	~ 10
Filling factor	$2 \times 10^{-3} - 7 \times 10^{-2}$
Luminosity (1200-11000Å;$L_\odot$)	0.1 - 1.4

The low ionization fraction immediately rules out photoionization as the source of excitation; the low filling factor is consistent with emission from a relatively thin cooling region behind a shock wave. Shock wave excitation is also indicated by molecular hydrogen line ratios where observed.

The association of HH objects and large scale molecular flows (e.g. Edwards and Snell, 1983, 1984) suggests that whatever powers these flows also may be responsible for the formation of HH objects. Infra-red data has shown that stars (or proto-stars) are the culprits. It also seems beyond doubt that some manifestation of stellar mass loss is the agency of energy or momentum transfer.

It is very important to establish the source of excitation for a given HH object or group of objects, not least because its determination can influence the choice of preferred formation mechanism. Cantó (1985) notes that there can be considerable doubt about the identification, as, for example, in the case of HH12, where three different identification criteria lead to three different excitation sources. HH1 and HH2 have provided a classic example where the obvious exciting candidate, the CS star, has turned out to be an innocent bystander (Pravdo et al, 1985).

The wide range of ionization state noted above implies a wide range of shock velocities within a given object. This can be caused by a mixture of shocks of different strengths and/or by the presence of curved shocks (Hartmann and Raymond, 1984). There is also evidence that some shocks may be very young (Dopita et al, 1982).

The radial and tangential velocities of HH objects can be large, as would be expected for a shock origin. The upper limits of the velocities are about 300 km s^{-1} from proper motion studies, and, for the case of the Orion HH objects, 450 km s^{-1} from line widths. The main kinematic features are discussed by Cantó (1985).

3.   WIND INTERACTIONS AND THE FORMATION OF HH OBJECTS

3.1   General Remarks

The impact of a hypersonic stellar wind (velocity V_*, mass loss rate $\dot{M}_*$

on surrounding gas (density n_0) sets up a two shock flow pattern in which an outer shock accelerates ambient gas and an inner shock decelerates the wind. The resultant dynamics is determined by the ratio of the cooling time in the shocked wind to the dynamical timescale. This ratio is greater than one if $V_* > V_c \equiv 250 (n_3 \dot{M}_6)^{1/9}$ km s^{-1} (Dyson, 1984), where $\dot{M}_6 \equiv \dot{M}_*$ $/10^{-6}$ $M_\odot$ yr^{-1} and $n_3 \equiv n_0/10^3$ cm^{-3}. The outer shock is then driven by the pressure of the shocked wind (Case A). If $V_* < V_c$, the shocked wind gas radiates well and the swept-up gas is accelerated by the wind momentum (Case B). This criterion assumes that there is no mixing of cool gas into the shocked wind gas.

Case A: the outer shock velocity $V_0 \sim (\dot{M}_* V_*^2/n_0)^{1/5} t^{-2/5}$ and the radius $R_0 \sim (\dot{M}_* V_*^2/n_0)^{1/5} t^{3/5}$. Cooling takes place behind the outer shock only, and the total luminosity per unit area of shock is $L_0 \simeq n_0 V_0^3$ $n_0^{2/5} t^{-6/5}$. Localized HH objects in principle could be identified with post-shock cooling regions as the outer shock encounters higher-than-average density condensations in the ambient gas. The luminosity of an HH object formed in this way would be $L_{HH} \simeq n_0 V_0^3 R_0^2 \Omega \simeq \dot{E}_* \Omega t^{-6/5}$ where Ω is the solid angle subtended at the star by the HH object, and $\dot{E}_*$ is the wind mechanical luminosity. The HH luminosity decays with time.

Case B: the outer shock velocity $V_0 \sim (\dot{M}_* V_*/n_0)^{1/4} t^{-\frac{1}{2}}$ and the radius $R_0 \sim (\dot{M}_* V_*/n_0)^{1/4} t^{\frac{1}{2}}$. Radiation is now produced behind both shocks and the ratio of the areal luminosities is $L_I/L_0 \simeq (V_*-V_0)/V_0 \simeq V_*/V_0$. The inner shock luminosity dominates. Localized HH objects again can be produced by high density concentrations of ambient gas and their luminosities would be time indepedent if they are so dense that the local V_0 is very low. As noted by Cantó (1979), the outer shock is not necessary in this interaction. It could have degenerated into a sound wave or the flow have reached pressure equilibrium with its surroundings. The HH luminosity would be $L_{HH} \simeq n_w V_*^3 r^2 \Omega$, where the inner shock is located distance r from the star and n_w is the wind density $(\equiv \dot{M}_*/4\pi r^2 V_*)$ at r. Obviously, again $L_{HH} \simeq \dot{E}_* \Omega$.

3.2   The Schwartz-Dopita Model

Schwartz and Dopita (1980) advanced essentially the Case B interaction above. Figure 1 sketches their model. As previously discussed, the bow-shock ($\equiv$ the inner shock) luminosity dominates the total luminosity, however, emission from behind the slow shock driven into the condensation (e.g. molecular or low ionization line) could have observable consequences. The post-shock temperature $T_s \sim \cos^2\psi$ (Fig. 1) and excitation is thus highest in the stagnation zone. Roughly speaking, the excitation would decrease with increasing distance from the star. HH43 (Schwartz, Dopita and Cohen, 1985) appears to be an example of this behaviour. This varying excitation is an important feature of this model and of all models where curved shocks are formed. Hartmann and Raymond (1984) have demonstrated that this mixed excitation emission is one plausible way to produce the wide excitation range demonstrated by optical and UV data.

   A variety of arguments can be stated in the context of this model - but which have much more general validity - to show that the wind must suffer a high degree of collimation.

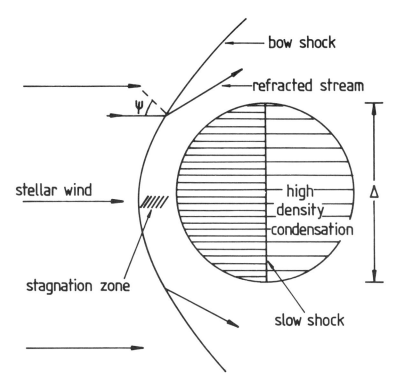

Figure 1.   The flow pattern for the impact of a stellar wind on a dense condensation.

   The cooling time of the post-shock flow must be less than the flow time around the condensation, otherwise the shocked wind gas will expand without cooling. This condition translates into the mass-loss rate requirement $\dot{M}_6 \gg 0.03 V_1{}^5 r_{0.1}{}^2 \Delta_1{}^{-1}$, where $V_1 \equiv V_*/100$ km s^{-1}, $r_{0.1} \equiv r/0.1$ pc and Δ_1 is the scale size Δ of the condensation in units of 1000 AU. Kahn's (1976) cooling approximation has been used. Very high mass loss rates are needed to satisfy this requirement for reasonable V_1.
   Secondly, the maximum post-shock compression is about $(V_*/C_0)^2$, where C_0 is the sound speed ($\simeq 10$ km s^{-1}) in the cooled emitting gas. A characteristic HH density of 10^4 cm^{-3}, say, requires $\dot{M}_6 \simeq 40 r_{0.1}{}^2/V_1$ (for a spherical wind).
   The final argument is well illustrated by HH43. The luminosity of HH43 is about $0.2 L_\odot$, whereas the luminosity of the exciting star (IRS 1) is about $5 L_\odot$ (Schwartz et al, 1985). Using the geometrical parameters given by Schwartz et al (1985), the stellar mechanical luminosity needed is $\dot{E}_* \simeq 40 L_*$ which, for $V_1 \simeq 2$, say, gives an implied mass-loss rate of $\dot{M}_6 \simeq 60$.
   There are other strong observational grounds which imply collimation of the wind, notably the association of bi-polar CO flows and HH objects. Liseau and Sandell (1986) have demonstrated convincingly the real association of these two phenomena.

A particular difficulty with this model is the production of HH objects which have a high proper motion. Hydrodynamic calculations (e.g. Nittmann, Falle and Gaskell, 1982) have shown that the maximum velocity which can be given to the condensation as a whole is about equal to the slow shock velocity $V_s \simeq (n_w/n_c)^2 V_*$, where n_w and n_c are respectively the pre-shock wind and condensation densities. In general, $V_s \ll V_*$ because of the high density contrast.

Many aspects of this model pose interesting and largely unanswered questions. It is known (.e.g Innes, 1985) that shocks of velocity greater than about 150 km s^{-1} are unsteady because of the thermally unstable post-shock cooling. The entire post-shock zone is likely to be unsteady and turbulent. Further, mixing in of cold condensation material via, for example, the process described by Hartquist et al (1986) may significantly affect the emitted spectrum as a result of charge exchange (Hartmann and Raymond, 1984). This mass addition can also strongly affect the dynamics of post-shock flow (Hartquist et al, 1986).

3.3   The Norman-Silk Model

Norman and Silk (1979) suggested that the break-up of a cocoon about a star by the action of a stellar wind would lead to the production of fast moving interstellar bullets which would plough through the interstellar medium driving bow shocks into the ambient gas. The cooling flows behind these shocks would be the HH objects. (The flow pattern in this model is essentially that of the S-D model in a different frame of reference). Three major observational differences between this and the S-D model are immediately apparent. Firstly, the bulk of the emission should occur at roughly the bullet speed and high proper motion HH objects are automatically produced (unless the object moves predominantly parallel to the line of sight). Secondly, the excitation sense is opposite to that of the S-D model; the highest excitation should be seen furthest away from the exciting source (e.g. HH1 and 2). Thirdly, shocked molecular emission could arise behind the more oblique parts of the shock if the interstellar gas contains molecules and could envelop the optically visible HH object. The remarks above regarding unanswered questions which can be addressed to the S-D model are equally applicable to the Norman-Silk model.

Cantó and Rodriguez (1986) have presented evidence in favour of this model, at least with regard to HH2. They find that the measured electron density in the components of this object fit the relationship $n_e \sim V_T$, where V_T is the total component velocity (radial velocity + proper motion velocity). This is most simply explained in terms of the motion of a shock of velocity V_T into a medium containing a magnetic field H_0 strong enough to dominate the pressure in the post-shock cooled gas. If this is the case, $H_0^2 \sim V_T^2$ and, for a 1-D field compression, $H \sim n_e$, thus giving the observed correlation.

There are serious difficulties with the formation mechanism for the bullets as originally proposed (see Section 3.2). In an attempt to circumvent the problems, Tenorio-Tagle and Rozyczka (1984) advanced a mechanism which depends upon the focussing of large scale wind or explosion driven shocks by obstacles in their path. A converging conical shock is produced which can lead to the formation of bullets provided that gas

shocked by the conical shock cools fast enough. An attractive feature of
this model is that, in principle, the bullets can outstrip the main shock,
and about 50% of HH objects seem to lie outside the boundaries of the associated molecular flows. However, a very serious difficulty with this
model is its critical dependence on the maintenance of strict geometrical
constraints. The converging shock must be conical and completely uniform.
It is very hard to see how these constraints can be satisfied in what is
undoubtedly an extremely irregular ambient medium.

   (The jet 'working surface' model (Section 4.2) is an extension of
this, but instead of bullets hurled by a one-off impulse, the bullets
have continuous momentum transfer to them).

3.4   The Cantó Model

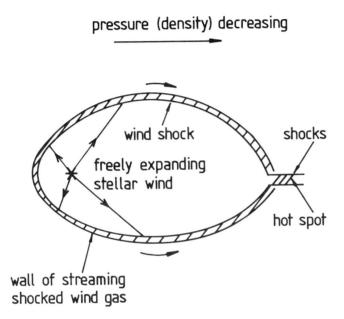

Figure 2.   The excavation of a cavity in an interstellar cloud by a stellar wind which cools on shocking.

Cantó (1980) recognised the severe energy problem associated with HH objects and suggested that if some means of focussing the winds could be
arranged, the difficulty could be removed. His suggestion was to use the
focussing properties of density gradients in the gas around the wind source (Fig. 2). Provided that the shocked stellar wind gas cools well, a
stationary state can be realised in which shocked wind gas is in pressure
equilibrium with the ambient gas. An ovoid cavity whose walls are defined by standing shocks in the stellar wind is excavated in the surrounding
gas. Because the ambient gas has a non-uniform pressure, the shocks are
oblique, the wind streamlines refract across them and the shocked gas
flows around the cavity walls. The flowing gas stream can converge to a

focus. Cantó (1980) suggests that HH objects can be identified either
with bright patches on the walls, or, most efficiently, with emission at
the focal point. The cavity shape is determined by the choice of density
distribution. A symmetric distribution leads to a two-lobe cavity. It
is very tempting to link this morphology to that of the bi-polar molecular
flows. Cantó (1985) has advanced a possible way of doing this. The flowing gas streams are supposed linked to the surrounding molecular gas by
viscous coupling. However, the physical details of the coupling mechanism remain to be elucidated.

The structure of the focal point depends critically on the obliquity
of the shocks through which the gas flows into this point. If these
shocks are more or less normal to the flow, a stationary HH object would
be produced which has no proper motion but which has a line width comparable to the velocity of the colliding streams. Very oblique shocks could
produce a similar (though presumably rather lower excitation) object,
again with no proper motion unless some means of re-exciting the cooled
gas occurs. The cooled gas could, given the right geometry (e.g. Tenorio-
Tagle and Rozyczka, 1985), take the form of a jet which could give rise
to HH objects as discussed in Section 4.1, with or without proper motions.
Cantó (1985) has also hypothesised that that gas injection into the focus
may be in the form of clumps which could drive bow-shocks ahead of themselves into surrounding gas and produce HH objects in the way described
by Norman and Silk (1979).

Although this model has the great virtue of efficiency, there are
some difficulties with it. Firstly a static configuration is set up in
a time-scale about equal to that for changes in the external density distribution to occur. Secondly, and perhaps most importantly, the external
density distribution must be extremely smooth.

3.5 The Königl Model

Königl (1982) considered the other extreme case of a wind blowing into an
inhomogeneous distribution, but where the wind does not cool after shocking. Here, the shocked wind expands and forms a De Laval nozzle which
points down the density gradient. HH objects are supposed to result from
the acceleration of clumps of material produced, for example, by the de-
tachment of portions of the wall. The collimation of the flow again produces some increased efficiency over the spherically symmetric case.
Problems with this model include the difficulty in acceleration of clumps
by gas streams, the necessity of having a smooth external density distribution, and finally, there may be stability problems with the subsonic
section of the nozzle.

4.   JET INTERACTIONS

4.1 General Remarks

Mundt and Fried's (1984) startling discovery of jet-like structures associated with T-Tauri stars has generated a new cottage industry for HH
production (see Cantó (1986) for a dissenting view). A review of the

jet properties – at least as far as is presently surmised – is given by Mundt (1985). We start off here with the basic premise that, somehow, stars produce high Mach number jets which are collimated at least down to a distance of about 1000 AU from the stars, and discuss general ways in which the interaction of jets with their surroundings can give rise to emission features which might be identified as HH objects.

Wilson and Falle (1985) have described how steady jets propagating into non-uniform surroundings set up internal shock structures. The jet tries to come into pressure equilibrium with the ambient gas, but cannot do so if $L_p < L_s$, where L_p is the length scale for pressure variations in the surrounding medium and L_s is the distance moved by the jet fluid in the internal sound crossing time in the jet. $L_s \simeq V_j R_j / C_j \equiv M_j R_j$, where V_j and C_j are respectively the jet velocity and internal sound speed and M_j is the internal Mach number. Shocks are set up if $L_p < M_j R_j$, and if the sense of adjustment to the pressure variation is to decrease the opening angle of the jet (or if it goes through a maximum). High Mach number jets are more susceptible to shock formation than low Mach number jets. In principle, this internal shock structure can contain oblique shocks and normal shocks (Mach discs). As a general rule, the shock obliquity increases and the Mach disc size decreases with increasing jet Mach number. This mixture of shocks should produce a wide range of excitation. Falle, Wilson and Innes (this meeting) have made the first attempt to match this type of structure to chains of HH objects, specifically to HH7-11.

In the steady case, HH objects which result from cooling behind internal shocks in jets cannot have high proper motions. Unsteady jets can also have internal shocks which can be set up in a variety of ways (Norman, Smarr and Winkler, 1984), and in this situation the shock pattern will move, perhaps then giving rise to proper motions.

Supersonic jets can entrain material from their confining surroundings (e.g. De Young, 1986). If internal shocks are present in the jet, this gas could be excited into emission by the hot jet material with similar spectral consequences to the mixing process suggested for HH2 (Hartmann and Raymond, 1984). An intriguing possibility is that this mixing process could lead to a supersonic turbulent boundary layer if, during the mixing process, the local cooling time becomes less than the sound crossing time for the mixing zone. Shock-shock collisions could dissipate kinetic energy ultimately leading to the relatively show collision of streams of dense gas and thus favour low excitation emission (Kahn, private communication). There is some evidence of boundary layer phenomena occurring in the HH46-47 system (Section 5.2).

4.2 The 'Working Surface' Model

At the head of the jet, the 'working surface', shocks occur in both the jet gas and the ambient gas. The structure is shown in Figure 3 (adapted from Smith et al, 1984). Dyson (1984) and Mundt (1985) independently proposed that HH objects could be produced in gas cooling behind either of these shocks. HH objects would trace the path of the working surface as the jet bores through the interstellar gas. The velocity of the working surface, V_s, is determined by momentum balance at the jet head and is

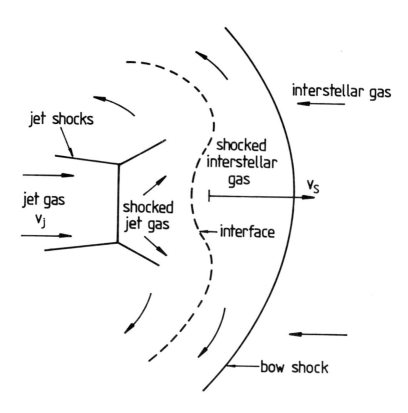

Figure 3. Schematic drawing of the working surface of a jet as it bores through ambient gas.

approximately (Dyson, 1984) $V_s \simeq (11/r_{0.1}\theta)(\dot{M}_6 V_j/n_3)^{\frac{1}{2}}$, where θ is the jet opening angle (in $°$) and V_s and V_j are expressed in units of 100 km s^{-1}. As a suitable example, $V_j = 3$, $n_3 = r_{0.1} = 1$, $\dot{M}_6 = 0.1$, $\theta = 10°$ gives $V_s \simeq 0.6$ and $V_j - V_s \simeq 2.4$. Cooling shocked ambient gas will produce a much lower excitation emission spectrum than cooling shocked jet gas. The wide range of excitation conditions observed in some HH objects can be produced in this way. The spatial distribution of the emission will be complex. Roughly, for emission behind either shock, the excitation will be highest furthest away from the star. This will be observed if emission from either shock dominates. However, the lower excitation shock is furthest from the star. The spatial distribution of excitation will not be so simple if emission comes from behind both shocks. It is likely that the emitting region will be clumpy because of thermal instab- ilities and/or Rayleigh-Taylor instabilities (cf. Allen and Hughes, 1983).

Provided, of course, that the angle of the jet to the line of sight is not too small, HH objects produced in this way will automatically pos- sess proper motions. Since $V_j > V_s$ always, very high proper motions (>300 km s^{-1}, say) require very high jet speeds. This requirement, to- gether with other evidence (e.g. the very high 450 km s^{-1} velocities

measured for the Orion HH objects - Section 5.1), suggests that at least
some HH phenomena involve extremely high wind or jet velocities, maybe as
high as 1000 km s^{-1}.

Reipurth et al (1986) have drawn together many of the ideas of Sections 4.1 and 4.2 to model the HH34 system. They argue that HH34 itself
is produced by the working surface. The short bright jet near the proposed exciting star could be produced by internal shocks in a jet confined
by a dense gas cloud around the star. The jet may originate on a stellar
or circumstellar scale. Alternatively they suggest that the jet is produced at the focal point of a flow collimated as in Cantó's (1980) model
(Tenorio-Tagle and Rozyczka, 1985).

5.   TWO PARTICULAR CASES

In this section we briefly discuss two associations of HH objects, the
Orion HH objects and the HH46-47 system, in the light of the more general
discussion above.

5.1   The Orion HH Objects

Axon and Taylor (1984) discovered nine high velocity condensations on the
front surface of OMC1 which had rather similar spectral characteristics
to HH objects. The investigation of the kinematics of these objects was
substantially extended by Taylor et al (1986) - henceforth TDAH. Very
high blue shifted line wings (up to 450 km s^{-1} from line centre) were observed in the [OI] 6300 Å lines. A very significant feature of this data
is the invariable accompaniment of these extended line wings (the HVC) by
narrow enhanced [OI] emission (the ZVC) at the systemic nebular [OI] velocity.

This latter feature, together with the extended spatial distribution
of the HH objects places severe constraints on possible models for their
production. It is, for example, hard to see how the Cantó model can produce several focal points. The bulk of the emission on the Silk-Norman
model should be produced at the bullet velocity and not at the systemic
nebular velocity. TDAH have discussed the relationship of the objects to
current models in some detail.

If the HH objects are produced by the cooling of a wind impacting on
dense condensations of ambient gas (cf. the Schwartz-Dopita model), arguments on the cooling time demand that the wind be collimated into a jet.
The necessity that the jet produce isolated HH objects simultaneously
visible over an extensive region of the sky led TDAH to propose a precessing jet model. HH objects are produced by the cooling of jet gas as it
shocks against isolated very dense condensations of ambient gas. This
precession might indicate that the likely excitation source, IRS2, is a
binary system. Interestingly, Lightfoot and Glencross (1986) have proposed a model for the HH7-11 system which also involves a precessing jet.

Although the HH objects seen in Orion emit strongly in [OI], this in
itself does not necessarily mean that they are shock excited. Strong [OI]
emission can be produced in the ionization front separating an HII region
from an HI region. In view of this, TDAH proposed an alternative model

for the Orion objects which utilizes a wind which needs some degree of
collimation, but must be spatially extended enough to power the HH objects
simultaneously. In this model, the stellar wind impacts on the rear (neutral) faces of dense intrusions in the ionization front which separates
OMC1 from the ionized Orion nebula. The ionization fronts on the faces
of these condensations illuminated by the exciting stars of M42 are the
sites of the ZVC. The stellar wind detaches small clouds of material from
the dense intrusions and accelerates them out of the 'shadows' of the intrusions into the stellar UV radiation field. Provided the small clouds
are optically thick in the Lyman continuum, they too have surface ionization fronts which emit in [OI]. This latter requirement essentially fixes the wind mass-flow rate. The fast moving clouds must be very small
and a large assemblage of such clouds (resembling an aerosol spray) must
be present. There are some serious unanswered questions involved with
this latter mechanism, for example relating to cloud acceleration and
survival in the hot shocked wind which, in this model, does not cool on
shocking.

TDAH note that objects formed by either model might occur in other
molecular clouds/HII region interface regions.

5.2   The HH46-47 System

Meaburn and Dyson (1986) have presented recent observational data on the
Hα and [SII] 6716, 31 Å line profiles along the emission line filament
HH47B which connects HH46 and HH47A. A remarkable result is that the Hα
profile across HH47B is broad ($\sim$100 km s^{-1}), and that the associated [SII]
profile shows a distinct splitting over the same velocity range. It appears that the emitting volume contains Hα emitting gas throughout, but
[SII] emitting gas at the boundary only.

Meaburn and Dyson (1986) have proposed a jet interaction model to
describe the system. They interpret HH46 as shocks in the jet nozzle,
HH47A as the working surface of the jet, and HH47B as being produced by
internal shocks in the jet. For an (admittedly arbitrarily chosen) inclination angle of the jet to the sky of 30°, they derive a jet mass
through-put rate of $\dot{M}_6 \simeq 0.02$ and a jet speed $V_j \simeq 1.8$. The ambient density ahead of the working surface is about 7 cm^{-3}, suggesting that the
jet has reached the outer low density regions of the cloud. They propose
that the [SII] emission results from surface phenomena on the jet, perhaps assoicated with entrainment of mass. The velocity separation of the
[SII] peaks implies that there is considerable deviation of the direction
of the gas velocity at the jet boundary from simple radial motion along
the jet.

As is well-known, HH47A and HH47B have extremely low excitation spectra. In the case of HH47B this could be due to pronounced obliquity of
the internal shocks and/or the mixing-in of neutral material at the jet
boundary. The estimated velocity of the working surface is $V_s \simeq 1.4$
(Meaburn and Dyson, 1986). Hence $V_j - V_s \simeq 0.4$. The low excitation
spectrum of HH47A may be due to the mixing in of partially ionized shocked jet gas with fully ionized shocked ambient gas. This system provides
an excellent example of the proposition that the kinematics cannot be
divorced from the spectral characteristics.

6. DISCUSSION

In spite of the large volume of observational and theoretical work of the last few years, there is no concensus of opinion about the way in which HH objects are formed. This is really not surprising if HH objects represent the cooling regions behind shocks. If stellar mass loss sets up any form of supersonic flow, shocks will inevitably appear somewhere in it. The only criteria which have to be satisfied are that somewhere, the post-shock cooling time is less than the dynamic timescale of the flow and that the shocks are fast enough to cause the necessary excitation. The necessary conclusion of this is that HH objects or systems of objects should be treated on an individual basis and that to look for a universal flow interaction to explain them all is not a profitable procedure.

The areas for future work are extensive. For example the calculation of the spectra of non-steady shocks with and without post-shock mixing of cool or hot gas is clearly necessary. Very little work has been carried out on the boundary layers and cocoons associated with jets as they traverse the ambient gas. The calculation of the structure of very high Mach number jets ($M_j > 10$) is another important area. Increasing evidence that precessing jets may be present also presents interesting possibilities. No mention has been made above of the possible role played by non-continuous stellar mass loss (e.g. the FU Ori phenomenon). The relationship of the bright jets, the HH objects and the large scale molecular flows remains largely a mystery. Perhaps stars have winds and jets simultaneously. A final, and in many respects, most fundamental area for future work, is to understand why stellar jets are there in the first place.

ACKNOWLEDGEMENT

I am grateful to the Scientific Organizing Committee of this Symposium for their invitation to present a discussion of largely unresolved questions to a captive audience.

REFERENCES

Allen, A. J. and Hughes, P. A.: 1983, Mon. Not. R. astr. Soc., 202, 935.
Axon, D. J. and Taylor, K.: 1984, Mon. Not. R. astr. Soc., 207, 241.
Böhm, K. H.: 1983, Rev. Mexicana Astron. Astrof., 7, 55.
Cantó, J.: 1979, Ph.D. Thesis, University of Manchester.
Cantó, J.: 1980, Astron. Astrophys., 86, 327.
Cantó, J.: 1985, in Nearby Molecular Clouds, ed. G. Serra, Lecture Notes
    in Physics, p.237 (Berlin; Springer-Verlag).
Cantó, J.: 1986, in Cosmical Gas Dynamics, ed. F. D. Kahn (Utrecht;
    VNU Science Press).
Cantó, J. and Mendoza, E. E.: 1983, Symposium on Herbig-Haro Objects,
    Rev. Mexicana Astron. Astrof., 7).
Cantó, J. and Rodriguez, L. F.; 1986, Rev. Mexicana Astron. Astrof., 13,
    57.

De Young, D. S.: 1986 (preprint).

Dopita, M. A., Binette, L. and Schwartz, R. D.: 1982, Astrophys. J., 261, 183.

Dyson, J. E.: 1984, Astrophys. Space Sci., 106, 181.

Edwards, S. and Snell, R. L.: 1983, Astrophys. J., 270, 605.

Edwards, S. and Snell, R. L.: 1984, Astrophys. J., 281, 237.

Haro, G.: 1952, Astrophys. J., 115, 572.

Hartmann, L. and Raymond, J. C.: 1984, Astrophys. J., 276, 560.

Hartquist, T. W., Dyson, J. E., Pettini, M. and Smith, L. J.: 1986, Mon. Not. R. astr. Soc., 221, 715.

Herbig, G.: 1951, Astrophys. J., 113, 697.

Innes, D.: 1985, Ph.D. Thesis, University of London.

Kahn, F. D.: 1976, Astron. Astrophys., 50, 145.

Königl, A.: 1982, Astrophys. J., 261, 115.

Lightfoot, J. F. and Glencross, W. M.: 1986, Mon. Not. R. astr. Soc., 221, 993.

Liseau, R. and Sandell, G.: 1986, Astrophys. J., 304, 459.

Meaburn, J. and Dyson, J. E.: 1986, Mon. Not. R. astr. Soc. (submitted).

Mundt, R.: 1985, in Protostars and Planets, eds. J. Black and M. Mathews (Tucson; Univ. of Arizona Press).

Mundt, R. and Fried, J. W.: 1984, Astrophys. J., 274, L83.

Nittmann, J., Falle, S. A. E. G. and Gaskell, P. H.: 1982, Mon. Not. R. astr. Soc., 201, 833.

Norman, M. L., Smarr, L. and Winkler, K. H.: 1984, in Numerical Astrophysics, ed. J. Centrella.

Norman, C. A. and Silk, J.: 1979, Astrophys. J., 228, 197.

Pravdo, S. H., Rodriguez, L. F., Curiel, S., Cantó, J., Torelles, J. M., Becker, R. H. and Sellgren, K.: 1985, Astrophys. J., 293, L35.

Reipurth, B., Bally, J., Graham, J. A., Lane, A. P. and Zealey, W. J.: 1986, Astron. Astrophys., 164, 51.

Schwartz, R. D.: 1983, Ann. Rev. Astron. Astrophys., 21, 209.

Schwartz, R. D. and Dopita, M. A.: 1980, Astrophys. J., 236, 543.

Schwartz, R. D., Dopita, M. A. and Cohen, M.; 1985, Astron. J., 90, 1820.

Smith, M. D., Norman, M. L., Winkler, K. H. and Smarr, L.: 1984, MPA Preprint 150.

Taylor, K., Dyson, J. E., Axon, D. J. and Hughes, S.: 1986, Mon. Not. R. astr. Soc., 221, 155.

Tenorio-Tagle, G. and Rozyczka, M.: 1984, Astron. Astrophys., 137, 276.

Tenorio-Tagle, G. and Rozyczka, M.: 1985, Proc. ESO-IRAM-Onsala Workshop on (Sub) Millimetre Astronomy.

Wilson, M. J. and Falle, S. A. E. G.: 1985, Mon. Not. R. astr. Soc., 216, 971.

TH 28: A NEW BIPOLAR HERBIG-HARO JET*

Joachim Krautter
Landessternwarte
Königstuhl
D-6900 Heidelberg
Germany

A bipolar Herbig-Haro jet system associated with the emission-line object Th 28 has been discovered from direct CCD-imaging, long-slit spectroscopy, and broadband infrared observations. The observations were carried out using the facilities of the European Southern Observatory, La Silla, Chile. A direct CCD-image taken through a narrow band H_α interference filter is shown in Figure 1.

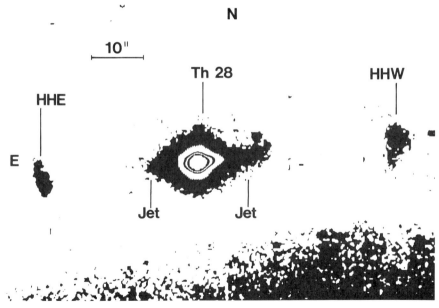

Figure 1. 120m H_α CCD-exposure of the Th 28 Herbig-Haro jet system.

---

* Based on observations collected at the European Southern Observatory, La Silla, Chile

I. Appenzeller and C. Jordan (eds.), Circumstellar Matter, 173–174.

Two oppositely directed jet-like structures (length 0.008
and 0.009 pc) are emanating from a central star-like source,
and two Herbig-Haro objects (Th 28-HHE and Th 28-HHW) are
located on both sides exactly on the axis defined by the bipolar jets at distances of 0.020 and 0.024 pc, respectively.
The system shows a remarkably simple structure. It exhibits
the highest degree of symmetry known yet for any comparable
system, indicating that the interstellar medium around Th 28
has a high degree of homogeneity. A strong variation of the
electron density by about two orders of magnitude along the
jets has been detected. The line-of-sight velocity for both
jets and HH objects is ±40 km s^{-1} only. A proper motion of
0."5/yr found for Th 28-HHE corresponds to a tangential velocity of 316 km s^{-1}.

The observations strongly suggest that Th 28 is a new
case of a bipolar, well-collimated high-velocity outflow
from a stellar object in a star-forming region. The low luminosity (L $\geq$ 0.015 L$_\odot$) of the powering source of Th 28 indicates the presence of a highly opaque disk-like circumstellar cloud orientated perpendicularly to the outflow of
material.

A more complete description of these results is published in Astron. Astrophys. <u>161</u>, 195 (1986).

HH 34: THE BOW SHOCK OF A JET

Th. Bührke and R. Mundt
Max-Planck-Institut für Astronomie
Königstuhl 17
D-6900 Heidelberg
FRG

ABSTRACT. Deep CCD imaging of HH 34 in H_α shows that the HH-object has a bow shock-like structure, of which the wings can be traced over about 1 arcmin ($\hat{=}$ 0.15 pc). A knotty jet is pointing towards the apex of the bow shock structure. Long-slit spectroscopy reveals that 1) the jet has approximately a constant radial velocity and electron density. 2). The spectrum of the jet is of a much lower excitation than that of HH 34. 3) HH 34 has a complex velocity and line excitation structure. The extended bow shock is interpreted by a jet of which the working surface is propagating with high velocity ($\approx$ 200 km/s) through a partially ionized medium.

1. CCD IMAGING

Deep CCD images of the HH 34 region has been obtained at the 2.2 m Telescope on La Silla through an H_α-filter (λ_C= 6565 Å, $\Delta\lambda$ = 67 Å, exposure time 1 h). The image is reproduced in Fig. 1. A 12" long, knotty jet points towards HH 34, which has a bow shock-like structure. The jet emanates from a highly reddened young star, which shows strong H_α-emission (Reipurth et al. 1986). A very similar morphology is observed in the case of HH 1 and HH 39 (Mundt, Brugel and Bührke 1986).

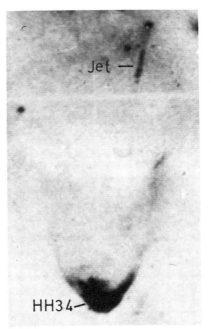

Fig. 1: H_α-frame of the HH 34 region

I. Appenzeller and C. Jordan (eds.), Circumstellar Matter, 175–176.

2. LONG-SLIT SPECTROSCOPY

We examined the HH 34 Jet by long-slit spectroscopy at the 2.2 m Telescope on the Calar Alto, Spain. The spectral resolution was 60 km/s (FWHM). In the knots of the jet the radial velocity (-85 ±5 km/s) was nearly constant as well as the electron density (N_e = 650 ±200 cm^{-2}), derived from the [SII] $\lambda\lambda$6716/6731 Å line ratio. In HH 34 we observed two components of the forbidden lines with a separation of 50 km/s, which converge to a single line near the leading edge of the bow shock. With increasing distance from the star, both components show a decrease in the radial velocity from -140 km/s to -60 km/s. The electron density varies between 60 cm^{-3} and 400 cm^{-3}. The line ratios imply for the jet a much lower excitation (shock velocity $v_S \approx 50$ km/s) than for HH 34 (v_S = 90 - 100 km/s). The different behavior is shown in Fig. 2:

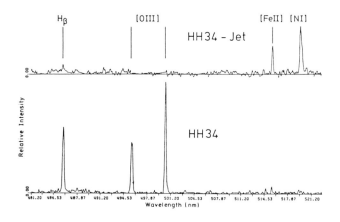

Fig. 2: Line intensities in the jet and HH 34

3. INTERPRATATION

The bow shock structure of HH 34 is explained by the working surface of a jet, which is propagating with a high velocity (200 km/s) through the ambient medium. The model calculations of Raga (1985) show that a partially ionized ambient gas is required in order to observe very extended strongly radiating bow shocks. A comparison of the observed radial velocities in HH 34 with the predicted ones of Raga's model shows significant differences (see Bührke and Mundt 1986 for details). The discrepancies are explained by additional emission from the jet gas shock-excited in or near the working surface, which is not included in the model. The fact that the jet cannot be traced all the way from the star to HH 34 can have various reasons, e.g. sections of free expansion.

REFERENCES

Bührke, Th. and Mundt, R. 1986, in prep.
Mundt, R., Brugel, E.W. and Bührke, Th. 1986, Ap.J., in press
Raga 1985, Dissertation, Univ. of Washington
Reipurth, B. et al. 1986, Astron. Astrophys. 161, 51

OBSERVATIONS OF JETS FROM YOUNG STARS

Edward W. Brugel
Center for Astrophysics and Space Astronomy
University of Colorado, Boulder, CO 80309 (USA)

Reinhard Mundt and Thomas Bührke
Max-Planck Institute for Astronomy
Heidelberg, West Germany

ABSTRACT. Optical jets, and collimated outflows, are now recognized as a common phenomena associated with young stars (Mundt 1985, Strom et al. 1986). Presented here are the results of new CCD imaging and spatially resolved spectroscopy for ten such objects. Using these and previously published data on twenty known jets, we compiled a set of observational criteria describing the phenomena. From this compilation we addressed several physical questions pertaining to the nature of collimated outflows associated with young stars.

1 INTRODUCTION

Morphologically, optical collimated flows are seen as radially projected single or bipolar elongations with an observed length to width ratio of ≈10:1. The radiated flux is dominated by the strong shock induced emission lines seen in HH objects, (e.g. [OI], [NII], [SII] and Hα). There is no optical continuum observed, and it is via the line emissions that jets are detected. Surface brightnesses are in general not uniform, but instead present a patchy or knotty structure. Regions of very weak or no emission along the outflow direction are common.

Spectroscopic data indicate that collimated flows have mean radial velocities of ≈ $100 - 400$ km/s. Both blue and redshifted jets are observed, indicative of their bipolar nature. Four outflows show evidence of radial velocity variations along the jet. The electron densities, from [SII] 6717/6731, range from 400-2000 cm^{-3}. There are cases (DG Tau, HH33/40) in which Ne decreases significantly along the jet.

To summarize, a jet from a young stellar object can be described by the following characteristics: bipolar optical morphology emanating from approxiamtely < 100 au from the driving source; lengths about 3×10^{17} cm; width $< 2 \times 10^{16}$ cm; radial velocity ≈100 km/s; velocity dispersion 80 km/s; Ne about 1000 cm-3; source luminosity 5-10 $L_{\odot}$; jet

I. Appenzeller and C. Jordan (eds.), Circumstellar Matter, 177–178.

and source system associated with a bipolar molecular outflow and/or a cometary reflection nebula.

2 DISCUSSION

This list of observed properties constitutes a useful definition or set of criteria for recognizing the phenomena of jets emanating from young stars. Analysis of these properties has lead us to the following general physical description and interpretation of these objects.

1. The emission line spectrum is created by shocks of $\approx 40 - 100$ km/s.

2. Bright knots probably represent the working surface of shocks.

3. The gaps of weak or no emission are probably regions of a freely expanding jet.

4. The surface brightness irregularities or knotty structures are due to internal shocks.

5. These outflows are efficient means of transporting kinetic energy over large distances, as the energy dissipation in the internal shocks is shown to be extremely small.

6. Physical parameters for a typical jet:

   $v_{jet} = 200 - 400$ km/s

   Mach number $= 10 - 40$

   $N_{jet} = 20\text{-}100$ hydrogen atoms cm^{-3}

   $\rho_{jet}/\rho_{ambient} = 1 - 2$

   $\dot{M}_{jet} = 0.05 - 2 \times 10^{-8} M_\odot/\text{yr}$

   $(L_{kin})_{jet} = .01 - 0.2 L_\odot$

   $2\dot{M}_{jet}/\dot{M}_{radio} = 0.1$

7. Time scales:

   statistical duration of outflow visibility: 2×10^4 years

   dynamical age of outflows : 200 - 3000 years

   age of driving sources : 10^5 years

   age of optical flow $\approx$ to the age of molecular flow.

A full report of this work will appear in the Astrophysical Journal.

REFERENCES

Brugel, E.W., Böhm, K.H. and Mannery, E. 1981, *Ap. J. Sup.* **47**, 117.

Mundt, R. 1985 in *Protostars and Planets II*, eds. D. Black and M. Mathews, University of Arizona Press, p.414.

Strom, K.M., Strom, S.E., Morgan, J.S., Wolff, S.C., and Wenz, M. 1986 (preprint).

DENSE CORES IN THE HH24-26 OUTFLOW REGION

K.M. Menten, C.M. Walmsley, R. Mauersberger
Max-Planck-Institut für Radioastronomie
Auf dem Hügel 69
D-5300 Bonn 1, F.R.G.

We have made observations of the (1,1) and (2,2) inversion lines of ammonia (NH_3) towards the dark cloud region containing the Herbig-Haro (HH) objects 24-27. These transitions are only excited at H_2 densities $>10^4$ cm^{-3}, and thus probe high density gas. From the observed hyperfine splitting one can calculate optical depths. The optical depth ratio can be used to determine the rotational temperature T_{21} which is equal to the kinetic temperature under dark cloud conditions.

With the 40" beam of the Effelsberg 100-m telescope we are able to detect a wealth of fine structure in the NH_3 distribution not seen in earlier lower resolution studies of the region (Matthews and Little 1983, Torrelles et al. 1983). Our map in the (1,1) line (see Fig. 1) shows a general elongation of the NH_3 emission in the N-S direction forming a ridge which connects the northern cluster of Herbig-Haro objects (HH24 A-D), the H_α emission line star no. 140, and HH25 and 26 with their exciting star, the embedded IR-source SSV59 (Strom et al. 1976). No enhancement in the NH_3 emission is seen towards HH27 which lies just outside our eastern map boundary.

Like HH objects, high velocity CO line wings are manifestations of outflow activity and Fig. 1 is an overlay of our NH_3 map and the CO high velocity emission detected by Snell and Edwards (1982) which suggests the existence of two distinct outflow centers in the region.

There is a conspicuous correlation of the ammonia emission peaks with the positions of the HH objects (except HH27) and the embedded IR-sources. There also is a local emission maximum near no. 140, the star which may be driving the northern outflow.

We have found evidence for an interaction of the outflow with the dense NH_3 clumps: A high S/N spectrum taken towards the position of SSV59, between HH25 and 26 shows evidence for broad line wings. Moreover, the linewidth towards the peak close to HH25 is significantly larger than towards neighbouring positions (1.0 km s^{-1} compared to typically 0.6 km s^{-1}). Also, the NH_3 rotational temperature at this position is higher than elsewhere in the cloud (15±1 K compared to 11-12 K).

A possible explanation for the enhanced temperature and broader lines towards HH25 is that these phenomena reflect the existence of an embedded heating source other than SSV59 which lies close to HH25. Evi-

I. Appenzeller and C. Jordan (eds.), Circumstellar Matter, 179–180.

dence for this comes from the fact that two IRAS point sources are found in this area. Also, the 100 μm emission observed by Cohen et al. (1984) is extended in the same direction as our NH₃ and peaks on HH25.

Finally, it should be noted that the geometry of the region around HH25/26 is rather complicated and an assignment of the outflows to one of the IR-sources is by no means straightforward: Although the appearance of the double peaked NH₃ structure around SSV59 and its elongation perpendicular to the blue and red lobes of the southernmost CO outflow detected by Snell and Edwards (1982, 1984) resembles the type of interstellar disk proposed by some authors to explain the bipolarity of molecular outflows, one should keep in mind that the HH-objects which are "normally" aligned with the blue-shifted gas are in this case situated along a line perpendicular to the CO outflow axis.

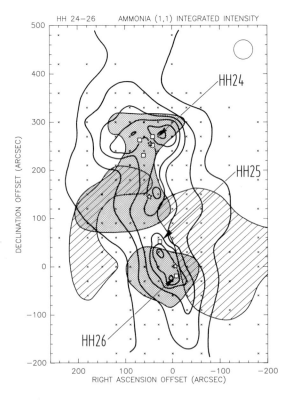

Fig. 1.: Map of the integrated $NH_3(1,1)$ main-beam brightness temperature. Coordinates of the $(0,0)$ position are $\alpha_{1950}=05^h43^m31^s.6$, $\delta_{1950}=-00^\circ15'23''$. Measured positions are marked by crosses and the circle indicates the 40" telescope beam. The lowest contour is 0.7 km s^{-1} and the contour increment 0.8 K km s^{-1}. The extent of red- and blueshifted CO-emission (Snell and Edwards 1982, 1984) is outlined by hatched and dotted areas respectively. The IR-sources SSV 63, H_α 140, and SSV 59 (top to bottom) are denoted by stars and HH-objects by squares.

REFERENCES

Cohen, M., Harvey, P.M., Schwartz, R.D., Wilking, B.A.: 1984, Ap.J. 278, 671
Matthews, N. and Little, L.T.: 1983, M.N.R.A.S. 205, 123
Snell, R.L. and Edwards, S.: 1982, Ap.J. 259, 668
Snell, R.L. and Edwards, S.: 1984, Ap.J. 281, 237
Strom, K.M., Strom, S.E., Vrba, F.J.: 1976, A.J. 81, 308
Torrelles, J.M., Rodriguez, L.F., Canto, J., Carral, P., Marcaide, J., Moran, J., Ho, P.T.P.: 1983, Ap.J. 274, 214

HERBIG-HARO EMISSION IN TWO BIPOLAR REFLECTION NEBULAE

H.J. Staude, Th. Neckel, M. Sarcander, K. Birkle
Max-Planck-Institut für Astronomie
Königstuhl 17
D-6900 Heidelberg
F.R.G.

ABSTRACT. CCD images show that the reflection nebula associated with PV Cep is bipolar. From spectroscopy of this object as well as of the bipolar Boomerang Nebula we find low excited Herbig-Haro emission and indications for collimated high velocity flows along the polar axes of both nebulae. The central star of the Boomerang Nebula is probably double.

1. BOOMERANG NEBULA

This bipolar nebula ($\alpha=12^h 42^m$, $\delta=-54^\circ 15'$) was discovered by Wegner and Glass (1979). At the 2.2 m telescope on La Silla we obtained CCD images and a red longslit CCD spectrum along its polar axis. The CCD rows corresponding to positions A and B in Figure 1 contain stellar spectra which are markedly different. We classify spectrum B as K0III-K2III, while spectrum A is definitely earlier. This is consistent with the UBVJ photometry by Wegner and Glass (1979) and the 12-100 μm fluxes of the coincident IRAS point source: these data can be fitted assuming a double star K0III+A0III, $A_V=2.9$ mag and d=900 pc. The components of the double star are oriented roughly NS, their angular separation is of the order of 1-2 arc sec. The spectra C and D contain a weak scattered continuum and the strong emission lines [OI] 6300, 6364 (also present in B), Hα, [NII] 6548, 6583 and [SII] 6716, 6731. These lines are absent north of D and south of B. They strongly recall the spectrum of the jet in L 1551 (Sarcander et al., 1985). We suggest that the K giant induces this HH emission by a collimated high velocity flow.

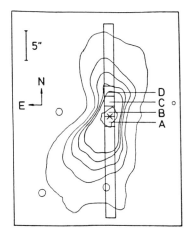

Fig. 1: The inner part of the Boomerang Nebula: red CCD contours with the position of our longslit spectrogram. Cross = central star.

181

I. Appenzeller and C. Jordan (eds.), Circumstellar Matter, 181–182.

2. THE PV CEP NEBULA

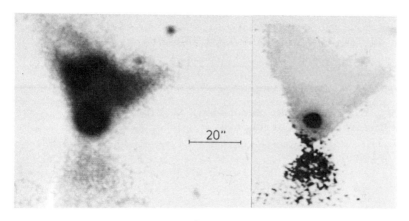

Fig. 2: CCD image of PV Cep in I (left) and the R-I colour index derived from the I and an R image. The reddening strongly increases from north to south.

The bipolar structure of the highly variable nebula associated with PV Cep (Cohen et al., 1981) became evident in R and I CCD pictures taken with the 3.5 m telescope on Calar Alto (Figure 2). Bipolar CO outflow is also present (Levreault, 1984). The southern lobe is deeply embedded in the associated dark cloud. Longslit spectroscopy in the red of PV Cep and the northern lobe revealed:

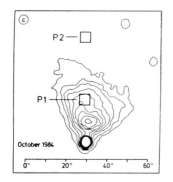

Fig. 3

The <u>star</u> shows chromospheric emission lines, broad Hα emission characteristic of a rotating or expanding shell, and blueshifted HH emission typical for T Tauri stars with circumstellar disks (Appenzeller et al., 1984). <u>The northern lobe</u> is a pure reflection nebula, with exception of positions P1 and P2 (Figure 3): here a blueshifted HH spectrum is emitted (v=-225km s^{-1}), again suggesting the presence of a highly collimated flow from the star along the polar axis of the bipolar nebula.

3. REFERENCES

Appenzeller, I., Jankovics, I., Östreicher, R.: 1984, Astron. Astrophys. **141,** 108

Cohen, M., Kuhi, L.V., Harlan, E.A., Spinrad, H.: 1981, Astrophys. J. **245,** 920

Levreault, R.M.: 1984, Astrophys. J. **277,** 634

Sarcander, M., Neckel, Th., Elsässer, H.: 1985, Astron. Astrophys. **288,** , L51

Wegner, G., Glass, I.S.: 1979, MNRAS **188,** 327

SHOCK EXCITED EMISSION KNOTS IN COMETARY REFLECTION NEBULA

Th. Neckel and H.J. Staude
Max-Planck-Institut für Astronomie
Königstuhl 17
D-6900 Heidelberg
FRG

ABSTRACT. Near infrared and Hα CCD images of the newly found cometary reflection nebula GN 20.18.3. led to the discovery of its illuminating star and of a bright Hα knot located on its axis of symmetry. The knot has a Herbig-Haro spectrum indicating the presence of a collimated outflow from the central star. Also in the Bok globule L 810 a red CCD image reveals a cometary reflection nebula surrounding the central star, from which a shock excited filament is emanating. These observations emphasize the common origin of cometary nebulae and collimated high velocity flows.

INTRODUCTION. Staude et al. (this volume) have presented observations of collimated outflows in two bipolar reflection nebulae. On the basis of CCD photographs and longslit spectrograms here we show the existence of collimated high velocity flows in two cometary reflection nebulae.

THE COMETARY NEBULA GN 20.18.3 (α=20h 18$\overset{m}{.}$3, δ=+37° 00') is associated with a small elongated dust cloud. An infrared CCD picture (Figure 1), taken with the 2.2 m telescope on Calar Alto, makes visible the fairly bright star #3 at the apex of the parabolic rim of the nebula. From this configuration and from its strong reddening (R-I ~ 2.6 mag) we argue that the nebula is illuminated by this star. A rough estimate of its distance yields d = 1 kpc. At this distance the R and I magnitudes of an early A star reddened by A_V = 11 mag equal those which we have observed.

The cometary nebula coincides with an IRAS point source. Integrating the IRAS fluxes between 12 and 100 µm and assuming a distance of 1 kpc, we obtain a contribution of 50 L_0 to the bolometric luminosity from this spectral range. This value agrees quite well with the assumption that the nebula is powered by an early A star.

An Hα frame of the nebula is shown in Figure 2. In its brightest part a nearly stellar knot becomes visible; here the ratio between Hα and Gunn r intensities is three times higher than throughout the nebula. From this we conclude that the radiation of the Hα knot is mainly emission.

I. Appenzeller and C. Jordan (eds.), Circumstellar Matter, 183–184.
© *1987 by the IAU.*

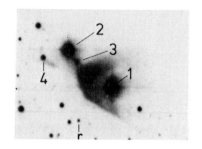

Figure 1. The cometary nebula
GN 20.18.3 in the infrared.
Star 3 illuminates the
nebula, stars 1 and 2 are
foreground objects.

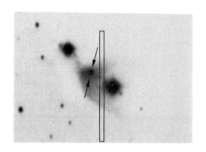

Figure 2. An Hα photograph of
GN 20.18.3 shows the bright Hα
knot near its center (arrows).

A spectrogram of the nebula taken with the slit oriented in N–S
direction about 4 arcsec west of the Hα knot shows besides Hα the
forbidden lines of [OI] λλ6300, 6363 Å. Both Hα and [OI] λ6300 Å
exhibit a component blueshifted by about 8 Å. Thus, it seems likely
that an emission feature at 6724 Å must be identified as blueshifted
[SII] 6731 Å line. The radial velocity of the blueshifted components
turns out to be ~350 km/s. Together with the intensity ratio of the
forbidden lines relative to Hα, which are typical for shock
excitation, we interpret our observations in terms of a collimated
high velocity flow emanating from the central star and colliding with
the ambient material at the position of the Hα knot.

THE CASE OF L 810. Neckel et al. (1985) have shown that in the
centre of the Bok globule L 810 a recently formed late B star is present. A shock excited filament is emanating from this star and points
towards a nearby H$_2$O maser. A new CCD picture (Figure 3), taken under
excellent seeing conditions, reveils interesting new details. First,
it shows that the central star of L 810, #7 in Figure 3, is embedded
in a cometary nebula with star #7 beeing located at its apex, the
usual configuration in cometary nebulae. Further, also the nearby
nebulous patch #8 exhibits a cometary shape. Possibly, this is a second cometary nebula with an additional young star in L 810. In con-

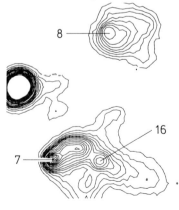

clusion, our observations substantiate the
tight connection between collimated high
velocity flows and bipolar or cometary
morphology.

Figure 3. The central part of the nebulae
embedded in the Bok globule L 810 with the
two cometary reflection nebulae. The
isophotes are constructed from a Gunn r
CCD frame taken with the 2.2 m telescope.

REFERENCES. Neckel, Th. Chini, R., Güsten,
R. and Wink, J.E.: 1985, Astron.Astrophys.
153, 253

Staude, H.J., Neckel, Th., Sarcander, M.,
Birkle, K.: This volume

BOW SHOCK MODELS OF HERBIG-HARO OBJECTS

A. C. Raga, K.-H. Böhm and M. Mateo
Astronomy Department
University of Washington, FM-20
Seattle, WA 98195
U.S.A.

ABSTRACT. It has recently been found that models of a radiating bow shock can explain qualitatively the strange emission line profiles observed in some Herbig-Haro (H-H) objects. It is also possible to compare directly the emission line intensity maps predicted from these models with CCD images of H-H objects. Such a comparison between our models and observations of HH 46/47 is presented, showing that the condensation HH 47A may tentatively be identified with a bow shock formed at the "head" of a jet.

1. INTRODUCTION

There now is relatively strong evidence that the emission line spectra of some H-H objects are formed in the recombination region behind a bow shock. Models of a bow shock formed around a "bullet" moving supersonically with respect to the surrounding medium predict emission line profiles (Raga and Böhm 1985, 1986) which are qualitatively similar to long-slit spectra obtained for HH 1 (Böhm and Solf 1985, see also Choe *et al.* 1985) and HH 32 (Solf *et al.* 1986). Hartmann and Raymond (1984) have also shown that a bow shock model successfully explains the emission line ratios observed in HH 1.

Given this quite convincing spectroscopic agreement between theory and observations, one would expect that bow shock models should also successfully predict the emission line intensity maps obtained from narrow-band CCD imaging of H-H objects. We have calculated such intensity maps from our bow shock models (Raga 1986). In this paper we attempt to compare our predictions with narrow-band images of H-H objects.

2. THE PREDICTED AND OBSERVED INTENSITY MAPS

We have developed bow shock models from which predictions of the emission line spectrum, line profiles, and spatial distribution of the emission can be obtained. If the size assumed for the seeing disk is small relative to the size of the bow shock, the emission line intensity maps predicted from these models show an arc-like shape (Raga 1986). We have also calculated line-ratio maps, which provide another possible observational test for our models.

A quite striking similarity is found between the Hα images of HH 34 obtained by Reipurth *et al.* (1986) and by Bührke and Mundt (1987) and the predictions from our models (Raga 1986). The observations of HH 34 show an intensity distribution which agrees quite well with the bow shock model predictions, but also shows a few low contrast in-

I. Appenzeller and C. Jordan (eds.), Circumstellar Matter, 185–186.

homogenieties ("condensations") which could in principle be due to the presence of other shocks, a time-dependent behaviour of the recombining gas, or the presence of inhomogenieties in the pre-shock gas. In other H-H objects with smaller angular diameter (for example, HH 32) the situation is less favourable, and comparisons of narrow band images of these objects with predictions from our models are inconclusive.

3. THE CASE OF HH 46/47

We have obtained narrow-band CCD images of the H-H object HH 46/47. This object morphologically appears to be jet-like, and proper motion studies (Schwartz *et al.* 1984) indicate that HH 47A might be the "head" of this jet. In figures 1 and 2 we show a comparison between a [S II] λ 6717 image of HH 47A (fig. 1) and the corresponding prediction from a model of a 72 km/s bow shock moving at an angle $\phi = 60°$ with respect to the plane of the sky (fig. 2). Although the predicted and observed images are qualitatively similar, a bow shock identification for HH 47A should be considered only tentative. Line ratio maps and high resolution spectroscopy should provide information needed for a more careful interpretation of this H-H object.

This work has been supported by NSF Grant AST-8519771.

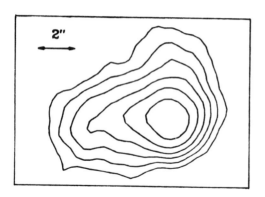 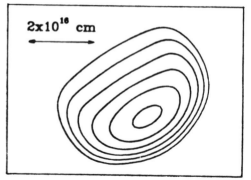

Figure 1 - [S II] λ 6717 image of HH 47A obtained with the CTIO 0.91 m telescope.

Figure 2 - [S II] λ 6717 image predicted from a 72 km/s model of a bow shock moving at an angle $\phi = 60°$ with respect to the plane of the sky.

REFERENCES

Bührke, T., and Mundt, R. 1987, these Proceedings.
Choe, S.-U., Böhm, K. H., and Solf, J. 1985, Ap. J. 288, 338.
Hartmann, L., and Raymond, J. C. 1984, Ap. J. 276, 560.
Raga, A. C. 1986, A. J. (September issue, in press).
Raga, A. C., and Böhm, K. H. 1985, Ap. J. Suppl. 58, 201.
Raga, A. C., and Böhm, K. H. 1986, Ap. J. (September 15 issue, in press).
Reipurth, B., Bally, J., Graham, J. A., Lane, A., and Zealy, W. J. 1986, Astr. Ap. 164, 51.
Schwartz, R., Jones, B. F., and Sirk, M. 1984, A. J. 89, 1735.
Solf, J., Böhm, K. H., and Raga, A. C. 1986, Ap. J. 305, 795.

OBSERVATIONAL TESTS OF THE BOW SHOCK THEORY OF HERBIG-HARO OBJECTS

K.-H. Böhm and A. C. Raga
Astronomy Department, Univ. of Washington, Seattle, WA 98195, U.S.A. and

J. Solf
Max-Planck-Institut f. Astronomie. Königstuhl, 69 Heidelberg, F.R.G.

ABSTRACT. We discuss four different tests of the bow-shock theory of Herbig-Haro objects, emphasizing especially tests based on position-velocity diagrams and on the appearance of "double layer" structures in the spatial maxima of the high- and low-velocity components of the emission lines. Though this latter effect is surprising. it is a fundamental consequence of the bow shock theory.

1. INTRODUCTION

It is now generally accepted that 1. typical Herbig-Haro (HH) objects move radially away from a young star (Herbig and Jones 1981). 2. HH objects often trace highly collimated bipolar outflows (Mundt 1986). 3. the HH emission line spectrum is formed in the recombination region behind a shock wave (Schwartz 1975). Recently there has been increasing evidence that in a number of cases the line emission is actually formed in bow shocks (as expected e.g. in front of an interstellar bullet or of the working surface of a jet). The first observational evidence came from (spatially integrated) flux ratios in the optical and ultraviolet range (Hartmann and Raymond 1984) and the study of spatially resolved high resolution emission line profiles ("position-velocity diagrams". see Böhm and Solf 1985).

2. THE PRESENT STATE OF OBSERVATIONAL TESTS OF THE THEORY.

Our computations (Raga and Böhm 1985, 1986; Raga *et al.* 1986; Raga 1986) show that there are at least four different tests of the bow shock theory available now, namely 1. the comparison of the observed and theoretical (spatially integrated) line fluxes (as done by Hartmann and Raymond 1984), 2. the comparison of observed and predicted postion-velocity diagrams. 3. the study of "double layers" of high and low velocity maxima which are seen, e.g., in the observations of HH 32 (Solf *et al.* 1986) and are also predicted by the theory. 4. the comparison of observed and predicted monochromatic images of HH objects. In this note we shall emphasize tests 2. and 3. . The results of these four different tests make it very probable that. at least in some cases. bow shocks are really responsible for the emission line formation in HH objects. As an illustration of an application of test 2, we show a comparison of the observed spatially resolved line profiles of Hβ and [O III] 5007 in HH 1 with the bow shock predictions for these lines. There are indications that the agreement can be further improved by a more sophisticated selection of parameters.

I. Appenzeller and C. Jordan (eds.), Circumstellar Matter, 187–188.
© *1987 by the IAU.*

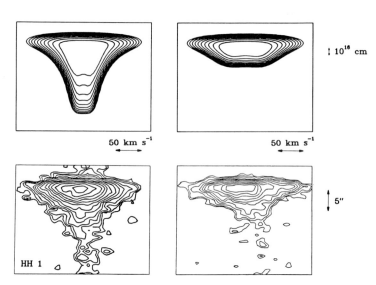

Fig. 1 - Comparison of the observed and predicted position-velocity diagrams for Hα (left) and O III 5007 (right) in HH 1.

3. THE DOUBLE LAYER TEST

In HH 32 one sees typically double-peaked lines (cf. Solf *et al.* 1986). It turns out that in all condensations (A, B, C and D) the spatial maximum of the low velocity peak of the line profile occurs about 0."5 - 1."0 farther away from the central star (AS 353A) than that of the high velocity peak. This was found for all slit orientations. Plotting these results on an image of the HH 32 complex one finds a surprising arrangement of "double layer" structures which seem to be rather enigmatic at first sight. Surprisingly it turns out that just such an effect is predicted by the bow shock theory (Raga *et al.* 1986) if all four condensations form individual bow shocks. Interestingly, the theory also predicts that the spatial separation of the two velocity components will be much larger in the bow shocks in which the bow shock axis forms only a small angle with the line of sight. This makes understandable why the effect is relatively easily detected in HH 32 but not in some other objects. We feel that this test is a rather strong indication of bow shocks in at least some HH objects.

This work has been supported by NSF Grant AST-8519771.

REFERENCES

Böhm, K. H., and Solf, J. 1985, Ap. J. 294, 533.
Hartmann, L., and Raymond, J. C. 1984, Ap. J. 276, 560.
Herbig, G. H., and Jones, B. F. 1981, A. J. 86, 1232.
Mundt, R. 1985, in *Protostars and Planets II* (ed. D. Black a. M. Matthews), p. 414.
Raga, A. C. 1986, A. J. 92, 637.
Raga, A. C., and Böhm, K. H. 1985, Ap. J. Suppl. 58. 201.
Raga, A. C., and Böhm, K. H. 1986, Ap. J. 308, (in press).
Raga, A. C., Böhm, K. H., and Solf, J. 1986, A. J. 92, 119.
Schwartz, R. D. 1975, Ap. J. 195, 631.
Solf, J., Böhm, K. H., and Raga, A. C. 1986, Ap. J. 305, 795.

FILTERED CCD IMAGES OF SOUTHERN HERBIG-HARO OBJECTS

B. Whitmore & D.H.M. Cameron
Dept. of Physics & Astronomy
University College London
Gower Street
London WC1E 6BT

R.F. Warren-Smith
Physics Department,
University of Durham,
South Road
Durham DH1 3LE

1. Introduction

It is currently believed that Herbig-Haro (HH) objects are a consequence of a high-velocity (up to at least 200 km s^{-1}) outflow of material from a young embedded star. These flows can often be detected by deep observations of optical emission lines using CCD cameras.

We obtained deep filtered CCD images of six southern star forming regions containing HH objects with the aim of detecting any ionised outflow which may be present. The filters used were centered on the H$_\alpha$ emission line (λ6563Å) and the combined [O I] lines at $\lambda\lambda$6300 and 6363Å, since they are amongst the strongest observed from HH objects. Infrared observations from the IRAS point source catalogue were used to search for evidence of an embedded protostar powering these HH objects.

2. Discussion

HH 46/47 lie within the Bok Globule ESO 210-6A and are associated with a well collimated optical outflow showing evidence of bipolarity (Dopita, Schwartz & Evans 1982). Our [O I] image shows the jet structure well, since H$_{alpha}$ emisson from the globule edge is not present. The exciting star of this outflow has a bolometric luminosity of 20 L$_{solar}$, based on a distance of 450 pc.

The H$_{alpha}$ image of HH 52, 53, 54A-E and 54X (Chamaeleon T2 association) shows that HH 54 has an extended complex morphology and a faint filament extending towards HH 54X. As can be seen from Figure 1 HH 52/53 are extended in the direction of HH 54, suggesting a possible association. A weak IRAS point source is situated near to the eastern edge of HH 54. This source is confirmed by IRAS AO data and another source is found to the east of HH 52 (see Figure 1).

HH 56 and 57 are associated with the small dark cloud Sandquist 178 in the Norma T1 association. A recently brighten star, identifed as a FU Orionis object, has appeared 20" W of HH 57 (Graham & Frogel 1985). Our H$_{alpha}$ image clearly shows that this new object is associated with HH 57, and has no connection with HH 56.

189

I. Appenzeller and C. Jordan (eds.), Circumstellar Matter, 189–190.
© *1987 by the IAU.*

Two IRAS point sources lie within the bounds of this image: the FU Ori
objects lies at the edge of the error box for one of these, the second
lies to the S of HH 57.   Their luminosities are 125 and 35 L_{solar}
respectively, based on a distance of 700 pc (Graham & Frogel 1985).
IRAS additional observations show that this first source is indeed the
FU Ori object, the likely driving source for HH 57.

The H_{alpha} image of HH 100 shows the extended nature of this
object and also shows a faint filamentary structure directed towards
HH 101.   The H_{alpha} image of HH 101 shows a spiral structure to its
emission knots, enveloped within a faint cone-like emission feature
pointing towards HH 100.   The morphological picture indicates that
these two sources are linked, a view also supported by the proper
motion studies which indicate that HH 100 and 101 are moving away from
the likely exciting star HH 100/IR.

The three images containing HH 48, 49/50 and 51 (Chamaeleon T1
association) all suffer some spurious reflections caused by the
telescope, making it difficult to detect any faint jet-like structure.
HH 48 appears star-like and is detected as a point source by IRAS.
There are no near-by IR sources in the vicinity of HH 49/50 and 51.
HH 55 (Lupus 2 dark cloud) appears star-like and shows no evidence for
any jet activity.   A T-Tauri star RU Lup lies 2 arcmin to the SW and is
the only IRAS source in the vicinity, thus suggesting it is the likely
exciting star for HH 55.

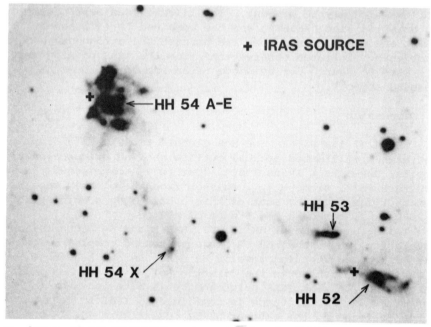

FIGURE 1. H_{α} image of HH 52/53/54:   crosses denote IRAS sources.

3.   References

Doptia, M.A., Schwartz, R.D.   & Evans, I, 1982, Ap.J.Letts., 263, L77.
Graham, J.A.   & Frogel, J.A., 1985, Ap.J., 289, 331.

DETECTION OF COLLIMATED BIPOLAR MASS FLOW IN HH24

J. Solf
Max-Planck-Institut für Astronomie, Heidelberg

ABSTRACT. High-resolution long-slit spectroscopy of HH24 has revealed the presence of collimated bipolar mass flow originating from the "central" infrared source SSV63. The bipolar system consists of the well known jet-like HH24C (mean V_{LSR} -190 km s^{-1} and a newly detected weak component, HH24E ($\sim$+170 km s^{-1}).

1. INTRODUCTION

High-resolution long-slit spectroscopy has been used successfully in studies of the detailed kinematics of Herbig-Haro (HH) objects flowing at supersonic velocities away from a young stellar source. In case of HH1 and HH32 the position-velocity diagrams deduced from the observed emission lines show remarkable similarities to those predicted from the hydrodynamic model of a radiating bow shock (Böhm and Solf 1985; Solf, Böhm and Raga 1986). Compared to these objects the anchor-shaped HH24 exhibits a considerably more complex morphology consisting of at least four subcomponents (Herbig 1974). The infrared source SSV63 (Strom, Strom and Vrba 1976) detected near the geometric center of the nebular complex has been considered to be the star "powering" HH24.

2. OBSERVATIONS AND RESULTS

Using the coudé spectrograph of the 2.2 m telescope on Calar Alto and a two-stage image intensifier tube a number of deep long-slit spectra in the red were obtained from various positions within HH24, one of them with the slit centered near the position of SSV63 at position angle 334°, thereby crossing HH24A and HH24C. The position-velocity diagram of the [SII]6716 line (Fig. 1) deduced from that spectrogram presents a bow-shaped, highly blues-hifted feature (mean V_{LSR} -190 km^{-1}) due to HH24C, a rather compact low-velocity feature ($\sim$+42 km s^{-1}) due to HH24A, and a weak third feature which is highly red-shifted ($\sim$+170 km s^{-1}) and obviously due to a rather unconspicuous nebular condensation between SSV63 and HH24A. This condensation, hereafter referred to as HH24E, was also observed at a different slit angle indicating that it is elongated pointing away from SSV63, similar to HH24C on the opposite

191

I. Appenzeller and C. Jordan (eds.), Circumstellar Matter, 191–192.

side of SSV63. Evidently, the high-velocity features HH24C and HH24E
form the counterparts in a bipolar flow originating from SSV63. The
jet-type morphology of both components indicates a rather high collimation of the flow with velocities above 200 km s^{-1}. Absorbing material
in a disk-like structure around SSV63 may be responsible that the farther HH24E is seen much fainter compared to the nearer HH24C. The inclination of the flow vector with respect to the line of sight and the
true flow velocity are not known. Proper motion measurements of HH24C
are required to answer these questions. The relations of the low-
velocity components HH24A (+42 km s^{-1}) and HH24B (-9 km s^{-1}) to the
bipolar flow system are unclear so far. (HH24D represents reflected
star light, probably from SSV63.) Since HH24A and HH24E are found at
the same position angle with respect to SSV63, it is tempting to inter-
prete HH24A as the fromt edge (or working surface) of the HH24E jet
which has been decelerated by interacting with intervening material.

3. REFERENCES

Böhm, K. H., Solf, J. 1985, Ap. J. **294,** 533,
Herbig, G. H. 1974, Lick Obs. Bull. No. 658,
Solf, J., Böhm, K. H., Raga, A. C. 1986, Ap. J. **305,** 795,
Strom, K. M., Strom, S. E., Vrba, F. J. 1976, A. J. **81,** 308.

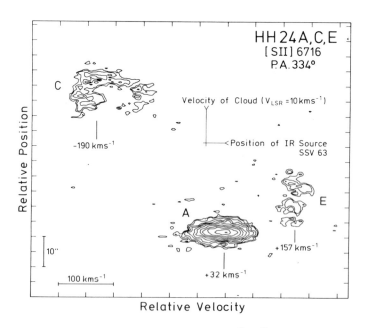

Figure 1. Position-velocity diagram of [SII] 6716 deduced from a
long-slit spectrogram crossing HH24A, C and E. Intensity contours are
spaced by factor of $\sqrt{2}$. Velocities are quoted relative to that of the
parent molecular cloud, presumably identical with that of the "central"
IR Source SSV63 (projected relative position marked).

A JET MODEL OF HERBIG-HARO OBJECTS

M.J. Wilson
Department of Applied Mathematics
The University
Leeds LS2 9JT

S.A.E.G. Falle and D.E. Innes
Max-Planck-Institüt für Kernphysik
Postfach 10 39 80,
D-6900 Heidelberg

ABSTRACT.   We present results of steady jet calculations in which the cooling and compression behind internal shocks leads to optical emission with an intensity pattern similar to the regular well-aligned emission knots characteristic of stellar jets.

There are, at present, known to be a number of well collimated 'jet-like' features associated with young stars or IR sources (Mundt & Fried 1983; Reipurth *et al.* 1986; Mundt *et al.* 1986). Many of these jets show knots of enhanced optical emission with spectra typical of low-excitation shock-heated gas (Mundt *et al.* 1986), a good example being the jet associated with HH-34 (Reipurth *et al.* 1986; Bührke & Mundt 1987).

It is well known (e.g. Prandtl 1952) that shocks are readily excited in a steady supersonic gas jet if it is initially out of pressure balance with its surroundings. Sanders (1983) proposed a sufficient condition for a non-uniform external pressure to produce shocks in a jet, while Falle & Wilson (1985) and Wilson & Falle (1985) derived a necessary and sufficient condition. This work indicates that the chances of shocks being induced in stellar jets by a pressure mismatch to its surroundings are very strong indeed.

Our assumptions are that the jet is axisymmetric, supersonic and steady (this being valid if the time-scale for changes in the external medium is greater than the flow-time down the jet). We can then use a Godunov scheme for steady supersonic flow (Glaz & Wardlaw 1985) to model the flow within the jet. The numerical scheme includes the dynamical effects due to radiative cooling and allows us to calculate the emission from any shock-heated material within the jet.

In this paper we present a calculation representative of our results to date. Shown in Figure 1 are contours of pressure for a jet which experiences an external pressure variation sufficient to excite a series of incident and reflected shock pairs, very much like those found in a perturbed adiabatic jet (e.g. Falle & Wilson 1985). The effect of the cooling is to increase the compression behind the shocks which, as can be seen, leads to a decrease in jet radius and shock strength.

I. Appenzeller and C. Jordan (eds.), Circumstellar Matter, 193–194.

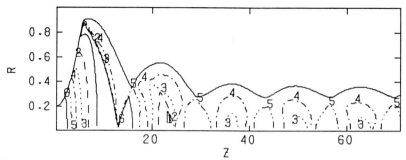

Figure 1. Log pressure contours for a steady jet with radius R and length Z (scaled units). The pressure increases from contour 1 to 6 with contour spacing Log $P = 0.38$.

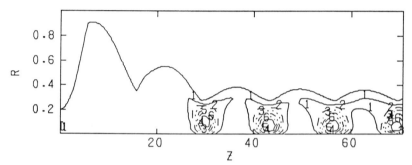

Figure 2. Linear [O I] $\lambda\lambda$ 6302,6365 intensity contours for a line-of-sight perpendicular to the jet axis.

The integrated emission from the jet in the [O I] $\lambda\lambda$ 6302,6365 line is shown in Figure 2. One finds that the optical emission from the jet is concentrated in a series of bright knots whose spacing is approximately $2M \times R$, where M is the jet mach number and R is the jet radius. Each knot is associated with the high density region behind a reflected shock. The density enhancement in an element of fluid having passed through an incident/reflected shock pair can be up to a factor of 100 depending upon the cooling length. Using the velocity structure and emissivity of the gas, line profile maps along the jet can be calculated for the optical forbidden lines. We find that those regions with the lowest velocity dispersion correspond to the bright knots.

References

Bührke, T. & Mundt, R, 1987, this volume.
Falle, S.A.E.G. & Wilson, M.J., 1985, *Mon. Not. Roy. Astr. Soc.*, **216**, 79.
Glaz, H.M. & Wardlaw, A.B., 1985, *J. Comput. Phys.*, **58**, 157.
Mundt, R., Brugel, E.W. & Bührke, T, 1986, preprint.
Mundt, R. & Fried, J.W., 1983, *Astrophys. J. Letters*, **274**, L83.
Prandtl, L., 1952, *Fluid Dynamics*, (Blackie: London).
Reipurth, B., Bally, J., Graham, J.A., Lane, A. & Zealey, W.J., 1986, *Astron. & Astrophys.*, in press.
Sanders, R.H., 1983, *Astrophys. J.*, **266**, 73.
Wilson, M.J. & Falle, S.A.E.G., 1985, *Mon. Not. Roy. Astr. Soc.*, **216**, 971.

CIRCUMSTELLAR SHELLS AND ENVELOPES

IRAS RESULTS ON CIRCUMSTELLAR SHELLS

H.J. Habing
Sterrewacht
Leiden.

ABSTRACT   The IRAS Point Source Catalog contains tens of thousands of stars with circumstellar shells. Most are cool stars surrounded by dust-rich envelopes, but many B and A stars are present with ionized circumstellar envelopes. IRAS and its main products are briefly discussed.

1. INTRODUCTION

In early 1983 a satellite, IRAS, was launched to make a survey of the sky in wavelength bands centered at 12, 25, 60 and 100μm. IRAS was the first astronomical infrared satellite and, probably, the first satellite to be cryogenically cooled. Its main instruments performed better than expected, and there were hardly any failures until on November 22, 1983 the last drops of coolant (super fluid Helium) were used and the instruments stopped functioning. During the 300 days of satellite operations 95% of the sky has been covered twice and 72% for a third time. The data were analysed and the first results were released in November 1984, less than a year after the last observations. One of the two prime "data products" is the Point Source Catalog (IRAS, 1984), containing 245,839 point sources with a reliable detection in at least one of the four wavelength bands. Most point sources have been detected at the two shortest wavelengths, 12 and 25μm, and most of these are stars; for approximately 83,000 objects enough information is available to conclude that the infrared emission is from a circumstellar shell. The IRAS Point Source Catalog may well contain the largest collection of circumstellar shells. Therefore it is highly appropriate that at this Symposium some time is spent on this collection and on the instrument that produced it.

I. Appenzeller and C. Jordan (eds.), Circumstellar Matter, 197–213.

TABLE 1.

IRAS -the InfraRed Astronomical Satellite

Project of the United States of America, the Netherlands and the
United Kingdom.

Primary Aim
A complete and reliable all-sky survey in broad bands at 12, 25, 60
and 100µm, plus spectrophotometry between 6 and 23µm of all isolated,
bright sources.

Instrument and Performance
A 60 cm cryogenically cooled telescope. Angular resolution between
0.8×4.5 arcmin ^{2} and 3.0 x 5.0
arcmin^2 (D/λ=50,000 to 6,000). Broadband spectral
resolution $\lambda/\Delta\lambda$=1.7 to 3.2. Spectrophotometry with $\lambda/\Delta\lambda$ varying from
14 to 35.

Operations
Three hundred days in 1983. Observations made at time t were repeated
at t+2 weeks and (for 72% of the sky) at t+6 months. Ninety six
percent of the sky has been observed at least twice.

Results
First generation "data products" released in November 1984. A second
generation is being prepared in the U.S.A. and in the Netherlands.
The Point Source Catalog has 250,000 entries, of which 2/3 are stars.
Approximately 81,000 stars have circumstellar shells.
Positions are (practically) always more accurate than 10 arcsec; the
photometric accuracy is expected to be better than 10%. The goals in
completeness and reliability have been met.

Some useful references
-IRAS, the Explanatory Supplement, 1984, eds. C. Beichman, G.
Neugebauer,
H.J. Habing, P.E. Clegg, T. Chester, U.S. Government Printing
Office (in press)
-"Light on dark matter", proceedings 1st IRAS conference,ed. F.P.
Israël
(Reidel publ., 1986)
-"IRAS Far-Infrared colours of normal stars"
L.B.F.M. Waters, J. Coté, H.H. Aumann, 1986, Astron. Astrophys. (in
press)
-"The brightest high-latitude 12 micron IRAS sources"
P. Hacking, G. Neugebauer and 11 other authors, Publ, Astron. Soc.
Pac. 97, 616

2. THE SATELLITE, ITS MISSION AND ITS MAIN PRODUCTS

In table 1 I give a short summary of the most important aspects of the IRAS project and what it produced. The project was planned and carried out as a survey mission: its goal was an index of the infrared sky and this was envisioned to be a catalogue of point sources. Consequently a strong emphasis was placed on this catalogue during the years of preparation and during the mission and the phase of data reduction. The goals were high: the catalogue should be as complete as possible (no sources missing) and reliable (all sources included should be real). Such criteria are somewhat contradictory, but nevertheless the resulting catalogue fulfills the two criteria quite well, except at low galactic latitudes (below about 2°; near the galactic centre below 4°); there the source density is very high and the detectors could no longer separate individual objects (confusion). For a thorough discussion of completeness, reliability and confusion and how much has been achieved I refer to the Explanatory Supplement (IRAS, 1984).

The IRAS Point Source Catalog is made available on magnetic tape or on microfiche. A printed version is foreseen. Mandatory for a good use is the Explanatory Supplement (IRAS, 1984), which describes the instrument, the observations and the data reduction.

Each point source is characterized by a position and by measured values (or upper limits) for the flux densities in the four wavelength bands. Very good accuracy (usually better than 10 arcsec) was obtained for the positions. Photometric accuracy was more difficult to achieve and, worse, the photometry can not be tested by an independent comparison: for most objects only IRAS data are available, except at 12μm. Nevertheless, the photometry is thought to be reliable to about 10%. A revised version of the Point Source Catalog with corrections will appear in late 1986. Also in late 1986 a new catalogue will appear that contains point sources found in small fields from the so called "Additional Observations" program. These fields were scanned by IRAS more often than during the regular survey and consequently much fainter point sources are recognized. The new catalogue will be of special interest for extragalactic studies but occasionally it may prove to be useful for a stellar astronomer.

The Point Source Catalog is reliable and complete and it contains a large number of new sources. Nevertheless the data are rather primitive for the demands of a modern astronomer: angular and spectral resolution are poor (see table 1). Identification of an IRAS source at other wavelengths is often necessary before any meaningful interpretation can be started. Luckily, the accuracy of the positions usually makes such an identification possible and unique.

To all who had been involved in the IRAS preparations it came as a happy surprise that the detectors worked so reliably and were so sensitive. It proved to be possible to construct meaningful maps of the brightness distribution - in jargon: "The DC behaviour was excellent". A first generation of sky maps has been made, a second generation is being prepared and the expectation is that here there will be a large improvement in quality. This second generation is

prepared in the U.S.A. (at IPAC, the IRAS Centre at the California
Institute of Technology) and in the Netherlands (at the Space
Research Laboratory of the University of Groningen). The two efforts
are independent, but the groups remain in contact. Results are to be
expected toward the Summmer of 1987. For the subject of this
Symposium sky flux maps are of less interest and I will not consider
them further.

3. THE CONTENTS OF THE IRAS POINT SOURCE CATALOG

The catalogue contains in total 245,839 point sources. The basic
data per source are position and flux density (or an upper limit) in
each of the four wavelength bands, but additional information is also
available (e.g. associations with objects in other catalogues; the
presence of confusing sources nearby). Some of the additional
information is in a separate file, the Working Survey Data Base (e.g.
flux densities from individual detections).

What are these sources? Because the positions are quite accurate
(see table 1) identification in other wavelength bands is often
possible. An example is the Edinburgh program for identification of
IRAS point sources near the South Galactic Pole, a first report of
which was given by Wolstencroft et al. (1986). (The first part of
the definitive report has appeared in Mon. Not. Roy. Astr. 223, 279).
Short wavelength and long wavelength sources appear to be very
different. Of the 153 sources with a detection at least at 12μm 149
(97%) are stars. Of the 161 sources detected at least at 60μm 145
(90%) are galaxies. At lower galactic latitudes the fraction of stars
will increase. The conclusion is that identifiable sources with a
12μm detection are by and large stellar. Are the 12μm sources all
identifiable? The brightest ones are as shown by Hacking et al.
(1985). They made a study of sources brighter than [12μ]=0.0[*], which
corresponds to 28.0 Jy; only the 269 sources at |b|>30° were
considered (and the Large Magellanic Cloud was avoided). Two of the
objects are galaxies, the others are all stars; only a few, rather
southernly located sources had no previous identifications.
Seventyseven percent of the sources are M-type stars, 9% are C-stars
and 16% are K-giants. An interesting point, missed by Hacking et al.,
is that practically all sources identified with a variable star have
a colour excess - and vice-versa. I will come back to this point in
section 6.

Another way to study the nature of the point sources is to
consider only the IRAS fluxes. Seven percent of the sources has been
detected in three bands, so that two colours can be constructed.
Figure 1, originally due to Chester (1986), shows the distribution of
all point sources detected at 12, 25 and 60μm in a colour-colour

---

[*] The square brackets stand for "magnitude". Magnitudes remain useful
for stellar studies although they are not used in the IRAS Catalog.
For their definition and the definition of colours see the appendix.

diagram of R_{21} (=log [f_ν(25μm)/f_ν(12μm)])and
R_{32} (= log [f_ν(60μm)/f_ν(25μm)]) ; f_ν is the flux density as given in
the catalogue.

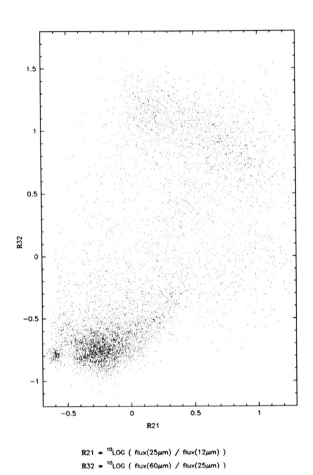

R21 = ^{10}LOG (flux(25μm) / flux(12μm))
R32 = ^{10}LOG (flux(60μm) / flux(25μm))

Figure 1: Colour-colour diagram of IRAS colours for
all objects in the Point Source Catalog with a
high-quality detection at 12, 25 and 60μm.
The flux densities have not been colour-corrected.

There is a sprinkling of points over the whole plane, but there is a
strong concentration at lower left and a weaker one at upper right.
The "swarm" at the lower left consists of stars with and without
circumstellar shells, the other concentration (upper right) consists
of galaxies and of "IR sources of galactic origin" (Chester, 1986).
    First I discuss the swarm of stars. If all stellar photospheres

were black bodies and their temperatures over 2000k, then all stellar
points would be close to R_{21} = -0.64 and R_{32} = -0.76 (the Rayleigh-
Jeans point). Indeed around that point a concentration is present
within the stellar swarm. The large swarm to the right (to the east)
of the Rayleigh-Jeans point (and well isolated from it) consists of
circumstellar dust shells with the 9.7µm band of silicate -thus of
cool, oxygen-rich giants with large mass loss rates. This follows
from many identifications and from inspection of the Low Resolution
Spectra -when available. The gap between the stars with and without
circumstellar shells is real; there are no observational limitations
known that could have produced the gap (Hacking et al.; 1985).
Presumably the presence of the gap means that there is a
discontinuity in the distribution of mass loss rates for late type
giants. I will come back to the point in section 6. Carbon rich
envelopes around C-stars tend to have lower R_{21} values and to lie in
the direction north by northeast from the Rayleigh-Jeans point.
Apparently R_{21} is a sensitive indicator for the presence of the 9.7µm
silicate band: this band goes into absorption and suppresses the 12µm
flux. As a consequence O-rich stars are weaker in the 12µm band than
C-stars.
Undoubtedly there remain many interesting and peculiar objects to be
found in Figure 1 -for example α Lyra (Vega) is at R_{21} = -0.52 R_{32} =
0.01. An interesting area is between R_{21}=0.3 and 1.0, and R_{32}=-0.6
and 0.0: many planetary nebulae are found there and several objects
that may be protoplanetary nebulae (v.d. Veen et al., 1987; Kwok et
al., 1987).

The concentration in the upper part of the diagram contains
practically all the galaxies with 12,25 and 60µm detections in the
Point Source Catalog. Here the two colour temperatures, one
corresponding to R_{21}, the other to R_{32}, are very different,
indicating that the sources contain both hot and cold dust. This type
of spectrum emerges from regions where considerable amounts of matter
are heated by hot stars like in starforming regions. Most of the "IR
components of galactic origin" in this concentration are probably
associated with molecular clouds and star formation processes; they
are almost always at very low galactic latitudes.

Qualitatively it thus appears that in all cases the IR point
sources consist of clouds of dust reemitting stellar photons. There
are two main concentrations in figure 1 because fundamentally two
different situations are involved: 1) evolved stars with small
amounts of mass (relative to the stellar mass) at very close distance
and 2) large clouds of gas and dust heated by embedded or bordering
stars; here the dust is at a somewhat larger distance from the star,
and the star may have ionized part of the cloud.

For only seven percent of the sources IRAS measured three
fluxes, from which two colours can be derived. What about the other
sources? Can useful information be extracted without further
identification? The answer is a partial yes for the 39% sources with
two flux density measurements: by using their distribution in the
sky, or rather, their distribution with respect to the Galaxy, quite
useful information can be extracted. At high declination galaxies

dominate the Point Source Catalogue at the longer wavelengths. Close
to the galactic plane (but just outside it) most 12 and 25μm sources
are circumstellar shell objects and R_{21} is an interesting and useful
piece of information: see section 6.

Even for sources detected in only one wavelength band it is
sometimes possible to derive their nature: a significant number of
point sources is seen only at 100μm; almost all are bright spots in
the interstellar cirrus, as is proven by the correlation between
their position in the sky and the distribution of HI (from 21cm line
surveys).

4. THE ATLAS OF LOW RESOLUTION SPECTRA

The low resolution spectrograph (see section 2) also functioned
without problems. It was a slitless instrument with its dispersion in
the direction of the scanning, so that a point source would cast its
spectrum gradually over the detectors while the telescope was
scanning the sky. This implies that only well isolated point sources
yielded useful spectra. Two overlapping wavelength bands, one from
7.7 to 13.4μm, the other from 11.0 to 22.6μm were scanned
simultaneously. In both cases the resolution increased from 20 at the
short wavelength side to 60 at the long wavelength side. The noise
varies from (typically) 1.5 Jy at the shorter wavelength to 3 Jy at
the longer wavelengths. The atlas of spectra has recently appeared
(IRAS, 1986); it contains 5425 spectra of isolated IRAS point sources
that are bright at 12 and 25μm; five objects are galaxies, the others
are (probably) all of galactic origin.

An analysis of the contents of the atlas has been made by Olnon
(1986). I mainly summarize his results. All the spectra have been
classified automatically by computer (a visual inspection afterwards
led to changes in only a few percent of the cases). The slope of the
spectrum ("red" or "blue") and the strength of spectral features
(9.7μm silicate band; 11.3μm SiC band; ionic lines) were used as
classification criteria; only information present in the spectrum was
used. Altogether there are 9 main classes of objects, each subdivided
in 10 subclasses, plus two extra subclasses. To what physical objects
do these classes correspond? Ninetyfour percent (4735 objects) appear
to be stars and most have circumstellar shells; some 1200 objects
have very little, if any, circumstellar material around them. The
spectra of the shells either contain the silicate bands at 9.7μm (in
emission of in absorption) and at 18.8μm (always in emission), or
they contain the SiC 11.3μm band indicative of a C-rich shell. The C-
spectra are four times less frequent than O-(Silicate) spectra. The
685 remaining galactic objects consist of planetary nebulae and of
compact HII regions (in about equal proportions) plus a small number
of "peculiar stars": proto planetary nebulae, emission line stars. In
Figure 2 a sample of spectra of the most common kinds is shown.

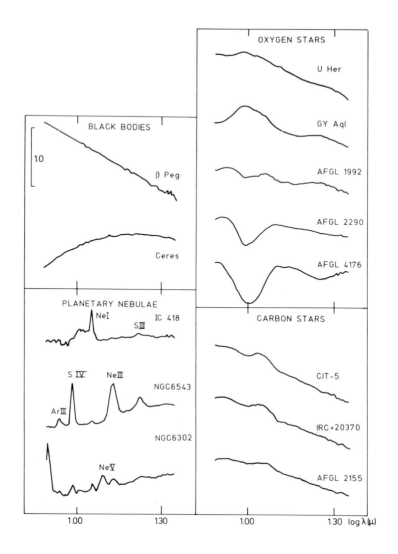

Figure 2: A sample of low resolution spectra from the Atlas. Among the "black bodies" β Peg is an M2II star with a temperature over a few thousand K, and the asteroid Ceres has a temperature around 300K. The blackbody spectra are given as a reference.

5. "NORMAL" STARS

This conference is on circumstellar matter. To make our topic more exclusive I will call stars "normal" if they have no circumstellar matter. IRAS was a poor detector of normal stars, which is not surprising: stellar photospheres are too hot to emit efficiently, even at 12μm, the shortest IRAS wavelength. A simple calculation confirms this: normal stars could be detected only when they are within a limiting distance D_{lim}. This distance depends on the detection limit at 12µm, on the luminosity L and on the effective temperature T_{eff}. For a detection limit of 1 Jy values of D_{lim} are shown in table 2 as a function of L and T_{eff}: Main sequence stars and normal giants (luminosity class III) are detected only when they are closer than 1kpc. The last column indicates the number of stars to be detected. The sum of these is much less than the total number of short wavelength sources in the IRAS catalogue. This confirms the conclusion drawn by Olnon (1986) from his analysis of the contents of the LRS-atlas (see section 4): most of the stellar objects have circumstellar shells, only a small fraction (Olnon estimates 1/4) are "normal".

The last colomn can not be very accurate -for example the number of B stars in the Bright Star Catalogue is much less than 3000. Nevertheless qualitatively it gives the right impression: the number of dwarfs decreases with advancing spectral type, and most bright G and K stars are giants.

TABLE 2.

Limiting distances and expected numbers of normal stars

Spectral type	V-[12µ][1]	V_{lim}[2]	M_v[3]	D_{lim} (parsec)	ρ[4] (parsec)$^{-3}$	expected[5] number
B5 V	-0.47	+3.93	-0.9	92	0.0009	3,000 (?)
A0 V	+0.05	+4.45	+0.7	56	0.001	117
A5 V	+0.53	+4.93	+2.0	39		
F0 V	+0.84	+5.24	+2.8	31	0.003	250
F5 V	+1.20	+5.60	+3.8	23		
G0 V	+1.50	+5.90	+4.6	18	0.004	80
G5 V	+1.84	+6.24	+5.2	16		
K0 V	+2.42	+6.82	+6.0	15	0.009	120
G5 III	+2.18	+6.58	+1.5	104	0.0008	3700
K0 III	+2.48	+6.88	+0.8	164	0.0001	22,000
K5 III	+3.73	+8.13	+0.0	423		

[1] from Waters et al. (1986)
[2] limiting V magnitude for inclusion in the Point Source Catalog, assuming that $[12µ]_{lim}$=+4.40
[3] from Allen, Astrophysical Quantities, 2nd edition, § 99
[4] local density of stars, from Allen, Astrophysical Quantities, 2nd edition, § 117
[5] assuming a spherical volume, except for the K giants, where a cylinder with a height of 700 pc was assumed and a radius of 320 pc

What are the colours of normal stars? An illuminating paper is
by Waters et al. (1986). They consider only IRAS point sources
identified with stars from the Bright Star Catalogue (Hoffleit and
Jaschek, 1982). This is a sensible approach: the PSC is complete at
12μm for magnitudes brighter than +4.4. After elimination of doubtful
cases Waters et al. have a sample of 6013 stars with a good 12μm
detection (and an increasingly smaller number of objects detected in
the other IRAS bands).

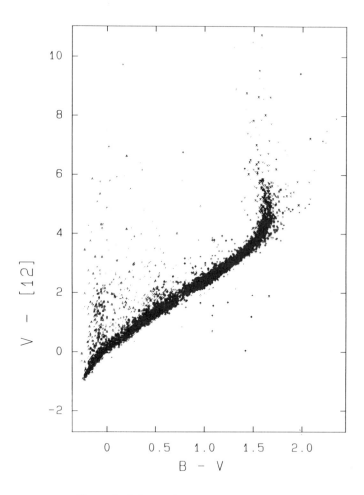

Figure 3: Colour-colour diagram for
objects from the Bright Star Catalogue
identified with objects in the
Point Source Catalog (Waters et al, 1986).

Figure 3 shows a diagram of B-V versus V-[12μ]; the 12μm flux density
is converted into a magnitude using 28.0 Jy as the zero point (see
appendix). Figure 3 shows at a glance which stars are "normal" and

where the deviations occur: a narrow, very well defined band runs in
the middle of the diagram. There is an obvious gap in the band at B-
V=0.75, which is due to the fact that the number of main-sequence
stars drops rapidly for spectral type later than G5. But at B-V=0.8
giant stars begin to fill in - this follows also from the last column
of table 2. That the stars in the narrow band are "normal" stars
follows from a comparison for main sequence O, B and A stars (B-
V<0.8) with colours expected from model atmospheres by Kurucz. The
theoretical curve has the same shape but is systematically lower by
10%. This shift may have been caused by the assumption of LTE in the
model computations.

    Almost all deviating cases in Figure 3 are above the band of
normal stars. The few cases below can probably be accounted for in
terms of duplicity, variability and misidentifications; they are not
further considered by Waters et al.. Being situated above the band
implies an excess flux at 12μm. First consider the excesses for B-
V<0.75. Most excesses occur around B-V=0.0, that is for spectral
types A0 or earlier. Especially among B and A type stars quite a few
show IR excesses: about ½ of the B stars and about ¼ of the A stars
have a V-[12μ] excess of 0.2 or larger. The percentage of stars with
an excess drops if one goes to later spectral types. Only 5% of the K
stars in Figure 3 (practically all giants) have a V-[12μ] excess
larger than 0.2. However, for M stars the situation is completely
reversed: Figure 3 shows a large number of excesses beyond B-V=1.5. I
will discuss this further in the next section.

    It thus appears that infrared excess occurs frequently in M
giants and in (main sequence) stars of spectral type earlier than A8.
Dwarfs and giants with spectral types F, G and K are mostly, but not
always free of an excess.

6. DGE STARS: COO- GIANTS WITH DUSTY ENVELOPES

    As discussed in the previous section only a few percent of the K
stars have an infrared excess, but this is not at all true for M
stars: Waters et al. state that 55% of the M0-M2 stars have a V-[12μ]
excess larger than 0.2, and they do not even discuss stars of later
spectral type (i.e. B-V>1.6). Clearly, if in a given sample 55% of
the stars have an excess, the question rises: what is the standard,
what is normal? Before I attempt to answer I will first try to answer
another question: what produces the excess emission?

    Already in the mid-sixties it became clear that late-type giants
emit more in the infrared than expected from a blackbody
extrapolation from shorter wavelengths. Solid particles in a
circumstellar envelope were thought to be the cause: they will
convert stellar photons very efficiently into infrared photons over a
broad range in infrared wavelengths. In oxygen-rich giants the
particles were recognized as a kind of silicates when Woolf and Ney
(1969) identified the 9.7μm absorption band. Later a band at 11.3μm
in the (less frequently found) carbon-rich giants was identified as
SiC (Treffers and Cohen, 1974). The same bands are found in the IRAS

atlas of low-resolution spectra (see section 4): most stars with
excess emission have the 9.7 band, a few have the 11.3 band. Some
stars have the silicate band in emission, others have it in
absorption: clearly the amount of circumstellar matter varies
strongly from one star to the other. Theoretical infrared spectra
have been calculated with increasing success, starting with the
pioneering work by Jones and Merrill (1976) and by Bedijn (1977). For
very recent work see Rowan- Robinson et al. (1986) and Bedijn (1986).
Theory and observation can be brought together very closely with a
minimum of assumptions: see Figure 4.

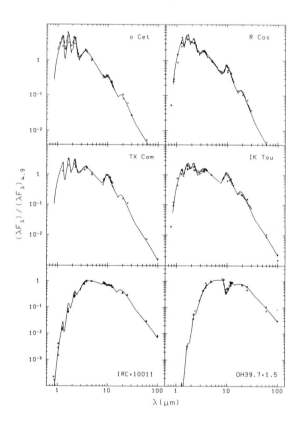

Figure 4: Infrared spectra of long period
variables; dots and crosses are observations
and the drawn line is a model fit, based on
the calculation of the transfer of radiation
through a dusty envelope (Bedijn, 1986).
The same basic model is used for all spectra,
only the optical depth (measured at 9.7μm)
varies from τ=0.4 for o Cet to τ=10 for OH39.7+1.5.

The main difference between various authors, and the main source of
uncertainty, is the precise absorption characteristics of the dust.
The IRAS Atlas of Low Resolution Spectra provides a large and
homogeneous data base that may clarify the absorption characteristics
much better - the work by Papoular, Pegourié and others here at this
Symposium is certainly a step in that direction. For a more extensive
and useful review see Kwok (1986).

At this point I like to propose some nomenclature. Stars with
strong infrared excesses are sometimes called CE-stars (circumstellar
envelope stars) and, sometimes, OH/IR stars. Both names are
inadequate. The first because it does not distinguish between stars
with ionized circumstellar shells, e.g. Be stars, and the cool giants
with dusty envelopes. The second is inadequate because only a
fraction of the stars show OH maser emission. I propose to call cool
giants with dust-rich envelopes "DGE-stars": dust-gas enveloped
stars. The OH/IR stars are then a natural subclass of the DGE-stars.
And many M giants will be DGE-stars.

I will now turn again to the question which M giants have IR
excess emission and which do not. Let me try to swim in deep water
and jump from the diving board of theory. Much work has been done on
the interior structure of well evolved stars (much more, so it
appears, than on the outer structure) - for a review see Iben and
Renzini (1983). This makes it clear that stars of very different
initial masses may become M giants; and they may become it more than
once. A sample of observed M giants must clearly be a mix of objects
with very different masses and internal structures. Two basic
structures can be distinguished: (i) stars with degenerate helium
cores and extensive hydrogen envelopes burning hydrogen into helium
at the boundary between the two; such stars are called First Giant
Branch stars (FGB stars); (ii) stars with a core of degenerate carbon
and oxygen, surrounded again by an envelope of hydrogen; the two
parts are separated by two thin layers, one in which hydrogen burns
into helium and one in which intermittently helium burns into carbon.
Such stars are called Asymptotic Giant Branch Stars (AGB stars). Most
AGB stars become much brighter than FGB stars.

The dichotomy between FGB and AGB stars reminds one of the fact
that there is a small group of M giants which are highly variable:
Mira variables. Their spatial density is only a few percent of that
of the "normal" M giants (Plaut, 1965). Wood (1974) has suggested
that Mira variables are AGB stars: their luminosities are high and
their spatial motions show that they have main sequence masses
$>1.0M_\odot$. With this in mind I speculate that the non-variable (i.e.
non-Mira) M giants are FGB stars, and that the DGE-phenomenon occurs
only in highly variable AGB stars. To support my case I will present
some evidence that pulsation and mass loss are closely connected. I
realize, however, that I cannot make a fully convincing case yet, or
answer all objections that might be raised.

The paper by Hacking et al. on the brightest 12μm IRAS sources
outside of the galactic plane contains 130 identifications with
variable stars. Of these 119 have a [2.2μ]-[12μ] index larger than
0.4 - value of 0.0 is expected for a blackbody, and, indeed, K-giants

in the Hacking et al. paper have low values (<0.2). I have selected
10 stars out of these 119 with index >1.0, checked their visual
amplitude and found in nine cases that this is larger than 1.6; six
stars are Mira's, 3 are SRb variables; one object I could not trace.
Next consider the 11 variables with a small index [2.2μ]-[12μ]. For
six I could find the variability properties; they are all SR or L
variables, with visual amplitudo's <1.0. This suggests that variables
with large excesses have large pulsation amplitudes, and those with
small (or no) amplitudes have small or no [2.2μ]-[12μ] excesses. The
Hacking et al. paper also contains 38 M giants with a name that does
not indicate variability. Checking them in the Bright Star Catalogue
I sometimes find the indication "var?" indicating at most small
amplitude variations. Of these 38 only 4 have a colour index larger
than 0.4 - the others donot. Again I find the suggestion that small
or no pulsations imply a low colour excess.

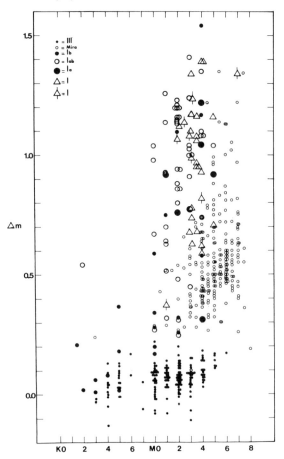

Figure 5: The [12μ]-[25μ] colour (called "ΔM")
of late type giants (luminosity class III) and
of supergiants as a function of spectral type
(see text). Courtesy of M. Raharto (Lembang).

The same suggestion is coming from a somewhat different direction. In Figure 5 is shown the [12μ]-[25μ] index (in the figure it is called "Δm") for some IRAS point sources for which a good spectral and luminosity classification is available; I owe this figure to M. Raharto from Lembang Observatory. It shows very clearly how different classes of objects have different [12μ]-[25μ] indices: M giants, mostly taken from the Bright Star Catalogue and without variability (or at most: variability with a small amplitude), have a small index. Mira variables and supergiants have a range of index values, but never a small index. There seems to be a gap of about Δm=0.15 between the two groups. It is probably the same gap that occurs in Figure 1 near the Rayleigh-Jeans Point (see section 3 for some discussion).

This is thus the suggestion: among M giants two fundamentally different types are present: FGB stars with no, or small amplitude variation in the visual and with low IR excesses and AGB stars pulsating with large visual amplitudes (there are mainly Mira variables) and with large IR excesses. The two types occupy well-separated areas in Figure 5; the separation between the two areas is probably real: AGB stars have always a significant excess, FGB stars do not.

In this section I have discussed the topic of the DGE-stars. Although the interest in this topic began when the first IR observations were made by the pioneers in the sixties, it is obvious to me that IRAS has given a very strong stimulus to the field and may well produce a breakthrough in our understanding of the late stages of evolution. This in itself is an important reason to be interested in DGE stars. There are, however, two more reasons.

The first is that the circumstellar envelopes are very rich in molecules. Carbon star envelopes are already well known for this richness -e.g. IRC+10216 has a large collection of different molecules (see e.g. Johansson et al., 1984), and oxygen-rich stars begin to give away their secrets as well (Guilloteau et al., 1986). To study chemistry under astrophysical conditions these envelopes may be more suitable than molecular clouds in interstellar space: in circumstellar shells one knows more accurately the physical conditions (radiation field, density, temperature).

The second additional reason why DGE stars are interesting is that they are very strong IR emitters (luminosities 2000 - 20,000 $L_\odot$, M_{bol} between -3.5 and -6.0) and that they do not suffer much interstellar extinction. Hence they can be studied at large distances in the Galaxy where optical observations fail. Plotted on a map of the sky DGE stars clearly outline the disk and the bulge of the Galaxy (Habing et al., 1985). The bulge and the disk of the Galaxy are two quite different components of the Galaxy, the first considered to be very old, the other young, or, at least, to be continuously rejuvenated. The IRAS data have given us new and very effective means to study the Galaxy "in a new light". Elsewhere I will review some of the means (Habing, 1987).

7. CONCLUSIONS

For the student of circumstellar matter the IRAS Point Source
Catalog contains a very large and new data base with ten thousands of
interesting objects. There are many envelopes of ionized material
surrounding B and A stars but most objects will be DGE-stars, i.e.
stars with dust-rich envelopes. The latest stages of stellar
evolution are heavily influenced by mass loss; it is likely that
analyses of the DGE-stars discovered by IRAS will shine much light on
this process - a possible breakthrough.

One final point. I do not believe in the superiority of large
numbers, despite my recent attachement to space research. Personally
I have always found (and still find) the single discovery of the
excess emission around VEGA (α Lyra) by Aumann and Gillett (see e.g.
Aumann et al., 1984) the most rewarding and interesting of the
discoveries by IRAS. Perhaps other discoveries of similar importance
are still hidden in the large IRAS database.

APPENDIX

Magnitudes and colours in the infrared

The magnitude of a point source is defined as -2.5 times
the   log ratio of the flux density and some standard flux density
(the zero point) both at the same wavelength. Magnitudes are
symbolized by [λ] , where λ is the wavelength. Traditionally the
spectrum of some "average" A0 V star was chosen to define zero points
at different wavelengths. In the IRAS context this habit has been
replaced by the marginally different habit to use a 10,000K blackbody
viewed at a solid angle of $1.57 \ 10^{-16}$ ster. This means that [12μ]=0.0
corresponds to 28.0 Jy, [25μ]=0.0 to 6.7 Jy.

A "colour index" (or "colour" for short) is the difference
between magnitudes at two different wavelengths, in the sense "blue
minus red". Redder objects have increasing colour indices. In the
infrared the colour index for "normal stars" is always very small and
increases only marginally with redness. This is a consequence of
being on the long wavelength side of the blackbody curve.

REFERENCES

Allen, C.W. 1963, Astrophysical Quantities (2nd ed.), the Athlone
    Press (London)
Aumann, H.H., Gillett, F.C., Beichman, C.A., de Jong, T., Hovck,
    J.R., Low, F.J., Neugebauer, G., Walker, R.G., Wesselius, P.R.
    1984, Astrophys. J. Lett. 278, L23
Bedijn, P.J. 1977, thesis, Leiden University
Bedijn, P.J. 1986, in "Light on Dark Matter", ed. F.P. Israel, Reidel
    Publ. Cy., Dordrecht p. 119
Chester, T.J. 1986, in "Light on Dark Matter", ed. F.P. Israel,

Reidel Publ. Cy., Dordrecht, p.3.
Guilloteau, S., Lucas, R., Nguyen-Q-Rieu, Omont, A. 1986, Astron. Astrophys. 165, L1
Habing, H.J. 1987, in "The Galaxy", ed. G. Gilmore, Reidel Publ. Cy., Dordrecht (in press)
Habing, H.J., Olnon, F.M., Chester, T., Gillett, F., Rowan-Robinson, M., Neugebauer, G. 1985, Astron. Astrophys. 152, L1
Hacking, P., Neugebauer, G., Emerson, J., Beichmann, C., Chester, T., Gillett, F., Habing, H.J., Helou, G., Houck, J., Olnon, F., Rowan-Robinson, M., Soifer, T., Walker, D. 1985, Publ. Astron. Soc. Pac. 97, 616
Hoffleit, D., Jaschek, C. 1982, "The Bright Star Catalogue", Yale University Observatory, New Have, Conn., USA
Iben, I., Renzini, A. 1983, Ann. Rev. Astron. Astrophys. 21, 271
IRAS, 1984, "The Explanatory Supplement", eds. C.A. Beichmann, G. Neugebauer, H.J. Habing, P.E. Clegg, T.J. Chester (The United States Government Printing Offices).
IRAS, 1986, Atlas of Low Resolution Spectra, eds. F.M. Olnon and E. Raimond, Astron. Astrophys. Suppl. Ser. 65, 607.
Johansson, L.E.B., Andersson, C., Elldér, J., Friberg, P., Hjalmarson,A, Hoeglund, B., Irvine, W.M., Olofsson, H., Rydbeck, G. 1984, Astron. Astrophys. 130, 227
Jones, T.W., Merrill, K.M. 1976, Astroph. J. 209. 509
Kwok, S.Y. 1986, preprint, to appear in "Physics Reports"
Kwok, S.Y., Hrivnak, B.J., Milone, E.F. 1986, Astrophys. J. (in press)
Olnon, F.M. 1986, in "Light on Dark Matter", ed. F.P. Israel, Reidel Publ. Cy., Dordrecht, p. 31.
Plaut, L. 1965, in "Galactic Structure" (Stars and Stellar Systems, vol. V), eds. A. Blaauw and M. Schmidt (The University of Chicago Press), p. 267
Rowan-Robinson, M., Lock, T.D., Walker D.W., Harris, S. 1986, Mon. Not. Roy. Astr. Soc. 222, 273
Treffers, R., Cohen, M. 1974, Astrophys. J. 188, 545
Van der Veen, W., Habing, H.J., Geballe, T. 1987, in "Planetary and Protoplanetary Nebulae", ed. A. Preite-Martinez, Reidel Publ.Cy., Dordrecht (in press)
Waters, L.B.F.M., Coté, J., Aumann, H.H. 1986, Astron. Astrophys. (in press)
Wood, P.R. 1974, Astrophys. J. 190, 609
Woolf, N.J., Ney, E.P. 1969, Astrophys. J. Letters 155, L181

1612 MHz OBSERVATIONS OF SOUTHERN IRAS SOURCES

M. E. Dollery and M. J. Gaylard
Hartebeesthoek Radio Astronomy Observatory
R. J. Cohen
University of Manchester, Nuffield Radio Astronomy Labs.,
Jodrell Bank

ABSTRACT. Eight of thirty-four previously unobserved IRAS sources were found to be relatively strong 1612 MHz OH emitters. Five of these emit at 1667 MHz. Of the eight half are high velocity range, population I type stars, the other half are low velocity range, population II type stars. The pump efficiencies are in the range $0.018 \leqslant e \leqslant 0.163$.

1. INTRODUCTION

The OH/IR candidates were chosen to fall within the region of variable OH/IR stars in the colour-colour plot of Olnon et al (1984), fig. 2. Assuming a 3σ detection limit of 3 Jy at 1612 MHz, for a 15 minute observation, a minimum maser pump efficiency of 0.02 would allow OH/IR stars with a non colour-corrected flux $S(25) \geqslant 300$ Jy to be detected.

2. OBSERVATIONS AND RESULTS

Observations were made at 1612 MHz with the 26m Hartebeesthoek antenna. The zenith system temperature was 45K. Only left circular polarization was accepted. The 3σ sensitivity limits were 1.5 Jy at 1612 MHz and 1.0 Jy in the main-lines. In the undetected sources, the upper limit on the intensity at 1612 MHz was 3 Jy. Results are listed in Table I. The eight OH/IR stars detected at 1612 MHz were re-observed at 1667 MHz, where maser emission was found in five, and at 1665 MHz where none were detected, with an upper limit of 1 Jy.

3. DISCUSSION

The infrared pump efficiency of 1612 MHz masers, taken from the ratio of the averaged OH flux of the peaks to the 35um flux (Evans & Beckwith 1977), is less than 0.25 (Elitzur et al, 1976). Here, derived pump efficiencies (Table II) were found to be in the range 0.018 to 0.163.

The peak separation, dv, distinguishes population I ($dv \gtrsim 29$ km/s) from population II ($dv < 29$ km/s) masers (Baud et al, 1981). Population I

215

I. Appenzeller and C. Jordan (eds.), Circumstellar Matter, 215–216.
© *1987 by the IAU.*

masers in this sample (Table II) lie within 3° of the galactic plane.
Three of the four population II masers lie at galactic latitudes well
outside the range of population I tracers.

Distances to the stars were estimated from the mean radial velocities of
the maser lines. They were obtained directly from a galactic rotation
model (R_o = 10 kpc), and by comparison with the velocity-longitude
relationships for HI, CO and HII regions. Estimates for population II
masers may be unreliable, if they depart greatly from circular rotation.

TABLE I. 18cm OH EMISSION FROM THE EIGHT DETECTED IRAS SOURCES

Object	IRAS ident.	1612 MHz		1667 MHz	
		v(peak) km/s	s(peak) Jy	v(peak) km/s	s(peak) Jy
OH259.8-01.8	11438-6330	-43.8	11.4		
		-16.9	8.5	-16.3	2.2
OH309.6+00.7	13442-6109	-73.2	3.1		< 1
		-24.8	1.8		
OH318.7-00.8	14582-5926	-57.5	4.5		< 1
		-18.5	11.1		
OH329.8-15.8	17319-6234	-20.7	8.6		
		+8.1	5.1	+8.7	2.4
OH338.1+06.4	16105-4205	-96.3	59.8	-97.6	6.2
		-70.3	59.4	-68.9	1.6
OH343.9+02.7	16460-4022	-46.6	10.7		
		-12.2	6.0	-10.9	13.8
OH348.2-19.7	18467-4802	-59.6	25.2	not observed	
		-35.4	20.8		
OH357.3-01.3	17411-3154	-39.4	134.5	-40.7	19.0
		-2.2	134.9	-3.9	13.4

TABLE II. OH/IR STARS - DERIVED PARAMETERS FOR THE DETECTED SOURCES

Object	S(OH)/S(35)	dv (km/s)	Pop.	Vlsr (km/s)	Distance (kpc)	Comments
OH259.8-01.8	0.047	26.9	II	-30.4	-	Large peculiar velocity
OH309.6+00.7	0.018	48.4	I	-49.0	3.2/ 7.6	Centaurus arm, tangent
OH318.7-00.8	0.076	39.0	I	-38.0	2.1/ 0.6	Centaurus arm
OH329.8-15.8	0.047	28.8	II	-6.3	0.3/ (4.4)	Carina arm / Local spur?
OH338.1+06.4	0.148	26.0	II	-83.3	5.4/(10.4)	Norma arm?
OH343.9+02.7	0.033	34.4	I	-29.4	2.6/(13.7)	Centaurus arm
OH348.2-19.7	0.163	24.2	II	-47.5	5 /(11.5)	3 kpc / Norma arm?
OH357.3-01.3	0.099	37.2	I	-20.8	6 /(11)	3 kpc / Norma arm

REFERENCES

Baud, B. et al, Astronomy & Astrophysics, 95, 156-170, (1981)
Elitzur, M. et al, Astrophysical Journal, 205, 384-396, (1976)
Evans, N. J. & Beckwith, S., Astrophysical Journal, 217, 729-740, (1977)
Olnon, F. M. et al, Astrophysical Journal, 278, L41-L43, (1984)

CIRCUMSTELLAR MATTER AS DETECTED BY IRAS--SOME SYSTEMATICS

W. P. Bidelman
Warner & Swasey Observatory
Case Western Reserve University
Cleveland, Ohio 44106 USA

ABSTRACT. Most of the sources noted by the Infrared Astronomical Satellite are, as expected, more or less normal cool giant stars, but a very large number of objects of earlier spectral type were also picked up. These cases must involve, in a variety of ways, circumstellar gas and dust. From a large-scale program of identification of the IRAS objects, some systematics emerge.

1. INTRODUCTION

The IRAS point-source catalogue gives values of flux density at effective wavelengths near 12, 25, 60, and 100 micrometers for almost a quarter of a million infrared sources over the entire sky. The complete identification of all of these sources, is, of course, a tremendous task, but I have been contributing towards it in two directions: (1) In about 100 25-square-degree areas mainly at higher galactic latitudes and in the northern sky identifications are being made for the stellar IRAS objects with the aid of infrared objective-prism plates taken with the Burrell Schmidt, and (2) Identifications are being attempted for all of the several thousand IRAS sources that have their maximum values of flux density at either 25 or 60 μm.

As a result, a large variety of objects have been identified. The large numbers of normal cool stars found are of little interest here, but the fact that a very substantial number of hotter objects have been detected should be of interest to this Symposium.

2. RESULTS

It appears most useful here to simply indicate the types of hot, or warm, objects that have been found in the IRAS data. Four classes may be noted:

Class A: Hot objects that have an abnormally large flux at 12 μm, declining at longer wavelengths. Most of these appear to be normal Be stars, though some unusual objects such as

I. Appenzeller and C. Jordan (eds.), Circumstellar Matter, 217–218.
© *1987 by the IAU.*

89 Her, υ Sgr, 3 Pup, and RY Sgr, as well as most of the RV Tauri stars, belong to this class.

Class B: Objects that have a higher flux at 25 μm than elsewhere in the infrared. Many of these have been termed "young, low-excitation, stellar" planetary nebulae, though in some cases the nebulae themselves are not seen. The interesting objects HD 161796 and AC Her belong to this class, as do many other emission objects of rather uncertain nature. This class contains several cool carbon Wolf-Rayet stars as well as many T Tauri stars. Also, most of the unidentified type II stellar masers are in this group.

Class C: Objects that have a higher flux at 60 μm than elsewhere in the infrared. This very diverse group contains many somewhat older planetaries, some T Tauri stars, such objects as η Car and AG Car, and even some galaxies. The enigmatic shell star β Pic, which probably has suffered envelope ejection, is a member of this group.

Class D: Objects that continue to increase in flux to 100 μm. Most of the galaxies detected by IRAS belong to this class, as do a few T Tauri stars and most of the so-called early-type "nebulous stars." It appears that many of the latter may be accidentally rather than genetically involved in the interstellar medium.

It is hoped to eventually publish in detail some of the results of this work. In the meantime, inquiries regarding identification of the IRAS sources are welcome. Finally, it is a pleasure to acknowledge that this work has been supported in part under NASA's IRAS Data Analysis Program and funded through the Jet Propulsion Laboratory.

CONTRIBUTION OF LINE EMISSION TO THE IRAS MEASUREMENTS: NGC 6853

C. Y. Zhang[1,2], A. Leene[1], S. R. Pottasch[1] and J. E. Mo[1,2]
1. Kapteyn Astronomical Institute,
   P.O. Box 800, 9700 AV Groningen, Holland
2. Purple Mountain Observatory, Academia Sinica, Nanjing, China

ABSTRACT.   The contribution from the line emission of ions to the radiation at four IRAS bands has been estimated. The dust grains are likely to be mixed with ionized gas.

1. INTRODUCTION

NGC 6853, also known as the Dumbbell Nebula, has been investigated by Hawley and Miller (1978), Pottasch, Gilra and Wesselius (1982) and Barker (1984). The IRAS observations of it extend the wavelength coverage from the ultraviolet and visual to the far infrared part of the spectrum. Although dust grains are very efficient infrared emitters, the line emission of some abundant ions can also contribute to the infrared fluxes in the IRAS bands. The available detailed measurements of visual and UV lines make it possible to estimate the contribution from ions to all four IRAS bands with reasonable certainty.

2. DATA

The Dumbbell Nebula was measured by the SUR-AO's (Additional Observations with the survey array) and the CPC-AO's (Additional Observations with the Chopped Photometric Channel). The data of the two SUR-AO's and the two of four CPC-AO's of this nebula have been analyzed. The maps at the four IRAS bands deduced from SUR-AO's and the co-added CPC-AO's maps at 50 and 100 μm bands are obtained, and compared with the visual and radio maps (Bignell, 1986). The flux density and in-band flux for each survey band are given in Table 1. The 50/60 μm maps from CPC/SUR and the two 100 μm maps from CPC/SUR look very similar to the visual map and radio map at 20 cm from the VLA observation, but there are significant differences between 25 μm map and the maps at the other bands. The 25 μm map has a position angle of about 76 which is quite different from the values of less than 33 at other bands. And the FWHM sizes of the 25 μm map in both major and minor axes are smaller than those of the maps at other IRAS bands.

I. Appenzeller and C. Jordan (eds.), Circumstellar Matter, 219–220.

3. CONTRIBUTION OF LINE EMISSION TO THE IRAS MEASUREMENTS

The infrared radiation from most planetary nebulae shows a very high
excess over the continuous radiation expected from free-free emission.
We have considered the possible contribution from line emission of ions
to the radiation detected in the IRAS bands. The electron density and
temperature in this nebula are chosen as $n_e \approx 200$ cm^{-3} and $T_e \approx 1.2 \times 10^4$ K
(Pottasch, 1984). The total flux of the H$_\beta$ line emission corrected for
interstellar extinction is 4.46×10^{-10} ergs cm^{-2} sec^{-1}
(Bussoletti et al., 1974).
We have taken into account the effect of the transmission function on the
predicted fluxes of line emission in the IRAS bands. The total contribution of the related lines to the four IRAS bands, the IRAS fluxes after
correction for line emission and the correction factors can be found in
the fourth, fifth and sixth columns of Table 1.
As can be seen, the 25 µm radiation is dominated by the O IV 25.87
µm line emission. In fact, unlike O III, O IV radiation comes from
a high ionization region, which must be in the inner part of the nebula.
It is therefore to be expected that the 25 µm map dominated by the
O IV 25.87 µm radiation would have a relatively smaller size and a
different shape.

TABLE I. Flux densities measured at IRAS bands and corrected for line
emission

λ (µm)	F (Jy)	F(in) (erg cm^{-2} sec^{-1})	F(line) (erg cm^{-2} sec^{-1})	F(corr) (Jy)	correction factor
12	8.2	$1.1\ 10^{-9}$	$7.5\ 10^{-10}$	2.6	68 %
25	41.1	$2.1\ 10^{-9}$	$1.5\ 10^{-9}$	11.0	75 %
60	135.2	$3.5\ 10^{-9}$	$9.6\ 10^{-10}$	97.7	28 %
100	206.3	$2.1\ 10^{-9}$	$4.6\ 10^{-10}$	160.0	22 %

REFERENCES

Barker T.: 1984, Astrophys. J., 284, 589.
Bignell C.: 1986.
Bussoletti E., Epchtein N. and Baluteau J.P.: 1974, Astron. Astrophys.,
    34, 141.
Hawley S.A. and Miller J.S.: 1978, Publ. Astron. Soc. Pacific
    90, 39.
Pottasch S.R., Gilra D.P. and Wesselius P.R.: 1982, Astron. Astrophys.,
    109, 182.
Pottasch S.R.: 1984, Planetary Nebulae, Reidel, Dordrecht.

IS THERE A SIGNATURE OF ICE IN THE IRAS LRS SPECTRA OF SOME MIRA VARIABLES?

M.S. Vardya
Tata Institute of Fundamental Research, Bombay, India

Some M Miras show the 9.7 and 20μm silicate emission. Recently, Vardya, de Jong & Willems (1986; hereafter VDJW) discovered using IRAS LRS spectra a weak broad emission feature ∿12μm; this new feature may be silicate (VDJW). Here we investigate whether this can be due to H_2O ice.

$H_2O(g)$ is important as a source of opacity in M stars (cf Vardya 1970), and has been detected in IR (cf Merrill & Ridgway 1979), in mm (Bowers & Hagen 1984) and the excess emission between 5-8μm in μ Cep and R Cas has been attributed to $H_2O(g)$ (Tsuji 1978).

Recently, de Muizon, d'Hendecourt & Perrier (1987) have explained excess <u>absorption</u> ∿12μm in some cool stars as due to ice, based on the presence of 3μm ice band. Hence, 12μm <u>emission</u> feature in M Miras can be signature of ice.

12μm emission feature is most prominent among 19 stars of VDJW in RU Her, R Aur and T Cas (Fig. 1); the light curves are fairly symmetric as f values indicate, defined as the ratio of the number of days between visual light minimum and the following maximum to period (P), and P∿>450d. RU Her also shows the 9.7μm emission. The visual light curves in Fig. 2 (Campbell 1955) show a hump or a change in gradient in the rise part, unlike smooth rise as in RR Aql, which shows no 12μm feature.

In the log S_{12}/S_{25} vs f plot of VDJW, where S_λ is the IRAS source flux density, there are two stars below the $400^{\circ}K$ line - Z Cyg of small period, a normal light curve but no 12μm feature, and WX Psc of very long period and 12μm emission.

Thus, it appears, that 12μm emission is more probable when the light curve is nearly symmetric with a hump or change in gradient in the rise part, the period is long, attains low temperature at minimum light and has low IR temperature.

The 9.7μm emission in M Miras is present when f<0.45 (VDJW); this is qualitatively understood if smaller the f is, stronger is the shock. This shock drives the mass loss and the condensates in the flow depend on the physical conditions. If f is small, ρ_g is high near the surface, freeze out occurs close to the surface at high temperature and silicates are the only condensates. As f increases, ρ_g is not that high and freeze out occurs further out at low temperature with formation of other

I. Appenzeller and C. Jordan (eds.), Circumstellar Matter, 221–222.

condensates as well.

The hump or change in gradient in the light curve indicates double shock. Hence ρ_g is not high and in these three stars, the freeze out occurs far out at low enough temperature for ice to form, if the condensation time, proportional to P, is long. WX Psc with low IR temperature, shows the 12μm feature; this strenghthens the above argument. Non-detection of gaseous H_2O in these stars (Bowers & Hagen 1984) may indicate depletion due to condensation.

Concluding, a case has been made that the 12μm emission feature in some M Miras may be due to ice. One needs to expand the list of stars with this feature and confirm the existence of ice by 3μm spectro-photometry.

References

Bowers, P.F. and Hagen, W. 1984, _Ap. J._ **285**, 637.
Campbell, L. 1955, Studies of Long Period Variables, (Cambridge: American Association of Variable Star Observers).
de Muizon, M., d'Hendecourt, L.B., and Perrier, C. 1987, IAU Symposium No.120, Goa (Abstract).
Merrill, K.M., and Ridgway, S.T. 1979, _Ann. Rev. Astr. Ap._, **17**, 9.
Tsuji, T. 1978, _Astr. Ap._, **68**, L23.
Vardya, M.S. 1970, _Ann. Rev. Astr. Ap._, **8**, 87.
Vardya, M.S., de Jong, T., and Willems, F.J. 1986, _Ap. J. (Lett.)_ **304**, L29.

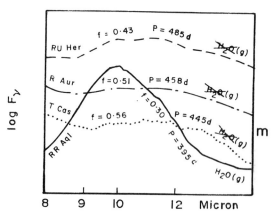
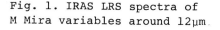
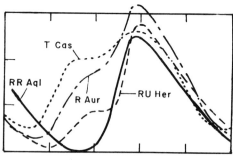

Fig. 1. IRAS LRS spectra of M Mira variables around 12μm.

Fig. 2. Visual light curves of M Mira variables (Campbell 1955)

A STATISTICAL ANALYSIS OF DUST FEATURES IN THE IRAS LOW RESOLUTION SPECTRA

O. Gal[1], M. de Muizon[2], R. Papoular[1], B. Pégourié[1]
1. Service d'Astrophysique, CEN Saclay,
   91191 Gif sur Yvette, France
2. Huygens Laboratory, Sterrewacht Leiden, The Netherlands
   and Observatoire de Meudon, France

ABSTRACT. Using the IRAS catalog of low resolution spectra (LRS), we have analyzed the silicate features in emission and absorption, the carbon-rich emission features and the featureless spectra, in the range 8-22 μm. The sample sizes are large enough to allow average properties to be established (e.g. energy distributions), as well as correlations between luminosities, excesses, colors and coordinates, and histograms and galactic distributions, all with a good degree of confidence.

1. METHOD

Select a group of spectra in the LRS data bank. Normalize the fluxes F_λ to 1 at a suitable w.l. λ_0 (say 10 μm) and obtain $\mathcal{F}_\lambda \equiv F_\lambda/F_{\lambda_0}$. For each λ, take the _geometric_ mean of all normalized fluxes, over the spectra of the sample. This gives $\bar{\mathcal{F}}_\lambda$, which is our "average profile" for the sample. This procedure improves S/N in proportion to the sample size. We also compute Σ_λ, the (geometric) std.dev. of $\mathcal{F}_\lambda$. It is an estimate of the dispersion of the spectral profiles in the selected sample. When the spectra of a given class are grouped in sub-classes according to their brightness (e.g. flux at 8 or 10 μm) so that Σ_λ, in each sub-class, remains reasonably small ($\leqslant 25\%$), the following general, gross trends are observed: _bright objects_ are more evenly distributed in longitude (l) and/or latitude (b), hence are closer to the Sun ; _weak objects_ are restricted to the bulge direction, hence are farther from the Sun ; also, they are generally redder.

2. PARTICULARS OF VARIOUS SPECTRAL CLASSES

2.1. Class 1 (featureless spectra)

On a colour-colour diagram, F_{12}/F_{25} vs. F_8/F_{12}, 3 distinct sub-classes are distinguished among the 445 brightest objects:
a) $2000 \leqslant Tc \leqslant 3000$ K : 212 objects with spectra $\propto \lambda^{-4}$ and nearly uniform distributions in l and b ; these are probably photospheres.

223

I. Appenzeller and C. Jordan (eds.), Circumstellar Matter, 223–224.

b) 500 ⩽ Tc ⩽ 1000 K : 199 objects grouped towards the bulge, with a Sic-like bump around 11.3 µ in their spectrum, indicating C-rich envelopes (which considerably increases the proportion of such shells as compared with silicate shells).

c) Tc ⩽ 500 K : 34 objects towards the inner galactic arms ($|b| ⩽ 10°$), with non-thermal spectra and SiC bumps.

2.2. Classes 2 and 6 (silicate features at 10 and 18 µm)

There is a clear change of spectral profile as a function of brightness: as the latter decreases, one notices a counter-clockwise tilt (reddening) of the spectra, the 10- and 18- µm excesses become stronger and narrower and the objects tend to cluster towards the galactic bulge. Objects with F_8 ⩽ 11 Jy and $|b|$ ⩽ 2.5° are roughly grouped along the tangents to the arms. 10- µm excesses larger than any observed from earth (⩾ 10) are found among the faintest objects. On the other hand, if bright objects (F_8 ⩽ 22 Jy, i.e. in the Sun's vicinity) are grouped according to their relative 10-µm excess, ε_{10}, and a histogram in b^{II} is drawn for each group, then it is found that the std.dev. of b^{II} from its mean decreases rather steadily from 24 to 8° as ε_{10} increases from 1 to 10 (the average $\varepsilon_{10} \simeq 1.5$). Thus, the distribution of ε_{10} in the Galaxy seems to follow the same general trends as the metallicity of stars.

In previous work, Pégourié and Papoular (1985) have derived empirical optical efficiencies (Q_{abs}) for silicate dust observed from Earth, with the help of suitable radiative transfer models. We used the same Q's to model the average spectra of subgroupsof classes 2 and 6. The optimum parameters (star and dust temp. T_*, T_d ; dust optical thickness at 10 µm, τ ; average grain radius, a ; star and internal shell radii, r_*, r_1) are given here for a) the brightest objects of class 2 (F_{12}⩾ 60 Jy), and b) class 6: a) $T_* = 3500$ K, $T_d(r_1) = 450$ K, $\tau(r_1/r_*)^2 = 29$, a = 1.6 µm ; b) $T_* = 1000$ K, $T_d(r_1) = 550$ K, $\tau.(r_1/r_*)^2 = 3.4$, a = 1.5 µm. The fits are satisfactory.

2.3. Class 4 (carbon-rich CS shells)

Assuming a power-law underlying continuum between 10 and 13 µm, the relative excess $\varepsilon(\lambda)$ can be extracted. Its average profile over the class is quite similar to Russel and Stephen's laboratory measurement of SiC extinction. (M. Cohen, 1984, MN 206), 137). However, if subclasses are formed again according to brightness, a clear trend appears: as F_8 decreases, the SiC feature strength $\varepsilon_{11.3}$ increases (up to 3, from an average of 0.4). Also, weaker features are broader; the change in energy distribution mainly takes the form of a growing bump (between 11.5 and 12 µm) upon the red wing of the SiC feature : ε_{12} varies by ≈ 30%. Finally, a bump grows between 8 and 8.5 µm, in rough correlation with $\varepsilon_{11.3}$, as F_8 decreases. The galactic distribution of class 4 is rather uniform in l but quite peaked about b = 0 ($\sigma_b \approx 5°$).

ON THE INFRARED EXCESS OF ALPHA LYRAE*

B.G. Anandarao[1] and D.B. Vaidya[1,2]
1. Physical Research Laboratory, Ahmedabad-380009, India
2. Presently at Indian Institute of Astrophysics,
   Bangalore-560034, India

SUMMARY. Recent measurements in far-infrared (12-200 μm) by IRAS (Aumann et al, 1984) and KAO (Harper et al, 1984) have shown evidence of infrared excess in the AO star Alpha Lyrae. This excess in the infrared radiation was interpreted as due to circumstellar dust constituted by grains much larger ($\gtrsim 10$ μm) than those of interstellar origin. However, there appears to be a difficulty in interpreting these observations in terms of a single or a combination of several temperatures of dust. We have formulated a simple model to explain the far-infrared data in 12-200 μm region available with IRAS and KAO observations in terms of multiple dust shells of different temperatures. We have shown that dust grains of size ~ 10 μm can explain the excess emission in the entire region 12-200 μm with temperatures of 130 ± 15 K and 40 ± 5 K. The far-infrared fluxes seem to vary as $\lambda^{-\beta}$ with $\beta = 3.0 - 3.8$, in quite contrast with small sized grains. This work lends further support to the interpretation of far-infrared excess in α Lyrae as emission from circumstellar shell with large (~ 10 μm) size grains. Our calculations show, however, that grains of size larger than 10 μm cannot explain the fluxes.

---

* Full paper has appeared in Astronomy and Astrophysics, 1986, 161, L9-L11.

I. Appenzeller and C. Jordan (eds.), Circumstellar Matter, 225.

IRAS OBSERVATIONS OF CLASSICAL CEPHEIDS

C. J. Butler
Armagh Observatory, Armagh, N. Ireland
H. P. Deasy and P. A. Wayman
Dunsink Observatory, Dublin, Ireland

ABSTRACT. IRAS observations of sources identified with cepheid variables are used to give estimates of observed mass-loss rates for those stars.

In this note we make a comparison, using the IRAS Point Source Catalogue (1985), between Cepheid variable stars and similar non-variable stars as they may be detected in the 12 micron, 25 micron, 60 micron and 100 micron bands of the IRAS detector system. 33 classical Cepheids and 38 non-variable supergiants were identified. The selection criteria required E(B-V) < 1.0 and 0.4 < (B-V)o < 1.0 in order to permit reasonable reduction, using an interstellar extinction curve (Savage and Mathis, 1979), of measured infrared colours to intrinsic values.
 The results for comparison of Cepheids with non-variable stars are shown in Fig.1. There is evidence that some long-period Cepheids have infrared excesses associated with them. The reddening correction to the infrared colour index $[12] - [25] = 2.5 \, \mathrm{Log} \, (F(25\mu)/F(12\mu))$ is only $\simeq$ 0.02 E(B-V), permitting the use of this ratio also, when available, for comparison with a supposed blackbody ratio.

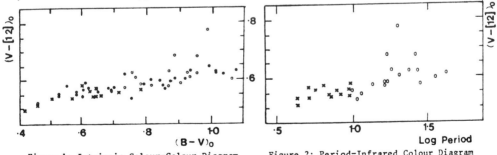

Figure 1: Intrinsic Colour-Colour Diagram   Figure 2: Period-Infrared Colour Diagram
Crosses - short period Cepheids; open circles - long period Cepheids;
filled circles - non-variable supergiants.

 The results show 'normal' values for short period and for most long period Cepheids with excess reddening for some Cepheids of longer period. It is suspected (Fig. 2) that processes that produce this emission are most prevalent at periods around 20 days. The values could be produced

227

I. Appenzeller and C. Jordan (eds.), Circumstellar Matter, 227–228.

either by circumstellar dust emission or be due to free-free emission
from a hot corona. Radiatively-driven mass loss is not indicated, because
this would be most effective for the largest stars and the suspected
dependence on period makes pulsation-driven processes more likely.

The few 60/100 micron values show, for δ Cephei, nearly blackbody
behaviour, but for RS Puppis and RV Scuti systematic departure is
indicated. For RS Puppis, associated with nebulosity detected optically
by Havlen (1972), a temperature around 40° K is indicated.

A principal interest of these observations was to obtain possible
values for the mass loss. This requires assumptions concerning the nature
of the stellar wind. Based on the theory of Gehrz and Woolf (1971) for
re-radiation of light in a thin dust shell, assumptions concerning gas
density conditions outside Cepheids and possible velocities led to
figures for mass loss by dust shown in Col.4 of the associated Table,
evaluated for those stars with significant excesses in the infrared
colours.

TABLE

Name	Period (days)	Excess (mag)	$\dot{M}$dust $(10^{-8} M_\odot y^{-1})$	$\dot{M}$ff	$\dot{P}$ $(10^{-6} y^{-1})$	DMdust $(M_\odot)$	DMff $(M_\odot)$
V350 Sgr	5.154	(0.18)	(0.32)	(2.2)			
RX Cam	7.912	0.43	0.84	2.8			
S Mus	9.659	(0.10)	(0.24)	(1.5)			
XX Cen	10.956	(0.39)	(1.1)	(3.8)			
RW Cam	16.413	0.57	2.4	8.3	<50 p	>.00048	>.0017
CD Cyg	17.071	0.76	3.8	20	3.73f	.010	.054
RU Sct	19.698	1.12	8.0	70	5.93e	.013	.12
X Pup	25.961	0.32	1.8	6.9	10.9 e	.0018	.0063
RY Vel	28.125	0.37	2.2	12			
RS Pup	41.388	0.22	1.7	10	29.5 e	.00056	.0034

Notes: Col.6: Rate of period change with sources indicated –
  p = Parenago (1956), e = Erleksova & Irkaev (1982), f = Fernie (1983).

Alternatively, mass-loss rates could be calculated on the basis that
the infrared excess is from free-free emission of hot ionised gas,
following the plasma equations of Wright and Barlow (1975), (Col.5).

Using information on likely crossing-times for the Cepheids in the
instability strip, based on observed rates of period change, the
integrated mass loss for these figures, given in Cols. 7 & 8, are
relatively small in comparison with total masses around 5 M$_\odot$.

REFERENCES

Erleksova, G.E. & Irkaev, B.N., 1982, Perem. Zvezd., 21, 715.
Fernie, J.D., 1983, IAU Symposium No. 105 (Reidel, 1984).
Gehrz, R.D., & Woolf, N.J., 1971, Astroph. J., 165, 285.
Havlen, R.J., 1972, Astr. Astroph., 16, 252.
IRAS Point Source Catalogue, Joint IRAS Working Group, 1985.
Parenago, P.P., 1956, Perem. Zvezd., 11, 236.
Savage, B.D. & Mathis, J.S., 1979, Ann. Rev. Astr. Astroph., 17, 73.
Wright, A.E. & Barlow, M.J., 1975, Mon. Not. R. Astr. Soc., 170, 41.

CIRCUMSTELLAR ENVELOPES OF OH-IR SOURCES

R.J. Cohen
Nuffield Radio Astronomy Laboratories
Jodrell Bank
Macclesfield
Cheshire   SK11 9DL
England

ABSTRACT.   This article reviews recent radio observations of maser emission from OH, H_2O and SiO molecules in the circumstellar envelopes of OH-IR sources. The different radio lines require different conditions for their excitation, and each therefore probes different regions in the circumstellar envelope. For some stars radio interferometer maps of several maser lines are now available, and a consistent picture of the envelope structure is beginning to emerge.

1.   INTRODUCTION

I will present an observational review of the circumstellar envelopes of OH-IR sources, highlighting the use of radio maser lines to probe their structure. The maser lines of OH, H_2O and SiO which I will discuss are very intense, with brightness temperatures of typically 10^{10}K, and each is emitted from different parts of the circumstellar envelope. The lines can now be studied using radio interferometers (VLA, MERLIN and VLBI) on angular scales from 1 arcsec down to 1 milliarcsec, corresponding to linear scales as small as one astronomical unit. These studies contribute to our understanding of the late stages of stellar evolution (M_* and $\dot{M}$), the physics of mass loss, the physics of the maser excitation processes, the three-dimensional structure of the circumstellar envelopes, circumstellar chemistry, and the effects of magnetic fields which can be investigated using Zeeman splitting of the OH lines. The OH-IR stars also provide a promising new technique for measuring distances by comparing angular and light-travel diameters. Finally as we heard yesterday from Prof. Habing, the OH-IR stars are a special group of stars which can be identified and studied throughout the Galaxy.

2.   BASIC IDEAS ABOUT CIRCUMSTELLAR MASERS

The OH-IR sources are believed to be long-period variable stars near the end of the asymptotic giant branch. Their slow pulsations drive material far enough above the stellar surface that dust grains can condense; and

I. Appenzeller and C. Jordan (eds.), Circumstellar Matter, 229–239.
© 1987 by the IAU.

these are then blown away from the star by radiation pressure, dragging the circumstellar gas with them (Goldreich & Scoville 1976). The mass loss rates are estimated to be $\sim 10^{-5} M_\odot$ yr^{-1}.

For the OH, H_2O and SiO masers the line excitation requirements, and also the shapes of the lines themselves, give clues as to the structure of the circumstellar envelope and the location of the masers within it (e.g. Olnon 1977, Elitzur 1981). The maser lines form a natural sequence of decreasing excitation. The SiO lines at 43, 86 GHz etc. are rotational transitions of vibrationally excited SiO. Lines up to V=3 have been observed (Scalise & Lépine 1978). The vibrational states have energy levels corresponding to temperatures of ~ 2000-5000K, and so appreciable population of these states is only likely close to the star. The H_2O 22 GHz line is produced by a transition between two rotationally excited states (6_{16}-5_{23}) each of which is some ~ 650K above the ground state. H_2O masers are therefore expected to occur near the star, but not so near as SiO masers. Finally the OH 1.6 GHz lines result from hyperfine splitting of the lambda doublet in the rotational ground state ($^2\pi 3/2$). These masers would generally be expected to lie furthest from the star.

The physics of cosmic masers has been reviewed recently by Elitzur (1982). The population inversion can be understood for the case of a two-level maser as a simple problem of rates. The populations are increased by so-called pump processes and decreased by loss processes, each of which can involve radiation, collisions or chemistry. Population inversion can be achieved if the pump or the loss rates for the two levels are unequal. Once inversion has been established the maser provides amplification by stimulated emission. The maser output is limited ultimately by the rate at which inversion occurs, and the maser saturates when the maser transition rate per unit volume approaches the inversion rate per unit volume. The maser can also be quenched when the collision rate dominates the other rates and the level populations thermalize.

3. OH1612 MHz MASERS

The OH1612 MHz masers nicely illustrate the physical principles just outlined. The pumping scheme for these masers is purely radiative, involving absorption of 35 μm photons by OH in the ground state, followed by a radiative cascade down the $^2\pi_1$ ladder and back to the $^2\pi 3/2$ ground state. Elitzur, Goldreich & Scoville (1976) showed that inversion of the 1612 MHz transition will occur provided the final transition to the $^2\pi 3/2$ ground state is optically thick. The maser will be quenched for $n_{H2} \gtrsim 4 \times 10^5$ cm^{-3}. The pumping scheme predicts a specific ratio of four 35 μm photons to one 1612 MHz photon for the case of saturation. Observations by Werner et al. (1980) have verified this. The proportionality between 35 μm emission and 1612 MHz emission was observed from source to source, and for each source individually as it varied throughout the stellar cycle. Monitoring of the OH 1612 MHz masers thus enables a purely radio determination of the stellar period to be made (Harvey et al. 1974; Herman 1983).

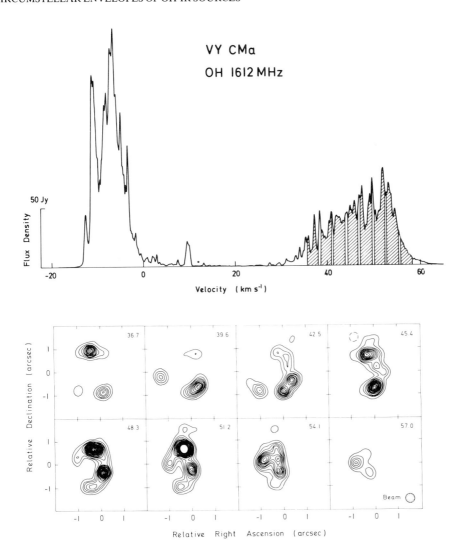

Fig.1. The OH 1612 MHz spectrum of VY CMa shows two main peaks of emission separated by some 50 km s^{-1}. Much fine structure is also evident at the 0.06 km s^{-1} resolution employed here. MERLIN maps of the emission over different velocity ranges (shaded) show different slices through the expanding envelope.

The OH 1612 MHz lines are almost without exception twin-peaked.
In the classical OH-IR sources such as IRC10011 the emission varies
smoothly with velocity. In other sources such as VY CMa there is much
fine structure in the emission profile, as shown in Fig.1. The two
emission peaks come from the near and far sides of the expanding circumstellar envelope, where the longest velocity-coherent paths for
maser amplification will be found if the outflowing gas has reached
terminal velocity (Olnon 1977; Reid et al. 1977). Radio maps made with
MERLIN and the VLA confirm that in most cases the 1612 MHz masers are
distributed in a uniformly expanding thin shell (Booth et al. 1981;
Bowers et al. 1983; Diamond et al. 1985). MERLIN maps of VY CMa are
shown in Fig.1. (Perry, unpublished). In this and in most other sources
we do not see complete shells of masers. Considering the special
excitation requirements and the effects of turbulence on the velocity-
coherent paths for maser amplification the lack of complete maser shells
is not surprising. Maps of the emission at different velocities show
different slices through the maser envelope with the velocities and
angular diameters being related in the way predicted by the simple
shell model. Some examples of this relation are shown in Fig.2, where
the dashed lines indicate least-squares fits to the data. By model-
fitting in this way it is possible to determine the shell radius even
in cases where the emission at the central velocity (which would show
the full shell size) is too weak to be mapped.

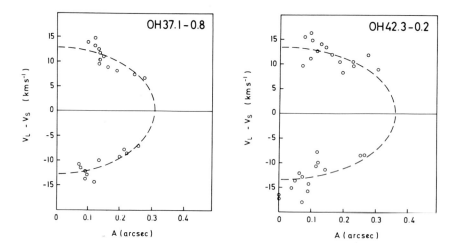

Fig.2 Radial velocities of OH 1612 MHz masers around two OH-IR sources
measured relative to the stellar velocity, are plotted against the
angular distance from the estimated stellar position. The dashed lines
show best fitting expanding shell models, from which the angular radii
and expansion velocities of the circumstellar shells are determined
(Chapman 1985).

The 1612 MHz emission is pumped by the pulsating 35 μm radiation field and varies in sympathy with it. Although the variations are synchronized with respect to the central star an outside observer sees a measurable phase-lag between the blue-shifted (near) and red-shifted (far) emission peaks. Accurate monitoring of the 1612 MHz masers thus enables the light travel time across the shell to be measured (Jewell et al. 1980; Herman 1983). The phase-lag measurement of the linear diameter can be combined with an interferometric measurement of the angular diameter of the shell to yield a geometrical estimate of the distance to the star (e.g. Baud 1981). These distance measurements could in principle be made accurate to 5%, and so might be of fundamental importance in helping to determine the galactic distance scale.

Phase-lags can also be measured between emission features at intermediate velocities in the OH spectrum. The uniformly expanding shell model predicts a linear change in phase-lag with velocity across the spectrum, and monitoring of this enables a check to be made on the spherical symmetry or otherwise of the envelope (e.g. Bowers & Morris 1984). By monitoring the variations with an interferometer it might be possible to determine phase-lags for individual regions throughout the envelope and so build up a fully three-dimensional picture of the maser shell. This is an experiment which could be started now if it was felt to be important enough.

The shell distribution of OH 1612 MHz masers reflects a true increase in OH abundance at a particular radius from the central star. The OH is produced by photodissociation of H_2O molecules in the external UV radiation field (Goldreich & Scoville 1976). The radius of maximum OH abundance depends in the mass loss rate from the star, and on the strength of the external UV radiation field (Huggins & Glassgold 1982). The dependence on mass loss rate has been confirmed observationally, but there is as yet no firm evidence for the weaker dependence on the UV radiation field.

4. OH MAINLINE MASERS

The OH mainline emission at 1665 and 1667 MHz also varies throughout the stellar cycle, but the variations are larger and less regular than those at 1612 MHz (Harvey et al. 1974; Fillit et al. 1977). These masers too are thought to be radiatively pumped. Infrared radiation from dust which is hotter than the gas can preferentially excite the upper half of the OH lambda doublet, and this inversion will be preserved in the cascade to the ground state, provided of course that the $^2\pi_1$ $J=\frac{1}{2}$ to $^2\pi_{3/2}$ $J=\frac{3}{2}$ transition is optically thin (Elitzur 1978). Pumping efficiences of 1% are predicted (Bujarrabal et al. 1980a), and these can be increased by line overlap (Bujarrabel et al. 1980b). OH mainline emission is expected to occur in warm optically thin parts of the circumstellar envelope, as compared with the cool optically thick regions which radiate at 1612 MHz.

Published mainline maps are available for only a small number of sources. They divide naturally into two groups. The first group of sources have a classical twin-peaked mainline spectrum, and the maps

show resolved shell structure with the mainline masers at comparable
radii to the 1612 MHz masers. Examples are OH 127.8 (Diamond, Norris &
Booth 1985) and IRC 10420 (Bowers 1984). OH 127.8 is particularly
interesting as the OH 1667 MHz emission at one velocity occurs at the
same position angle as a pronounced gap in the 1612 MHz emission at the
same velocity. This must reflect an asymmetry in the envelope structure
and the mass loss. IRC 10420 is also interesting in that the mainline
emission region is larger than the 1612 MHz region, contrary to the
pumping models. The second group of stars has more variable line shapes,
with emission right across the velocity range. They frequently have
large degrees of circular polarization. MERLIN maps show many unresolved
masers distributed over a compact region some 10^{15}cm in extent.
Examples of such sources are U Ori (Chapman & Cohen 1985) and VX Sgr
(Chapman & Cohen 1986). These masers occur in warmer denser regions of
the envelope than have been considered in the mainline pumping models.

5. H₂O MASERS

Circumstellar H₂O masers follow the stellar cycle even more irregularly
than do OH mainline masers, showing strong variations in amplitude from
cycle to cycle, and variations in the line shape (Schwartz et al. 1974;
Cox & Parker 1979). Gómez Balboa & Lépine (1986) have attempted to
model the amplitude variations in terms of periodic shock waves propagating into the circumstellar envelopes.
     The H₂O maser regions are typically 10^{15} cm in extent, and their
structure has been shown to vary on a timescale of a year (Johnston,
Spencer & Bowers 1985). The most detailed information is available for
the supergiant S Per, which has been studied by trans-Atlantic VLBI
(Diamond et al. in press). Two dozen individual maser spots were
detected over a region some 300 AU in extent. Individual maser spots
were resolved, and the spot sizes measured to be a few AU. The kinematics of the H₂O masers can be interpreted in terms of either an
expanding thin shell with turbulence, or an expanding thick shell with
acceleration.
     The pumping of circumstellar H₂O masers is thought to be set up
via collisions. Collisional excitation followed by radiative decay can
lead to an overpopulation of the so-called "backbone" rotational levels
(de Jong 1973) relative to other levels of comparable energy, and in this
way population inversions will be set up for several rotational transitions of the H₂O molecule (Cooke & Elitzur 1985). The inner boundary
of the H₂O maser region is determined by collisional quenching. The
model predicts that the inner radius r_1 should increase with mass loss
rate according to

$$r_1 \propto \dot{M}^{2/3} V^{-1}$$

‥oke & Elitzur 1985), and hence that the size of the maser region should
increase with mass loss rate. The rather sparse data presently available
are plotted in Fig.3. A correlation in the predicted sense is evident.
However it depends entirely on the difference between the supergiants
and the Mira-type variables, and more data are needed to establish the
result firmly.

6. SiO MASERS

Circumstellar SiO masers show even more irregularity than H_2O masers in their line shapes, in their cycle-to-cycle variations, and in the

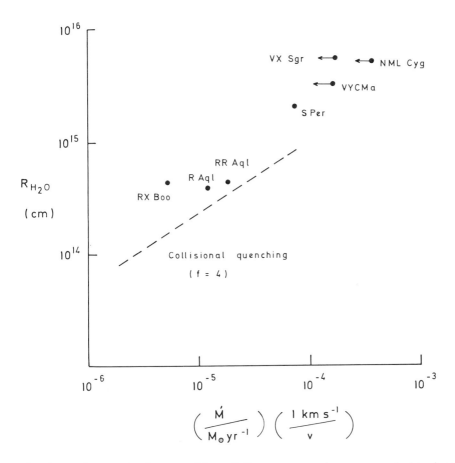

Fig.3. Sizes of circumstellar H_2O maser regions are plotted against mass loss rates for seven stars. The dashed line shows the predicted inner boundary set by collisional quenching of the maser, for an assumed $H_2O:H_2$ abundance of 4×10^{-4} (Cooke & Elitzur 1985). Sizes of the H_2O maser regions are taken from Johnston, Spencer & Bowers (1985), Cohen & Chapman (1986), Diamond et al. (in press), and unpublished MERLIN data on VY CMa. Mass loss rates are from Gehrz & Woolf (1971) and Bowers, Johnston & Spencer (1983). Mass loss for the supergiants VY CMa, VX Sgr and NML Cyg are shown as upper limits because of asymmetries in the circumstellar envelopes.

variations from star to star. Correlations between SiO maser intensity
and stellar light have been seen in most but not all of the sources
studied to date (e.g. Lane 1982; Nyman & Olofsson 1986). These correlations do not in themselves constitute proof that the pumping is
radiative, since the SiO masers are believed to lie close to the star
in hot dense regions where the thermal time constants are very small
compared to the stellar period. Theoretical considerations suggest
that collisional pumping is most likely (Elitzur 1982). The J=1-0
rotational transitions of different vibrationally excited states show
strong similarities in their velocity structure. For one star VX Sgr
near-simultaneous VLBI maps of the V=1 and V=2 J=1-0 masers have been
made. The two lines exhibit strong similarities in their spatial and
velocity distributions, which tends to confirm the suggestion that they
are generated in the same circumstellar regions despite their different
excitations (Lane 1982). The maser region is six stellar radii in
extent. Only one other star, R Cas, has been mapped to date, and it
shows a distribution of SiO masers some four stellar radii in extent
(Lane 1982, and refs therein). This is an area where more observations
are urgently needed.

7. VX SGR: A CASE STUDY

The supergiant VX Sgr is the first star for which radio maps of all the
classical maser lines are available (Lane 1982; Chapman & Cohen 1986).
The data have been combined by Chapman & Cohen, subject to the assumptions given in their paper, to provide a comprehensive radio picture of
the circumstellar envelope. A summary of the results is shown in Fig.
4. The SiO, H_2O and OH 1612 MHz masers lie at successively larger
distances from the star, as expected from their line excitation requirements. However the OH mainline masers lie at similar distances to the
H_2O masers, which seems to indicate some density and/or temperature
asymmetries in the envelope structure. Indeed asymmetries and multiple
shell structure have been noted in the maser envelopes of most other
supergiants studied to date (e.g. Bowers 1984; Diamond et al. 1984).
     The kinematics of the masers around VX Sgr are consistent with
radial outflow in which the flow velocity increases systematically with
distance from the star. The SiO masers show the most irregularity in
their velocity pattern, but their kinematics are broadly consistent
with expansion (Lane 1982). However the SiO masers are all moving at
less than escape velocity. The OH mainline and H_2O masers form a thick
shell region in which most of the acceleration appears to take place.
By the time the outflowing gas leaves this zone it has achieved escape
velocity. Further out the OH 1612 MHz masers are excited in a well-
defined thin-shell region. At this stage the outflowing gas is close
to terminal velocity. From a more detailed consideration of the velocity
field Chapman & Cohen conclude that the driving force per unit mass
acting on the outflowing gas must increase outwards to at least fifty
stellar radii. An increase in opacity with distance out to twenty
stellar radii was in fact incorporated by Goldreich & Scoville (1976)
in their model of OH-IR envelopes in order to account for infrared

occultation data (Zappala et al. 1974). It will be interesting to see
if observations of other OH-IR sources support this.

     An observation relevant to the dynamics and the envelope asymmetries
is the strong circular polarization of the OH mainline emission. Cohen
& Chapman identify two Zeeman groups in the OH mainline spectra of VX
Sgr. The line splitting implies a magnetic field of 2 mG in each case.
Such a field has an energy density comparable to that of the outflowing
gas, and may therefore be expected to influence the dynamics. In this,
as in other respects, it remains to be seen whether VX Sgr is typical
of the majority of OH-IR sources.

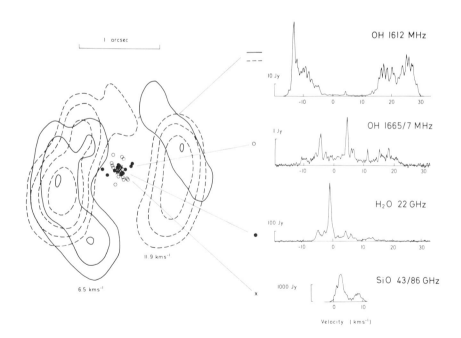

Fig.4 Maser emission from the circumstellar envelope of VX Sgr (from
Chapman & Cohen 1986). Spectra of the OH, H$_2$O and SiO lines are plotted
on the right, and the locations of the masers are indicated schematically on the left. For the OH 1612 MHz masers only the emission near
the stellar velocity is shown, to make the shell structure clear.

REFERENCES

Baud, B., 1981. Astrophys.J., 250, L79.
Booth, R.S., Kus, A.J., Norris, R.P. & Porter, N.D., 1981. Nature,
    290, 382.
Bowers, P.F., 1984. Astrophys.J., 279, 350.
Bowers, P.F., Johnston, K.J. & Spencer, J.H., 1983. Astrophys.J.,
    274, 733.
Bowers, P.F. & Morris, M., 1984. Astrophys.J., 276, 646.
Bujarrabal, V., Destombes, J.L., Guibert, J., Marlière-Demuynck, C.,
    Nguyen-Q-Rieu & Omont, A., 1980a. Astr.Astrophys., 81, 1.
Bujarrabal, V., Guibert, J., Nguyen-Q-Rieu & Omont, A., 1980b.
    Astr.Astrophys., 84, 311.
Chapman, J.M., 1985. Ph.D. Thesis, University of Manchester.
Chapman, J.M. & Cohen, R.J., 1985. Mon.Not.R.astr.Soc., 212, 375.
Chapman, J.M. & Cohen, R.J., 1986. Mon.Not.R.astr.Soc., 220, 513.
Cooke, B. & Elitzur, M., 1985. Astrophys.J., 295, 175.
Cox, G.C. & Parker, E.A., 1979. Mon.Not.R.astr.Soc., 186, 197.
Diamond, P.J., Norris, R.P. & Booth, R.S., 1984. Mon.Not.R.astr.Soc.
    207, 611.
Diamond, P.J., Norris, R.P. & Booth, R.S., 1985. Mon.Not.R.astr.Soc.,
    216, 1P.
Diamond, P.J., Norris, R.P., Rowland, P.R., Booth, R.S. & Nyman, L-A.,
    1985. Mon.Not.R.astr.Soc., 212, 1.
Elitzur, M., 1978. Astr.Astrophys., 62, 305.
Elitzur, M., 1981. "Physical Processes in Red Giants", p.363, eds.
    Iben, I. & Renzini, A., Reidel, Dordrecht, Holland.
Elitzur, M., 1982. Rev.Mod.Phys., 54, 1225.
Elitzur, M., Goldreich, P. & Scoville, N., 1976. Astrophys.J., 205, 384.
Fillit, R., Proust, D. & Lépine, J.R.D., 1977. Astr.Astrophys., 58, 281.
Gehrz, R.D. & Woolf, N.J., 1971. Astrophys.J., 165, 285.
Goldreich, P. & Scoville, N., 1976. Astrophys.J., 205, 144.
Gómez Balboa, A.M. & Lépine, J.R.D., 1986. Astr.Astrophys., 159, 166.
Harvey, P.M., Bechis, K.P., Wilson, W.J. & Ball, J.A., 1974.
    Astrophys.J.Suppl., 27, 331.
Herman, J., 1983. Ph.D. thesis, University of Leiden.
Huggins, P.J. & Glassgold, A.E., 1982. Astr.J., 87, 1828.
Jewell, P.R., Webber, J.C. & Snyder, L.E., 1980. Astrophys.J., 242, L29.
Johnston, K.J., Spencer, J.H. & Bowers, P.F., 1985. Astrophys.J.,
    290, 660.
de Jong, T. 1973. Astr.Astrophys., 26, 297.
Lane, A.P., 1982. Ph.D. Thesis, University of Massachusetts.
Nyman, L-A. & Olofsson, H., 1986. Astr.Astrophys., 158, 67.
Olnon, F.M., 1977. Ph.D. Thesis, University of Leiden.
Reid, M.J., Muhleman, D.O., Moran, J., Johnston, K.J. & Schwartz, P.R.,
    1977. Astrophys.J., 214, 60.
Scalise, E. & Lépine, J.R.D., 1978. Astr.Astrophys., 65, L7.
Schwartz, P.R., Harvey, P.M. & Barrett, A.H., 1974. Astrophys.J.,
    187, 491.
Werner, M.W., Beckwith, S., Gatley,I., Sellgren, K., Berriman, G. &
    Whiting, D.L., 1980. Astrophys.J., 239, 540.

Zappala, R.R., Becklin, E.E., Mathews, K. & Neugebauer, G., 1974.
     Astrophys.J., 192, 109.

DETECTION OF THE FIRST EXTRA-GALACTIC OH/IR STAR

P. R. Wood, M. S. Bessell[1] & J. B. Whiteoak[2]
Mount Stromlo and Siding Spring Observatories[1]
The Australian National University, Canberra and
CSIRO Radiophysics[2], Epping.

ABSTRACT. A search has been made for 1612 MHz OH maser emission from OH/IR stars in the Magellanic Clouds. Candidate objects were selected mainly on the basis of their 25μm flux densities and the 25 to 12μm flux ratio as given in the IRAS point source catalog; two known long-period variables and two HII regions (30 Doradus and N159) were also examined. One OH source (IRAS 04553-6825) was detected, this being the first OH/IR star found in the Magellanic Clouds. Upper limits were placed on the flux for 17 other sources. The expansion velocity of the circumstellar material surrounding IRAS 04553-6825, as indicated by the OH peak separation of 11 km s^{-1}, is surprisingly small compared to Galactic sources of similar bolometric and OH luminosity. The OH intensity of IRAS 04553-6825, and the upper flux limits placed on many of the other objects examined, indicate that Magellanic Cloud OH/IR stars do not emit OH as strongly as their Galactic counterparts of similar 25μm/12μm flux ratio. Both the low expansion velocity of IRAS 04553-6825 and the low OH intensity of the Magellanic Cloud infrared sources may be explained by the low metal abundance in the Clouds.

I. Appenzeller and C. Jordan (eds.), Circumstellar Matter, 241.

OH MASERS IN ENVELOPES OF LATE TYPE STARS

A.M. Le Squeren[1], P. Sivagnanam[1], F. Tran Minh[1],
M. Dennefeld[2], F. Foy[1]
[1] : Observatoire de Paris, section de Meudon, 92190-Meudon
[2] : Institut d'Astrophysique de Paris, 75014-Paris
France

ABSTRACT. Two studies of OH maser emission in envelopes of late type
stars -miras (3) and OH/IR objects- have been performed with the Nançay
radio-telescope. Mainly, the OH miras are found with thicker dust
envelopes than the non OH ones. A sample of unidentified IRAS point
sources selected on their colors has been observed. We have detected 46
new OH sources. The OH detection rate is a function of the galactic longitude and of the IRAS spectral classification.

1. MIRA STARS

1.1. The sample

The Mira type red giants are long period variables, with an important
mass loss ($\sim 10^{-6}$ $M_\odot/y$) and then a cold (some 10^2K) gas and dust envelope
($\sim 10^3$ stellar radius) where OH masers are often observed. We have
selected all known oxygen-rich miras within 1 kpc of the sun (~ 230 objects) for a high sensitivity search (0.08 Jy) of 18cm OH maser lines
(1612, 1665 and 1667 MHz).

1.2. Results

From Lockwood (1) we got the M spectral type at optical maximum. OH
emission is not possible below M5.5, and is very common above this limit.
The OH detection rate only increases continuously with increasing
period without clear limit between OH and non OH stars. Near the galactic plane, this rate is better and an OH star has a higher probability
to present type II (1612 MHz line) emission ; likely a stronger UV
intensity in this plane increases photodissociation of H_2O in OH.
 The colors (logarithmic ratio of $\nu S\nu$) between the IRAS flux densities
S_ν are a little different from blackbody colors. The mean [25-12] color
increases from non OH to type I OH, and to type II OH miras, but the
[60-25] color does not. All miras follow the same relation between the
normalized flux densities at 1 kpc, $Fv=Sv^*d^2$. So the colors are functions
of only one flux. OH masers (specially type II) need bright envelopes,

243

I. Appenzeller and C. Jordan (eds.), Circumstellar Matter, 243–244.
© *1987 by the IAU.*

i.e, log [Fv(12)]>1.2 (Fv in Jy), and are very common above this limit ;
moreover the OH miras mainly belong to IRAS LRS class 2 (silicate band
in emission), the non OH miras to class 2 (silicate band in emission),
the non OH miras to class 1 (no band) : OH miras have thicker dust
envelopes than non OH ones. So their colors differ because their mean
fluxes strongly differ (at least from 1 to 300) ; it does not reveal
other differences.

2. NEW OH/IR OBJECTS

2.1 Selected sources

A color-color plot between the 12, 25 and 60μm IRAS fluxes shows a
sequence from the bluest objects (non OH Miras, mainly class 1) (3) to
the reddest OH/IR objects with thick envelope (mainly class 3 -silicate
band in absorption- or 2) (2). We have searched for the OH counterpart
of the IRAS point sources located in the delineated box 0.20<12/25<0.45,
1.87<25/60<6.61, corresponding to OH/IR objects, for which a LRS spectrum (class 1-4, 4 : SiC band in absorption) had been obtained (202
sources). Presently we have observed 120 objects at 1612 MHz. The sensitivity limit is 0.15 Jy. 66 OH/IR stars were detected (46 new detections)

2.2. Results

Selected IR objects are mainly located in the galactic arms. A histogram
of these objects shows peaks in the direction of the spiral arms. Moreover OH detection rate is higher between 350° and 70° (70%) than between
70° and 250° (30%). This part of longitude corresponds to the galactic
arms where ultraviolet radiation density in the strongest, and could
produce OH molecules in envelopes of the stars by photodissociation of
H_2O molecules.
      The detection rate clearly increases from class 1(17%) to 3(68%),
consequently with the envelope thickness, furthermore the detection
rate of class 4 is peculiarly high : 50 %. The OH flux in classes 3-4
is on average stronger than in classes 1-2. 88% of the classes 1-2 sources have a mean peak flux < 2Jy ; 66% of the classes 3-4 sources have
a mean peak flux > 2Jy.
      If we draw in the selected bow a line parallel to the locus of
blackbodies the OH detection rate is higher below than above this
line. This fact may be due to differences of the envelope mean thickness,
or of the 35μm flux, which pumps the 1612 MHz masers.

REFERENCES
(1) Lockwood G.M., : 1972, Ap.J. Suppl., 24, 75.
(2) Olnon F., Baud B., Habing H.J., De Jong T., Harris S., and Pottasch
    S. : 1984, Ap.J., 278, L41
(3) P. Sivagnanam, AM. Le Squeren, F. Foy (in preparation)

INFRARED SPECKLE INTERFEROMETRY OF OH-STARS

Jessica M. Chapman[1] & R.D. Wolstencroft[2]
[1]Jodrell Bank, Macclesfield, Cheshire   SK11 9DL
[2]Royal Observatory, Blackford Hill, Edinburgh   EH9 3HJ

We have begun a co-ordinated programme of high angular-resolution radio and infrared measurements to study the physical structure of the circumstellar envelopes surrounding high mass-loss OH-stars. Here we give near-infrared (NIR) angular diameters for 5 stars. For each of these stars the spatial distribution of the OH maser emission at 1612 MHz or 1665 MHz has been previously mapped[1,2,3].

The observations were taken in September 1983 using the 3.8m UKIRT telescope on Mauna Kea with the speckle-slit system and broadband UKT5 photometer in the K(2.2μm), L'(3.8μm) and M(4.8μm) bands. The NIR speckle observing technique used was similar to that described by Dyck & Howell (1982)[4]. For each source, between 6 and 18 visibility profiles were obtained giving the source visibility in the north-south direction as a function of spatial frequency. Averaged visibility functions and model fits to the data are shown in Fig.1.

The circumstellar envelopes of VX Sgr, NML Cyg and OH39.7+1.5 were partially resolved between 2.2μm and 4.8μm. For these sources we have obtained angular diameters by fitting a two-component model to our visibility data. The model assumes a point source contribution to the NIR emission from the stellar photosphere and a Gaussian distribution of the extended emission from the circumstellar envelope. Results are given in Table 1 where the columns are:
1)   source name
2)   stellar classification
3)   near-infrared wavelength (λ)
4)   best-fit percentage stellar contribution (A)
5)   best-fit FWHM of the extended emission (α)
6)   adopted stellar distance (D)
7)   linear diameter of the infrared emission (d_{IR})
8)   ratio of the OH-1665 MHz to infrared diameters (d_{1665}/d_{IR})
9)   ratio of the OH-1612 MHz to infrared diameters (d_{1612}/d_{IR})

I. Appenzeller and C. Jordan (eds.), Circumstellar Matter, 245–246.

References
1. Chapman, J.M. & Cohen, R.J., 1985. Mon.Not.R.astr.Soc., <u>212</u>, 375.
2. Chapman, J.M. & Cohen, R.J., 1986. Mon.Not.R.astr.Soc., <u>220</u>, 513.
3. Diamond, P.J., Norris, R.P., Rowland, P.R., Booth, R.S. & Nyman, L-A.,
   1985. Mon.Not.R.astr.Soc., <u>212</u>, 1.
4. Dyck, H.M. & Howell, R.R., 1982. Astr.J., <u>87</u>, 400.

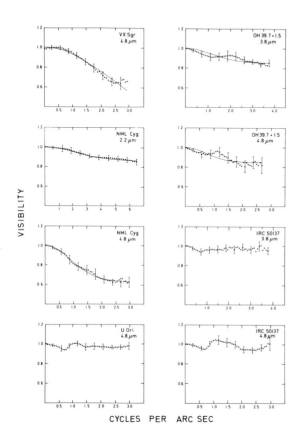

<u>Figure 1</u>. Near infrared
visibility curves
obtained for 5 OH-stars
using speckle interfero-
metry.

SOURCE	CLASS	λ (μm)	A (%)	α (arcsec)	D (kpc)	d_{IR} (10^{15} cm)	$\dfrac{d_{1665}}{d_{IR}}$	$\dfrac{d_{1612}}{d_{IR}}$
VX Sgr	M4eIa+M9.5	4.8	11^{+23}_{-11}	0.15±0.03	1.7	3.8±0.8	0.4+1.3	11.1±2.4
NML Cyg	M6Ia	2.2	86±10	0.17±0.01	2.0	5.1±0.2	-	12.4±1.2
NML Cyg	M6Ia	4.8	62±10	0.42±0.01	2.0	12.5±0.3	-	5.0±0.5
OH39.7+1.5	OH-IR	3.8	85±10	0.32±0.05	0.8	3.7±0.6	-	9.2±1.5
OH39.7+1.5	OH-IR	4.8	85±10	0.52±0.13	0.8	6.2±1.6	-	5.5±1.5
IRC 50137	M10	3.8	*	<0.08	0.8	<1.0	-	>25.0
IRC 50137	M10	4.8	*	<0.10	0.8	<1.2	-	>20.0
U Orionis	M6e+M9e	4.8	*	<0.10	0.3	<0.5	>4.0	-

Table 1

*Unresolved

OH0739-14: An old star blowing bubbles

Bo Reipurth
European Southern Observatory
Casilla 19001, Santiago 19
Chile

ABSTRACT. Two large bipolar bubbles emanating from the OH/IR star OH0739-14 have been discovered. Interference filter images show that the bubbles are emission-line objects, and longslit spectra reveal that the bubbles are expanding, with a radial velocity-difference between the bubblefronts of over 200 km/sec.

The OH/IR star OH0739-14, also known as OH231.8+4.2, has a faint optical counterpart, a small elongated nebula, which reflects the light of an embedded M9 giant or supergiant (J.Cohen and Frogel 1977,M.Cohen 1981). A bright near-infrared nebulosity is oriented along the same axis (Allen et al.1980). Furthermore, a detailed spatial and velocity mapping of the 1667 MHz OH maser emission shows a bipolar velocity gradient also along this axis (Bowers and Morris 1984,Morris et al.1982). Recently, Cohen et al.(1985) made longslit spectroscopy along the flow axis, and discovered blue- and red-shifted shock-excited regions outside the optical nebula.

Deep CCD images of OH0739-14 through various interference filters have been carried out at the Danish 1.5m telescope at ESO,La Silla. Fig.1 shows an H-alpha image, on which three basic features of the object can be identified. Firstly, the optical reflection nebulosity is seen as two bright, rather narrow lanes. Secondly, an obscuring disk cuts across these two lanes, and thirdly, two large shock-excited bubbles surround the reflection lanes. An image taken through a continuum-filter, otherwise identical to the H-alpha filter, shows only the reflection lobes and the obscuring disk, while yet another image, through a filter transmitting the [SII] 6717/6731 lines, again shows the bubbles, although weaker than in H-alpha.The total extent of the bubble-system is 50 arcseconds.

Longslit spectroscopy obtained with the ESO 2.2m telescope yield a heliocentric radial velocity for the

I. Appenzeller and C. Jordan (eds.), Circumstellar Matter, 247–248.

northern bubblefront of -42 km/sec, and +180 km/sec for the
southern front, both with a standard deviation of 27 km/sec.
The inclination of the system is about 47 degrees to the
plane of the sky, on the basis of near-infrared polarimetry
(Tielens,Werner and Capps, cited in Cohen et al.1985). The
space velocities of the northern and southern bubblefronts
relative to the central maser are then about 110 and 190
km/sec. At a distance of 1.2 kpc and with this inclination,
the physical extent of the system is around 0.42 pc (0.17
and 0.25 pc for the northern and southern part). A lower
limit to the dynamical age of the bubbles is thus 1500
years.

The outflow is probably generated by the wind from a
hot accretion-disk surrounding a companion white dwarf. The
companion accretes matter from the mass-losing red giant,
which is at the very end of AGB evolution. We are most
likely here witnessing the rapid evolution immediately
preceding the formation of a bipolar planetary nebula.

REFERENCES
Allen,D.A.,Barton,J.R.,Gillingham,P.R.,Phillips,B.A.:1980,
    Mon.Not.Roy.Astr.Soc.190,531
Bowers,P.F.,Morris,M.:1984,Astrophys.J.276,646
Cohen.J.G.,Frogel,J.A.:1977,Astrophys.J.211,178
Cohen,M.:1981,Publ.Astron.Soc.Pacific 93,288
Cohen,M.,Dopita,M.A.,Schwartz,R.D.,Tielens,A.G.G.M.:1985,
    Astrophys.J.297,702
Morris,M.,Bowers,P.F.,Turner,B.E.:1982,Astrophys.J.259,625

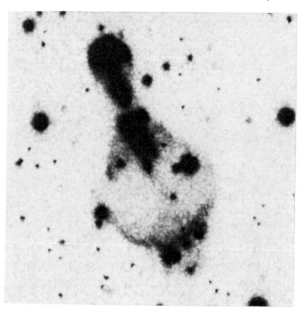

Fig.1. A deep H-alpha CCD image of OH0739-14.

A MODEL FOR MASER LINE PROFILES OF LATE-TYPE STARS

J.R.D.Lépine
Instituto Astronomico e Geofisico -U S P
Caixa Postal 30627
01051 Sao Paulo
Brazil

ABSTRACT. The profiles of maser lines from spherically symmetric expanding shells are computed with the help of a simple model.It is shown that a beaming effect must be taken into account in order to correctly account for the OH profiles of type II OH stars.

1.INTRODUCTION

According to the models of Elitzur et.al.(1976) and of Reid et.al.(1977), the two OH emission features of type II OH stars originate in the near and far side of an expanding circumstellar shell. We present here a more complete derivation of the profiles of maser lines which takes into account the whole contribution of a spherical shell. The profiles depend only on the expansion velocity law V(r) and on the population difference across the maser transition n(r) . We remark that since the equation for radiation transfer takes similar forms in the cases of optically thin thermal emission and of saturated maser emission (in both cases the intensity of radiation increases linearly with column density),the same "thermal" profiles are obtained in both cases. The typical profiles with two peaks can only be obtained if the maser intensity depends on a power m>1 of the amplification path length; this occurs in the case of unsaturated masers, or in the case of saturated maser emission with a beaming effect.

2.BASIC ASSUMPTIONS

The first step in the calculation of the line profile is to obtain the profile I(p,v) observed in a line-of-sight that passes at a distance p from the center.The length of the amplification path contributing to the intensity at a velocity v in the spectrum, limited by the Doppler shift due to the velocity gradient along the path,is $\Delta l(z)= \Delta v/(dv/dz)$,where z is the distance along the line-of-sight, v the projection of V(r) , and Δv the thermal width of the line.
 We assume that the observed intensity in a given line-of-sight and at a given velocity is proportional to some power of the length $\Delta l(z)$ and

249

I. Appenzeller and C. Jordan (eds.), Circumstellar Matter, 249–251.

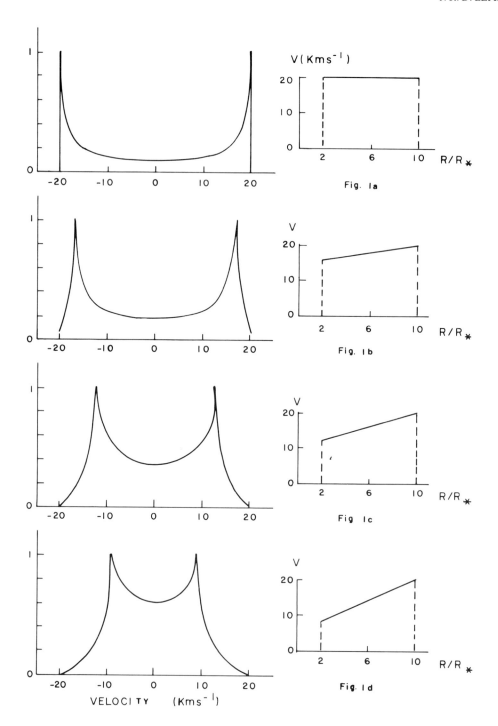

Figure 1:Computed maser profiles for several expansion velocity
laws.The velocity laws are indicated at the right of each profile.

to some power of $n(r)$. A convenient expression for $n(r)$ is $n(r) = A\,r^{-q}$ in a shell $(R_1 < r < R_2)$ and $n(r) = 0$ outside this range. Therefore a convenient expression for the intensity due to a line-of-sight is:

$$I(p,v) = r^{-q}\left[dv(p,z)/dz\right]^{-m} \qquad (1)$$

where m, q are parameters which allow us to describe different situations. The profile from the whole envelope should be obtained by integrating $I(p,v)$ over p. We substitute the integration over p by an integration over r using the relations:

$$p^2 = r^2 - z^2 \qquad (2)\;; \qquad v(p,z)/V(r) = z/r \qquad (3)$$

$$\frac{dv(p,z)}{dz} = \frac{V(r)}{r}\left[1 - \frac{v^2}{V(r)^2} - \frac{v^2}{V(r)^2}\frac{d\ln v}{d\ln r}\right] \qquad (4)$$

The result is:

$$I(v) \propto \int_{R_1}^{R_2} v(r)^{-m}\left[1 - (v/V)^2 - (v/V)^2\,(d\ln v/d\ln r)\right]^{1-m} r^{m+1-q}\,dr \qquad (5)$$

3. RESULTS AND DISCUSSION

When $V(r) = V0$ (constant), which occurs at large distances from the star, the expression (5) reduces to:

$$I(v) = \left[1 - (v/V0)^2\right]^{1-m}$$

For $m=1$ we obtain $I = $ constant, which corresponds to the rectangular profiles observed in the case of optically thin thermal emission.

For $m=2$ (intensity proportional to the square of the amplification path length) we obtain the same expression $I(v) = 1/(V0^2 - v^2)$ which was derived by Reid et.al.(1977) without stating the condition $m=2$.

The integral (5) must be computed numerically for other velocity laws $V(r)$. We show in figure 1 the normalized profiles that are obtained with linear increase of the expansion velocity $(V(r) = A + B*r)$ between 2 and 10 stellar radii, with $m=2$ and $q=4$. This choice of q corresponds to both density and pumping radiation flux decreasing like $r(-2)$. We remark that the profiles obtained with moderate acceleration resemble the OH profiles of visible Miras.

The additional dependence of the intensity on the amplification path length $(m=2)$ is a beaming effect. A large path length produces a large coherence area at the output, so that the radiation is concentrated in beam of smaller solid angle and the observer receives more radiation (the coherence area acts like a large emitting antena).

REFERENCES
Elitzur,M.,Goldreich,P.,Scoville,N.,1976,Astrophys.J.205,384.
Reid,M.J.,Muhleman,D.O.,Moran,J.M.,Johnston,K.J.,Schwartz,P.R.,
1977,Astrophys.J.214,60.

SHOCK WAVES - THE TRIGGER MECHANISM OF SIO MASERS IN CIRCUMSTELLAR
ENVELOPES OF COOL GIANTS AND SUPERGIANTS

Astrid Heske
Hamburger Sternwarte
Gojenbergsweg 112
2050 Hamburg 80, F.R.G.

Circumstellar envelopes of cool giants and supergiants are composed
of different parts which can be observed through characteristic spectral features. One is SiO-maseremission which is typically found in
Miras and some semi-regular variable supergiants. These masers are
thought to originate in those parts of the upper atmosphere or lower
circumstellar envelope where still no mass-loss takes place. Location
and trigger mechanism was still not definitely clear.

A well defined sample of about 50 cool giants and supergiants being
selected by their high (I-K)-value and covering a wide range of spectral types (G8-M8) was observed, among other spectral features, in
the Hα-line (1.4m CAT, La Silla (Chile); 2.2m Calar Alto (Spain)) and
the SiO (J=1-0)-line (100m-telescope in Effelsberg). The resolution
was 0.03 Å or 1.5 kms^{-1} and 0.7 kms^{-1} respectively.

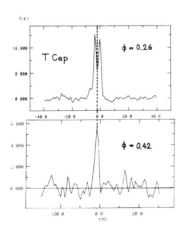

Fig.1

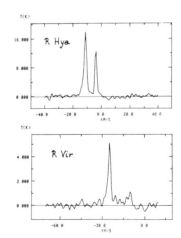

Fig.2

Figures 1 and 2 show three examples of double-peaked maser profiles.
As these profiles look similar to most of the OH-maser profiles and in
T Cep the single peak (observed at phase 0.42) falls exactly in the

253

I. Appenzeller and C. Jordan (eds.), Circumstellar Matter, 253–254.

centre of the double-peaked profile (observed at phase 0.26) this
might be taken as evidence that in these examples rear and front of
a shell-type outflow of matter where SiO-masers are located is observed. Thus half of the velocity separation may be assigned to the
outflow velocity. This value was found to be proportional to the ratio
of the blue- to the redshifted peak intensity, I_{blue} / I_{red}. It can
be due to a geometrical effect, the star obscuring part of the rear
emitting area. If radial amplification of the masers and the peak flux
being proportional to the emitting area provided, the intensity ratio
I_{blue} / I_{red} thus is proportional to $1/R^2$ as the rear emitting area
grows with distance R from star. The SiO-masers are though found to be
located between 2 and 6 stellar radii (0.3 ... $1 * 10^{14}$cm) moving out
with velocities between 8 and 1 kms^{-1}.

From the whole sample comparison of the Hα- and SiO-spectra led to the
correlations that SiO-masers are only observed when the giant or supergiant shows variable Hα-emission and when the star is cooler than
2500 K, if it is a giant. In supergiants SiO-maser seem to occur also
at earlier types (e.g. μ Cep, M2Ib).

The observational results show evidence for SiO-masers being triggered
by shock waves where Hα-emission originate. The masers may then be
located in the post shock material several stellar radii above the
star, moving with velocities less than 8 kms^{-1}. This is consistent
with the velocity field of Mira-atmospheres derived by Willson and
Bowen (1986). Their calculated density distribution implies that the
particle density can still be high enough (N $\lesssim 10^{12}$ cm^{-3}) at several
stellar radii for SiO-masers to be excited (Elitzur 198o). Shock waves
as triggers for SiO-masers imply that they could be collisionally
pumped as the kinetic temperature rises immensely in the shock front.
The more complex maser profiles would also fit in this picture as it
should be taken into account that during one cycle more than one shock
wave may be expelled, and that not in every giant or supergiant the
rear emitting region may be observed.

References:

M. Elitzur 198o, Astrophys. J. <u>24o</u>, 553

L.A. Willson and G.W. Bowen 1986, Third Trieste Workshop on the
    Relationship between Chromospheric/Coronal Heating and
    Mass-Loss; eds. R. Stalio and J. Zirker, pp. 127

THEORY OF CIRCUMSTELLAR ENVELOPES

R. Wehrse
Institut f. Theoretische Astrophysik
Im Neuenheimer Feld 561
D 6900 Heidelberg
Federal Republic of Germany

ABSTRACT. After a comparison of conventional photospheres with circumstellar envelopes the radiation in spherical shells is considered. We discuss the transfer equation and a new quasi-exact solution in term of the transition matrix. Various methods used for the numerical evaluation of the specific intensities are summarized. The general properties of the radiation fields and some recent detailed model calculations are briefly reviewed.

1. INTRODUCTION

In the usual nomenclature "circumstellar envelope" or "circumstellar shell" designates the space around a star, which is outside the star's atmosphere, but where the matter is still connected to the star. Although there is evidently no strict separation between outer layers of the atmosphere and the circumstellar envelopes, both have quite different characteristics, the most important being the different geometry and the different energy as well as momentum balance. In all cases this means a much more complicated behaviour of the circumstellar envelopes. Therefore, it has not yet been possible to construct realistic models for circumstellar envelopes from first principles and without severe simplifications.

In Table 1 we list some details, but it is beyond the scope of this paper to present a complete review of the field, since recently excellent and detailed papers have appeared summarizing the present knowlegde e.g. on the physical and chemical state of matter in the envelopes around cool stars (Omont, 1985 ; Gail and Sedlmayr, 1986), the mass loss mechanisms for hot and cool stars (Hearn, 1987, Holzer, 1987), etc..

We will focus on the radiative transfer in circumstellar

I. Appenzeller and C. Jordan (eds.), Circumstellar Matter, 255–266.

T a b l e   1

	conventional photosphere	extended envelope
geometry	plane parallel, i.e. extension of photosphere $\Delta r \ll$ stellar radius r	spherical (first approximation only)
density stratification	hydrostatic equilibrium	hydrostatic equilibrium or steady hydrodynamic equilibrium
velocity fields	turbulence, no macroscopic fields	in- or outflow and turbulence
temperature stratification	radiative + convective equilibrium	equilibrium of hydrodynamic + radiative losses and gains (incl. e.g. dissipation of acoustic and Alfvén waves)
state of matter	gas, mostly in or near LTE	gas and dust, mostly far away from LTE
outer boundary conditions	no incident radiation negligible pressure at τ_{out} (small)	no incident radiation negligible pressure at τ_{out} (small) velocities $\to$ const. for $\tau \to \tau_{out}$
inner boundary conditions	diffusion approximation for radiation	prescribed energy flux continuity of density
constraint		mass infall/outflow rate = const.
parameters	effective temperature gravity chemical composition turbulence velocity	luminosity mass radius chemical composition (effectively: runs of density, velocity and temperature)

shells, which requires a more sophisticated treatment than in stellar atmospheres, because (i) in addition to turbulence velocity fields with large scales (in most cases radial flows) have to be taken into account; (ii) the geometrical extension of the configuration has to be considered explicitly; (iii) radiative processes dominate over collisional processes so that the absorption and scattering coefficients are coupled directly to the radiation field (and not only via the energy equation).

In Section 2 the radiative transfer equation for spherical configurations with radial velocitiy fields is given and various methods for its solution are described. The resulting characteristics of the radiation fields are summarized in Section 3. Finally, in Section 4 we discuss briefly some recent calculations modeling the envelopes of cool giants, Be stars, and supernovae.

2. THE RADIATIVE TRANSFER EQUATION FOR SPHERICALLY EXTENDED CONFIGURATIONS AND ITS SOLUTION

We will consider subsequently spherical configurations, because for them methods for the evaluation of the radiation field are well established, although no standards have emerged yet (cf. Beckman and Crivellari, 1985). This restriction to spherical geometry implies that we cannot discuss polarisation effects and that we have to exclude binary stars, discs, and jets. On the other hand, speckle interferometry (Roddier,Roddier,and Karovska, 1985) indicates that the distribution of matter around most single stars is approximately spherical and we therefore expect that the majority of lines originating in the shells of these objects are well described by this approximation. For exceptions at highest luminosities see Wolf's contribution at this conference. In addition we will neglect time dependent and stochastic effects (for the latter see e.g. Traving, 1975, Gierens, Traving, and Wehrse, 1986, or Albrecht, 1986).

The equation for the stationary transport of unpolarized radiation can be written
(i) in the Eulerian or observer's frame (see e.g. Mihalas, 1978, or Cannon, 1985):

$$\left(\mu \frac{\partial}{\partial r} + \frac{1}{r}(1-\mu^2)\frac{\partial}{\partial \mu} \right) I = -(\varkappa + \sigma) I + (\varkappa + \sigma) S \tag{1}$$

with

μ = cosine between the normal and the ray direction
r = radial coordinate
I = specific intensity

$\varkappa$ = absorption coefficient
σ = scattering coefficient
S = source function

or
(ii) in the Lagrangian or comoving frame

$$\left\{ \mu_0 \frac{\partial}{\partial r} + \frac{1}{r}(1-\mu_0^2)\frac{\partial}{\partial \mu_0} - \frac{v_r v}{cr}\left[(1-\mu_0^2)+\mu_0^2 \frac{\partial \ln v}{\partial \ln r}\frac{\partial}{\partial v_0}\right]I_0 = (\varkappa_0+\sigma_0)(S_0-I_0) \right.$$

(2)

with

$\mathbf{v}$ = radial velocity
ν = frequency.

The subscript 0 indicates quantities in the comoving frame.
For comparison with observations they have to be transformed
to the rest frame.

The absorption and the scattering coefficients as well as
the source function depend on radius, the angle coordinate,
the frequency, the temperature, mean intensities, and
particle occupation numbers. Whereas in comoving frame calculations the phase function is the only source of the
direction dependence and can in most cases be neglected, in
observer's frame computations all these quantities are
strongly anisotropic because in addition the Doppler effect

$$\nu = \nu_0 \left(1 \pm \mu v/c \right)$$

(3)

has explicitly to be taken into account. On the other hand,
this is at least partly compensated by the fact that the
observer's frame equation is considerably simpler.
The choice of the coordinate frame is determined by
numerical (and perhaps personal) conveniance; the physics
to be considered (e.g. complete or partial redistribution)
must in both cases be identical.

There is no general analytical solution of the partial
integro differential equations (1) or (2) known. However,
they can easily be solved if they are transformed to a
system of ordinary differential equations by discretizing
the angle-frequency space, i.e. by considering the specific
intensity I and the source function S as vectors

$$I = (I_1, I_2, I_3, \ldots, I_n)^t$$
$$S = (S_1, S_2, S_3, \ldots, S_n)^t$$

(4)

where n is the numbers of angles times the numbers of
frequencies. There are presently two schemes
used for the discretisation of the angle space (Fig. 1):
(i) rays in the configuration space, i.e. the integration
follows the propagation of the light. This discretisation
has the advantage that the peaking effect (see below) is

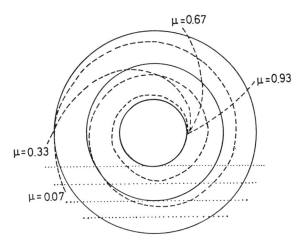

Fig. 1. Discretisation schemes for spherical radiative transfer equations. Dotted lines indicate light rays with "impact" parameters p= 1.0, 1.5, 2.0, and 2.5. The logarithmic spirals represent the discretisation $\mu(r)$=const for a Gaussian 4 point division.

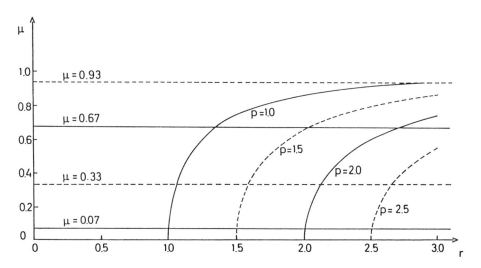

Fig. 2. Representation of the same discretisation schemes as in Fig. 1, but in the r x μ space. The horizontal lines give $\mu(r)$= const for the Gaussian division and the curves indicate the light rays.

well represented. It has the disadvantage that the angles to
the radial direction vary with r and therefore makes divergence-free formulations impossible.
(ii) constant angles to the normal direction. The corresponding curves in the configuration space are logarithmic
spirals. With this discretisation a divergence-free form of
the transfer equation can be constructed (Grant and
Peraiah, 1972, Peraiah, 1984) and for a Gaussian division
in μ the numerical accuracy is already very high for a
rather small number of number of angles, if the radiation
does not peak too much. Such a discretisation is therefore
particularly advantageous for problems of high optical
depths and small to moderate geometrical extensions.

Note that both discretisations must be considered equally
"natural", since it depends only on the mapping whether they
are represented by straight lines (the first one in the
configurations space, the second one in the r x μ space,
see Fig. 2) or by curves (r x μ space for the first one,
configuration space for the second one).

Both forms of the transfer equation (eqs. 1 and 2) can
now be written in matrix notation

$$\frac{d}{dr} I_r = A_r I_r + Q_r \tag{5}$$

where the coupling matrix A contains all terms proportional
to I and Q describes the photon sources.
Now the solution can be expressed in terms of the transition
matrix

$$I_r = \phi_{r,0} I_0 + \int_0^r \phi_{r,r'} Q_{r'} \, dr' \tag{6}$$

The transition matrix ϕ is defined by the matrix
differential equation (see Bronson,1970)

$$\frac{d}{dr'} \phi_{r,r'} = A \, \phi_{r,r'} \tag{7}$$

with the initial condition

$$\phi_{r',r'} = 1 \tag{8}$$

This formalism for the solution of a system of differential
equations is frequently used in quantum electrodynamics (
cf. e.g. Louisell, 1973), but to my knowledge has not yet
been employed in radiative transfer theory. The main reason
seems to be that eq. 6 hardly can be used in numerical

calculations because in the form given above it contains
terms that increase exponentially with the optical depth and
can hardly be manipulated by a computer (however, see below). Therefore, either the transfer equation or the
solution has first to be transformed analytically into a
form suitable for numerical evaluations. In the literature a
large variety of such transformations are found ; they can
be summarized in the following way (see also Kunasz,
1985):
a) one-step-solutions
They are possible if the source function is known or it can
be expressed as a linear function of the specific intensity.
This is e.g the case for line radiation from a
two-level-atom, where

$$S = \varphi^{-1} \left\{ (1-\varepsilon) \iint R\, I\, d\mu'\, d\nu' + \varepsilon B \right\} \tag{9}$$

(Mihalas, 1978). Then the differential equation for the
transition matrix can be transformed into differential
equations for the transmission and reflection matrices
(Schmidt and Wehrse, 1987). The system may be stiff, but
the corresponding solutions contain only decreasing and
weakly inreasing terms. Although this method in many cases
may not be economical due to the large matrices involved and
the small integration step-size required it provides a way
for obtaining quasi-exact solutions of the general problem.
If A and/or Q have a simple structure or - on physical
grounds can be simplified - the solution of eq. can be obtained by quadratures, often even analytical expressions can
be given. This is e.g. the case if the depth dependence of
the extinction coefficients and of the source function can
be approximated by polynomials (Schmid-Burgk, 1975) or if
the line profile is narrow and the radial velocity field has
a large gradient (Sobolev, 1960).

b) iterative methods
These methods, which may be used in parallel with the
solution of the rate equations, may be divided into the
following groups:
(i) moment methods, in which differential equations
for the angle-moments of the specific intensity are
solved with closure relations that are updated
iteratively. The most well-known method of this group
is the Feautrier-methods with variable Eddington factors
(Mihalas, 1978).
(ii) Newton-Raphson iterations, in which the transfer
equation and the equation determining the source function
are paramterized (or discretized) with respect
to depth and solved by linearisation (i.e. a Newton Raphson
method).
(iii) operator perturbation methods, in which the matrix A
(which is considered as an operator) is written as

$$A = A_0 + A_1 \tag{10}$$

and A_1 is considered to be a perturbation. The solution then proceeds in complete analogy to the Dyson expansion in quantum electrodynamics (Dyson, 1949 a,b). Note that for the continuous depth space the iteration converges, if only all coefficients are bounded, but that in the discrete space in general the differentiation operator is also perturbed and then for convergence also the condition

$$\| 1 - A_1^{-1} A_0 \| < 1 \tag{11}$$

is required (Kalkofen, 1985). This method has recently become very popular because decompositions of A have been devised (Scharmer, 1984) in which the zero order term contains already the largest part of the information and still can be very quickly evaluated.

(c) Monte Carlo methods
In this method the paths of individual photons are followed and the specific intensity is determined a posteriori by photon statistics; for an example see Lee and Meier, 1980. Complicated redistribution functions and/or deviations from sphericity can easily be included, but as for the one-step-methods the source function must begiven explicitly or by an expression linear in the specific intensity. For reaching a high accuracy this method is in many cases very time consuming.

In addition to the "pure" methods summarized above, various combinations e.g. in perturbation approaches are possible and have been attempted.

3. SOME CHARACTERISTICS OF SPHERICAL RADIATIVE FIELDS

The radiation fields in spherical configurations differ from those in plane-parallel geometry in many respects:
(i) The primary difference is the dilution of the field, i.e. the fact that for a conservative system the flux is proportional to r^2 , whereas for a plane-parallel medium it is constant. In numerical calculations this effect can easily be taken into account by replacing the intensity I by $I' = 4 \pi r^2 I$, since the transfer equation keeps its form under this transformation. Although it is often rather small (e.g. in red giant atmospheres with $\Delta r/r \approx 1.e-2$)it may have severe consequences, for instance by "switching on" the formation of molecules like H_2O, which absorb strongly and change the temperature structure in the optically thin layers.

(ii) If a photon in a sphere travels outwards (without interactions) its angle to the normal decreases contiuously ("peaking effect", Fig. 1). It is particularly important, if the medium is optically thin and geometrically very extended. In the transfer equation the second term on the left hand side takes care of this effect. Since an accurate discrete representation of this term is inhibited by the requirement that no photons should be generated or lost by the numerical evaluation of the source term, it mainly causes the complications of spherical radiative transfer (for details see Schmidt and Wehrse, 1987).

(iii) The escape probability of photons travelling in directions other than the normal one is larger since for them the optical depth to the surface is lower than in plane parallel geometry. This implies that the mean intensity cannot build up as high in spherical geomtries and therefore the radiation pressure (in particular from lines) and fluorescence efficiences are smaller.

(iv) In a sphere all radial velocity fields lead to velocity gradients (at least in the transversal direction) and affect directly and in a depth dependent way the radiation field whereas in slabs $v(r)$ = const. only means a global Doppler shift.

(v) Since the continuous absorption is usually weak in circumstellar shells, for small velocities often the lines are extremely optically thick and therefore deviations from complete redistribution become significant (Hubeny, 1985).

4. SOME EXAMPLES FOR MODEL CALCULATIONS

The most significant effects of spherical radiative transfer are found for cool giants and supergiants since many of these stars have extended photospheres (Watanabe and Kodaira, 1978, 1979; Schmid-Burgk, Scholz and Wehrse, 1981) and are surounded by huge envelopes. In the photospheres of luminous M stars the geometrical extension "switches on" the formation of water vapor which makes the outer parts several hundert degrees cooler than expected from plane parallel models. This temperature decrease shows up in the depths of the molecular lines and bands.

Main diagnostic tools for these outer layers are CO lines, since a) they show up in all cool giants; b) they form both in the outer photosphere (Δv = 2 lines mainly with $\lambda \approx$ 2.3μ , see e.g Höflich et al., 1986) and in the envelope (Δv = 1 lines with $\lambda \approx$ 4.5μ, cf. Sahai and Wannier, 1985) and in both spectral ranges lines of rather different excitation potential are visible; c) accurate transition probabilities are available (Tipping, 1976); and d) the profiles can be well observed by means of Fourier transform spectroscopy (Maillard, 1974, Hinkle, 1978).

By fitting such CO $\Delta v = 2$ profiles from Her Höflich et al.
(1986) can show that the lines are formed in LTE and that
the photosphere and the envelope must be separated by a
chromosphere in which the CO molecule is destroyed. The $\Delta v =$
1 emission lines from the envelope of the carbon rich object
IRC +10216 have been studied in detail by Sahai and Wannier
(1985). Using radiative transfer calculations in the
Sobolev approximation (1960) and statistical equilibrium
level populations for several rotation lines of the P and
the R branch and taking into account different apertures
they are able to derive from the observed line strengths
reliable information on the temperature distribution and the
mass loss rate.

Whereas for these cool stars the photosphere and the envelope can be calculated separately due to the small Rosseland
opacity of the shell, this is no longer possible for hotter
stars in which the envelope is ionized: By physically
consistent NLTE models for Be stars Höflich (1986) demonstrates that level occupations and the temperature structure
in the photosphere are strongly influenced by the density in
the shell. If he takes this effect fully into account, he is
able to reproduce the line strenghts and profiles, Balmer
jumps etc. of all Be stars, for which reliable data have
been published, with a smaller number of free parameters
than previously considered to be necessary.

An example for an unexpected result is provided by the
atmosphere of a supernova type II atmosphere during the
coasting phase, which may be considered as an extreme
circumstellar envelope: Although the density is very low and
Thomson scattering by far prevails over absorption, the
continuum is formed in LTE (Höflich, Wehrse,and Shaviv,
1986). The cause for this unusual behaviour is that under
these conditions the electron scattering is so strong that
the radiation field becomes essentially local again, as
could be shown by test calculations in which the Thomson
cross-section was artificially decreased and the level departure coefficients immediately increased.

These few examples show that spectral features calculated
from present day models for spherical envelopes can
successfully be used to interpret observations. On the other
hand, our knowledge on the physics of such regions (in
particular the hydrodynamics and its interaction with the
thermodynamics and the radiation field) is still rather
limited since simple concepts are missing and even with a
large computer it is today just possible to calculate for a
given velocity distribution the radiation field and the
occupation numbers of a few levels ($\lesssim$ 100) consistently.
Fortunately, this also means that much better models and new
effects can be expected in near future, when a new
generation of machines becomes available.

Acknowledgement: This work was supported by the Deutsche Forschungsgemeinschaft (SFB 132).

References

Albrecht, M. : 1986, Ph.D. Thesis, Frankfurt University.
Beckman, J.E., Crivellari, L., eds. : 1985, Progress in Stellar Spectral Line Formation Theory, Reidel, Dordrecht.
Bronson, R. : 1970, Matrix Methods, Academic Press, New York.
Cannon, C.J.: 1985, The Transfer of Spectral Line Radiation, Cambridge University Press, Cambridge.
Dyson, F.J. : 1949a, Phys. Rev. 75, 486.
Dyson. F.J. : 1949b, Phys. Rev. 75, 1736.
Gail, H.P., Sedlmayr, E. : 1986, Proc. Irsee Conf. on Interstellar Matter, in press.
Gierens, K.M., Traving, G., Wehrse, R.: 1986, J. Quant. Spectrosc. Radiat. Transfer, in press.
Grant, I.P., Peraiah, A. : 1972, Mon. Not. R. astr. Soc., 160, 321.
Hearn, A.G. : 1987, this volume.
Hinkle, K. : 1978, Astrophys. J. 220, 210.
Höflich, P.A.: 1986, Ph.D. Thesis, Heidelberg University.
Höflich, P.A., Wehrse, R., Shaviv, G.: 1986, Astron. Astrophys. 163, 105.
Höflich, P.A., Lowe, R.P., Moorhead, J., Scholz, M., Wehlau, W., Wehrse, R. : Mon. Not. R. astr. Soc. 220, 377.
Holzer, T. : 1987, this volume.
Hubeny, I. : 1985, in: Progress in Stellar Spectral Line Formation Theory, Beckman, J.E., Crivellari, L., eds., Reidel, Dordrecht, p. 27.
Kalkofen, W. : 1985, in: Progress in Stellar Spectral Line Formation Theory, Beckman, J.E., Crivellari, L., eds. , Reidel, Dordrecht, p. 153.
Kunasz, P.B. : 1985, in: Progress in Stellar Spectral Line Formation Theory, Beckman, J.E., Crivellari, L., eds., Reidel, Dordrecht, p. 319.
Lee, J.-S., Meier, R.R. : 1980, Astrophys. J. 240, 185.
Louisell, W.H. : 1973, Quantum Statistical Properties of Radiation, J. Wiley & Sons, New York.
Maillard, J.P. : 1974, Highlights of Astronomy 3, 269.
Mihalas, D. : 1978, Stellar Atmospheres, W.H. Freeman, San Francisco.
Omont, A. : 1985, in: Mass Loss from Red Giants, Morris, M., Zuckerman, B., eds., Reidel, Dordrecht, p. 269.
Peraiah, A.: 1984, in: Methods in Radiative Transfer, Kalkofen, W., ed., Cambridge Universty Press, Cambridge, p. 281.
Roddier, F., Roddier, C., Karovska, M. : 1985, in: Mass Loss from Red Giants, Morris, M., Zuckerman, B., eds.,

Reidel, Dordrecht, p. 63.

Sahai, R., Wannier, P.G. : 1985, Astrophys. J. **299**, 424.

Scharmer, G.B.: 1984, in: Methods in Radiative Transfer,
    Kalkofen, W., ed., Cambridge University Press,
    Cambridge, p. 173.

Schmid-Burgk, J. : 1975, Astron. Astrophys. **40**. 249.

Schmid-Burgk, J., Scholz, M., Wehrse. R. : 1981, Mon.
    Not. astr. Soc. **194**, 383.

Schmidt, M., Wehrse, R. : 1987, in: Numerical Methods in
    Radiative Transfer, Kalkofen, W., ed., Cambridge
    University Press, Cambridge, in press.

Sobolev, V.: 1960, Moving Envelopes of Stars, Harvard
    University Press, Cambridge/Mass.(Russian Edition 1947).

Tipping, R.H. : 1976, J. molec. Spectrosc. **61**, 272.

Watanabe, T., Kodaira, K. : 1978, Publ. Astr. Soc. Japan,
    30, 21.

Watanabe, T., Kodaira, K. : 1979, Publ. Astr. Soc. Japan,
    31, 61.

STUDIES OF VARIABILITY OF CIRCUMSTELLAR H_2O MASERS

G. M. Rudnitskij
Sternberg Astronomical Institute
Moscow State University
Moscow V-234, 119899
USSR

From March 1980 to December 1983, the author took part in regular observations of variability of maser radio emission in the H_2O line at 22 GHz. The observations were carried out at the 22-meter radio telescope of the P. N. Lebedev Physical Institute (USSR Academy of Sciences) in Pushchino (Moscow Region). The interval between consecutive observational sessions was usually 1.5-2 months. The observational program included 21 late-type variable stars (Miras and SRs): R Aql, RR Aql, RT Aql, SY Aql, U Aur, NV Aur, RX Boo, VY CMa, S CrB, KY Cyg, NML Cyg, U Her, W Hya, X Hya, R Leo, U Lyn, U Ori, UU Peg, VX Sgr, RS Vir, RT Vir. The results for eight stars ending June 1982 were published by Berulis et al. (1983). A comparison was made between the time dependences of the H_2O line radio flux F and the curves of visual and near-infrared brightness of the stars. Miras (R Aql, R Leo, U Ori, U Aur), as a rule, have a rise in F connected with the visual maximum (phase 0), the maximum F occurring at phases 0.1-0.2 (see figure for an example). Not all visual maxima (only one out of each two or three) are accompanied by H_2O flares. This Miras' behaviour was also noted earlier in the H_2O line by Berulis et al. (1984), Gómez Balboa and Lépine (1986), as well as in the SiO maser line v=1, J=2-1 by Nyman and Olofsson (1986).

Two models of H_2O line variability, connected with propagation of periodic shock waves in the inner layers of circumstellar shells (where H_2O maser emission is generated), are suggested. Model 1 connects the H_2O flux rise with non-saturated amplification at the H_2O line frequency of free-free radio continuum emission, originating in hot ionized gas behind the shock front. Model 2 explains H_2O maser bursts by fast dissipation of the shock-wave energy in the region of H_2O line generation. As a test, parallel observations of the H_2O line, $H\alpha$ emission, and cm-wave continuum can be proposed. In Model 1, there must be net

267

I. Appenzeller and C. Jordan (eds.), Circumstellar Matter, 267–268.

correlation between F H_2O, on one hand, and radio continuum (yielding the background input for the maser) and, accordingly, Hα, on the other. In Model 2, the H_2O flare must follow in time the moment of Hα and radio continuum extinction, when the shock enters the region of maser generation.

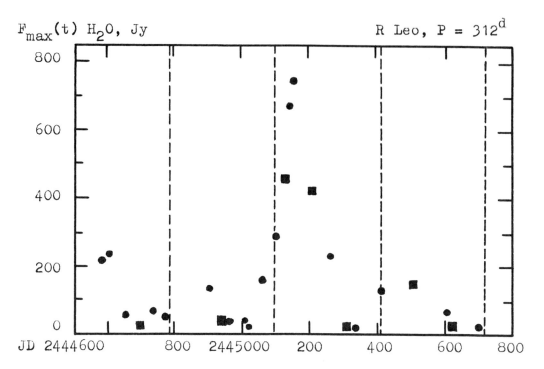

Time dependence of H_2O maser emission for the Mira-type variable R Leo in 1981-1983. $F_{max}(t)$ H_2O is the flux density in the maximum of the main H_2O emission peak. Optical maxima of the star (taken from the Bulletin de l'AFOEV) are marked by vertical dashed lines. Dots - this work, squares - data from Nyman and Olofsson (1986).

REFERENCES

Berulis, I.I., Lekht, E.E., Pashchenko, M.I., Rudnitskij, G.M. 1983. Soviet Astr., 27, 179.

Berulis, I.I., Gladyshev, A.S., Lekht, E.E., Pashchenko, M.I., Rudnitskij, G.M., Sorochenko, R.L., Khozov, G.V. 1984. Sci. Inf. Astr. Council Acad. Sci. USSR, 56, 92.

Gómez Balboa,A.M., Lépine,J.R.D. 1986. Astr. Ap., 159, 166.

Nyman, L.-Å., Olofsson H. 1986. Astr. Ap., 158, 67.

MASS-LOSS FROM COOL STARS

WHAT IS THE ESSENTIAL PHYSICS OF MASS LOSS FROM LATE-TYPE STARS?

Jeffrey L. Linsky[1]
Joint Institute for Laboratory Astrophysics, University
of Colorado and National Bureau of Standards, Boulder,
Colorado 80309-0440

ABSTRACT

In this review I consider what clues the data are providing us concerning the mass loss from late-type stars. I consider in turn the
major classes of mass-loss mechanisms (thermally-driven winds, radiatively-driven winds, and wave-driven winds), and consider whether the
empirical mass loss rates and other data are consistent with any of
these mechanisms acting alone. It is likely that several mechanisms
act together to produce the large mass loss rates in the Mira and nonpulsating M supergiants. Studies of the solar atmosphere suggest that
thermal bifurcation driven by molecular condensation instabilities may
play a critical role in cooling the atmospheres of luminous cool stars
and forming silicate dust. It is possible that several metastable
modes of atmospheric structure may exist for a given set of stellar
parameters.

1. INTRODUCTORY REMARKS

Historically mass loss has been investigated because of the roles it
can play in stellar evolution and because it is one important process
(the other being stellar explosions) that chemically enriches the
interstellar medium. In a recent review Iben (1985) pointed out that
for stars less massive than 10 $M_\odot$, mass loss does not appreciably affect evolution on the main sequence, but when stars exhaust the hydrogen and helium fuel in their cores rapid mass loss from the asymptotic giant branch stars leads to rapid evolution to white dwarfs
(M < 1.4 $M_\odot$) rather than supernovae, which would have occurred had the
stellar mass remained above 1.4 $M_\odot$. This example calls attention to

---

[1]Staff Member, Quantum Physics Division, National Bureau of Standards.

271

I. Appenzeller and C. Jordan (eds.), Circumstellar Matter, 271–287.

the intimate role mass loss plays in stellar evolution scenarios.
Cassinelli (1979), Dupree (1981,1986), Dupree and Reimers (1987),
Wannier (1985), Linsky (1985), Goldberg (1985), Knapp and Morris
(1985), Drake (1986), and Lada (1985), among others, have reviewed
empirical estimates of mass loss rates which employ a broad range of
techniques. While these techniques are being pursued vigorously,
uncertainties in the empirical estimates remain large, especially at
the critical pre-main sequence and the asymptotic giant branch stages
of evolution. Stellar evolution calculations, therefore, are generally based on parameterized scaling laws such as the so-called
Reimer's law $\dot{M} = cL/gR$, where $c \approx 4 \times 10^{-13}$ in solar units (Reimers
1975). Such laws are often applied to ranges of stellar parameters
far beyond the range for which they were derived. In particular, Iben
(1985) called attention of the failure of this law to account for the
low rate of supernovae in the galaxy and thus mass loss during the
asymptotic giant branch. Furthermore, the use of such parameterized
scaling laws implicitly assumes that mass loss depends only on the
stellar parameters (i.e., L, g, R), independent of stellar age, interior structure, and history. None of these assumptions may be valid.

In reading through the literature I sense a strong, almost philosophical, predilection toward a clean and simple approach to the topic
of mass loss involving:

(1) the use of various diagnostics to measure mass loss rates for
stars in different regions of the H-R diagram and stages of evolution;

(2) the determination of semi-empirical scaling laws that relate
mass loss rates to such stellar parameters as L, T_{eff}, R, M, g, and
chemical composition;

(3) the identification of plausible mass loss mechanisms for different regions of the H-R diagram consistent with the scaling laws;

(4) the prediction of observables on the basis of the mass loss
mechanisms and the critical test of the mechanisms by the comparison
of predictions with observations.

This straightforward approach may, unfortunately, ignore some of
the essential physics of mass loss. For example, stars are likely
spatially complex. The Sun may provide us with a useful prototype in
that the mass loss rate is negligible in magnetically closed regions
so that all the mass loss occurs in the magnetically open regions
primarily at the poles. Second, mass loss can be transient as is
likely the case for pre-main sequence stars (Mundt 1984; Herbig 1977).
Third, several mechanisms may operate together (examples will be given
below) so that no one mechanism may explain mass loss for each type of
star. Finally, certain instabilities and nonlinearities may be
important such that the mass loss rate may depend nonuniquely on the
stellar parameters. Thus history may be important and stars could
have "individuality." If so, then the mass loss rate for a given star
at any given time may be overconstrained and thus not predictable.

My purpose is not to disparage the study of mass loss from stars,
but rather to point out the need to identify the essential physics of
mass loss and to confront theory with observations continuously. In
this review, I will attempt to do so by discussing each proposed mass
loss mechanism and the relevant data together.

2. MASS LOSS MECHANISMS: GENERAL CONSIDERATIONS

There are a number of excellent reviews of mass loss mechanisms relevant to late-type stars including reviews by Cassinelli (1979), Holzer (1980,1987), Castor (1981), and Holzer and MacGregor (1985). Conceptually, the three broad classes of mechanisms may be distinguished by the relative importance of different terms in the momentum equation for steady-state radial flows,

$$u \frac{du}{dr} + \frac{GM}{r^2} + g_T + g_R + g_W = 0 \quad , \tag{1}$$

where u is the mean flow spread, G is the gravitational constant, M is the stellar mass, and r the radial distance. The term $g_T = (1/\rho)dP/dr$ represents thermal pressure acceleration, g_R is the radiative acceleration, and g_W is the wave pressure acceleration. We distinguish three regimes:

(1) When $g_T > g_R,g_W$, the wind is thermally-driven. This is the mechanism first proposed by Parker (1958) and commonly thought responsible for the solar wind, but as discussed by Holzer (1979) this mechanism may not be responsible for all of the solar wind acceleration.

(2) When $g_R > g_T,g_W$, the wind is radiatively-driven. In hot stars radiation pressure on resonance and subordinate lines of abundant ions in the ultraviolet generally has been assumed to be an important acceleration mechanism. Below we consider radiation pressure on grains as a mechanism for accelerating winds in M supergiants.

(3) When $g_W > g_T,g_R$, the wind is wave-driven. Below we describe two variants of this mechanism, acceleration by Alfvén waves and by periodic shock waves.

This classification says nothing directly about the energy equation, the temperature distribution, the geometry, the role magnetic fields play in channeling the flow, or the heating mechanism. These aspects of the stellar wind problem implicitly determine the magnitudes of g_T, g_R, and g_W, and are important in determining the asymptotic flow speed and mass loss rate. They also provide all of the complexity and subtlety to the stellar mass loss problem.

3. THERMALLY-DRIVEN WINDS

Parker (1958) first presented the solution to the momentum equation for an isothermal, steady-state, radial flow that satisfies boundary conditions. In this solution the flow goes through a critical point between subsonic flows ($r < r_{crit}$) and supersonic flow ($r > r_{crit}$). At the critical point, the temperature, T_{crit}, is

$$T_{crit} = 8 \times 10^6 \text{ K } \left(\frac{M}{M_{sun}}\right) \left(\frac{r_{sun}}{r_{crit}}\right) \quad . \tag{2}$$

It is important to recognize the <u>inverse relationship</u> between T_{crit} and r_{crit}. The mass loss rate is

$$\dot{M} = 4\pi \; r^2 \rho v \quad , \tag{3}$$

and when hydrostatic equilibrium is valid

$$\rho(r) = \rho_0 \; e^{-r/H} \quad , \tag{4}$$

$$H = \frac{kT_{cor}}{\mu g} \; . \tag{5}$$

For the Sun, empirically $T_{cor} \approx 2 \times 10^6$ K, so that the density scale height $H \approx 0.15 \; r_{sun}$ and $r_{crit} \approx 4 \; r_{sun}$. Thus for the Sun

$$\dot{M} = 4\pi \; (4 \; r_{sun})^2 \; \rho_0 \; e^{-25} \; v \quad ,$$

which is a very small number ($\approx 10^{-14}$ M_{sun} yr^{-1}).

This very simple calculation is instructive because it highlights the roles played by T_{cor}/T_{crit} and by the hydrostatic equilibrium assumption. When T_{cor}/T_{crit} approaches unity, r_{crit} approaches the photosphere where the densities are large and the mass loss rate becomes large. Conversely, for stars with $T_{cor}/T_{crit} \ll 1$, the mass loss rate is negligible due to the exponential decrease in density out to the distant critical point. However, if one can greatly increase the density at the critical point either by dynamical events, turbulent motions, or by the input of momentum by waves, then the mass loss rate will increase in proportion to this density increase. An essential point is therefore to investigate conditions for which the effective density scale height can exceed the thermal value. The large photospheric linewidths in α Ori (cf. Goldberg 1979) imply that scale heights in M supergiants can be far larger than thermal.

3.1.   Empirical Estimates of Coronal Temperatures

Whether or not a thermally-driven wind can produce significant mass loss depends on the temperature of the hot gas in the stellar corona, the density at the radial distance of the critical point corresponding to $T_{crit} = T_{cor}$, and whether or not the hot plasma is confined by closed magnetic loops. The latter two questions are difficult to answer, but the <u>Einstein</u> X-ray Observatory has provided us with valuable information on which types of stars emit X-rays and thus have coronal gas hotter than 1×10^6 K (cf. Rosner, Golub and Vaiana 1985 for a recent review). The X-ray data for late-type stars indicate that dwarfs have coronae with 10^6-10^7 K plasma as do giants earlier than about spectral type K2 III (Ayres <u>et al</u>. 1981; Haisch and Simon 1982). No bright single giants and supergiants have been detected, except for Canopus (F0 Ib-II), β Dra (G2 Ib-II), and α TrA (K4 II) (Brown 1986), and the upper limits are often much lower than for the Sun. Pre-main sequence stars are often detected as bright X-ray sources as are the "naked T Tauri" stars, which are as young as the classical T Tauri stars but without evidence for circumstellar gas (Walter 1986). Each of the stellar types detected as X-ray sources could have thermally-driven winds if the hot plasma is not magnetically confined.

Plasma as hot as 150,000 K can be detected by IUE as emission in the C IV 1550 Å and N V 1240 Å lines. The IUE observations of late-type stars (see Linsky and Jordan 1987 for a recent review) are consistent with the findings from Einstein. In particular, dwarfs, giants earlier than spectral type K2 III, and pre-main sequence stars emit spectral lines indicative of plasma at least as hot as 150,000 K. One of the important results from IUE was the discovery of a fundamental change in atmospheric structure as one proceeds to the right in the H-R diagram from the yellow giants (spectral types earlier than K1 III) to the red giants. Whereas the yellow giants have spectra with all of the high-temperature lines present, the spectra of red giants contain none of these lines (Linsky and Haisch 1979; Simon, Linsky and Stencel 1982). Instead red giant spectra contain such low temperature species as O I, Si II, Fe II, and S I, as well as blue-shifted absorption features in the Mg II and Ca II lines indicative of cool winds (Stencel and Mullan 1980). Dupree (1986) and Dupree and Reimers (1987) have summarized the evidence for the onset of cool winds in the red giants. The hottest plasma in the well-studied red giant α Boo (K2 III), as indicated by the Ly-α, C II 2325 Å multiplet, and Si III] 1892 Å lines (Ayres, et al. 1987), is probably cooler than 20,000 K. These IUE data indicate that the red giants cannot have appreciable thermally-driven winds.

IUE observations also led to the discovery of a new class of stars called the "hybrid" stars (Hartmann, Dupree and Raymond 1980) which are K bright giants and G supergiants with C IV and N V emission, indicative of plasma at least as hot as 150,000 K, and blue-shifted Mg II absorption features, indicative of cool winds. The highest temperature plasma in these stars is not known because only one member of the class, α TrA (K4 II) has been detected as an X-ray source (Brown 1986). The winds in these stars could be radiatively-driven if the 10^5 K gas is located at a critical point near 3 r_* (see Table 1) and the blue-shifted Mg II absorption features are formed in overlying gas that has cooled to ~6000 K. If this picture is valid, then one should see blue-shifted emission in the C IV 1548 Å resonance line and C III 1909 Å intersystem lines at roughly half the expansion velocities seen in Mg II (100-200 km s^{-1}) but no expansion has been detected in either C IV or C III (Hartmann et al. 1985; Brown, Reimers and Linsky 1986). Thus the winds in hybrid stars are probably accelerated by another mechanism and the hot plasma may be magnetically confined.

The IUE spectra of late-type giants and supergiants either show emission lines of essentially all ions up to the highest temperature lines (C IV and N V) observable by IUE or they show no lines formed at temperatures above 10,000-20,000 K. No post-main sequence star has yet been detected with a maximum temperature between 20,000 and 150,000 K, although some pre-main sequence stars may be contrary examples.

The existence of this apparently forbidden range in T_{cor} may be a simple consequence of thermal instability. McWhirter, Thonemann and

Table 1. Predicted critical temperatures for thermally-driven winds

Class	Example	Spectral Type	$\frac{M_*}{M_\odot}$	$\frac{r_*}{r_\odot}$	T_{crit} (K)		
					$r_{crit}=r_*$	$r_{crit}=3r_*$	$r_{crit}=10r_*$
Main Sequence	Sun	G2 V	1	1	8×10^6	2.7×10^6	8×10^5
Pre-Main Sequence	T Tau	K1	~1	~4	2×10^6	6.7×10^5	2×10^5
Red Giant	α Boo	K2 III	~1	25	3.3×10^5	1×10^5	3.3×10^4
Hybrid	α Aqr	G2 Ib	~5	~100	4×10^5	1.3×10^5	4×10^4
M Supergiant	α Ori	M2 Iab	~10	~1000	8×10^4	2.7×10^4	8×10^3

Wilson (1975), among others, have computed the radiative power loss, P_{rad} (ergs cm^3 s^{-1}) of an optically thin solar abundance plasma in steady-state ionization equilibrium. They find that P_{rad} rises steeply with increasing temperature until T = 15,000 K, but then is roughly constant for 15,000 K $\lesssim$ T $\lesssim$ 5 × 10^5 K. Thus if the heating rate is sufficient to force the plasma to be hotter than about 15,000 K, the maximum plasma temperature will then run away to 10^6 K where cooling by thermal conduction and wind expansion can balance the energy input. If the wind expansion is inhibited by closed field lines, then one would expect an energy balance at higher temperatures and pressures as cooling would be dominated by conduction and radiation only.

Castor (1981) proposed a somewhat different explanation for the simultaneous disappearance of hot plasma and onset of cool winds in the early K giants. He called attention to two important time scales: the radiative cooling time

$$t_c = E \left(\frac{dE_R}{dt}\right)^{-1} = \frac{nkT}{n_e^2 P_{rad}(T)} \quad , \qquad (6)$$

and the expansion time for sonic flows

$$t_{exp} = \frac{r_*}{v_{sound}} \quad . \qquad (7)$$

The radiative power loss $P_{rad}(T)$ is roughly constant over the temperature range 15,000 K $\lesssim$ T $\lesssim$ 5 × 10^5 K, but it is roughly a factor of 10^2 lower for 10^6 K $\lesssim$ T $\lesssim$ 10^7 K. To the left of the boundary, $T_{cor} \gtrsim 10^6$ K so that $P_{rad}(T)$ is small and $t_c \gg t_{exp}$. Thus the wind remains hot as it leaves the star. To the right of the boundary, $T_{cor} < 5 \times 10^5$ K so that $P_{rad}(T)$ is a factor of 10^2 larger and $t_{exp} \gg t_c$. The wind thus rapidly cools if it started hot. As a result of the radiative instability there may be no outer atmospheres with $15,000 \text{ K} \lesssim T_{cor} \lesssim 5 \times 10^5 \text{ K}$.

Recently, Antiochos and Noci (1986) and Antiochos, Haisch and Stern (1986) have found that for low gravity stars magnetic loops with $T < 10^5$ K appear to be thermally stable. This work points out the complexity of thermal stability analyses when closed magnetic fields are included, but it does not alter the previous arguments that should be valid for open or no fields, which is likely when winds are present.

4. RADIATIVELY-DRIVEN WINDS

Historically the second mechanism considered for the acceleration of winds in late-type stars was radiation pressure on circumstellar dust grains. Cassinelli (1979), Zuckerman (1980), Castor (1981), and Drake (1986) have reviewed the empirical evidence for large mass loss rates in luminous cool stars. The specific values proposed by different authors for individual stars and the functional dependence of the mass loss rate on stellar parameters are, unfortunately, in a highly con- fused state (cf. Goldberg 1979) and can provide only rough guidance concerning the mass loss mechanism.

Woolf and Ney (1969) first discovered broad emission features at 10-14 μm in M supergiants (but not carbon stars), which they argued could not be photospheric or chromospheric in origin. Instead, they argued that these emission features must be circumstellar and are probably due to thermal emission from silicates as the wavelength de- pendence of the emission feature is similar to the opacity of silicate grains like olivine. Gilman (1969) showed that the likely constitu- ents of grains that condense out of circumstellar gas are refractory silicates for oxygen-rich stars (M stars), carbon grains in carbon- rich stars (C stars), and silicon-carbide grains in stars for which the O/C ratio is close to unity (S stars). Subsequently, Gilman (1972) showed that the important physical processes in radiatively- driven mass loss are first momentum transfer from the radiation field to the grains, and then momentum transfer to the gas by collisions. Gehrz and Woolf (1971) presented infrared observations of many late- type stars and estimated mass loss rates and terminal velocities.

Subsequent development of the radiatively-driven wind theory con- sisted of treating in detail grain condensation and growth, momentum deposition on the grains, coupling of the grains to the gas, and prop- erties of the flow itself. Castor (1981), Kwok (1980), Cassinelli (1979), Nuth and Donn (1982) have summarized this work at length. Menietti and Fix (1978) showed that the flow does pass through the sonic point at the radial distance where the grains condense. Their models are consistent with $\dot{M} \approx \Delta L/v_\infty c$, where ΔL is the total power radiated by the grains in the 10 μm feature (and presumed equal to the total radiative power absorbed by the gas), v_∞ is the flow velocity far from the star, and c is the speed of light.

One can speak of winds as radiatively driven only if radiation pressure on the dust results is most of the momentum deposition to the gas <u>and</u> if this momentum deposition has occurred before the gas

achieves sufficient outward velocity to escape the star. The first
question is whether there is enough dust opacity in the wind. Assuming good momentum coupling of dust and gas, the radiative acceleration
on the gas will exceed the gravitational acceleration when

$$g_R = \frac{kL_*}{4\pi r^2 c} > \frac{GM_*}{r^2} \qquad (8)$$

(cf. Holzer 1987 for an opacity correction term to this equation).
Jura (1986a,b) argues that most of the luminosity L_* for M supergiants
is in the 1-2 μm region and for $M_* = 2 M_\odot$ and $L_* = 10^4 L_\odot$, $k(1-2 \ \mu m)$
need only be as large as 3 cm^2 g^{-1}. Since for gas/dust ratios typical
of the interstellar medium $k(1-2 \ \mu m) \approx 30$ cm^2 g^{-1}, there is probably
sufficient dust opacity to produce mass loss in M supergiants. Note,
however, that Hagen, Stencel and Dickinson (1983) conclude on the
basis of a similar calculation that there is insufficient dust opacity
for radiation pressure alone to account for the observed mass loss
rates of M supergiants.

The second question is whether there is sufficient momentum in
the stellar radiation field to explain the observed mass loss, i.e.

$$L_*/c \geq \dot{M} \ v_{exp} \ . \qquad (9)$$

This inequality is satisfied for all M supergiants (Jura 1986a),
except for a few rapidly evolving stars for which L_* could have been
ten times larger than its present value as recently as 1000 years ago.
However, the high velocity molecular (CO) outflows for many pre-main
sequence stars violate this condition (Lada 1985); thus radiation
pressure from the central objects cannot explain these outflows.

These two arguments make radiation pressure on dust a possible
candidate to explain mass loss from post-main sequence stars that exhibit evidence for circumstellar dust grains -- the M supergiants and
the C and S stars. Even for these stars, however, there are several
problems that appear to rule out this mechanism acting <u>alone</u> as the
likely cause of mass loss.

(1) Radiation pressure on grains cannot initiate the flows. The
measured properties of pure silicates like olivine, the so-called
clean grains, are such that they absorb mainly near 10 μm and very
little in the near infrared where most of the photospheric radiation
is located. As a result the grains act as inverse greenhouses so that
$T_{grain} < T_e$ and $T_{grain} < T_{rad}$. These grains can condense close to a
star, roughly 1.06 r_* for α Ori (Draine 1981), where densities are
high. However, they absorb only a small portion of the stellar light
and the resulting mass loss rates are low. By comparison, dirty
silicates absorb well throughout the near infrared and near 10 μm, so
that they evaporate close to a star. Draine (1981) calculates that
they cannot exist within 4.5 r_* of α Ori, for example. This estimate
is confirmed by the 11 μm heterodyne interferometry measurement (cf.
Sutton <u>et al</u>. 1977) that the inner radius of the dust shell is about
12 r_*, and Low's (1979) interferometric measurement that the inner

radius is at least 10 r_*. At these distances the density must be low (and thus the mass loss rate small) unless the flow of gas out to 4.5 r_* is produced by a different mechanism.

(2) Escape velocities are reached before the grains form. For α Ori, Goldberg (1979) cites evidence for $v \approx v_{esc}$ already deep in the chromosphere. Also if grains form typically at 10 r_*, then some other source of momentum deposition has already provided 90% of the work needed to lift the gas out of the stellar gravitational potential.

(3) The gas and dust column densities are uncorrelated in M supergiants (Hagen 1978; Hagen et al. 1983), contrary to expectation if radiation pressure on dust is driving the mass loss.

We conclude that radiation pressure on dust by itself is not the cause of significant mass loss anywhere in the H-R diagram. However, Jura (1986a,b) noted the excellent correlation of 12 μm excess, indicative of circumstellar dust, with pulsation for M supergiants. This suggests a two-step process in which pulsations raise material to large distances above the photosphere where grains can condense and radiation pressure can contribute to the mass loss.

4.1.   Does Dust Formation Quench Chromospheres?

Jennings and Dyke (1972) called attention to an empirical inverse correlation between Ca II H and K line emission and 9.7 μm dust emission. They concluded that chromospheres disappear just as dust appears in the early M supergiants. Jura (1986a) reexamined this inverse correlation using IRAS data with a similar conclusion, and Hagen et al. (1983) found that M supergiants with large gas/dust ratios have no apparent Ca II emission. These data all support Jenning's (1973) speculation that dust formation "quenches" chromospheres in that energy that would otherwise heat chromospheres to temperatures ($T_e > 5000$ K) where Ca II could be collisionally excited is instead radiated in the infrared by dust mixed with cool ($T_e < 1000$ K) outflowing gas. In this scenario there appear to be two stable regimes for a circumstellar envelope -- either it is completely warm (chromospheric) and not dusty, or it is completely cool and dusty.

To test this scenario Stencel, Carpenter and Hagen (1986) used IUE to observe 15 K and M giants and supergiants (excluding Miras) with different gas/dust ratios. They found that all of their sample stars, including giants as late as M5 III, have Mg II, Fe II, Aℓ II], and C II] emission features indicative of plasma at chromospheric temperatures, but the radiative losses in these lines (indicative of the heating rates in the chromosphere) in the dusty stars are an order of magnitude smaller than those for the stars with large gas/dust ratios. They concluded that dust formation can alter the outer atmospheric structure but not eliminate the presence of matter at chromospheric temperatures. These data reinforce the concept of thermal bistability within a given atmosphere (see below).

5. MASS LOSS BY PERIODIC SHOCK WAVES

Many late M giants, like Mira (gM6e), are long period variables that
show evidence of large mass loss (10^{-6}-10^{-7} M_{sun} yr^{-1}) and low terminal velocities (≈ 10 km s^{-1}). Willson and Hill (1979), Wood (1979),
Willson and Bowen (1985), and Bertschinger and Chevalier (1985) have
presented numerical calculations of the dynamic response of a Mira
star atmosphere to a periodic train of upward propagating shocks
driven by a piston located at the base of the atmosphere. These calculations may also be useful in understanding the essential physics in
other pulsating stars including the semi-regular variables, Cepheids,
and RR Lyrae stars. Nonpulsating cool giants and supergiants
generally have wide line profiles, implying turbulent velocities of
20-30 km s^{-1} (Reimers 1987). These stars may also be pulsating but
with many radial or nonradial modes.
     Willson and Hill (1979) showed that mass loss is inevitable for
a periodic train of waves by the following argument. A star has a
natural gravitational period which is the gravitational return time,
$P_0 \approx 2r_0/v_0$, for a particle with velocity v_0 to return to its radial
position r_0. When the atmosphere is driven with a pulsation period
$P = P_0$, then particles are forced into periodic ballistic orbits in
which they return to their initial positions and there is no mass
loss. However, if $P < P_0$, then particles do not have sufficient time
to return to their initial location but find themselves further from
the star when the next wave arrives. Mass loss is thus inevitable.
Furthermore, P_0 increases with increasing r such that there must be a
critical radius, r_{crit}, where the condition $P = P_0$ is satisfied. The
mass loss rate depends on the density at r_{crit} and can be very large
for stars with small values of r_{crit}/r_*. However, the outflow velocity
at r_{crit} is typically much less than the escape velocity at this point,
contrary to the situation for thermally-driven winds.
     They also called attention to several important effects. First,
particles can accumulate kinetic energy from successive shocks and
shocks can catch up to previous shocks and combine to enhance the mass
loss. Second, contrary to intuition, the mass loss far from the star
is essentially a steady flow rather than a series of discrete events
produced by individual shocks. Third, the ratio of the cooling to expansion time scales is a crucial parameter. Wood (1979), for example,
showed that in the isothermal limit (rapid cooling) there is no continuous mass loss, but rather occasional ejections of matter with a
time-averaged mass loss rate of $\sim 10^{-12}$ M_{sun} yr^{-1}. In the adiabatic
limit, however, he computed unrealistically high mass loss rates
(0.02 M_{sun} yr^{-1}). Real flows should be an intermediate case with
nearly isothermal shocks near the base where the densities are highest
and nearly adiabatic shocks at the top where the densities are lowest.
The inclusion of heating near the top of the atmosphere in the calculations of Willson and Hill (1979) results in higher pressures and
enhanced mass loss. In effect, these models begin to resemble
thermally-driven winds but with enhanced densities at the thermal
critical point due to the shock wave forces. Wood (1979) discussed
another mixed acceleration flow in which the addition of period shock

waves into a Mira atmosphere with a pre-existing wind driven by radiation pressure on grains enhances the mass loss rate by a factor of 40,
while the terminal velocity of the flow is not significantly changed.

More recently Willson and Bowen (1985) presented calculations for
periodic waves in Miras and other pulsating stars in which the adiabatic or isothermal approximations are relaxed. One important result
is that the atmospheres can become very distended, especially for long
period waves. In other words, the dynamical density scale height can
become very much larger than the static (i.e., thermal) scale height
leading to orders of magnitude increases in density. A second important point is that the radiative relaxation time τ_{rad} increases with
decreasing density and thus increasing radial position. The condition $\tau_{rad} = P$ determines the inner radius (r_{ad}) of an adiabatic zone
since beyond this point the radiative relaxation time is too long for
a shock to radiate its internal energy before the next shock appears.
For stars with $r_{ad} \ll r_{crit}$, the gas from r_{ad} out to r_{crit} is heated
and the wind is thermally driven. For stars with $r_{ad} > r_{crit}$ the gas
below r_{crit} is cool and the thermal pressure gradient is a small
contributor to the wind acceleration.

Willson and Bowen's (1985) calculations suggest that for large
mass loss rate long period variables (i.e., Miras) the winds are
driven by pulsations and radiation pressure on dust (cf. Jura 1986a),
but for small mass loss rate Miras the winds are only driven by pulsations. They speculate that the winds for RR Lyrae and short period
Cepheids are thermally driven. However, the phenomenology of Miras is
exceedingly complex and such important observations as a stationary
layer detected in CO data (Hinkle, Hall and Ridgway 1982) are not yet
explained by the theory.

6. ALFVÉN-WAVE-DRIVEN WINDS

Hollweg (1974) reviewed the extensive in situ measurements of hydromagnetic waves in the solar wind made by spacecraft. The existence of
these waves has led several authors (e.g. Belcher 1971; Parker 1975) to
suggest that undamped Alfvén waves can impart momentum to the solar wind
and thereby affect the flow properties. Recent work has concentrated
on explaining both the wind and heating of the solar corona by these
waves, but Leer and Holzer (1980) have pointed out that if Alfvén
waves deposit most of their energy beyond the critical point, then
the asymptotic flow speeds will tend to be unreasonably large.

Given that momentum deposition by Alfvén waves in the solar
corona has many attractive features, it was natural to consider this
mechanism for stars in general. An important consideration is whether
the Alfvén waves are damped or not beyond the critical point. Belcher
and Olbert (1975) assumed that the waves are adiabatic (undamped) on
the basis that Alfvén waves in astrophysical plasmas tend to be very
difficult to damp. They pointed out that winds accelerated by such
waves could be cool or hot if heated by another mechanism. Since densities are likely to fall off faster than r^{-2}, while field strengths
should be proportional to r^{-2}, the Alfvén speeds and field fluctua-

tions can be very large far from the star. Their solutions also exhibit a cutoff Alfvén flux below which there is no mass loss.

Hartmann and MacGregor (1980) applied the Alfvén wave mechanism
to late-type giants and supergiants (cf. Castor 1981). They considered Alfvén waves of low amplitude ($\delta B \ll B$, $\delta v \ll v_A$) with wavelengths small compared to the pressure scale height or variations
in any stellar parameters. They also assumed radial fields with $B = B_0(r_0/r)^2$. Since they did not consider closed loops, tension in the
field lines is negligible and they implicitly considered only regions
analogous to solar coronal holes. They found that solutions to the MHD
equations assuming no damping result in terminal velocities ≈ 300 km s^{-1},
which are unrealistically large for late-type supergiants, but not very
much larger than for the hybrid stars. Conversely, if the Alfvén waves
are highly damped (dissipation scale lengths much less than a stellar
radius), then the wave flux would be dissipated as heat in the high
density portion of the corona close to the star and there would be
negligible mass loss. Instead, they make the ad hoc assumption that
the dissipation scale length is a stellar radius and found that winds
are cool (T $<$ 10^4 K) for luminous (log g $<$ 2) stars and hot (T $>$
10^5 K) for giants and dwarfs (log g $>$ 2) with reasonable values of
terminal velocities and mass loss rates.

A number of important details must still be investigated. For
example, mass loss rates of 10^{-5}-10^{-6} M$_{sun}$ yr^{-1} are predicted for a
star like α Ori only for coronal base fields of 10 Gauss. It is hard
to imagine how dynamos in extremely slowly rotating M supergiants
could produce fields this large. Clearly the dissipation scale length
plays a critical role in determining mass loss rates and terminal
velocities (Holzer, Fla and Leer 1983) and must be calculated realistically. Finally the field lines are assumed radial so the solutions
cannot be valid for those portions of a stellar corona where the field
lines are closed. Thus the hybrid stars might be hybrid in the sense
that the cool wind originates in open field regions while the hot gas
is confined to closed loop structures. In any case Alfvén-wave-driven
winds are an attractive possibility for explaining the cool flows in
the nonpulsating K and M giants and supergiants as well as the hybrid
stars (Hartmann, Dupree and Raymond 1981) and perhaps the T Tauri
stars.

7. SIMULTANEOUS METASTABLE ATMOSPHERIC MODES AND MASS LOSS

The preceding discussion provides a general picture of a non-Mira cool
giant or supergiant atmosphere consisting of a turbulent photosphere,
a warm (5000-8000 K) chromosphere extending out to roughly 10 r$_*$ that
is distended either by turbulence or a complex interplay of radial and
nonradial pulsations, and a dusty circumstellar envelope beyond 10 r$_*$.
The mass loss mechanism may be a three-step process (i.e., Jura 1986a)
involving "levitation" by wave momentum deposition, dust formation,
and the final removal of the gas and dust by a mixture of the wave
pressure, radiation pressure on the dust, and thermal pressure terms.

The Miras may differ only in that the pulsations are easily observed because they are primarily in one radial mode and there may be no permanent chromosphere but rather transient heated gas behind the shocks. For both Miras and non-Miras, the feedback of dust formation on the existence of chromospheric gas is unclear.

Is this all of the essential physics implied by the data? The answer is probably no and the clue as to what is missing comes from an unlikely source -- the Sun. Ayres and Testerman (1981) and Ayres, Testerman and Brault (1986) showed that the infrared solar spectrum in the CO vibration-rotation bands is inconsistent with a homogeneous atmosphere but suggests instead thermal bifurcation into discrete structures (perhaps magnetic flux tubes) with steep chromospheric temperature rises and cool regions containing CO with no chromospheric temperature increases with height. Subsequent work by Ayres (1981), Kneer (1983), Muchmore and Ulmschneider (1985), and Muchmore (1986) has explored how cooling in the CO vibration-rotation bands can produce a condensation instability or molecular "catastrophe" in which the initial formation of CO, say by compression, radiatively cools the gas which produces more CO (since the association rate is highly temperature-dependent) and thus more radiative cooling. The thermal bifurcation of the solar atmosphere is thus driven by the destabilizing effect of the steep temperature dependence of the CO formation and radiative loss rate. Analogously, the interstellar medium has at least two stable thermal regimes (Field, Goldsmith and Habing 1969).

Stencel, Carpenter and Hagen (1987) and Stencel (1986) have proposed that the CO condensation instability is an example of the essential physics that occurs in the chromospheres of M supergiants. Other molecules like SiO, CS, OH and H_2O (Muchmore, Nuth and Stencel 1986) can behave in a manner similar to CO. One plausible scenario is that a chain of molecular "catastrophes" can occur in which cooling by CO and the resultant pressure perturbation produce conditions ripe for SiO condensation that triggers formation of other molecules and eventually silicate dust and perhaps also SiO maser emission. A schematic outline of this scenario is shown in Figure 1.

If detailed calculations and observations give credence to this new picture of a cool supergiant atmosphere, we must recognize that the essential physics of these stars includes thermal instabilities, dynamic phenomena, the presence of very different thermal regimes in close proximity, and several mass loss mechanisms working together. The atmospheres of these stars thus appear to be highly complex and even chaotic. We must even consider the possibility that several metastable modes of atmospheric structure may exist for a given set of stellar parameters. The theory of mass loss from these stars must properly include all of this essential physics.

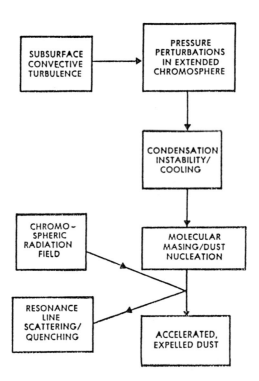

Fig. 1. A flow chart summarizing a mass scenario including the condensation instability and molecular cooling (from Stencel, Carpenter and Hagen 1986).

8. ACKNOWLEDGMENTS

I wish to thank the sponsors of IAU Colloquium No. 122 for their hospitality and financial support. This work is supported in part by NASA grants NGL-06-003-057 and NAG5-82 to the University of Colorado. I am indebted to T. R. Ayres, A. Brown, and R. Stencel for discussions and comments on this manuscript.

9. REFERENCES

Antiochos, S. and Noci, G.: 1986, Astrophys. J. 301, p. 440.
Antiochos, S., Haisch, B.M., and Stern, R.A.: 1986, Astrophys. J.
    (Letters), in press.
Ayres, T.R.: 1981, Astrophys. J. 244, p. 1064.
Ayres, T.R., Judge, P., Jordan, C., Brown, A., and Linsky, J.L.: 1987,
    Astrophys. J., in press.
Ayres, T.R., Linsky, J.L., Vaiana, G.S., Golub, L., and Rosner, R.:
    1981, Astrophys. J. 250, p. 293.

Ayres, T.R. and Testerman, L.: 1981, Astrophys. J. 245, p. 1124.
Ayres, T.R., Testerman, L., and Brault, J.W.: 1986, Astrophys. J. 304, p. 542.
Belcher, J.W.: 1971, Astrophys. J. 168, p. 509.
Belcher, J.W. and Olbert, S.: 1975, Astrophys. J. 200, p. 369.
Bertschinger, E. and Chevalier, R.A.: 1985, Astrophys. J. 299, p. 808.
Brown, A.: 1986, Advances in Space Research, in press.
Brown, A., Reimers, D. and Linsky, J.L.: 1986 in "New Insights in Astrophysics," ESA SP-263, p. 169.
Cassinelli, J.P.: 1979, Ann. Rev. Astron. Astrophys. 17, p. 275.
Castor, J.: 1981, in "Physical Processes in Red Giants," eds. I. Iben, Jr. and A. Renzini (Boston: Reidel), p. 285.
Draine, B.T.: 1981, in "Physical Processes in Red Giants," eds. I. Iben, Jr. and A. Renzini (Boston: Reidel), p. 317.
Drake, S.A.: 1986, in "Cool Stars, Stellar Systems, and the Sun," eds. M. Zeilik and D. M. Gibson (Berlin: Springer-Verlag), p. 369.
Dupree, A.K.: 1981, in "Effects of Mass Loss on Stellar Evolution," eds. C. Chiosi and R. Stalio (Dordrecht: Reidel), p. 87.
Dupree, A.K.: 1986, Ann. Rev. Astron. Astrophys. 24, in press.
Dupree, A.K. and Reimers, D.: 1987, in "The Scientific Accomplishments of the International Ultraviolet Explorer," eds. Y. Kondo et al. (Dordrecht: Reidel), in press.
Field, G., Goldsmith, D., and Habing, H.: 1969, Astrophys. J. (Letters) 155, p. L149.
Gehrz, R.D. and Woolf, N.J.: 1971, Astrophys. J. 165, p. 285.
Gilman, R.C.: 1969, Astrophys. J. (Letters) 155, p. L185.
Gilman, R.C.: 1972, Astrophys. J. 178, p. 423.
Goldberg, L.: 1979, Quart. J. Royal Astron. Soc. 20, p. 361.
Goldberg, L.: 1985, in "Mass Loss from Red Giants," eds. M. Morris and B. Zuckerman (Dordrecht: Reidel), p. 21.
Hagen, W.: 1978, Astrophys. J. Suppl. 38, p. 1.
Hagen, W., Stencel, R.E., and Dickenson, D.F.: 1983, Astrophys. J. 274, p. 286.
Haisch, B.M. and Simon, T.: 1982, Astrophys. J. 263, p. 252.
Hartmann, L., Dupree, A.K., and Raymond, J.C.: 1980, Astrophys. J. (Letters) 236, p. L143.
Hartmann, L., Dupree, A.K., and Raymond, J.C.: 1981, Astrophys. J. 246, p. 193.
Hartmann, L., Jordan, C., Brown, A., and Dupree, A.K. 1985, Astrophys. J. 296, 57.
Hartmann, L. and MacGregor, K.B.: 1980, Astrophys. J. 242, p. 260.
Herbig, G.H.: 1977, Astrophys. J. 217, p. 693.
Hinkle, K.H., Hall, D.N.B. and Ridgway, S.T.: 1982, Astrophys. J. 252, p. 697.
Hollweg, J.V.: 1974, Publ. Astron. Soc. Pacific 86, p. 561.
Holzer, T.E.: 1979, in "Solar System Plasma Physics," eds. C.F. Kennel, L.J. Lanzerotti, and E.N. Parker (Amsterdam: North-Holland), p. 101.
Holzer, T.E.: 1980, in "Cool Stars, Stellar Systems, and the Sun," ed. A.K. Dupree, Smithsonian Astrophysical Observatory Special Report No. 389, p. 15.

Holzer, T.E.: 1987, this volume.

Holzer, T.E., Fla, T. and Leer, E.: 1983, Astrophys. J. 275, p. 808.

Holzer, T.E. and MacGregor, K.B.: 1985, in "Mass Loss from Red
     Giants," eds. M. Morris and B. Zuckerman (Dordrecht: Reidel), p.
     229.

Iben, I., Jr.: 1985, in "Mass Loss from Red Giants," eds. M. Morris
     and B. Zuckerman (Dordrecht: Reidel), p. 1.

Jennings, M.: 1973, Astrophys. J. 185, p. 197.

Jennings, M. and Dyke, H.: 1972, Astrophys. J. 177, p. 427.

Jura, M.: 1986a, Irish Astron. J. 17, p. 322.

Jura, M.: 1986b, Astrophys. J. 303, p. 327.

Knapp, G.R. and Morris, M.: 1985, Astrophys. J. 292, 640.

Kneer, F.: 1983, Astron. Astrophys. 128, p. 311.

Kwok, S.: 1980, J. Roy. Astron. Soc. Canada 74, p. 216.

Lada, C.J.: 1985, Ann. Rev. Astron. Astrophys. 23, p. 267.

Leer, E. and Holzer, T.E.: 1980, J. Geophys. Res., 85, 4681.

Linsky, J.L.: 1985, in "Mass Loss from Red Giants," eds. M. Morris and
     B. Zuckerman (Dordrecht: Reidel), p. 31

Linsky, J.L. and Haisch, B.M.: 1979, Astrophys. J. (Letters) 229, p.
     L27.

Linsky, J.L. and Jordan, C.: 1987, in "The Scientific Accomplishments
     of the International Ultraviolet Explorer," eds. Y. Kondo et al.
     (Dordrecht: Reidel), in press.

Low, F.J.: 1979, in "High Angular Resolution Stellar Interferometry"
     (IAU Colloq. No. 50), eds. J. Davis and W.J. Tango (Sydney: Univ.
     Sydney), p. 15.

McWhirter, R.W.P., Thonemann, P.C., and Wilson, R.: 1975, Astron.
     Astrophys. 40, p. 63.

Menietti, J.D. and Fix, J.D.: 1978, Astrophys. J. 224, p. 961.

Muchmore, D.: 1986, Astron. Astrophys. 155, p. 172.

Muchmore, D.O., Nuth, J.A. III, and Stencel, R.E.: 1986, submitted to
     Astrophys. J. (Letters).

Muchmore, D. and Ulmschneider, P.: 1985, Astron. Astrophys. 142, p.
     393.

Mundt, R.: 1984, Astrophys. J. 280, p. 749.

Nuth, J. and Donn, B.: 1982, Astrophys. J. (Letters) 257, L103.

Parker, E.N.: 1958, Astrophys. J. 128, p. 664.

Parker, E.N.: 1975, Space Sci. Rev. 4, p. 666.

Reimers, D.: 1975, in "Problems in Stellar Atmospheres and Envelopes,"
     eds. B. Boschek, W. H. Kegel, and G. Traving (Berlin: Springer-
     Verlag), p. 229.

Reimers, D.: 1987, this volume.

Rosner, R., Golub, L., and Vaiana, G.S.: 1985, Ann. Rev. Astron.
     Astrophys. 23, p. 413.

Simon, T., Linsky, J.L., and Stencel, R.E.: 1982, Astrophys. J. 257,
     p. 225.

Stencel, R.E.: 1987, this volume.

Stencel, R.E., Carpenter, K.G., and Hagen, W.: 1986, Astrophys. J.
     308, p. 859.

Stencel, R.E. and Mullan, D.J.: 1980, Astrophys. J. 238, p. 221.

Sutton, E.C., Storey, J.W.V., Betz, A.L., Townes, C.H., and Spears, D.L.: 1977, Astrophys. J. (Letters) 217, p. L97.

Walter, F.M.: 1986, Astrophys. J. 306, p. 573.

Wannier, P.G.: 1985, in "Mass Loss from Red Giants," eds. M. Morris and B. Zuckerman (Dordrecht: Reidel), p. 65.

Willson, L.A. and Bowen, G.H.: 1985, in "The Relationship Between Chromospheric/Coronal Heating and Mass Loss," eds. R. Stalio and J. Zuker (Trieste: Astronomical Observatory), p. 127.

Willson, L.A. and Hill, S.J.: 1979, Astrophys. J. 228, p. 854.

Wood, P.R.: 1979, Astrophys. J. 227, p. 220.

Woolf, N.J. and Ney, E.P.: 1969, Astrophys. J. (Letters) 155, p. L181.

Zuckerman, B.: 1980, Ann. Rev. Astron. Astrophys. 18, p. 263.

THEORY OF WINDS FROM COOL STARS

Thomas E. Holzer
High Altitude Observatory
National Center for Atmospheric Research[*]
P. O. Box 3000
Boulder, CO 80307

ABSTRACT. The goal of this paper is to provide a framework for thinking about the various physical processes that may play significant roles in driving the massive winds of cool, low-gravity stars. First, some general theoretical considerations involving mass, momentum, and energy balance are discussed. Next, the value of the solar wind as an analog for these late-type stellar winds and for related astrophysical flows is briefly examined. Finally, four specific mass-loss mechanisms are discussed, and the possible importance of each of these mechanisms for massive winds from cool, low-gravity stars is evaluated.

1. INTRODUCTION

The two principal purposes of this review are to complement the related observational and theoretical papers presented at this conference concerning the use of the solar wind as an analog for other astrophysical flows and concerning physically meaningful ways of approaching the general mass loss problem. It is hoped that we can develop here a framework for thinking about the various physical processes that may be important in driving massive winds from cool, low-gravity stars: a framework that is useful both to researchers developing theoretical models and to those interpreting observations of these massive stellar winds. Because an extensive review of this subject was published a little more than a year ago (Holzer and MacGregor 1985), the present paper has simply been adapted from that review, with some restructuring to make it more appropriate to this conference and with some extension of the material covered to address specific questions raised at this conference. There is no attempt made here to provide a complete reference list, as that is readily available to the reader upon reference to the aforementioned review (Holzer and MacGregor 1985) and to other recent reviews (e.g., Cassinelli and MacGregor 1986; Goldberg 1984). The interested reader can find more detailed discussion of each of the four mass-loss mechanisms considered here in the review of Holzer and MacGregor (1985) and in papers referenced therein.

---

[*]The National Center for Atmospheric Research is sponsored by the National Science Foundation.

I. Appenzeller and C. Jordan (eds.), Circumstellar Matter, 289–305.

2. GENERAL THEORETICAL CONSIDERATIONS

Let us consider the steady, radial, spherically symmetric expansion of a stellar atmosphere. Although non-steady flow phenomena and substantial departures from spherical symmetry can be expected to occur in all stellar winds, a study of steady, spherically symmetric expansion provides us with a basis for understanding the more complex flow systems in real stellar atmospheres. The equations describing mass, momentum, and energy balance in our simple flow system can be written

$$ - \dot{M}_* = 4\pi\rho u r^2 = \text{constant} , \tag{1}$$

$$ u \frac{du}{dr} = - \frac{1}{\rho} \frac{dp}{dr} - \frac{GM_*}{r^2} + D , \tag{2}$$

$$ \frac{1}{\gamma-1} pu \frac{d}{dr} \ln \left(\frac{p}{\rho^\gamma} \right) = - \frac{1}{r^2} \frac{d}{dr} (qr^2) + Q , \tag{3}$$

where ρ, u, p, and q are the mass density, flow speed, thermal pressure, and heat flux density, γ is the ratio of specific heats, ρD and Q are the net volume rates at which momentum and heat are added to the flow, and G, M_*, and $-\dot{M}_*$ are the gravitation constant, stellar mass, and stellar mass loss rate. Depending on the physical processes giving rise to momentum addition ($D \neq 0$) and heat addition ($Q \neq 0$), we may also need Maxwell's equations, a radiative transfer equation, and other auxiliary equations to close our system of equations.

It will prove useful at times to write the energy balance description in the form [derived from eqns. (2) and (3)]

$$ F = (-\dot{M}_*) \left[\frac{1}{2} u^2 + \frac{\gamma}{\gamma-1} \frac{p}{\rho} - \frac{GM_*}{r} \right] + 4\pi qr^2 = F_o + F_A , \tag{4}$$

where the "o" subscript refers to a reference level, r_o, and

$$ F_A = \int_{r_o}^{r} dr' \left[(-\dot{M}_*) D + 4\pi r^2 Q \right] . \tag{5}$$

Evidently, F is the energy flux of the wind associated with advection of flow energy, enthalpy, and gravitational potential energy and with the conduction of heat. Note that the energy added to the wind as it expands [cf. F_A in eqns. (4) and (5)] results from both momentum addition and heat addition. It is most important to distinguish momentum addition from heat addition and to avoid considering energy addition and heat addition to be synonymous. These distinctions can be understood through consideration of equations (2)-(5). Such a consideration reveals that momentum addition corresponds to the addition of flow energy to the fluid through the application of a body force, while heat addition corresponds to the addition of internal energy to the fluid (and an increase in entropy of the fluid), generally through the dissipation of some energy flux (such as an energy flux associated with an electromagnetic radiation field or hydromagnetic waves).

2.1 Energy requirements

The energy-per-unit-mass of a stellar wind can be defined by [cf. eqn. (4)]

$$E = \frac{1}{2} u^2 + \frac{\gamma}{\gamma-1} v_t^2 - \frac{1}{2} v_g^2 + \frac{1}{2} v_q^2 , \tag{6}$$

where the characteristic speeds associated with internal energy (v_t), gravity (v_g), and thermal conduction (v_q) are defined by

$$v_t^2 = p/\rho , \tag{7}$$

$$v_g^2 = 2GM_*/r , \tag{8}$$

$$v_q^2 = 8\pi qr^2/(-\dot{M}_*) . \tag{9}$$

In the present context, it is most appropriate to interpret these characteristic speeds in terms of the energy available to the flow that resides in internal energy, gravitational potential energy, and a conductive energy flux. The gravitational contribution to the available energy is, of course, negative, because it reflects the work that must be done to lift the wind out of the gravitational field.

If we consider only stellar winds that are strongly gravitationally bound at the atmospheric base and whose energy flux is almost entirely in the form of directed flow energy at large distances from the star, then we can evaluate equation (6) at the atmospheric base (r_o) and at a suitably large distance from the star (r_∞) as follows

$$E_o \approx -\frac{1}{2} v_{go}^2 + \frac{1}{2} v_{qo}^2 , \tag{10}$$

$$E_\infty \approx \frac{1}{2} u_\infty^2 . \tag{11}$$

Evidently, the difference between E_∞ and E_o represents the energy-per-unit-mass that must be added to the stellar atmosphere above the atmospheric base in order to drive the wind. This energy requirement can be expressed in terms of the energy flux F_A by making use of equations (5) and (9)-(11):

$$F_{A\infty} = \int_{r_o}^{r_\infty} dr \ [(-\dot{M}_*) \, D \, + \, 4\pi r^2 Q]$$

$$\approx \frac{1}{2} \, (-\dot{M}_*) \, (u_\infty^2 \, + \, v_{go}^2) - 4\pi q_o \, r_o^2 \, .$$

(12)

If we include the conductive flux as a component of the energy flux said to be driving a stellar wind, then we can conclude that the driving energy flux, F_{do}, that must be supplied at the atmospheric base (r_o) and dissipated in the atmosphere to produce a wind characterized by $-\dot{M}_*$ and u_∞ is given by

$$F_{do} \geq F_{A\infty} + 4\pi q_o \, r_o^2 \approx \frac{1}{2} \, (-\dot{M}_*) \, (u_\infty^2 \, + \, v_{go}^2) \, .$$

(13)

Note that F_{do} may be larger than the nominal estimate of the driving energy flux, because some part of the energy flux supplied to the wind through heating can be radiated away (i.e., Q is a net heating function, and the dissipation of the driving energy flux must account for the actual heating rate, which is given by the sum of Q and the radiative cooling rate).

The information contained in equation (13) can be simply stated in words: the energy flux required to drive a stellar wind is that necessary to lift the mass transported by the wind out of the stellar gravitational field, to accelerate the wind to its asymptotic flow speed, and to supply the energy radiated away by the wind. Let us, for the moment, ignore the energy radiated by the wind, as it is generally observed to be small for the winds we consider (see, however, section 3 on thermally driven winds). In massive winds from early-type stars, $u_\infty^2 \gg v_{go}^2$, and most of the driving energy goes into accelerating the flow to its asymptotic speed. In solar-type winds (and some hybrid-star winds), $u_\infty^2 \approx v_{go}^2$, and comparable parts of the wind's driving energy go into lifting the expanding atmosphere out of the stellar gravitational field and into accelerating it to its asymptotic speed. In the massive winds from low-gravity, late-type stars (except for some hybrid stars), $u_\infty^2 \ll v_{go}^2$, and almost all of the driving energy of the wind goes into lifting the expanding atmosphere out of the stellar gravitational field.

It is possible to use equation (13) to estimate the energy flux density emanating from the atmospheric base that is required to drive a stellar wind, provided we know the values of the stellar mass and radius $(M_*$ and $R_*)$ and the wind mass loss rate and asymptotic flow speed $(-\dot{M}_*$ and $u_\infty)$. Assuming $r_o = R_*$, and expressing $-\dot{M}_*$ in units of solar masses per year, equation (13) can be written

$$\frac{F_{do}}{4\pi r_o^2} \gtrsim 3.3 \times 10^3 \left(\frac{-\dot{M}_*}{10^{-7}} \right) \left(\frac{M_*}{M_\odot} \right) \cdot$$

$$\cdot \left(\frac{400 \, R_\odot}{R_*} \right)^3 \left(1 + \frac{u_\infty^2}{v_{go}^2} \right) \ erg \ cm^{-2} \ s^{-1} \, .$$

(14)

The solar wind, which has a very low mass loss rate $(-\dot{M}_\odot \approx 2 \times 10^{-14} \, M_\odot \ y^{-1})$, is found to have a driving energy flux density requirement of about 10^5 erg cm^{-2} s^{-1}. (Note that the value of 5×10^5 erg cm^{-2} s^{-1} normally quoted for high-speed solar wind streams takes account of the non-spherically symmetric expansion of high speed streams.) For a typical K5 supergiant $(M_* \approx 16 M_\odot$, $R_* \approx 400 R_\odot$), which has a much larger mass loss rate $(-\dot{M}_* \approx 10^{-7} M_\odot \ y^{-1})$ than the sun, the driving energy flux density requirement is about 6×10^4 erg cm^{-2} s^{-1}, which is comparable to the solar case. Similarly, a typical Mira variable $(M_* \approx M_\odot$, $R_* \approx 400 R_\odot$), which has an even larger mass loss rate $(-\dot{M}_* \approx 2 \times 10^{-6} M_\odot \ y^{-1})$, has a driving energy flux density requirement of only about 5×10^4 erg cm^{-2} s^{-1}, again comparable to the solar case. Why is it that these very massive winds have atmospheric base driving energy flux density requirements comparable to that of the very tenuous solar wind? First, the radii of these stars with massive winds are so large that the gravitational potential energy barrier ($\propto R_*^{-1}$) is much smaller than that of the sun; second, the large stellar radii correspond to a stellar surface area ($\propto R_*^2$) from which the driving energy flux is supplied that is much larger than the solar surface area. In other words, despite the fact that $\dot{M}_*/\dot{M}_\odot \approx 10^7$, this is more than made up for by the fact that $(R_*/R_\odot)^3 \approx 6 \times 10^7$ [cf. eqn. (14)].

2.2. Momentum requirements

Consideration of the wind momentum flux has been found to place important constraints on the mass loss mechanism of for massive winds from hot stars, and for this reason some workers assume that momentum flux considerations should be equally useful in the study of massive winds from cool stars. Yet, in the case of cool stars, the usefulness is severely limited. Consider, for example, a wind driven purely by radiation pressure (above some level $r = r_o$ in the atmosphere), for which momentum balance is described by

$$\frac{1}{r^2} \frac{d}{dr} \left[\rho u^2 r^2 + (1-\eta) L / 4\pi c \right] = - \frac{\rho v_{go}^2 r_o}{2 r^2} . \tag{15}$$

Here, $\eta(r)$ measures the fraction of the asymptotic stellar radiation momentum flux (L/c) received by the wind during its transit from the atmospheric reference level (r_o) to a distance r. Integrating equation (15) from the reference level (r_o) to a point in the asymptotic flow regime (r_∞) yields

$$-\dot{M}_* \, u_\infty = \frac{\eta_\infty L}{c} \left(1 + \frac{v_{go}^2}{u_\infty^2} \frac{u_\infty}{2 <u>} \right)^{-1} , \tag{16}$$

where $<u>^{-1} = \int_{r_o}^{r_\infty} dr \, r_o / u \, r^2$. When the asymptotic flow speed is much larger than the gravitational escape speed at r_o (more specifically, when $v_{go}^2 \ll 2 u_\infty <u>$), then equation (16) reduces to the more familiar form

$$-\dot{M}_* \, u_\infty \approx \eta_\infty L / c . \tag{17}$$

Note that η_∞ can be greater than unity when there is multiple scattering of photons and less than unity when only a fraction of the stellar radiation field couples to the wind, but it is frequently assumed that $\eta_\infty \approx 1$. Equation (17) is applicable to many hot star winds, in which the flow accelerates rapidly ($<u> \gtrsim u_\infty/2$) and the asymptotic flow speed is much larger than the atmospheric base gravitational escape speed (viz., $u_\infty^2 \gg v_{go}^2$, with $r_o = R_*$). Its only possible applicability to cool stars, however, appears to exist when one sets the reference level far enough out in the atmosphere ($r_o \gg R_*$) that $v_{go}^2 \ll u_\infty^2$ (also, it is necessary that $u_o^2 \ll u_\infty^2$). Then it is required that some mechanism (e.g., waves) extend the atmosphere in $R_* \lesssim r \lesssim r_o$ and that radiation pressure abruptly take over from the atmospheric extension mechanism, near $r = r_o$ and accelerate the wind to its asymptotic flow speed. Even in this very special case, it is a mistake to imagine that the wind from a star for which $v_{g*}^2 \ll u_\infty^2$ is driven by radiation pressure (as might be inferred from eqn. (17)), for most of the driving energy is supplied by the atmospheric extension mechanism.

The preceding discussion indicates why one should generally avoid trying to apply momentum flux arguments to winds from cool stars, even though the same arguments have proved useful in considerations of winds from hot stars. Up to this time, it appears that such arguments have led to more confusion than clarification of our understanding of cool star winds.

2.3. Mass loss rate and asymptotic flow speed

There are some rather simple concepts relating to the mass flux of a stellar wind that can help us interpret observations of mass loss rates and asymptotic flow speeds. Consider first an isothermal, thermally driven wind: from equations (1) and (2), with $D = 0$, we obtain

$$(u^2 - v_t^2) \frac{1}{u} \frac{du}{dr} = \frac{2}{r} (v_t^2 - \frac{1}{4} v_g^2) . \tag{18}$$

Integrating equation (18) from r_o to the point where $u = v_t$ [i.e., the critical point of eqn. (18), $r_c = r_o \, v_{go}^2/4v_t^2$ (Parker 1958): cf. section 2.4)

$$-\dot{M}_* \propto u_o = v_t \left(\frac{v_{go}^2}{4v_t^2} \right)^2 \exp \left(- \frac{v_{go}^2}{2v_t^2} + \frac{3}{2} \frac{u_o^2}{2v_t^2} \right) . \tag{19}$$

For stellar atmospheres that are strongly gravitationally bound at the atmospheric base (i.e., $v_{go}^2 \gg 4v_t^2$), it is clear from equation (14) (cf. leading term in argument of exponential) that the mass loss rate increases exponentially with increasing atmospheric temperature. This strong temperature dependence of $-\dot{M}_*$ results from the atmospheric extension produced by increasing the temperature and thus the density scale height, and it is evidently directly related to an equally strong temperature dependence of the radiative flux emanating from a thermally driven wind (cf. section 3).

Note that in the above discussion of mass loss rate, only the properties of the atmosphere inside the critical point (where $u = v_t$ in the isothermal case) are considered. More generally, we can say that the mass loss rate of a stellar wind is normally primarily determined by the properties of the atmosphere inside the sonic point of the flow (the sonic point corresponds to the critical point in the isothermal case; see, however, section 2.4), because beyond this point compressive information cannot propagate upstream in the wind, and thus cannot reach the atmospheric base so as to modify u_o. This statement is also normally applicable to winds with driving forces other than the thermal pressure gradient force $(D \neq 0)$, for which equation (19) can be generalized to

$$-\dot{M}_* \propto u_o = v_t \left(\frac{r_s}{r_o} \right)^2 \exp \left[-\frac{v_{go}^2}{2v_t^2} \left(1 - \frac{2}{v_{go}^2} \int_{r_o}^{r_s} dr \; D - \frac{r_o}{r_s} \right) - \frac{1}{2} + \frac{u_o^2}{2v_t^2} \right], \quad (20)$$

where r_s is the location of the sonic point, at which the flow speed equals the sound speed. From equation (20) we can infer that energy addition by either heating (which increases v_t) or by momentum addition (corresponding to an increase in D) is effective in increasing the mass loss rate $(-\dot{M}_*)$ only if it occurs in the region of subsonic flow. This inference is also significant for the asymptotic flow speed of the wind, as we see below.

Recall from equation (13) that the asymptotic flow speed is simply related to the (net) energy added to the wind and to the mass loss rate:

$$u_\infty^2 \approx 2 \left(F_{A\infty} + 4\pi q_o \, r_o^2 \right)/(-\dot{M}_*) - v_{go}^2 \, . \quad (21)$$

Obviously [cf. eqn. (12)], $F_{A\infty}$ is increased regardless of whether energy addition to the subsonic or supersonic region of flow is increased, but (as seen above) $-\dot{M}_*$ is increased only if energy addition to the subsonic region is increased. As it happens, energy addition to the subsonic region tends to produce such a large increase in $-\dot{M}_*$, relative to the increase in $F_{A\infty}$, that u_∞ is either unaffected or slightly decreased by subsonic energy addition. In general, we conclude that the addition of energy to the region of subsonic stellar wind flow has the effect of increasing the mass loss rate $(-\dot{M}_*)$ but has relatively little effect on the asymptotic flow speed (u_∞), while the addition of energy to the region of supersonic flow increases u_∞ but has relatively little effect on $-\dot{M}_*$ (e.g., Leer and Holzer 1980). For a wind in which $u_\infty^2 \ll v_{go}^2$, most of the driving energy of the wind must, therefore, be supplied in the region of subsonic flow, and it is to this region where we must direct most of our attention when discussing the driving mechanisms of massive winds from cool stars.

2.4. Critical points

Often (e.g., in discussions on the fourth day of this conference), much is made of the critical point(s) of the stellar wind equation [e.g., eqn . (18)] in discussions of the physics of stellar winds. Although the critical point is most important from a mathematical standpoint, its usefulness in gaining physical insight is limited. We have just seen that in an isothermal, thermally driven wind, the critical point marks the point beyond which compressive disturbances cannot be transported upstream (i.e., towards the star). It is no coincidence that the critical point plays this role in the isothermal case. It will, in fact, play such a role whenever information is transported along flow lines by longitudinal waves in which the restoring force has the same acceleration dependence as does the driving force of the wind. There are, however, many circumstances in which such a condition is not met. For example, if information is transported along flow lines by high frequency sound waves that propagate adiabatically in the background isothermal flow, then the sonic point in the flow (the physically relevant point in this case) is located where $u^2 = \gamma v_t{}^2$, but the critical point is still located at $u = v_t$. Even when the sonic point corresponds to the critical point, information relevant to the acceleration of the wind may be transported upstream beyond this point. For example, if Alfvén waves play a significant role in accelerating the wind, and if the Alfvén speed exceeds the sound speed, then Alfvén waves generated in the region of supersonic flow can transport relevant information upstream into the subsonic region. In a thermally driven wind, electron thermal conduction can modify the wind's driving force by transporting heat upstream in the supersonic flow (in the presence of a positive radial temperature gradient), and in a radiatively driven stellar wind, the transport of electromagnetic radiation (at the speed of light) may have a similar effect.

In general, then, the critical point cannot be expected to mark the point in the flow beyond which information relevant to the acceleration of the wind can be transported upstream. The sonic point, which does not necessarily correspond to the critical point is more likely to fill this role. It is probably safest to view the critical point as a mathematical singularity with limited physical meaning but considerable mathematical significance. Moreover, when we hear a statement like "the wind cannot be thermally driven because the critical point would have to be too near the star" (or "inside the stellar surface", or "too far from the star", etc.), we can be relatively certain that the speaker has avoided an available physical discussion of the problem by appealing to a mathematical platitude.

3. THE SOLAR WIND: A HOT, TENUOUS WIND FROM A RELATIVELY COOL STAR

The solar wind is of interest to us, in part, because it is the best observed of all stellar winds. We must keep in mind, though, that its very existence would be difficult to infer from remote-sensing observations alone--i.e., from the sort of observations that are all we have available to us in our studies of other stellar winds. The solar wind is also of interest because it seems to represent a wind driven by two of the mechanisms we consider as candidates for driving massive cool-star winds and because it is frequently cited (appropriately or otherwise) as an analog by theorists attempting to support their ideas for driving mechanisms of a variety of astrophysical flows. In this conference, it has been referred to as an example of a bipolar flow and as an example of a wind that illustrates the combined effects of rotation and magnetic field in driving a flow along the rotation axis. Unfortunately, although we do have much to learn from the solar wind, neither of these particular references is accurate.

We cannot in this space even begin to present a full discussion of the solar wind (cf. the early work of Parker (1958, 1963, 1964a,b), as well as the review by Leer et al. 1982 and references therein), but we can use the solar wind to illustrate some of the general principles outlined in section 2, and we can also correct the two misconceptions referred to in the preceding paragraph. First, let us consider the concepts introduced in section 2.3. The best observed and most uniform type of solar wind flow is the high speed solar wind stream. These streams are best observed because they are the only type of solar wind for which a coronal source has been unambiguously identified, so that observation of the wind acceleration region is possible. In high speed streams, the asymptotic flow speed of the wind is comparable to the gravitational escape speed at the coronal base: i.e., $u_\infty^2 \approx v_{go}^2$. We can thus infer from the results of section 2.3 that comparable amounts of energy must be supplied to the flow in the subsonic and the supersonic regions. Since the solar atmosphere is strongly gravitationally bound at the coronal base

(viz., $v_{go}^2 >> 5\, v_{to}^2/2 + u_o^2/2$) the energy flux supplying the required energy to the flow
must be transported outward from the coronal base by some mechanism other than
advection. Thermal conduction is one such transport mechanism that can be expected
to be important, because the corona is so hot ($T \gtrsim 10^6$ K). Yet, in order for conduction
to supply adequate energy to the region of supersonic flow to produce $u_\infty \approx v_{go}$ without
the temperature in the subsonic being so high as to produce a larger than observed mass
flux, it would be required that radially outward thermal conduction lead to a solar wind
temperature that increases radially outward (Leer et al. 1982). Suggestions that this
may be the case (e.g., Olbert 1983, Scudder and Olbert 1983) appear at present to have
little basis (Lallement et al. 1986, Shoub 1986). It appears, therefore, that some form of
mechanical energy transport must supply energy to the region of supersonic flow, and a
prime candidate for such a transport mechanism is Alfvén waves (e.g., Hollweg 1978 and
references therein). Hence, the solar wind appears likely to be a wave-modified thermally
driven wind, with the thermal pressure gradient force lifting the wind out of the solar
gravitational field and the waves helping to accelerate the wind to its asymptotic flow
speed (through both heating and momentum addition in the supersonic region). This
concept of two driving mechanisms working together is one to remember in the following
discussions of massive winds from cool stars.

Now, what about the idea that the solar wind is some sort of analog of bipolar flows
from other stars? The idea that the solar wind is a bipolar flow apparently comes from
the attention received by high-speed solar wind streams and their solar source, coronal
holes. During the declining and minimum phases of the solar cycle, well-developed
coronal holes usually exist near the solar poles (e.g., articles in Zirker 1977). Hence, during these periods there is a tendency for two (very broad in solar latitude) regions of
well-organized high-speed solar wind to dominate the structure of interplanetary space,
and the centers of these two wind regions tend to be nearly 180° apart. It is essential to
realize, however, that the solar wind always fills all of interplanetary space (not, for
example, just the solar polar regions), and that, if anything, the mass flux density of
high speed streams is less than (by as much as 30% to 50%) that of the lower speed, less
well organized wind. In any case, the particle density in the high-speed streams is much
lower (a factor of two or more) than it is in the lower speed wind. Hence, if we were
able to obtain remote, density-sensitive observations of the solar wind (comparable to
observations of winds from other stars), we might well infer that most of the solar wind
is coming from the equatorial region, but certainly not that it is a bipolar flow from the
two polar regions. Unless we substantially modify the usual concept of bipolar flows in
astrophysical systems, it seems inappropriate to use the solar wind as an analog for such
flows, except insofar as it represents a typical atmospheric expansion from a central
gravitating body.

Finally, we may ask whether the solar wind provides supporting evidence for the
speculation that bipolar flows can be driven and self-focused by the effects of a magnetic
field frozen into a rotating body lying between the oppositely directed flows. The answer
is simply that the solar wind has little relevance to this speculation, because beyond a
few solar radii the outflowing solar atmosphere begins to take control of the solar magnetic field, stretching it into the classic spiral structure characteristic of a nearly radial
fluid flow and a "frozen-in" magnetic field that is tied to the rotating central star (e.g.,
Parker 1963). The magnetic energy density in the solar wind at the orbit of the earth is
only about 1% of the flow energy density, so the magnetic field has little effect of the
flow in the asymptotic region. This particular misuse of the solar wind as an analog for
the rotational/magnetic driving of a bipolar flow probably has resulted from the

inappropriate use of cartoons in place of sound, physically based calculations. Such calculations, if performed, might give us some feeling for whether such a driving mechanism has any hope of explaining some bipolar flows. The absence of such calculations will continue to give free reign to the unbridled imaginations and unsupported speculations of our cartoon-astrophysicists.

4. MASSIVE, THERMALLY DRIVEN WINDS FROM COOL, LOW-GRAVITY STARS

In the terminology we are using, a thermally driven wind is one in which the only significant outward force on the expanding atmosphere is exerted by the thermal pressure gradient [viz., $D \approx 0$ in eqn. (2)]. The energy supplied to the atmosphere to maintain the temperature required to produce a sufficiently large pressure gradient force can, in principle, be transported by a variety of mechanisms, including thermal conduction from a hot atmospheric base, a mechanical wave flux, and a radiative energy flux. In practice, however, the massive winds from cool stars seem themselves to be so cool that thermal conduction cannot play a significant role. Furthermore, a radiative equilibrium atmosphere for such cool stars is not hot enough to drive a massive wind through a thermal pressure gradient force alone. Thus, we are left with a mechanical energy flux as the only source (considered up to this time, at least) of energy for heating a thermally driven wind. We must remember, however, that a mechanical energy flux not only heats a wind through its dissipation, but also directly accelerates the wind through the action of a wave pressure gradient force (i.e., $D \neq 0$). If this wave force is not negligible in comparison with the thermal pressure gradient force, then we are not dealing with a strictly thermally driven wind, but rather with a wave-modified thermally driven wind, a thermally modified wave-driven wind, or simply a wave-driven wind. This point is touched on again below.

In examining the viability of a thermally driven wind, we can, at first, ignore the mechanism maintaining the atmospheric temperature and simply consider an isothermal (in the subsonic region) wind that exhibits the mass loss rate of an observed wind and a suitable atmospheric base pressure. Then we can calculate the electromagnetic radiation flux (associated with radiative cooling) emanating from such a wind and ask whether such a radiative flux is consistent with observations. Carrying out such an exercise for the range of physical parameters relevant to massive winds from cool, low-gravity stars, it has been found (Holzer and MacGregor 1985) that all such thermally driven winds would have exceedingly large radiative energy flux densities--orders of magnitude larger than the solar radiative energy flux densities at comparable temperatures. Almost all of this radiation arises from a region that extends a few hundredths to a few tenths of a stellar radius above the atmospheric base. The radial extent of this radiating region is orders of magnitude larger than that of the comparable region in the solar atmosphere, and this, of course, accounts for the magnitude of the radiative energy fluxes. Such large radiative energy fluxes would be readily detectable (they are, after all, often comparable in magnitude to the stellar luminosity and occurring at a temperature much higher than the stellar photospheric temperature), but they are not observed.

Watanabe (1981) has tried to get around this difficulty by postulating that the region of elevated atmospheric temperature required for a massive, thermally driven wind is quite narrow, so that the total radiative flux from this region is quite small. Although one can formally obtain thermally driven winds for such an atmospheric temperature structure, Watanabe's proposal has a number of difficulties associated with it (Holzer and MacGregor 1985). One of the most serious is that if such an atmospheric temperature structure were produced by the dissipation of any currently known mechanical energy flux, the wave pressure force would dominate the thermal pressure gradient force and the correctly described flow would not be thermally driven.

It seems, therefore, that the thermal pressure gradient force acting alone does not provide a viable mechanism for driving massive winds from cool stars. In addition, the preceding discussion indicates that thermal driving is a very inefficient wind acceleration mechanism when the wind density is sufficiently high that radiative energy losses become substantial. For example, in the study of isothermal winds by Holzer and MacGregor (1985), the energy required to keep the atmosphere sufficiently warm (i.e., to counter radiative losses) was generally several orders of magnitude larger than that required to lift the wind out of the gravitational field of the star and to accelerate it to its asymptotic flow speed.

5. WAVE-DRIVEN WINDS

The difficulties encountered in trying to drive massive winds from cool stars with only a thermal pressure gradient force lead us to consider the other type of driving mechanism that may be important in the acceleration of the solar wind: i.e., small to moderate amplitude mechanical waves, specifically hydromagnetic waves (e.g., Hartmann and MacGregor 1980). These waves may be either compressive (e.g., acoustic waves or magnetoacoustic waves: more generally, slow or fast hydromagnetic waves) or non-compressive waves (intermediate hydromagnetic--i.e., Alfvén--waves). Hartmann and MacGregor (1980) have pointed out that the compressive waves would only be able to extend an atmosphere at its base, but not able to lift a massive wind out of the stellar gravitational field, because such waves steepen rapidly into weak shocks and are dissipated within a few pressure scale-heights of the atmospheric base. The non-compressive Alfvén waves, in contrast, are not so readily damped and can do work on an expanding stellar atmosphere over a large enough radial range to drive a quite massive wind.

The momentum addition to a wind associated with a flux of Alfvén waves arises from the wave pressure gradient and is expressed by [cf. eq. (2)]

$$D = -\frac{1}{\rho}\frac{d}{dr}\left(\frac{<\delta B^2>}{8\pi}\right) = -\frac{1}{\rho}\frac{d}{dr}\left(\frac{1}{2}\rho<\delta v^2>\right), \tag{22}$$

where $<\delta B^2>$ and $<\delta v^2>$ are the mean-square magnetic field and velocity field of the wave. The net atmospheric heating rate associated with Alfvén wave dissipation and with radiative cooling of the atmosphere can be written in the form [cf. eqn. (3)]

$$Q = f/\lambda - L_R, \tag{23}$$

where the Alfvén wave energy flux density, f, is given by

$$f = \rho<\delta v^2>\left(v_a + \frac{3}{2}u\right), \tag{24}$$

$v_a = B/\sqrt{4\pi\rho}$ is the Alfvén speed, and λ is the damping length for the waves. The damping length in the winds of interest is likely to be determined primarily by ion-frictional damping and the radiative loss rate is probably adequately expressed in terms of an optically thin radiative loss function (Hartman and MacGregor 1980). Because of the strength of radiative losses, the energy balance of a massive wind at chromospheric temperatures ($T \approx 10^4$ K) can frequently be well approximated by using $L_R \approx f/\lambda$ [cf. eqn. (23)] in place of equation (3).

If momentum balance in the flow is dominated by the Alfvén wave pressure gradient force, then (Holzer et al. 1983) the sonic point of the flow occurs near 1.75 R_*, and the mass loss rate is

$$-\dot{M}_* \approx 1.8 \times 10^{-13} \left(\frac{f_o}{10^6 B_o} \right) \left(\frac{R_*}{R_\odot} \right)^{7/2} \left(\frac{M_\odot}{M_*} \right)^{3/2} M_\odot \, y^{-1} , \tag{25}$$

where f_o (in erg cm^{-2}s^{-1}) and B_o (in gauss) are both evaluated at the atmospheric base ($r_o \approx R_*$). When the Alfvén waves are undamped (i.e., all the wave energy is transferred to the flow through direct acceleration by the wave pressure gradient force) the asymptotic flow speed of the wind is approximately (Holzer et al. 1983)

$$u_\infty \approx v_{go} \left[\frac{8}{7} \left(\frac{v_a}{u} \right)_s + \frac{5}{7} \right]^{1/2} , \tag{26}$$

where the subscript s refers to the sonic point. Evidently, an Alfvén wave driven wind will have a mass loss rate comparable to those observed for massive, cool-star winds (cf. section 2.1) even for relatively modest wave energy flux densities at the atmospheric base of $f_o \approx 10^5 B_o$ erg cm^{-2} s^{-1}. (Note that if the waves are to be small amplitude in the region of subsonic flow, and thus remain undamped by nonlinear processes, then it is required that $B_o \gtrsim 10$ gauss and that $(v_a/u)_s \gg 1$). The asymptotic flow speed, however, is much too large ($u_\infty^2 \gg v_{go}^2$) for a wind driven by undamped Alfvén waves. Hartmann and MacGregor (1980) noticed this difficulty and pointed out that damping of Alfvén waves in the region of supersonic flow would significantly reduce the asymptotic flow speed without affecting the mass loss rate. Such damping might be provided by the frictional coupling of atmospheric ions with neutral atoms. Yet, the range of damping rates that produce massive winds with low asymptotic flow speeds is found to be quite small (Holzer et al. 1983), so the wave-driven wind mechanism probably requires some sort of atmospheric self-regulation process to be operative if it is to be the dominant mechanism in driving a substantial number of massive cool star winds. Such a self-regulation mechanism, which would limit the effective wave damping length to an appropriate, narrow range, is still being sought.

6. RADIATIVELY DRIVEN WINDS

Our understanding of radiatively driven winds comes primarily from extensive studies of massive, very high speed ($u_\infty^2 \gg v_{go}^2$) winds from early-type stars (e.g., review by Cassinelli and MacGregor 1986, and references therein). The radiative flux in these hot stars exhibits a significant energy flux in the ultraviolet portion of the electromagnetic spectrum, and since the atmosphere has many strong resonance lines in this same part of the spectrum, the radiation field is well coupled to the expanding atmosphere and can add a substantial amount of momentum to a stellar wind. The situation is quite different for cool stars, in which the radiative energy flux peaks in the red or near-infrared, and the strong atmospheric resonance lines occur in the visible and ultraviolet. The resulting poor spectral match implies that the potential effectiveness of radiative driving of massive winds from these cool stars can be realized only if some other opacity source can provide for efficient coupling of the radiation field to the wind. Two possible sources of opacity are provided by molecules and dust in the stellar atmosphere, but at present only the effects of dust have been considered extensively (e.g., review by Goldberg 1984, and references therein). We can illustrate the sort of role dust (or molecules) might play in driving a massive, cool-star wind with a very simple analysis, from which we can also infer some of the problems that need to be addressed before we understand the actual roles dust and molecules play.

The force per unit mass associated with radiative driving of a stellar wind can be written in the form [cf. eqns. (2) and (8)]

$$D = \frac{v_{go}^2 r_o}{2r^2} \, \Gamma(r) \, . \tag{27}$$

Evidently, Γ is just the ratio of the magnitudes of the local gravitational force and the local radiative force. In general, Γ must be considered to be a function of the various macroscopic (and perhaps microscopic) parameters of the expanding stellar atmosphere, but for the purpose of the present simplified discussion, we take Γ to be a function of radial distance. This approach allows us to draw some general conclusions about radiative driving of winds, but it would not be appropriate, for example, to apply such an approach to the study of phenomena like the radiative amplification of hydromagnetic waves.

In the models of dust-driven winds that might be appropriate to those cool stars with dusty atmospheres, the atmospheric dust and gas are generally so strongly frictionally coupled that radiative momentum transfer to the dust can be treated as momentum addition to the atmosphere as a whole (i.e., to both dust and gas, the latter of which comprises most of the stellar wind mass). For these models, therefore, the radial variation of Γ is determined by effects such as grain condensation and destruction, spectral redistribution of the emergent stellar radiation field through interaction with the dust, and radial variation of the dust density through atmospheric expansion. Let us consider a particularly simple case, in which there is no dust near the star, and dust is formed in a narrow radial range centered at radial distance $r = r_c$. The reason for labelling this distance r_c is that it also marks the critical point of the flow in our simple representation, as can be seen by considering a modified form of the isothermal wind equation (18), which includes the effects of radiative momentum addition:

$$(u^2 - v_t^2) \, \frac{1}{u} \, \frac{du}{dr} = \frac{2}{r} \left[v_t^2 - \frac{1}{4} \, v_g^2 \, (1{-}\Gamma) \right] . \tag{28}$$

In the case at hand, we assume $4v_t^2 < v_g^2$ and $\Gamma \ll 1$ in $r < r_c$, and $\Gamma > 1$ in $r > r_c$. Evidently, then, the zero of the right side of equation (18) occurs at $r = r_c$ and this is where the flow must be sonic. An examination of equation (20) indicates that moving the critical point inward from its position r_{ct} for a thermally driven wind to the position r_{cd}

where dust is formed leads to an increase in the mass loss rate by a factor of

$$\frac{\dot{M}_{*d}}{\dot{M}_{*t}} = \left(\frac{r_{cd}}{r_{ct}} \right)^2 \exp\left[2\left(\frac{r_{ct}}{r_{cd}} - 1 \right) \right]. \tag{29}$$

The magnitude of the radiatively driven mass loss can be expressed simply in terms of critical point parameters:

$$-\dot{M}_{*d} = 4\pi \rho_c \, v_{tc} \, r_c^2 . \tag{30}$$

The exponential increase of the mass loss rate $(-\dot{M}_{*d})$ with decreasing r_{cd} [cf. eqn. (29)] reflects the exponential falloff of density in the subsonic region of the wind and the consequent exponential increase of critical point density (ρ_c) as the critical point is moved inward [cf. eqn. (30)].

If we try to argue that the principal driving mechanism of winds from cool, low-gravity stars with dusty atmospheres is radiation pressure on dust grains, then we encounter the difficulty that this mechanism tends to produce relatively small mass loss rates, as long as the dust formation region is at relatively large radial distances (i.e., $r_{cd} \approx 10 \, R_*$). This difficulty could be overcome if a region of sudden grain formation were to occur very near the star (viz., $r_{cd} \lesssim 2R_*$), but the bulk of observational evidence seems to weigh against such a hypothesis, at least at present (e.g., Goldberg 1984). Alternatively, if grains were formed over a broad radial region near the star, or if "clean" grains near the star were to become "dirtier" with increasing radial distance (Draine 1981), resulting in a gradual increase of grain opacity over a broad radial region, then the atmospheric scale height in the subsonic region $(r < r_c)$ could be significantly increased, and a substantial mass loss rate could be radiatively driven even if the critical point were to occur relatively far from the star $(r_c \lesssim 10 \, R_*)$. Unfortunately, we will not be in a position to evaluate such ideas until our understanding of grain formation in stellar atmospheres is substantially improved.

Another difficulty that is encountered in trying to attribute the driving of winds from cool, low-gravity stars to radiation pressure alone is that the atmospheric opacity structure must have a very special configuration to avoid producing large asymptotic flow speeds. To illustrate this point, consider again the simple case in which grain formation occurs in a narrow radial region. The asymptotic flow speed can be calculated from equation (12), if we first note that [cf. eqn. (27)]

$$\int_{r_o}^{r_c} dr \, D - 0 , \tag{31a}$$

$$\int_{r_c}^{\infty} dr \, D \approx \frac{v_{go}^2 r_o}{2 \, r_c} \left(\frac{\Gamma_c}{1+\beta_c} \right), \tag{31b}$$

where $\beta_c = \langle -d\,(\ln \Gamma)/d\,(\ln r)\rangle$ is an average over the region just beyond r_c. Combining equations (12) and (31),

$$u_\infty^2 \approx v_{go}^2 \left(\frac{r_o}{r_c} \right) \left(\frac{\Gamma_c}{1+\beta_c} - 1 \right).$$ (32)

If we wish to have a substantial mass flux (i.e., r_c/r_o is not too large), to ensure that the mass flux is lifted out of the gravitational field (i.e., $u_\infty^2 > 0$), and to have a small asymptotic flow speed (i.e., $u_\infty^2 \ll v_{go}^2$), then we see that r_c, β_c, and Γ_c must be very finely tuned. Is there some sort of atmospheric self-regulation mechanism operating that can produce such a fine tuning? This, of course, is just the question that we had to ask in the consideration of wave-driven winds and that we must ask again in the context of shock-driven winds.

7. SHOCK-DRIVEN WINDS

Many of the stars which are surrounded by enough dust to provide strong dynamical coupling between the radiation field and the atmosphere also exhibit regular pulsations that can drive large-amplitude shock waves outward through the atmosphere. Because such shock waves deposit both heat and momentum in the stellar atmosphere, they may play an important role in driving stellar winds from pulsating stars, especially the Miras (e.g., Willson and Hill 1979; Wood 1979; references in Holzer and MacGregor 1985). Indeed, because of the frequent coexistence of dust and pulsations, it is possible that in some stars radiation-driving and shock-driving combine to produce the observed stellar winds (e.g., Wood 1979; Jura 1984). Let us first, however, consider the shock-driving mechanism in isolation.

Although a shock-driven wind is inherently non-steady, we can gain a feeling for the basic physical effects important in such a wind by considering a time-averaged view, which involves casting an approximate theoretical description in the steady flow frame-work that we have used up to now. Following our usual procedure, we ask first how we might formulate momentum and heat addition terms arising from the outward propagation of quasi-periodic shocks. When a shock front passes through a fluid element in a stellar atmosphere, that fluid element experiences rapid heating and acceleration. In a relatively strong shock, the heating is almost entirely dissipative (i.e., non-adiabatic or irreversible), which is to say that compressive (adiabatic) heating is small in comparison with the total heating associated with shock front passage. Hence, most of the heat produced will not be lost in the expansive cooling of the fluid element throughout the broad rarefaction region behind the shock front. The average heating rate owing to shock passage is thus

$$Q_s \approx \rho \, (\delta u)^2 / \tau,$$ (33)

where δu is the shock speed in the local average rest frame of the atmosphere (moving at speed $<u>$), and τ is the pulsation period. We can similarly approximate the momentum addition rate by

$$D_s \approx \delta u / \tau.$$ (34)

Of course, in the dense stellar atmospheres we are considering, radiative cooling behind the shock front can play an important role in counteracting shock heating. In the limit of very strong radiative cooling, the shocks can be described as isothermal (the above description was for adiabatic shocks) and the atmosphere is then taken to be in radiative equilibrium. In this case, although the shock heating effects become negligible, the momentum addition [eqn. (34)] still plays a role in accelerating the wind. This contrasts sharply with the adiabatic case, in which (as long as $\delta u / <u>$ is large) heating is the major source of energy addition to the wind.

Theoretical descriptions have been developed for the two extreme cases of adiabatic and isothermal shocks, and the results of the two descriptions are drastically different. Using an adabatic shock description (in which the wind is largely thermally driven) Wood (1979) calculated a mass loss rate some four orders of magnitude larger than the observationally inferred rate for the Mira variable being modelled, and some ten orders of magnitude larger than the mass loss rate calculated using an isothermal shock description. One finds in an *a posteriori* evaluation of these calculations that that radiative cooling rate far exceeds the shock heating rate in the wind driven by adiabatic shocks, while just the opposite is true for the isothermal case. Thus, neither description comes close to providing an energetically consistent description of the wind. Evidently, there is much work to do before a clear statement can be made about the importance of shock-driving in the winds of pulsating stars, and at least some of this work (involving a more realistic treatment of energy balance) needs to be carried out before one can make a suitable evaluation of the idea that shock-driving and radiation-driving may work in concert to produce the massive winds from pulsating stars with dusty atmospheres.

8. SUMMARY

The goal of this paper has been to provide a framework for thinking about the various physical processes that may play significant roles in driving the winds of cool, low-gravity stars. We have concentrated on four mass loss mechanisms and described them in the context of steady, radial, spherically symmetric flow, but the basic physical character of these mechanisms remains essentially the same when they are invoked in the context of asymmetric, nonsteady mass loss, and when they act in concert with each other. Hence, an understanding of the relatively simple models described here should provide a sound basis for understanding the relatively complex phenomena that are actually observed.

The general discussion of mass, momentum, and energy balance (cf. section 2), on which the descriptions of the simple models were based, provides us with the ability to articulate one of the most difficult problems encountered in trying to understand the driving of massive winds from cool, low-gravity stars. Observations seem to indicate that the asymptotic flow speeds of these winds are generally much less than the gravitational escape speed at the base of the atmosphere (i.e., $u_\infty^2 \ll v_{go}^2$). If this inference is correct, then we can conclude from the arguments given in section 2 that most of the energy required to drive the wind goes into lifting the wind out of the stellar gravitational field, and that most of this energy is supplied to the region of subsonic flow. The small fraction of the energy supplied by the region of supersonic flow is tightly constrained by the requirement that the asymptotic flow speed be small, but positive. This tight constraint, which seems to apply to a broad range of stellar winds, implies the existence of some sort of atmospheric self-regulation (associated with the wind driving mechanism(s)) that finely tunes the amount of energy deposited in the region of supersonic flow. One of the difficult problems we face is, thus, to explain how a driving mechanism or combination of driving mechanisms can lead to this sort of atmospheric self-regulation in a broad range of massive winds from cool, low-gravity stars.

ACKNOWLEDGMENTS. I am grateful to Keith MacGregor for useful discussions and for comments on the manuscript. I would also like to thank Lorraine Hori for preparing the camera-ready manuscript.

REFERENCES

Cassinelli, J.P. and MacGregor, K.B. 1986, in *Physics of the Sun*, ed. P.A. Sturrock, T.E. Holzer, D. Mihalas, and R.K. Ulrich (Chicago: University of Chicago Press), p. 47.

Draine, B.J. 1981, in *Physical Processes in Red Giants*, ed. I. Iben and A. Renzini (Dordrecht: Reidel), p. 317.

Goldberg, L. 1984, in *CNRS-NASA Monograph Series on Nonthermal Phenomena in the Stellar Atmosphere* (Washington, D.C.: NASA), Ch. 7.

Hartmann, L. and MacGregor, K.B. 1980, *Ap. J.*, **242**, 260.

Hollweg, J.V. 1978, *Rev. Geophys.*, **16**, 689.

Holzer, T.E., Flå, T., and Leer, E. 1983, *Ap. J.*, **275**, 808.

Jura, M. 1984, *Ap. J.*, **282**, 200.

Lallement, R., Holzer, T.E., and Munro, R.H. 1986, *J. Geophys. Res.*, **91**, 6751.

Leer, E. and Holzer, T.E. 1980, *J. Geophys. Res.*, **85**, 4681.

Leer, E., Holzer, T.E., and Flå, T. 1982, *Space Sci. Rev.*, **33**, 161.

Olbert, S. 1983, in *Solar Wind 5, NASA Conf. Publ. 2280*, 149.

Parker, E.N. 1958, *Ap. J.*, **128**, 664.

_____. 1963, *Interplanetary Dynamical Processes* (New York: Interscience).

_____. 1964a, *Ap. J.*, **139**, 72.

_____. 1964b, *Ap. J.*, **139**, 93.

Scudder, J.D. and Olbert, S. 1983, in *Solar Wind 5, NASA Conf. Publ 2280*, 163.

Shoub, E. 1986, in preparation.

Watanabe, T. 1981, *Publ. Astron. Soc. Japan*, **33**, 679.

Willson, L.A. and Hill, S.J. 1979, *Ap. J.*, **228**, 854.

Wood, P.R. 1979, *Ap. J.*, **227**, 220.

Zirker, J. (Ed.) 1977, *Coronal Holes and High Speed Solar Wind Streams*, Colorado Associated University Press, Boulder.

WHAT DO BINARIES TEACH US ABOUT MASS-LOSS FROM LATE-TYPE STARS ?

Dieter Reimers

Hamburger Sternwarte, Universität Hamburg
Gojenbergsweg 112, 2o5o Hamburg 8o
Federal Republic of Germany

ABSTRACT. It is shown that the binary technique - a B star companion is used as a light source which probes the wind of the red giant primary - has yielded accurate mass-loss rates and wind velocities for 8 G to M (super)giants and (in some cases) estimates of wind temperature.
  Eclipsing binary systems have in addition revealed that G and K supergiants possess extended chromospheres which could be detected outwards to $\sim$ 1 R_* (stellar radius) above the photospheres. Electron temperatures T_e and hydrogen ionization n_e/n_H seem to increase with height up to at least o.5 R_* ($n_e/n_H = 1o^{-2}$, $T_e = 1o^4$ K at o.5 R_*), and the winds start to be accelerated at heights above $\sim$ o.5 R_*.
  Mass-loss rates appear to increase steeper than linearly with $L/g \cdot R$. It is shown that the observed mass-loss rates are consistent with stellar evolution constraints for both Pop. II and Pop I stars.

1. INTRODUCTION

The study of mass-loss from stars began in 1956 - before the detection of the solar wind - with Deutsch's spectroscopic observations of the α Her visual binary system.

  Deutsch (1956) noticed that α Her (M5II), like other M (super) giants, has blue-shifted cores of strong resonance lines and that the same blue-shifted lines are seen in the spectrum of its G giant companion 5" apart. Since single G giants do not show sharp, blue-shifted resonance lines, Deutsch concluded that the M giant has a vast expanding envelope that encloses the companion, and that the line shifts in the line of sight of the companion are above the local escape velocity, i.e. circumstellar matter escapes the system.

  As will be shown below, the binary technique of studying mass-loss of G to M giants and supergiants is a powerful tool to obtain accurate mass-loss rates and wind velocities as well as estimates of the wind temperature and information about the wind acceleration region. In particular, Zeta Aurigae binary systems are the only stars - besides the Sun - where the structure of the extended chromosphere and wind can be observed with high spatial resolution. For single stars, on the other

I. Appenzeller and C. Jordan (eds.), Circumstellar Matter, 307–318.

hand, it has turned out to be very difficult to determine mass-loss
rates quantitatively from CS lines. The reason is that while it is
possible to measure ion column densities N_{ion} and wind velocities v_w
from a theoretical analysis of P Cyg type profiles superimposed upon
the cores of strong resonance lines in stars like α Ori (c.f. Bernat
and Lambert, 1976), it is not possible to infer from spectroscopic
observations where in the line of sight the CS lines are formed, and
since $\dot{M}$ = const $\cdot$ N_{ion} $\cdot$ v_w $\cdot$ R_i , where R_i is the inner shell radius,
accurate mass-loss rates cannot be determined from optical observations
of M giants and supergiants. The same difficulty - in addition to others -
arises if one tries to determine mass-loss rates from circumstellar
dust emission.

The only technique for measuring $\dot{M}$ at optical wavelengths appears
to be spatially resolved imaging of CS shells in scattered resonance
line photons like KI 7699 Å or NaD (Mauron et al. 1984, 1986). However,
since the lines observed up to now are from minor ionization species,
good knowledge of the ionization of metals and of the formation of CO
is required. In particular, nonequilibrium effects (flow time $\sim$ recombination time scale) have to be taken into account. Reliable mass-loss
rates are available now for α Ori and μ Cep (cf. Table 1).

2. VISUAL BINARIES

Deutsch (1956) invented the technique to determine the rate of mass-
loss of a cool giant from CS absorption lines of a predominant ionization stage seen in the spectrum of a visual companion. By this
technique one avoids the difficulty one has in single stars of locating
the shell.

The visual binary technique has been applied to α Her (Reimers,
1977), α Sco (Kudritzki and Reimers, 1978) and - with less accuracy -
to o Cet (Reimers and Cassatella, 1985).

Besides mass-loss rates and wind velocities (see Table 1) as
inferred from the strengths of absorption lines seen in the spectrum
of the companions, some additional information can be obtained from a
study of excitation. In case of α Her, the observed absence of lines
from excited fine structure levels of TiII ($N(a^4F_{5/2})/N(a^4F_{3/2})$ < o.15)
implies that the hydrogen particle density n_H < 7.5 $\cdot$ $10^4 cm^{-3}$ for T =
5o K and n_H < 1.5 $\cdot$ 10^4 cm^{-3} for T >> 1oo K. With a typical density of
n_H $\approx$ 2 $\cdot$ 10^4 cm^{-3} as found from column densities, one can exclude
T >> 1oo K at a typical distance of 3oo M giant radii. This is roughly
in accordance with adiabatic cooling of the wind at large distances
($T \sim r^{-4/3}$).

In case of α Sco, the observed population of fine structure levels
within the ground state of TiII which is due to electron collisions
permits an independent estimate of the electron density in the wind
(Kudritzki and Reimers, 1978).

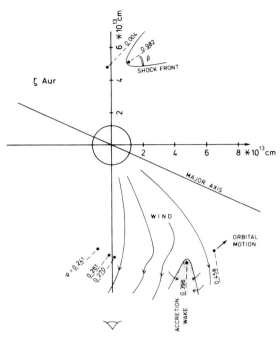

Figure 1. A roughly to-scale presentation of the accretion shock front
and wake as observed for ζ Aur (from Che and Reimers, 1986).

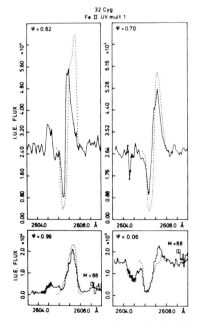

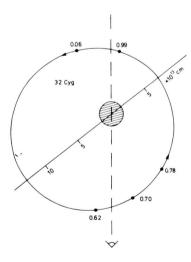

Figure 2. Location of B star relative to K supergiant at observed
phases

Figure 3. Dependency of a wind line (Fe II UV Mult.1 resonance line) on
phase (see Fig.2) in comparison with theory (....).

3. ζ AUR/VV CEP TYPE BINARIES

With IUE, the optical separation of red giants with hot companions can be replaced by a separation through complementary energy distributions of the components. At IUE wavelengths, in particular in the short-wavelength range, one observes a pure B star spectrum upon which numerous CS P Cygni type lines formed in the extended wind and chromospheric absorption lines (near eclipse) of the red giant are superimposed.

The B star serves as an astrophysical light source (a "natural satellite") which moves around in the wind of the red giant. However, compared to widely separated visual binaries, a number of additional difficulties arise
- a non-spherical, 3-dimensional line transfer problem has to be solved since the light source (B star) is excentric from the wind symmetry center. Computer codes that solve this problem have been developed in the 2-level approximation by Hempe (1982, 1984) and for the multilevel case by Baade (1986)
- the wind is disturbed in the immediate surrounding of the B star as it moves supersonically through the wind and formes an accretion shock front (Chapman, 1981). However, a detailed study of the accretion shocks has shown that their geometrical size is very small compared to the CS shell (Fig.1) and can be neglected in line transfer calculations (Che-Bohnenstengel and Reimers, 1986)
- the hot B star ionizes the wind, i.e. an HII region is formed within the red giant wind. In 31 Cyg and α Sco, in particular, the size of the HII region is larger and it has to be taken into account quantitatively since ions like SiII and FeII which are used for the mass-loss rate determination may be doubly ionized within the HII regions.

On the other hand, ζ Aur binaries are the only stars besides the Sun where the winds and extended chromosphere can be studied with spatial (height) resolution.

3.1 Wind lines

The wind is visible at all phases in P Cyg type profiles (during total eclipse of B star pure emission lines) of ions like FeII, SiII, SII, MgII, CII, AlII, and oI. These lines are formed by scattering of B star photons in the wind of the red giant. A few wind lines like FeII Mult.9 ($\sim$1275 Å) are seen in pure absorption due to the branching ratios of the upper levels which favour reemission as FeII UV Mult. 191 photons (Hempe and Reimers, 1982; Baade, 1986).

Theoretical modelling of wind line profiles and of their phase dependency has yielded accurate mass-loss rates and wind velocities for a number of systems (Table 1). It has turned out that a good mass-loss determination requires both phases with the B star in front of the red supergiant (which yields wind turbulence v_t) and phases with the B star behind the red supergiant(which yield the wind velocity v_w). Typically, $v_w \approx 2 v_t$. Further details can be found in Che et al. (1983). It turned out that it was possible to match the circumstellar line profiles at all phases with one set of parameters v_w , v_t and - within a factor of 2 - one mass-loss rate $\dot{M}$ (Figs.2-5). This means that at least in the orbital

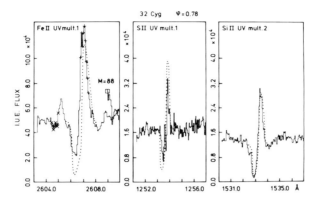

Figure 4. Comparison of theoretical (....) and observed wind line profiles for three ions (as for Fig. 2 and 3 from Che et al. 1983).

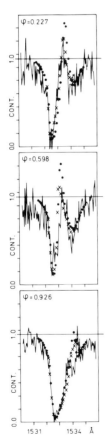

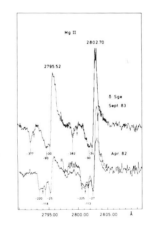

Figure 9. High velocity clouds in δ Sge (M2II, mean wind velocity ⋏3o km/s) seen in the spectrum of its B star companion.

Figure 5. Observed and theoretical SiII UV Mult. 1 line at different phases for 22 Vul (G3Ib-II), from Reimers and Che-Bohnenstengel (1986).

plane the envelope asymmetries (in density) are within a factor of 2 on
a length-scale of several K giant radii. The example of α Sco shows
that the influence of interstellar (IS) lines has to be considered carefully. As demonstrated in Fig.6 with SiII 1526 Å, all circumstellar o.oo
eV lines in the spectrum of α Sco B suffer from IS absorption on the
longward side of the profile (IS velocity is +5 km/s, Kudritzki and
Reimers, 1978), i.e. the reemission part of the CS P Cyg profile appears
absorbed at the resolution of IUE, while all other lines (like SiII
1533 Å) have the expected P Cyg profiles. Both v.d. Hucht et al. (198ο)
and Bernat (1981) used only the pure absorption lines to determine mass-
loss rates which consequently came out too high by one order of magnitude with a scatter of a factor of $\sim$ 3o among rates from different ions.
The correct treatment of the P Cyg profiles with Hempe's (1982) non-
spherical line transfer code yields a mass-loss rate of $\sim$ 10^{-6} $M_\odot$/yr
(Hagen, 1984, Hagen et al. 1986), consistent with Hjellming and Newell's
(1983) rate from radio emission from the HII region within the wind and
with the rate from optical observations of TiII lines in α Sco B
(Kudritzki and Reimers, 1978).

In case of 32 Cyg and 22 Vul, the wind electron temperature T_e
could be estimated from the observed population of excited FeII levels.
For 32 Cyg, at distances of more than 5 K giant radii, Che-Bohnenstengel
(1984) found T_e = 48oo K for n_e/n_H = o.o1 and $T_e \stackrel{\sim}{} 10^4$ K for smaller
electron densities. The LTE value would be 42oo K. In 22 Vul, popula-
tion of excited FeII levels is much higher, and a wind electron temperature of 3o $\pm$ 1o·10^4 K was estimated with the assumption of pure
electron collision excitation (Reimers and Che-Bohnenstengel, 1986).
Such a high wind excitation is particularly remarkable in a star like
22 Vul which has common characteristics with 'hybrid atmosphere' stars
like α Aqr : a high wind velocity, location in the HR diagram, it is a
young intermediate mass star, and it has an extended chromosphere observed to 1 R_* above the photosphere during its 1985 eclipse (Schröder
and Che-Bohnenstengel, 1985). However, since radiative excitation via
high levels cannot be excluded at present, the high wind temperature
of 22 Vul needs to be confirmed by improved theoretical techniques.

3.2 Chromospheric lines

The extended chromosphere - where the wind starts to expand - could be
studied by means of the technique applied by O.C. Wilson, H.G. Groth, K.O.
Wright and others in the 195os; and IUE data are a major advance in
several respects: The B star provides a smooth continuum on which one
observes numerous absorption lines up to heights (projected binary sepa-
rations) of more than one supergiant radius above the photosphere. In
addition to absorption lines, the wavelength and time dependence of
totality in the UV (Fig.7) can be used for a density model of the inner
chromosphere (Schröder 1985a, 1985b, 1986). Chromospheric densities
could be represented by power laws of the form $\rho \sim r^{-2}$. h^{-a} with
a $\stackrel{\sim}{} 2.5$ where r is the distance from the center of the star and h
is the height above the photosphere. The empirical density distribution
shows that after a steep decrease in the inner chromosphere already in
the upper chromosphere (height > 1/2 to 1 giant star radii R_*) expansion

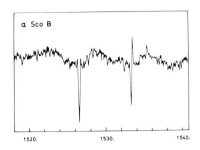

Figure 6. Interstellar absorption at SiII 1526 Å (o.oo eV) in comparison with the undisturbed wind line SiII 1533 Å (o.ol eV).

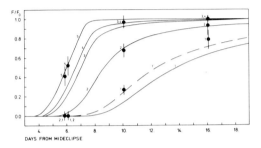

Figure 7. Light curves at 135o Å (1), 1513 (2), 1783 Å (3), 196o Å (4)
and 2992 Å (5) during eclipse of 32 Cyg. Solid line represents best fit
with model chromosphere (from Schröder, 1986).

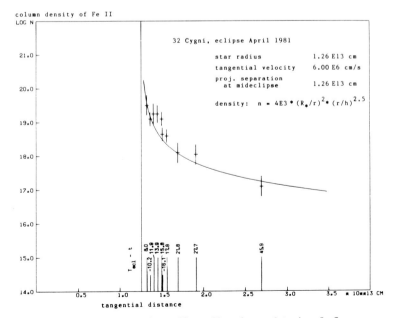

Figure 8. Chromospheric density distribution obtained from curve of
growth analysis of ultraviolet FeII lines (Schröder, 1985a,b).

starts ($\rho \sim r^{-2}$). A typical density at a height of 2 R_* is 10^7 cm^{-3} (Fig.8).

Since one observes total particle densities in the expanding chromosphere up to h $\overset{\sim}{\sim}$ 1.5 R_K , and in addition the wind density and velocity outside of $\sim$ 5 R_K , one can try to look for consistency by assuming a steady wind, i.e. to apply the equation of continuity. Using $\dot{M} = 4 \pi r^2 \cdot \rho(r) \cdot v(r)$ and $\rho(r) = \rho_0 \cdot (R_*/r)^2 \cdot (v/(r-R_*))^a$ for the chromosphere, we find $v(r) = \dot{M} (4 \pi \rho_0 R^2)^{-1} \cdot (1 - R_*/r)^a$ and a wind terminal velocity $v_\infty = \dot{M} (4 \pi \rho_0 \cdot R^2)^{-1}$ which can be checked with observed values for ρ_0, $\dot{M}$ and v_∞ for consistency. For 32 Cyg and 31 Cyg Schröder (1985) found consistency, which means that the empirical density distribution (when extrapolated by the equation of continuity to the outer wind) yields the correct mass-loss rate. In case of ζ Aur, the chromospheric density distribution was far too steep - at least at that particular limb position during eclipse - to give the mean mass-loss rate, which might be stellar analogue to a solar coronal hole.

Another stellar-solar analogue is the prominence detected during egress of the 1981 eclipse of 32 Cyg (Schröder, 1983). After egress from eclipse, an additional 'dip' in the light curve was seen at wavelengths λ < 2ooo Å for at least 6 days. Since the observed "prominence" was optically thin, the observed frequency dependence of optical depths $\tau_\nu \sim \nu^{5.5}$ could be used to identify the opacity as Rayleigh scattering at HI ground state. A linear extension - perpendicular to the line of sight - of about 1/6 K giant radii ($\overset{\sim}{\sim}$ 3o $R_\odot$) and an apparent height of $\sim$ 15 $R_\odot$ above the limb (at 135o Å) was estimated from the light curves. The small observed velocity of +2o km/s as measured from a few absorption lines like VII 311o.7 or TiII 3o72 A seen in addition to the normal chromospheric lines indicates a slowly moving cloud. Also, even a moderate velocity perpendicular to the line of sight, e.g. a slow prominence moving upwards with the wind velocity of $\sim$ 6o km/s can be excluded, since within the 6 days the cloud was seen it would have moved by 45 $R_\odot$ (3 times the observed height above the limb). The density in the observed prominence was of the order of 10^{12} cm^{-3}, about a factor of 1o higher than in the surrounding chromosphere. If the excess pressure was balanced by magnetic fields, a field strength of $\sim$ 4 Gauss would have been necessary.

Another finding important for our understanding of chromospheres and for the mass-loss mechanism operating in K supergiants is that the chromospheric electron temperature is $\sim$ 10^4 K with a tendency to increase with height, e.g. from T_e $\overset{<}{\sim}$ 85oo K at o.2 R_K above the photosphere to about 11 ooo K at o.5 R_K height in 32 Cyg. Hydrogen ionization appears to increase over the same range from about $n_e/n_H = 10^{-3}$ to 10^{-2} (Schröder, 1986). Both findings mean that extensive nonradiative heating occurs in heights above the photosphere where wind acceleration has started. At least in these stars, wind acceleration through radiation pressure on dust grains can be excluded as a possible mass-loss mechanism.

Although the extensive study of ζ Aur/VV Cep type stars with the IUE has brought a major advance in determining mass-loss rates of G to M supergiants and in understanding the structure of chromospheres, wind acceleration regions and winds of these stars, a number of difficulties

and problems must be overcome before wind acceleration can be understood in detail: (i) The S/N of single IUE spectra and the spectral resolution are not good enough for studying the wind acceleration region
(h $\stackrel{\sim}{\sim}$ 1 to 3 R$_*$). This difficulty can be overcome with the Hubble Space
Telescope. (ii) The 3-dimensional line transfer code uses the Sobolev
approximation for the source function which is not appropriate since
v_w/v_t is not >> 1. (iii) Our wind temperature estimates neglected radiative transitions which may be important for FeII. Although a multilevel NLTE code has been developed (Baade, 1986), calculations involving 3-dimensional radiative transfer are not yet possible with so
many levels as for FeII. (iv) Simple geometries have been used while inhomogeneities and departures from spherical outflow due to interactions
with the companion have been neglected. In at least two stars (δ Sge,
32 Cyg) time variable high velocity components can be seen in the
strongest wind lines. So, e.g. the M2II giant δ Sge with a mean wind
velocity of $\stackrel{\sim}{\sim}$ 3o km/s, shows MgII 28oo components with velocities up to
nearly 4oo km/s (Fig.9). While the high velocity clouds seem to contri-
bute little to the total envelope mass, they may contribute to the
energy budget of the wind. (v) A further difficulty arises through the
presence of an early B star in the wind of the red giant. Within the HII
region around the B star, Fe, e.g., is ionized a second time. Since in
a system like 31 Cyg with a period of $\sim$1o yrs, where these effects are
observed (Che et al. 1983), the HII region moves with the B star, and
since the recombination time scale in the HI region is large compared
to the orbital period, the FeII/FeIII ionization balance is probably
not in detailed equilibrium.

4. MASS-LOSS RATES

Table 1 summarizes the results for 8 binary systems studied so far. In
addition, mass-loss rates obtained for α Ori and μ Cep by means of the
resonance line imaging technique by Mauron et al. (1984, 1985, 1986)
are included since about the same accuracy as with the binary technique
is achieved. Mass-loss rates for more stars and in particular more
accurate values cannot be expected in the near future for normal giants.
I shall therefore briefly discuss the implications for stellar evolution. Nearly all stellar evolution calculations which included mass-
loss in the red giant stage have applied the semiempirical scaling law
$\dot{M}\left(M_\odot/yr\right) =\eta\cdot 4.1o^{-13}$ L/g$\cdot$R (1) proposed on dimensional arguments
and calibrated with empirical mass-loss rates for a number of Pop I
giants and supergiants (Reimers, 1975). The dimensionless factor
η (1/3 $\leq \eta \leq$ 3) has been introduced in order to take into account the
then considerable uncertainty of mass-loss rates.
     For this reason we test the simple scaling law with the improved
empirical mass-loss rates.
     In Fig.1o, the mass-loss rates from Table 1 are plotted versus
L/g$\cdot$R . Two conclusions are evident
1) All observed mass-loss rates can be represented within realistic
error bars (factors of 2 in both $\dot{M}$ and L/g$\cdot$R to either side) by a
relation $\dot{M}$ = 5$\cdot$1o^{-13} L/g$\cdot$R .   (η = 1.25)
2) The new empirical rates seem to indicate, however, a tendency for

Table 1: Compilation of accurate mass-loss determinations

Star	Spectral Type	Technique	$\dot{M}(10^{-8} M_\odot/yr)$	v_w (km/s)	v_t (km/s)	References
α^1 Her	M5II	Vis.bin.	11	8	4	Reimers (1977)
α Sco A	M1Iab	Vis.bin.	70 +)	17		Kudritzki & R. (1978)
		Radio	200 +)			Hjellming & Newell (1983)
		IUE	1oo	17	8	Hagen et al. (1986)
HR8752	GoIa	Radio	1ooo	3o		Lambert & Luck (1978)
ζ Aur	K4Ib	IUE	o.6	4o	3o	Che et al. (1983)
32 Cyg	K5Iab	IUE	2.8	6o	25	"
31 Cyg	K4Ib	IUE	∿ 4	8o	2o	"
δ Sge	M2II	IUE	2	28	2o	Reimers & Schröder (1983)
22 Vul	G3II-Ib	IUE	o.6	16o	5o	Reimers & Che-B. (1986)
α Ori	M1.5Iab	KI Image	2oo	1o		Mauron (1985)
μ Cep	M2Ia	NaD Image	3ooo	47		Mauron et al. (1986)

+) With a mean molecular weight μ = 1.4

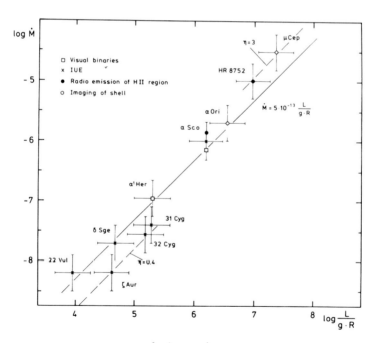

Figure 1o. Mass-loss rates $\dot{M}$ $\left(M_\odot/yr\right)$ for stars from Table 1 versus
L/g·R (in solar units).

a steeper dependence of mass-loss on luminosity, resp. $L/g \cdot R$.

A comparison of observed late stages of stellar evolution with evolutionary calculations applying red giant mass-loss in the simple parametrized form imposes constraints on η

i)   Horizontal branch star masses and the observed upper luminosity limit of globular cluster AGB stars require $\eta \approx 0.4$ (Renzini, 1977) for Pop.II stars

ii) The observed maximum initial mass for white dwarf progenitors M_{WD} as determined from white dwarf members of the intermediate age cluster NGC 2516 is $M_{WD} \approx 8 M_\odot$ or even higher (Reimers and Koester, 1982; Weidemann and Koester, 1983)

iii) Mass-loss rates of advanced AGB stars (OH-IR stars) seem to be higher by about an order of magnitude than predicted by the scaling law with $\eta = 1$ (Baud and Habing, 1983). Similarly, PN demand higher mass-loss rates if interpreted as excited winds of red giant progenitors.

Inspection of Fig.1o shows that for luminosities typical at the tip of the Pop.II AGB ($\log L \approx 3.2$, $\log L/g \cdot R \approx 5.3$), $\eta = 0.4$ does not contradict observations within realistic errors.

Similarly, for high luminosity $\eta = 3$ gives a good fit to observations. According to Iben and Renzini (1983), $\eta = 3$ would yield $M_{WD} \approx 1o$. However, observations at high luminosities in Fig.1o are from massive stars and should not be extrapolated deliberately to advanced evolutionary stages of intermediate mass stars. For such stars, Baud and Habing have proposed a parametrization of observations of mass-loss from OH-IR stars in the form

$$\dot{M}(t) = \mu \cdot \frac{L \cdot R}{M_e(t)} \quad \text{where} \quad \mu = \frac{M_e}{M} 4 \cdot 1o^{-13} \quad \text{and } M_e(t) \text{ is the envelope mass}$$

instead of the total mass. This modified relation gives about the same rates for most of the AGB lifetime as relation (1), while at the end of the AGB phase the reduced envelope mass leads to a steep increase of mass-loss rates, in accordance with observational requirements.

In conclusion, mass-loss rates accurately determined with the binary technique are in full agreement with stellar evolution constraints. However, a semiempirical fit of mass-loss rates with $\dot{M} \sim L/g \cdot R$ (or $\dot{M} \sim (L/g \cdot R)^{1.3}$ as might be derived from Fig.1o) should not be extrapolated to stars of distinctly different properties like OH-IR stars with thick dust shells, F stars, carbon stars etc.

ACKNOWLEDGEMENT: This paper is based on the work of the 'binary wind team' at Hamburg Observatory, R. Raade, A. Che-Bohnenstengel, K. Hempe, K.-P. Schröder and K. Schönberg. The Deutsche Forschungsgemeinschaft supported the binary project with several grants to the author.

REFERENCES

Baade,R. 1986, Astron.Astrophys. 154, 145
Baud,B., Habing,H.J. 1983, Astron.Astrophys. 127, 73
Bernat,A.P. 1981, Ap.J. 252, 644
Chapman,R.D. 1981, Ap.J. 248, 1o43
Che-Bohnenstengel,A. 1984, Astron.Astrophys. 138, 333
Che,A., Hempe,K., Reimers,D. 1983, Astron.Astrophys. 126, 225
Che,A., Reimers,D. 1983, Astron.Astrophys. 127, 227
Che-Bohnenstengel,A., Reimers,D. 1986, Astron.Astrophys. 156, 172
Deutsch,A.J. 1956, Ap.J. 123, 21o
Hagen,H.J. 1984, Diplomarbeit Universität Hamburg
Hagen,H.J., Hempe,K., Reimers,D. 1986, Astron.Astrophys. in preparation
Hempe,K., 1982, Astron.Astrophys. 115, 133
           1984, Astron.Astrophys.Suppl. 56, 115
Hempe,K., Reimers,D. 1982, Astron.Astrophys. 1o7, 36
Hjellming,R.M., Newell,R.T. 1983, Ap.J. 275, 7o4
Iben,I., Renzini,A. 1983, Ann.Rev.Astr.Ap. 21, 271
Kudritzki,R.-P., Reimers,D. 1978, Astron.Astrophys. 7o, 227
Lambert,D.L., Luck,R.E. 1978, M.N.R.A.S. 184, 55
Mauron,N. 1985, Doct.Thesis, Univ. Toulouse
Mauron,N., Fort,B., Querci,F., Dreux,M., Fauconnier,T., Lamy,P. 1984
     Astron.Astrophys. 13o, 341
Mauron,N., Cailloux,M., Prieur,J.L., Pilloles,P., Lefèvre,O. 1986,
     Astron.Astrophys. in press
Reimers,D. 1975, Mem.Soc.Roy.Sci. Liège 6e Ser. 8, 369
Reimers,D. 1977, Astron.Astrophys. 61, 217 (Erratum 67, 161)
Reimers,D., Koester,D. 1982, Astron.Astrophys. 116, 341
Reimers,D., Schröder,K.-P. 1983, Astron.Astrophys. 124, 241
Reimers,D., Cassatella,A. 1985, Ap.J. 297, 275
Reimers,D., Che-Bohnenstengel,A. 1986, Astron.Astrophys. in press
Renzini,A. 1977, In "Advanced Stages of Stellar Evolution" (eds.
     P. Bouvier, A. Maeder), Geneva p. 151
Schröder,K.-P. 1983, Astron.Astrophys. 124, L16
Schröder,K.-P. 1985a, Astron.Astrophys. 147, 1o3
               1985b, Ph.D.Thesis University of Hamburg
               1986, Astron.Astrophys. in press
Schröder,K.P., Che-Bohnenstengel,A. 1985, Astron.Astrophys. (Letters)
     151, L5
v.d.Hucht,K., Bernat,A.P., Kondo,Y. 198o, Astron.Astrophys. 82, 14
Weidemann,V., Koester,D. 1983, Astron.Astrophys. 121, 77

CHROMOSPHERIC DENSITY DISTRIBUTION, OPACITY, IONIZATION AND
WINDACCELERATION OF 3 K SUPERGIANTS IN ζ AURIGAE SYSTEMS

Klaus-Peter Schröder
Hamburger Sternwarte, Universität Hamburg
Gojenbergsweg 112
D-2050 Hamburg 80
Federal Republic of Germany

Results from 58 chromospheric eclipse spectra of three ζ Aurigae type
binary systems – with a K supergiant primary and a B star companion –
taken with IUE at high resolution are presented. Curves of growth have
been constructed at 20 phases using selected chromospheric absorption
lines superposed upon the B star spectrum. In order to fit average density models to the chromospheres, 3 samples of column densities (for
ζ Aur., 32 Cyg. (fig.1) and 31 Cyg. K giants) have been used.

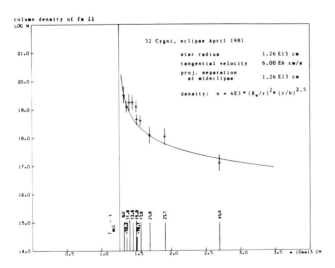

fig. 1: Observed column densities of Fe II from ingress and egress
of chromospheric eclipse versus tangential distance from the center
of the giant (vertical line indicating the limb of the giant photosphere). Comparison is made to a track of theoretical column densities, calculated by numerical integration over the density function
mentioned in the plot. Time intervals from mid eclipse are indicated.

Assuming continuous outflow of matter and knowledge of $\dot{M}$ (Che et al,

I. Appenzeller and C. Jordan (eds.), Circumstellar Matter, 319–320.

1983), wind velocities can be derived. While ζ Aurigae's wind acceleration region shows temporarily a deficiency in density and mass loss at
the 1979 eclipse, density gradients and wind acceleration of 31 Cygni
and 32 Cygni are very similar in the corresponding eclipses in 1981 and
1982. Their density and velocity structure can be represented by power
laws (except for the inner layers):

$$n(r) = n_0 \cdot (R_* / r)^2 \cdot (r/h)^a \quad \text{and} \quad v(r) = v_\infty \cdot (1 - R_* / r)^a$$

with a = 2.5 ± 0.5 and v_∞ = 55 ± 15 km/s. h = $r - R_*$ = height.

 If pure Rayleigh scattering by neutral hydrogen is assumed for the
continuum absorption coefficient, a density model of the inner chromosphere of 32 Cygni is derived. A non-pointlike B star and improved geometric parameters are applied . The density model of the inner chromosphere is in agreement with both the absorption line data and the continuum opacities which have been observed from IUE spectra near total
eclipse at 5 selected wavelengths (i.e. from wavelenth dependent eclipse
demonstrated by fig. 2).

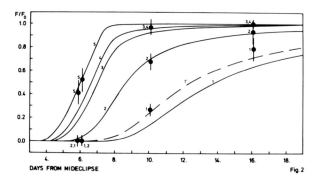

DAYS FROM MIDECLIPSE   Fig.2

fig. 2: 32 Cygni normalized fluxes at partial eclipse, at 1350 Å (1),
1513 Å (2), 1783 Å (3), 1960 Å (4) and 2992 Å (5). Solid lines: best
model of a Rayleigh scattering chromosphere, using the density distri-
bution $n_H(h) = 2.8 \cdot 10^8 cm^{-3} \cdot (R_*/r)^2 \cdot (r/h)^{1.8}$. Geometric parameters
are: R_* = $1.26 \cdot 10^{13}$cm, R_B = $2.3 \cdot 10^{11}$cm, projected separation at mid
eclipse D = $1.223 \cdot 10^{13}$cm, tangential velocity v_T = $6.0 \cdot 10^6$cm/s. Line 1'
uses $\sigma_R' = 0.7 \cdot \sigma_R(1350Å)$. $\sigma_R(\lambda)$ have been taken from Mihalas (1978).

 Using 16 high resolution IUE spectra of ζ Aurigae, 32 Cygni and
31 Cygni taken during chromospheric eclipse, the FeI/FeII ionization
equilibrium has been determined empirically as FeI/FeII ≃ $10^{-3.5}$.
Fe (Ti and V also) is ionized by UV radiation from the B star companion
while recombination is controlled by electrons from collisionally
ionized hydrogen. An attempt is made to estimate average electron densities n_e and electron temperatures T_e consistent with both FeI/FeII
ratios and the necessary hydrogen ionization. Typical values as taken
from 32 Cygni are about n_e ≃ $5 \cdot 10^6$/cm³, T_e ≃ 10500 K at h ≃ $5 \cdot 10^{12}$cm ≃
0.4 R_* (where n_H is about 10^9/cm³). For Details see Astron.& Astrophys.
147, p. 103ff (1985) and Astron.& Astrophys., in press (1986), by
K.-P. Schröder.

δ ANDROMEDAE (K3 III) : A HYBRID GIANT IN AN EXTENDED DUST SHELL

C.Jordan, P.G.Judge   and   M.Rowan-Robinson
Dept. of Theoretical Physics   Dept. of Mathematics
Oxford University   Queen Mary College
1, Keble Road   Mile End Road
Oxford, OX1 3NP, U.K.   London, E1 4NS, U.K.

ABSTRACT. Observations made with the Infrared Astronomical Satellite (IRAS) showed the existence of a set of K stars with excess emission at 60 μ. The brightest star in the group is δ And (K3 III). Spectra of δ And have been obtained with the International Ultraviolet Explorer (IUE) in both the long and short wavelength regions. These spectra show several features unusual in a K giant as cool as δ And, in particular, emission from CIV, MgII h and k lines with blue wings stronger than the red and MgII absorption blue shifted to up to 300 km s^{-1}. Overall δ And is similar to the known hybrid bright giant α TrA (K4 II). We have discovered the first 'hybrid' giant star. The IRAS observations are interpreted in terms of a cool (~100 K) dust cloud surrounding δ And - a spectroscopic binary system- and a third component at 1200 A.U.

1. INTRODUCTION

Low resolution spectra of giant stars led Linsky and Haisch (1979) to propose that these could be divided into two types, 'solar' and 'non-solar' according to whether or not they showed CIV emission. The 'dividing line' proposed occurs between K0 III and K2 III. Whilst X-ray observations (e.g. Haisch and Simon, 1982) support a similar division into hot and cool coronae, some authors have doubted the existence of a sharp dividing line (e.g. Dupree, 1981). Later observations of K bright giants, e.g. α TrA (K4 II), showed that several have a combination of CIV emission and MgII absorption indicative of high velocity stellar winds. Since massive winds are thought to develop to the cool side of the dividing line these bright giants are known as 'hybrids' (e.g. Reimers, 1977, 1982; Hartmann, Dupree and Raymond, 1981).

Observations with the IRAS satellite showed a set of K-stars with an unusual 60μ excess, indicating the presence of cool dust (Rowan-Robinson et al. 1984, Rowan-Robinson, private communication). In view of the unusual properties of the K bright giants we have observed two of these IRAS sources with IUE, including the brightest, δ And (K3 III).

I. Appenzeller and C. Jordan (eds.), Circumstellar Matter, 321–322.

2. OBSERVATIONS AND RESULTS

A paper giving full details of the IRAS and IUE observations and their
analysis is in press (Judge, Jordan and Rowan-Robinson, 1986). A long-
wavelength, high dispersion spectrum of δ And shows MgII emission line
profiles with a stellar self-reversal such that the blue emission wing
is stronger than the red. This is characteristic of 'solar' type
chromospheres. However, absorption further to the blue, extending to
$\sim$300 km s^{-1} is also observed, showing the presence of a wind. The short-
wavelength low dispersion spectra show transition region lines, including CIV emission. Thus δ And (K3 III) is a 'hybrid' giant, the first
to be discovered. The surface fluxes are similar to those of the known
hybrid giant α TrA (K4 II). We can demonstrate that the emission is
unlikely to come from the companion in the spectroscopic binary. Moreover, δ And is a photometric standard for its class and is most unlikely
to be an earlier type star misclassified. The second star, HD 129456
(K3 III) shows MgII profiles similar to those of δ And.

The dust cloud extends around both δ And and the third member
(M2 V) of the triple system at 1200 A.U. It may be analogous to an
Oort cometary cloud or a disk system, in which case the position of the
stars may lead to interesting dynamic interactions.

Given the long period of the spectroscopic binary (41 years) mass
transfer is not expected and δ And does not appear to have evolved
abundances. The rotation rate is uncertain, but could be faster than
average perhaps as a consequence of formation in a multiple system.
Several of the hybrid bright giants are also possible binaries (Reimers,
1982). At present there does not seem to be a *direct* link between the
dust and the hybrid properites, both could be consequences of the
multiple system.

The discovery of δ And as a hybrid giant casts doubt on the concept
of a simple and sharp dividing line for giants.

REFERENCES

Dupree, A.K. 1981, in *'Effects of Mass Loss on Stellar Evolution'*
      (Eds. C.Chiosi and R.Stalio) D.Reidel, p.87.
Haisch, B.M. and Simon, T. 1982, *Astrophys. J.*, **263**, 252.
Hartmann, L., Dupree, A.K. and Raymond, J.C. 1981, *Astrophys.J.*,**246**,
      193.
Hartmann, L., Jordan, C., Brown, A. and Dupree, A.K. 1985, *Astrophys.J.*
      296, 576.
Judge,P.G., Jordan, C. and Rowan-Robinson, M. 1985, *Mon. Not. R. astr.
      Soc.* In Press.
Linsky, J.L. and Haisch, B.M. 1979, *Astrophys. J.* **229**, L27.
Reimers, D. 1977, *Astron. Astrophys.*, **57**, 395.
Reimers, D. 1982, *Astron. Astrophys.*, **107**, 292.
Rowan-Robinson, M. et al. 1984, *Astrophys. J.*, **278**, L7.

MODELLING THE OUTER ATMOSPHERES AND WINDS OF K GIANT STARS

P.G. JUDGE
Department of Theoretical Physics
1, Keble Road
Oxford, OX1 3NP
England

ABSTRACT. It is shown how empirically derived constraints affect models of the outer atmospheres and winds of K giants, taking α Boo (K 2III) as an example. The importance of *empirical* approaches prior to making *semi-empirical* models is stressed. The reliability of recent wind models is assessed.

1. INTRODUCTION

The majority of cool, luminous stars (i.e. cooler and brighter than the sun) are giants (luminosity class III) between spectral types ~ G5-K5. For such stars, the only observational "evidence" for substantial mass loss ($\dot{M} \geq 10^{-10}$ $M_\odot yr^{-1}$) is the redward asymmetries of resonance lines of e.g. MgII and CaII (e.g. Ref. 1.), since narrow 'circumstellar' absorption components are observed in giants only later than ~ K4. The interpretation of redward asymmetries is known to be ambiguous (e.g. Ref. 1.), and the mass-loss interpretation is an extension of work by Hummer and Rybicki (Ref. 2.).It has never been demonstrated that the outflowing material has $v > v_{esc}$ in a K giant. The present paper assesses this interpretation and the reliability of recent semi-empirical modelling based on the asymmetric profiles in the light of recently derived empirical constraints (Ref. 3.).

2. EMPIRICAL CONSTRAINTS

An important result from studies of UV emission lines observed with IUE in the spectra of cool giants is that the regions where the line fluxes are *created* are close to hydrostatic equilibrium (Refs. 3,4). The conclusion is based on measurements of linewidths, electron densities, column densities and emission measures. The emitting regions are supported largely be the pressure of non-thermal motions (e.g. waves) leading to geometrical thicknesses $\geq$ those of early hydrostatic models (Ref. 5)., with $\Delta h_{emission} \leq 0.3$ R_* (and not ~ R_* as suggested in Ref. 6). The success of the hydrostatic models implies that the large-scale ordered flows inferred from line symmetries are unimportant (to first order) in the momentum balance of the *emitting* regions: the asymmetries

I. Appenzeller and C. Jordan (eds.), Circumstellar Matter, 323–324.

in MgII k must be associated with flowing regions overlying the emitting
regions which scatter the resonance radiation. A further important result
for α Boo is the empirical evidence for strong temperature and density
inhomogeneities in the chromosphere (Ref. 7, 3) implied by CO absorption
and CII] emission lines.

3. EMPIRICAL CONSTRAINTS: HOW DO THESE AFFECT SEMI-EMPIRICAL MODELS?

Semi-empirical models are required for the interpretation of line profile asymmetries, since assumptions concerning the atmospheric structure
(e.g. spherical symmetry) must be made to perform radiative transfer
calculations. Recently, Drake (Ref. 8) has attempted to model the 'wind'
of α Boo using the calculations of MgII k profiles in spherically symmetric outflowing wind models.

Spherical symmetry restricts flowing material to one degree of freedom
only: with steady-state outflows the matter *must* eventually reach the
local escape velocity (at distances $\gg R_*$). The empirical evidence of
inhomogeneities in the flux creation regions shows that spherical
symmetry cannot be correct. At worst, the extra degrees of freedom will
allow *circulation* of matter in the chromosphere, leading to very small
mass loss rates. This must be regarded as a serious possibility given
that the flow velocities implied from the line asymmetries are $\ll v_{esc}$
(R_*). Comparisons show that the model of Drake (Ref. 8) has electron
densities which are substantially lower than those derived empirically,
and that the model is too extended geometrically in the regions where
the UV line fluxes are created. The validity of the model where the
line asymmetries are produced must therefore also be in question. The
interpretation of radio fluxes in terms of spherically symmetric winds
(Ref. 9) must also be re-examined in the light of the empirical constraints. It may be possible to reproduce observed asymmetries and radio
fluxes with circulating models radically different from the wind models.

4. CONCLUSIONS

In conclusion, sufficient *empirical* evidence exists to question the use
of *semi-empirical* methods in interpreting line profiles in one-component
models of K giants. Current mass loss estimates may be seriously in
error. Direct imaging of CS shells in e.g. MgII k with HST (suggested
in Ref. 8) is required to confirm or reject current wind models.

REFERENCES

1. Stencel R.E. and Mullan D.J., 1980, Astrophys. J. 238,221.
2. Hummer D.G. and Rybicki G.S., 1969, Astrophys. J. 153, L107.
3. Judge P.G., 1986.Mon. Not. R. astr. Soc. 221, 119.
4. Judge P.G., 1986.Mon. Not. R. astr. Soc. in press.
5. Linsky J.L., 1980. Ann. Rev. Astron. Astrophys. 18, 439.
6. Carpenter K.G., Brown A. and Stencel R.E.,1985,Astrophys. J. 289,676.
7. Heasley J.N. et al.,1978. Astrophys. J. 219, 970.
8. Drake S.A., 1986,"Progress in Spectral Line Formation Theory",Reidel:
Dordrecht, p. 351.
9. Drake S.A. and Linsky J.L., Astron. J. 91, 602.

ACOUSTIC WAVE DRIVEN MASS LOSS IN LATE-TYPE GIANT STARS

M.Cuntz[1], L.Hartmann[1,2] and P.Ulmschneider[1]
[1]Institut für Theoretische Astrophysik, Im Neuenheimer Feld 561
6900 Heidelberg, Federal Republic of Germany
[2]Harvard-Smithsonian Center for Astrophysics, 60 Garden Street
Cambridge, MA 02138, USA

ABSTRACT. Mass loss generated by radiatively damped acoustic waves is investigated. We find that a persistent wave energy flux leads to extended chromospheres. Mass loss is quite likely produced if the wave field retains a transient character and if large wave periods are used.

1. INTRODUCTION

Three facts suggest that acoustic waves may be important for the generation of mass loss in late-type giant stars: Pulsation appears to be a viable mechanism for driving mass loss from Mira stars. For late-type giant stars one expects nonradial oscillations of angular quantum number l lower than ten and large wave periods (compared with the acoustic cut-off period), approaching the pulsation period. Short-period acoustic wave calculations show that with increasing period more wave energy is fed into mass motion than into heating.

2. EXTENDED CHROMOSPHERES DUE TO DISSIPATION OF ACOUSTIC WAVE ENERGY

With an Eulerian spherical time-dependent hydrodynamic code which treats radiation losses with a Cox and Tucker type law we have computed models for α Boo. Our aim was to investigate the acoustic wave spectrum between the periods $0.1\ P_{co}=1.4\ 10^4$ s and the pulsation period $6\ 10^6$ s, P_{co} being the acoustic cut-off period. In our first model we introduced a monochromatic acoustic shock wave with an initial amplitude of 0.2 Mach and a period of $0.1\ P_{co}$ in an atmospheric shell of 20 percent of the stellar radius. The result was that the scale height greatly increased; at the upper boundary of the shell the pressure scale height is more than a factor of six higher than in the undisturbed atmosphere. This behaviour agrees well with the observation that the chromosphere in α Boo is extended (Carpenter and Brown 1985). The larger scale height is primarily due to the increased mean temperature caused by shock heating and to a lesser degree due to wave pressure caused by wave momentum dissipation in the shocks. As the dynamical steady state resulted in a very low mean flow velocity it was concluded that monochromatic short-period acoustic waves lead mainly to extended atmospheres and not to mass loss.

I. Appenzeller and C. Jordan (eds.), Circumstellar Matter, 325–326.

3. GENERATION OF MASS LOSS BY ACOUSTIC WAVES

To satisfy the requirement for spatial resolution of the waves, and the boundary condition at infinity a compromise must be made as to the spatial extent of the atmosphere. For our preliminary results we consider three criteria to suggest the presence of mass loss in our finite atmospheric shell. The <u>first criterion</u> looks for the time-averaged flow velocity in the upper part of our model. After passage of the initial switch-on effect and after the damping of the atmospheric oscillation the mean flow velocity $v_m \gg 0$ approaches a steady state. This steady mean flow we take as indication of mass loss. If (<u>second criterion</u>) in the upper parts our model the flow speed v becomes comparable or larger than the sound speed c mass loss is suggested. When in the upper parts of our atmosphere the mean flow speed becomes both larger than the sound speed and the escape speed v_{esc} we take this as our <u>third criterion</u> for mass loss. By various degree these criteria suffer from the unknown influence of the overlying atmosphere. In our second model we take a short-period Gaussian acoustic frequency spectrum of the same energy centered at the period of $1.4\ 10^4$ s with a standard deviation of $2.0\ 10^4$ s and cut off at $5\ 10^3$ s to avoid negative periods. In the upper part of the model for a given time we now find mass loss after our first two criteria. At another time however we have v<c. Unlike to the monochromatic case the atmosphere is now unable to find a state with $v_m=0$. Long period waves from the spectrum or generated by overtaking shocks again and again lead to supersonic flows which suggests episodic mass loss. Our third model assumes long period ($3.0\ P_{co}$) monochromatic adiabatic acoustic shock waves with an initial amplitude of 0.1 Mach in an atmosphere which extends over 11 stellar radii. We find $v_m > v_{esc}$, c which fulfills our third criterion and mass loss rates of about 10^{-11} M_o per year which are in rough agreement those by Drake and Linsky (1984). Yet it is known that adiabatic calculations overestimate the mass loss (Wood 1979).

4. CONCLUSIONS

We find that short-period acoustic waves are able to produce extended chromospheres. Monochromatic waves with periods much below the cut-off period do not lead to appreciable gas flows. Short-period acoustic wave spectra possibly generate episodic mass ejections. Adiabatic waves with periods of several times the cut-off period produce reasonable mass loss rates. We suggest that both long and short period waves are needed for an effective acoustic mass loss mechanism, the short period waves to produce extended chromospheres and the long period waves to push off the mass.

REFERENCES

Carpenter, K.G., Brown, A.: 1985, Astrophys. J. <u>289</u>, 676
Drake, S.A., Linsky, J.L.: 1984, Proc. 3rd Cambridge Workshop on Cool
        Stars, Stellar Systems and the Sun, S.L. Baliunas, L. Hartmann
        Eds., Springer, Berlin, p. 350
Wood, P.R.: 1979 Astrophys. J. <u>227</u>, 220

CIRCUMSTELLAR SHELLS OF A-K LUMINOUS SUPERGIANTS

Kenneth H. Hinkle
Kitt Peak National Observatory
National Optical Astronomy Observatories[1]
P. O. Box 26732
Tucson, Arizona, U.S.A.   85726

ABSTRACT.   Infrared vibration-rotation bands of CO are ideal probes of the circumstellar environment of yellow supergiants.   Results for a sample of stars are reviewed.

1.   INTRODUCTION

The most luminous stars of all spectral types appear to be losing mass. Theoretical calculations (Choisi, Nasi, and Sreenivasan 1978) indicate that the yellow supergiants form an interesting subset of the most luminous stars with mass loss rates that can exceed $10^{-3}M_\odot yr^{-1}$.   The spectra of some yellow supergiants have conspicuous spectral features indicating the presence of thick circumstellar shells (Sargent 1961). Interestingly, the circumstellar environment of yellow supergiants is not conducive to the formation of dust.   A survey of IRAS measurements of infrared excesses of G supergiants found that only 4% have detectable circumstellar emission (Odenwald 1986).   Stothers (1975) has suggested that these dust shells are "fossils" left from when the stars were on other portions of the H-R diagram.
    A program of observing the infrared spectra of yellow supergiants at high resolution was begun a few years ago using the Kitt Peak 4 meter telescope and Fourier transform spectrometer.   Of particular interest to this program are spectral lines which originate entirely in the circumstellar shell and are not contaminated by an underlying photospheric profile.   For some yellow supergiants we have discovered that CO $\Delta v=2$ bands are present in the 2.3μm spectrum.   CO is formed entirely in the circumstellar envelopes of yellow supergiants since it cannot exist in warm photospheres.   CO vibration-rotation lines are formed in LTE under the low density conditions expected in a circumstellar shell (Hinkle and Lambert 1975).   The entire CO $\Delta v=2$ band, consisting of dozens of lines, may be observed in a single spectrogram and simply modeled to produce excitation temperatures and column densities.

[1]Operated by the Association of Universities for Research in Astronomy, Inc., under contract with the National Science Foundation.

327

I. Appenzeller and C. Jordan (eds.), Circumstellar Matter, 327–328.

2. VARIABLE CIRCUMSTELLAR SHELLS

Two yellow supergiants studied in detail are V509 Cas and ρ Cas. Lambert, Hinkle and Hall (1981) give first results from infrared spectroscopy of these stars. Both stars are massive and have no infrared excess but large mass loss rates. Time series near-infrared spectroscopy (Sheffer 1985; Sheffer and Lambert 1986) has revealed that both stars are pulsating with periods of 500 to 600 days. This pulsation is of large amplitude with a shock travelling outwards through the atmosphere during each cycle. Both stars have strong 2.3μm CO lines. The CO samples a region of excitation temperature $\sim$1500 K, implying a distance of about 2 R$_*$ above the photosphere. New results show that the CO line profiles undergo large changes during the pulsation cycle, indicating that the circumstellar CO line forming region is not decoupled kinematically from the photosphere. Such an association between the circumstellar shell and the photosphere may result in discrete mass loss events tied to the pulsation cycle. Discrete mass loss events complicate measurements of the time averaged mass loss rate (Hinkle 1983) and could reconcile the very large mass loss rates derived by Lambert, Hinkle, and Hall with more canonical values.

3. BINARY SYSTEMS WITH DUST

ε Aur and 3 Pup are both yellow supergiants with infrared excesses. Both of these systems are binaries. In the case of ε Aur the dust is associated with the companion to the F Ia primary (Backman et al. 1984); 3 Pup may be a similar system (Lambert, Tomkin and Hinkle 1986). Both 3 Pup and ε Aur (when eclipsed) have weak CO $\Delta v=2$ lines present. The CO in ε Aur is formed in a gas torus surrounding the dust-enshrouded companion. Hinkle and Simon (1986) find that the CO spectrum demands the yellow supergiant in the ε Aur system have a mass $\lesssim 7 M_\odot$, implying it is a low mass, post asymptotic giant branch star.

REFERENCES

Backman, D. E. et al. 1984, Ap. J., **284**, 799.
Choisi, C., Nasi, B., and Sreenivasan, S. R. 1978, Astr. Ap., **63**, 103.
Hinkle, K. H. 1983, P.A.S.P., **95**, 550.
Hinkle, K. H. and Lambert, D. L. 1975, M.N.R.A.S., **170**, 447.
Hinkle, K. H. and Simon, T. 1986, Ap.J., submitted.
Lambert, D. L., Hinkle, K. H. and Hall, D. N. B. 1981, Ap.J., **248**, 638.
Lambert, D. L., Tomkin, J., and Hinkle, K. H. 1986, in preparation.
Odenwald, S. F. 1986, Ap.J., **307**, 711.
Sargent, W. L. W. 1961, Ap.J., **134**, 142.
Sheffer, Y. 1985, M.A. Thesis, University of Texas at Austin.
Sheffer, Y. and Lambert, D. L. 1986, in preparation.
Stothers, R. 1975, Ap. J. Letters, **197**, L25.

CIRCUMSTELLAR ENVELOPE OF THE SUPERGIANT 89 HERCULIS

J.L. Climenhaga[1], J.Smoliński[2],J. Krempeć-Krygier[2],
B. Krygier[3], and S. Krawczyk[3]
[1] Department of Physics, University of Victoria, Canada
[2] Polish Academy of Sciences,
   N. Copernicus Astronomical Center, Toruń, Poland
[3] N. Copernicus University, Toruń, Poland

A bright enigmatic supergiant, 89 Her (HD 163506; F2Ia, $M_{bol}=-7.5$),exhibits the light variations (Percy et al. 1979, Fernie 1981, 1983) as well as the changes of radial velocities and structures of Balmer and sodium lines. Sargent and Osmer (1969) have given some evidences for the existence of expanding circumstellar envelope around 89 Her and have discovered 24 emission lines of neutral metals in its spectra.

We have undertaken the complex study of the nature of 89 Her basing on our spectroscopic observations carried out at the Dominion Observatory starting from 1970. The aim of the present paper is a discussion of narrow emission lines originated in the envelope of 89 Her. We have discovered 10 new narrow emission lines, as follows: $\lambda\lambda$ 4952.36 (NiI), 5587.36 (FeI), 5591.322 (ScI, FeII), 5702.666 (TiI), 5711.0735 (MgI), 5796.757 (CrI), 5846.306 (VI), 6270.238 (FeI), 6303.41 (EuII), 6325.22 (TiI) and probably FeII 6247.562 and CrI 6657.54 Å. The equivalent widths of the all 34 emission lines measured on our spectrograms, taken in 1970, 1975, 1977 and 1978, exhibit the significant (up to a factor of eight) irregular time variations. The profiles of these lines are usually single but sometimes some of them are splitted into two components.

We have measured the radial velocities of these emission lines using the "Arcturus" oscilloscope machine in Victoria as well as those of absorption lines in order to explain a general nature of 89 Her. The changes of the average radial velocities of absorption and emission lines with the phase of binary system estimated from absorption lines are shown in Fig.1. The radial velocities of absorption lines undergo long-term periodic changes indicating the binary nature of 89 Her. The best fitting of all our measured radial velocities reveals 221.93 day orbital period and $\gamma =-27.99 \pm 0.56$ km/s (Smoliński et al. 1980). The binary nature of 89 Her was supported by Arellano Ferro (1984) who obtains a longer orbital period, i.e. 285.8 days. On the other hand, the daily average radial velocities of the emission lines seem to exhibit only random fluctuations with the range of 6 km/s. It is worthwhile to underline a fact that the average value of all measured radial velocities of the emission lines, i.e. $V_r^{em}=-27.94 \pm 2.28$ km/s, is equal

I. Appenzeller and C. Jordan (eds.), Circumstellar Matter, 329–330.

to the γ-value of the binary system. Therefore, the emission lines are the recombination lines formed in a common envelope of the binary system. Probably they originate due to the density enhancement and adiabatic cooling appearing in the process of mass loss from the star.

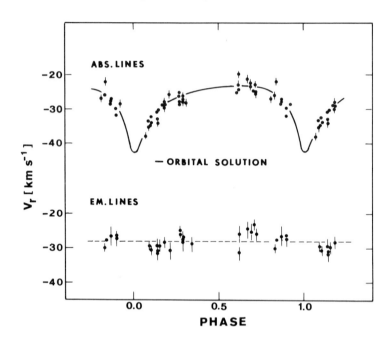

Fig.1 The changes of radial velocities of absorption and emission
lines with the phase of binary system.

Their steady occurence indicates the significant mass loading into the envelope supported by a behaviour of the circumstellar lines of hydrogen and sodium ($D_1 D_2$) and by existence of the infrared excess (Humphreys and Ney 1974).

REFERENCES

Arellano Ferro, A.: 1984, Pub. Astron. Soc. Pacific, 96, 641.
Fernie, J.D.: 1981, Astrophys. J., 243, 576
Fernie, J.D.: 1983, Astrophys. J., 265, 999
Humphreys, R.M., Ney, E.P.: 1974, Astrophys. J. Letters 187, L75
Percy, J.R., Baskerville, I., Trevorrov, D.W.: 1979, Pub. Astron.
  Soc. Pacific, 91, 368
Sargent, W.L.W., Osmer, P.S.: 1969, in Mass Loss from Stars,
  ed. M. Hack, Reidel, Dordrecht, Holland, p.57
Smoliński, J., Climenhaga, J.L., Krempeć-Krygier, J., Krygier, B.,
  Krawczyk, S.: 1980, in Fifth European Regional Meeting: Variability
  in Stars and Galaxies, Liege, Belgium

STELLAR CORONAE, CHROMOSPHERES OF COOL STARS

OBSERVATIONS OF STELLAR CORONAE

G. S. Vaiana[†] and S. Sciortino
Osservatorio Astronomico di Palermo
Palazzo dei Normanni
00134 Palermo, ITALY

ABSTRACT.

We present an overview of recent stellar X-ray observations, with some discussion of the requirements for future observations. We argue that solar observations indicate that coronal X-ray emission is strongly related to surface magnetic field activity; we show that the interpretation of X-ray stellar coronal emission from late-type stars within the framework of models analogous to those developed for the solar corona is viable, and it is supported by many experimental results. The extension of this solar analogy to the early-type stars is quite questionable and remains an unsolved problem, while the working hypothesis of an X-ray phase, related to phenomena of magnetic field-related activity, as contrasted to a wind phase during the PMS evolutionary stage is suggested by the present status of observations.

1. INTRODUCTION

Over the past two decades, the study of the very hot component of solar and stellar atmospheres has widened in its scope, range and perspective. In particular, over the past decade a new astronomical discipline has emerged: stellar X-ray astronomy. It is set at the crossroad of several major areas of astrophysical research. In the solar physics context it provides not only a major testing ground for theories but also motivates major readjustment in the perspective of the solar discipline. In the more general astrophysical context, stellar X-ray astronomy is a testing ground for several plasma processes and magnetic field-related mechanisms (magnetohydrodynamics and plasma physics of activity in flares and transients, energy release and wave propagation, magnetic field generation in self gravitating bodies and evolutionary effects of activity). In the stellar context, X-ray emission is one of the most sensitive monitors of activity, and relates to most of the major area of stellar research, covering almost all stellar masses and evolutionary stages.

In this paper, we will review and summarize the work in progress, with an emphasis on the experimental and on the observational perspective; for other aspects not covered here, we refer the interested reader to the specific literature and to some other recent reviews (Rosner, Golub and Vaiana 1985; Linsky 1985; Serio 1985; and reference therein).

---

† also Harvard-Smithsonian Center for Astrophysics

I. Appenzeller and C. Jordan (eds.), Circumstellar Matter, 333–345.

Prior to the *EINSTEIN* launch (1978), the domain of stellar X-ray astronomy was limited to only a few "odd" stars: compact binaries, as well as relatively larger samples of dwarf novae, cataclysmic variable and RS-CVns (Catura *et al.* 1975; Mewe *et al.* 1975; Nugent and Garmire 1978). If the Sun had really been prototypical, based on the 10^3 increase of *EIN-STEIN* Observatory sensitivity over previous spacecrafts, we expected to detect few solar-like stars at 10-20 pc with exposure times of $\sim 10^3$ s, while, with exposure time of $\sim 10^4$ s, we expected to detect thousands of times more intense sources up to distances of few kpc. The Sun turned out, not surprisingly, to be prototypical in the sense that a majority of solar-like stars emitted nearly at the same level of the Sun. Yet a major element of surprise was that the Sun lies near the bottom end of the observed range of X-ray luminosities of late-type stars, which spans over three decades. A second element of surprise was the ubiquity of the X-ray emission from all kinds of stars (cf., Fig. 1) independent of their mass and stage of evolution (Vaiana *et al.* 1981, Helfand and Caillault 1982).

As X-ray emission emerged as a general phenomena in all types of stars and an exceptional emission levels (with respect to the Sun) was detected for a substantial fraction of stars, a numbers of obvious questions arose: Why do most stars chose to put a fraction of their energy in high energy photons? Why do some stars chose to emit thousand times more high energy photons than others of identical mass and luminosity class? Why does this sort of behavior seem to be relatively independent of gravity (at least up to some spectral type)?

In the following we first review and update the general properties of stellar X-ray emission as deduced from the observations; and then discuss the specific characteristics of group of stars, with a particular emphasis on late spectral type and pre-main sequence stars.

2. WHAT HAVE WE LEARNED FROM THE SUN ?

In order to answer the questions posed above we need a closer look at the phenomena of the solar corona, with the aim of understanding if the physics of these phenomena can guide us in explaining stellar X-ray emission, at least for the case of late-type stars.

Here we limit ourselves only to a brief summary of the results from early rockets flights and from the X-ray telescopes on board *SKYLAB*; for a detailed discussion, we refer the reader to the extensive review of Vaiana and Rosner (1978) and the references therein quoted:

-   The outer atmosphere of the Sun is hot, structured and magnetized. The spatial contrast is highest at X-ray wavelengths, with clear evidence of magnetic confinement of plasma (loops). Within these loops the temperature T and the electron density n_e attain higher values than in the surroundings. The X-ray images of the Sun reveal regions of open magnetic topology: the coronal holes; less luminous in X-ray, they are found to be the sources of high speed solar wind streams. In general, higher values of the magnetic flux are correlated with higher values of L_x and higher values of L_x/L_v.

-   The atmosphere is dynamic, with variations on a remarkably wide range of time scales (even excluding flare variability).

-   The magnetic field has also an active role in *in situ* heating of plasmas within loops.

3. GENERAL PROPERTIES OF OBSERVATIONS

The stellar X-ray data have added many more new problem areas to those derived from the solar realm alone. We now have to contend with emission from early-type stars, from pre-

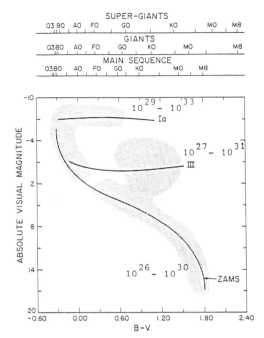

Figure 1. Schematic illustration of where stellar X-ray emission has been detected along the H-R diagram. Ranges of observed luminosities are also indicated.

main sequence and young cluster stars, from giants and WD stars, and finally from the host of low-mass dwarf stars. At the spectral type transition region between early and late type stars, the observations are so well defined that we can address the question of the onset of convection along the main sequence . A summary of the prototypes of the activity runs as follows:

- Dwarfs in the spectral type range from F to M are X-ray emitters. L_x is, to first order, independent of T_{eff} and surface gravity. The values of L_x range between 10^{26} and 10^{30} erg s^{-1}, and scale with the stellar surface angular velocity Ω (cf. Fig. 2) as $L_x \sim \Omega^n$, where $1 < n < 2$ (Vaiana et al. 1981, Walter 1982, Pallavicini et al. 1981).

- Stars earlier than B5 are X-ray emitters, at emission levels ranging between 10^{29} to 10^{34} erg s^{-1}, L_x is independent of surface gravity and scales with L_{bol} as $L_x \sim 10^{-7} L_{bol}$ (Harnden et al. 1979; Seward et al. 1979; Long and White 1980, Pallavicini et al. 1981).

- In the spectral type range from B8 to A5 detailed analysis has shown that, contrary to earlier reports, there is no credible evidence of X-ray emission from normal main sequence stars and Am stars (Schmitt et al. 1985a); however, there is some evidence for emission from Ap stars (Cash and Snow, 1982).

- Late giants and supergiants show a cutoff in X-ray emission levels as one moves to later spectral type (cf. Fig. 3), in particular, the M giants and the G and M supergiants have not been detected (Ayres and Linsky 1980, Ayres et al. 1981: Vaiana et al. 1981; Haisch and Simon 1982).

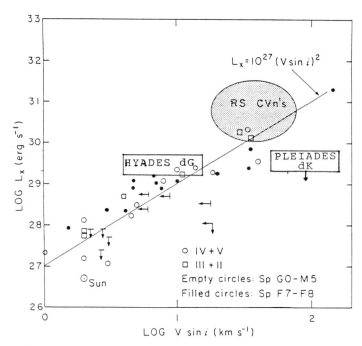

Figure 2. Scatter plot of soft X-ray luminosity versus rotation rate (adapted from Pallavicini *et al.* 1981) and extended to newly analyzed *EINSTEIN* and *EXOSAT* data. Note that independent of spectral type, the correlation is quite good, except for the data points of the Pleiades fast rotating K stars.

- The X-ray luminosity is a function of stellar age, the dependence is not simple and, in general terms, is consistent with the decline of Ω with stellar age (Vaiana 1983, Stern 1983, Micela *et al.* 1985, Caillault and Helfand 1985).

- Pre-main sequence stars have X-ray luminosities more than a thousand times the X-ray luminosities of normal main sequence stars (Feigelson 1984 and reference therein).

- The X-ray spectra of late-type stars are thermal, with single temperature components in the range 0.2-2.0 keV (10^6-10^7 K); however, many spectra show evidence for multi-temperature plasma (Holt *et al.* 1979; Swank *et al.* 1981; Vaiana 1983; Mewe *et al.* 1982; Schrijver *et al.* 1984; Majer *et al.* 1986), consistent with a continuous emission measure distribution in temperature; such emission measure distributions have been considered more appropriate for loop modeling analysis by some authors (Majer *et al.* 1984; Mewe 1984; Schmitt 1984; Schmitt *et al.* 1985b; Stern *et al.* 1986).

- Solar-like transients have been detected in late dwarfs (Haisch *et al.* 1980; Haisch 1983, Kahler *et al.* 1982) and in evolved stars which are members of close binary systems (e.g., RS-CVns; Walter *et al.* 1980; Agrawal *et al.* 1983); moreover, more energetic events named superflares, in which the release of energy as total soft X-ray luminosity is 10^2-10^3 times that in the solar case (Stern 1983, Montmerle *et al.* 1983; Stern *et al.* 1983; Caillault and Helfand 1985), have been detected in a few young dwarf stars and in pre-main sequence stars (cf. Fig. 4).

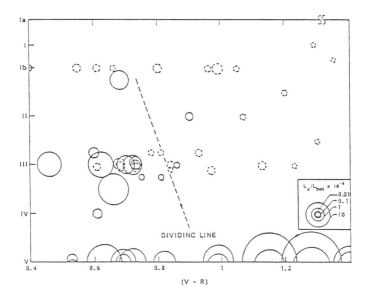

Figure 3. H-R diagram of a sample of single late-type stars observed with the *EIN-STEIN* Observatory (adapted from Antiochos *et al.* 1986), showing detections (solid circle) or upper limits (broken circle). The size of a circle is proportional to the bolometric luminosity. Note the lack of detections to the right of the dividing line (dashed).

- X-ray emission from close binary system, such as the RS-CVn (Walter *et al.* 1980) and the W UMa stars (Cruddace and Dupree 1984), in which the accretion should not play a major role, are substantially more intense than those from single stars of similar stellar structural characteristics.

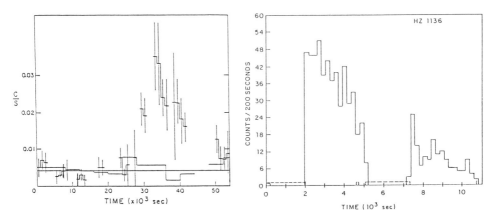

Figure 4. (right) X-ray light curve of a complete flare-like event seen in AS205 within 4 hours (Walter and Kuhi 1983). (left) Superflare seen in the K Pleiades stars Hz 1136 within 2 hours (Caillault and Helfand 1985).

4   EARLY TYPE STARS

Because the solar analogy is not immediately applicable to the early type stars, the problem of the physical mechanism responsible for their emission is more intriguing, and our understanding is quite primitive as compared to the case of late type stars. Two points are essential for model-building, and have not as yet been definitely established:

-   The X-ray source location, especially within the context of the stellar wind geometry; here spectral studies will play an essential role.

-   The spatial geometry/distribution of the emitting plasma (i.e., is the emitting gas diffuse, or does it occur in the form of fairly well-defined hot "bubbles" or shock structures); here variability studies will be of considerable interest.

In addition to the constraints already cited above, other observational constraints for modelling the X-ray emission from OB stars can be summarized as follows:

-   There is some evidence that L_x does not scale as M (Ramella et al. 1986);

-   The spectra are not absorbed at 0.5 keV (Cassinelli and Swank 1983);

-   There is some indication of long-term (Snow, Cash and Grady 1981) and short-term (Collura et al. 1986) variability of the X-ray emission.

The model originally proposed for OB star X-ray emission, namely that a hot corona lies near the stellar surface, underlying a far cooler high-speed wind (Hearn 1975; Cassinelli and Olson 1979; Waldron 1984), cannot easily reconciled with the absence of absorption features at 0.5 keV (cf., Fig. 5). More recent models attempt to explain the X-ray emission by considering that the massive radiatively-driven wind might be unstable to density perturbations, with the blobs of enhanced density produced by the instability shocking, and thereby producing material distributed throughout the wind, which in turn leads to the observed X-ray emission (Lucy and White 1980; Lucy 1982). However, Cassinelli (1985) has interpreted the evidence for very hot plasma ($T > 1.5 \ 10^7$ K) as a requirement for confinement of this hot gas, and hence hypothesizes that this confinement may be magnetic in character.

Much work is now in progress on the problem of understanding the origin of X-ray emission from early-type stars. For example, Ramella et al. (1986) are reinvestigating the correlation of L_x with stellar and wind parameters, based on a larger sample of stars than in the original EINSTEIN investigations; and Collura et al. (1986) are investigating the short-term variability of a sample consisting of a dozen OB stars.

5.   LATE GIANTS AND SUPERGIANTS

The major problem raised by the observations is whether there is really a change of character of X-ray activity in evolved stars with respect to main sequence stars, i.e., whether the applicability of the magnetically-confined corona prototype to the atmospheres of late giants and supergiants can be maintained (e.g., Dupree 1982, Linsky 1982). Beyond the cutoff in the X-ray emission level at later spectral type (cf. Fig. 3), the scaling of the X-ray emission for some of the detected giants with Ω^2 is the major observational constraints to date. This latter experimental result supports the notion that magnetic activity is the driving mechanism of the chromosphere-corona activity in evolved stars as well. Indeed, if this is the case, we will expect (as in the case of the main sequence stars) a good correlation of X-ray luminosity level with stellar parameters such as the rotational surface velocity, which are related to dynamo activity. Unfortunately, the present status of data at hand does not allow us to study these correlations in detail.

An ongoing program to survey the total sample of late giants and supergiants listed in the bright stars catalogue (Hoffleit 1982) and observed with the *EINSTEIN* IPC is in

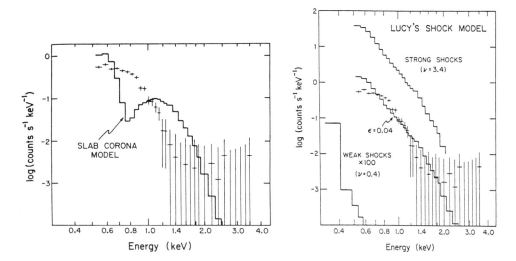

Figure 5. (right) SSS spectrum of ε Ori, with the best fit based on the slab corona model plus cool wind of Cassinelli and Olson (1979). Note the lack of the absorption feature at ~ 0.5 keV in the experimental data. (left) Same spectrum with the best fit derived from Lucy (1982) model in which both weak shocks and few isolated stronger shocks are included (adapted from Cassinelli and Swank 1983).

progress (Maggio *et al.* 1986). The preliminary results support the previous finding of the existence of a dividing line also in the X-ray emission.

Present modeling of X-ray emission from evolved stars addresses the problem of the weakening of X-ray emission at later spectral type. Some authors have suggested that the coronae of the evolved stars are more analogous to solar coronal holes than to solar loops (Linsky and Haisch 1979); Linsky and Haisch noted that the appearance of cool winds (as seen in UV lines) coincides with the disappearance of hot coronae (as seen in soft X-rays). A different description (Ayres *et al.* 1981), originally suggested to explain the presence both of coronae and cool stellar wind in the so-called hybrid-spectrum supergiants (Hartmann, Dupree and Raymond 1980), assumes that the X-ray emission derives from magnetic loops extending to some given distance above the star's surface, and that above that height a flow of cool wind develops; this wind will significantly absorb the softer X-rays emitted by the compact coronal structures. More recently, Antiochos *et al.* (1986) have proposed that because of the low surface gravity of the stars without X-ray emission, a hot (T > 10^6 K) corona is thermally unstable, and must cool down to lower temperatures, thus explaining the (apparent) lack of coronae.

However, due to the drastic reduction of the *EINSTEIN* IPC sensitivity as one goes from 0.25 keV to 0.10 keV, which results in poor IPC sensitivity to X-ray emission when the coronal temperature is less than 10^6 K, the actual status of our knowledge does not rule out the existence of warm coronae (T ~ 10^5 k) in supergiants; this problem will remain for future telescopes, which are sensitive in a softer energy band than the *EINSTEIN* Observatory was.

6.   SOLAR-TYPE STARS

Turning to the subject of solar-like stars, the fundamental question which arises is related to the extrapolation of solar modelling to the stellar case: how appropriate is the solar analogy? In addition to this central problem, other questions emerge; for example:

-    How do the parameters of coronae (L_x, T, n_e) vary with stellar parameters such as M, T_{eff}, Ω, Z?

-    Is there any evidence for a dependence of stellar X-ray activity on stellar age?

-    What happen to activity in the very low mass stars and in the close binaries?

-    At what spectral type does the onset of surface convection occur?

All of these questions are related to larger underlying problems, such as the dependence of stellar surface activity on the stellar dynamo, the nature of surface activity in itself, the structure of the convection zone in low-mass stars, and the general problem of flaring in astrophysics. Many of these questions can now be addressed with the aid of new (at least within the astronomical realm) statistical techniques which allow one to study unbiased (or, in general, volume-limited) samples of stars with the use of survival analysis; these techniques, which permit the inclusion of upper limits to detections, have enabled one to construct maximum likelihood integral X-ray luminosity functions. For a detailed description of these techniques, we refer the reader to Schmitt (1985), Feigelson and Nelson (1985), Isobe *et al.* (1986), and extensive applications of these techniques can be found in Schmitt *et al.* (1985a), Micela *et al.* (1985), Maggio *et al.* (1987), Bookbinder (1985), Micela *et al.* (1986). In the following, we limit our discussion to a summary of some of the major results, including new data not yet published.

6.1. X-ray Emission and Rotation

A dependence of X-ray emission level on rotation rate is expected on the basis of stellar dynamo theory; indeed, the observed correlation between X-ray luminosity and stellar rotational velocity is quite striking, notwithstanding the typical scatter of approximately one order of magnitude in the correlation. However, larger deviations are evident, as in the case of the Pleiades dK stars (which are rapidly rotating, and lie much below the trend line connecting X-ray emission levels with rotation rate; Micela *et al.* 1984, 1985). Both the observed scatter and the behavior of Pleiades dK stars seem to indicate that other parameters are relevant in determining X-ray emission levels, such as for example stellar age; this indeed seems to be the case for the dK Pleiades stars (cf., Fig. 2)

6.2. Age Dependence of X-ray Emission

The dependence of X-ray emission on stellar age is effectively studied by surveying open clusters of distinct age, whose members should be in the same evolutionary stages (at least within each individual spectral type) and have similar chemical composition. X-ray surveys of the Pleiades (Micela *et al.* 1985; Caillault and Helfand 1985), of the Hyades (Stern *et al.* 1983, Micela et al 1986), of Orion (Smith *et al.* 1983, see also Caillauit 1987) have been carried on. The results of these surveys show a clear decline of X-ray emission with increasing stellar age, with a possible saturation toward the age of the Pleiades (cf., Fig. 6). Recently Maggio *et al.* have surveyed the sample of the dG stars within 25 pc falling in all the *EINSTEIN* IPC fields, and have found that a similar trend is present if one considers the individual star ages (cf., Fig. 6).

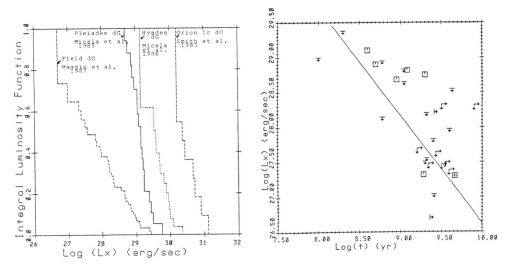

Figure 6. (right) Integral X-ray luminosity functions for four samples of stars well characterized in age: the nearby dG stars (Maggio *et al.* 1987), dG Hyades cluster members (Micela *et al.* 1986), dG Pleiades cluster member (Micela *et al.* 1985), and the slow rotating dG stars of Orion Ic (Smith, Pravdo and Ku 1983) (adapted from Micela *et al.* 1986). (left) Log-Log scatter plot of X-ray luminosity and stellar Lithium age; the straight line shows the best fit power law relation. Note the decline of X-ray luminosity with increasing stellar age (adapted from Maggio *et al.* 1987).

6.3. X-ray Emission in Low Mass Stars

There is some evidence of weakening of X-ray emission for stars later than dM5 (Book-binder 1985). This phenomenon is easily explained in terms of a change of character of the dynamo mechanism in fully convective stars; in fact the dynamo mechanism is likely to be confined to the boundary layer between the radiative core and the convective envelope (Schmitt and Rosner 1983), where the magnetic field can be stored for the amplification mechanism to work effectively, before the buoyancy force brings the field to the stellar surface (Schussler 1983; Rosner 1983). When the star become fully convective (~ at M5, according to the current models), such a mechanism cannot work, thus explaining the weakening of X-ray emission at the low end of the main sequence.

6.4. The Onset of Convection

Since dynamo activity relies essentially on convection, we expect that X-ray emission should be a sensitive diagnostic of the switch from radiative energy transport to convective energy transport in the outer layers of main sequence stars as one progresses from hot B stars to cooler F stars. The survey of Schmitt *et al.* (1985a) confirms these expectations, and indicated that in the B-V range 0.1 to 0.5, the X-ray luminosity increases rapidly with B-V, the stars in the range 0.1 to 0.3 being weaker X-ray emitters than the stars in the B-V range 0.3 to 0.5 (cf., Fig. 7). Moreover, the X-ray emission is virtually absent (at the sensitivity level of typical *EINSTEIN* images) in stars with B-V ~ 0.0.

7. PRE-MAIN SEQUENCE STARS

During the final phase of star formation, a contracting protostar lies within an envelope of accreting gas, so as to be invisible optically. However, it has been suggested that in this phase such an object might be visible as an X-ray source. Assuming that the outer

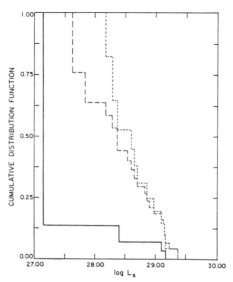

Figure 7. Integral X-ray luminosity functions for single stars in the color ranges 0.1≤B-V≤0.3 (solid line), 0.3≤B-V≤0.5 (long- dashed line), and 0.3 ≤ B-V ≤ 0.42 (short dashed line). It is evident the decline of X-ray luminosity level at lower value of B-V (adapted form Schmitt *et al.* 1985a).

atmospheres of T Tauri stars expand outward (as a stellar wind) and are thermally driven, Bisnovatyi-Kogan and Lamzin (1977) suggested the existence of a hot (10^6 K) corona near the base of the outflow region, with an expected X-ray luminosity of ~ 10^{34} erg s^{-1}. The collision between this outflow and the ambient interstellar medium would generate further X-ray emission (Schwartz 1978). In contrast, models of T Tauri star atmospheres which are dominated by accretion from the ambient interstellar matter predict X-ray emission due to the impact of the accreting matter on the stellar surface (Ulrich 1976; Mundt 1981); the temperature of the heated matter is determined essentially by the infall speed. All these predictions were disproved by the pre-*EINSTEIN* observations: only one single source detected with ANS and SAS-3 could be associated with a PMS object (den Boggende *et al.* 1978), but at an X-ray luminosity level much lower than the predictions, and with a spectrum harder than that predicted on the basis of the accretion models.

The *EINSTEIN* Observatory detected several hundred sources with typical exposure time of 10^3 s. The majority of the PMS stars have been detected in the Orion nebula (Ku and Chanan 1979; Ku *et al.* 1982), in the Taurus-Auriga complex (Gahm 1980; Feigelson and DeCampli 1981; Walter and Kuhi 1981), in the ρ Oph (Montmerle *et al.* 1983), in Chamaleon cloud (Feigelson and Kriss 1981) and in S Mon (NGC 2264) complex (Simon *et al.* 1985). The X-ray luminosity levels range from 10^{29} to 10^{31} erg s^{-1}, i.e., thousands of times more vigorous than that of the corresponding main sequence stars. Essentially all the sources are characterized by extreme variability of the emission (cf. Montmerle *et al.* 1983), and the level of X-ray emission is correlated with the stellar optical magnitude but not with other classical indicator of stellar PMS activity (such as the intensity in H$_\alpha$); furthermore, the X-ray spectra of PMS stars are intrinsically harder than those of normal main sequence stars and show some cut-off at low energy due to the absorption of the matter in the star-forming region (cf. the extensive review of Feigelson 1984).

New material has been presented at this colloquium on the basis of ongoing work. For example, Caillault (1987) has analyzed 19 IPC and HRI *EINSTEIN* fields covering the central

$2° \times 2°$ of the Orion region, and reports more than 200 distinct X-ray sources, 24 of them having no known optical counterpart at $M_v > 16^{mt}$, and 69 out 200 X-ray sources showing evidence for variability. Walter (1987) (cf. also Walter 1986) has undertaken a search for X-ray sources identified with T Tauri stars in Taurus, Ophiucus, and Corona Australis, surveying ~ 45 square degrees which are very crowded with X-ray sources, and has detected a total of 206 X-ray sources, 85 of which have known optical counterparts. In this latter survey, only 15 out of the 45 T Tauri stars present in the combined field-of-view have been detected. Moreover, the optical study of 56 objects reveals that 32 are "naked" T Tauri stars, i.e., stars coeval with the T Tauri, but differing mainly in the lack of a cir-cumstellar envelope.

The X-ray observation of PMS stars raise three major questions, namely: a) the origin of the X-ray emission, which is 10^3 times stronger than for most MS stars; b) the origin of the observed variability; and c) the time-dependent relation between X-ray emission levels and stellar parameters (such as angular momentum, structure of convective interior etc.) as star evolution proceeds.

The first question can be rephrased in terms of the capability of the solar analogy to explain the observed emission; that is, are there quasi-steady activity centers responsible for the observed emission (as happens on the sun) or is the X-ray emission instead due to a superposition of continual flaring (a possibility directly relevant to the second question as well); the present observational status does not allow us to discriminate between these two alternatives, and more high quality spectral and temporal data are needed to answer the question. The third question must also remain unresolved for the present: it requires more extensive surveys in a variety of star formation regions in order to acquire unbiased data on larger samples and in order to study the possible parameter correlations within these samples. This is a program for the future.

8. CONCLUSIONS

We can summarize the results discussed above as follows:

For late spectral type stars:

- Evidence points to activity connected to dynamo action (i.e., coupling of rotation and convection with ambient magnetic field);
- Shallow convection zone stars show a steep decline in activity;
- Fully convective stars, i.e., dwarf M later than M5, show an apparent decrease in X-ray detection;
- The level of X-ray luminosity scales as Ω^2 for all the field stars and the Hyades stars; however, this phenomenological relation cannot be reconciled with the emission levels of the Pleiades K stars.
- The level of X-ray emission depends on stellar age, younger stars being more intense emitters.

For pre-main sequence stars:

- The lack of direct correlation between T-Tauri with strong emission lines and X-ray PMS suggest a different phase for wind phenomena and X-ray phenomena;
- The luminosity range, the variability and the spectral signatures point in the direction of magnetic field-related phenomena for the X-ray phase.

ACKNOWLEDGMENTS

We thank our colleagues who have provided material prior to publication; in particular we are gratefully to R. Rosner, J. Schmitt, S. Serio, A. Maggio and G. Micela for their many discussions, comments, and suggestions. This work was supported by Piano Spaziale Nazionale and Ministero Pubblica Istruzione.

REFERENCES

Agrawal, P. C., Rao, A. R., Riegler, G. R., Stern, R. A. 1983, *Proc. Int. Cosmic Ray Conf., 18th, Bangalore*, Vol. 1
Antiochos, S. K., Haisch, B. M., Stern, R. A. 1986, *Ap. J. (Letters)*, **307**, L55.
Ayres, T. R., Linsky, J. L. 1980, *Ap. J.*, **235**, 76.
Ayres, T. R., Linsky, J. L., Vaiana, G. S., Golub, L., Rosner, R. 1981, *Ap. J.*, **250**, 293.
Bisnovatyi-Kogan, G. S., Lamzin, S. A. 1977, *Sov. Astron. AJ*, **21**, 720.
Bookbinder, J. 1985, PhD Thesis, Harvard University.
Caillault, J. 1987, this colloquium.
Caillault, J., and Helfand, D. J. 1985, *Ap. J.*, **289**, 279.
Cash, W., Snow, T. 1982, *Ap. J. (Letter)*, **263**, L59.
Cassinelli, J. P., Swank, J. H. 1983, *Ap. J.*, **271**, 681.
Cassinelli, J. P., Olson, G. L. 1979, *Ap. J.*, **229**, 304.
Cassinelli, J. P. 1985, in *The Origin of Nonradiative Heating/Momentum in Hot Stars*, ed. A. B. Underhill and A. G. Michalitsianos, NASA Conference Publ. 2358, p. 2.
Catura, R. C., Acton, L. W., Johnson, H. M. 1975, *Ap. J. (Letters)*, **196**, L47.
Collura, A., Sciortino, S., Serio, S., Vaiana, G. S., Harnden, F. R., Jr., Rosner, R. 1986, in preparation.
Cruddace, R. G., Dupree, A. K. 1984, *Ap. J.*, **277**, 263.
Dupree, A. K. 1982, in *Advances in Ultraviolet Astronomy: Four Years of IUE Research*, NASA Conference Publ. 2238, p. 3.
den Boggende, A. J., Mewe, R., Gronenschild, E. H., Grindlay, J. E. 1978, *Astron. Astrophys.*, **62**, 1.
Feigelson, E. 1984, In *Cool Stars, Stellar System, and the Sun*, ed S. Baliunas, L. Hartmann, New York, Springer-Verlag, p. 27.
Feigelson, E. D., DeCampli, W. M. 1981, *Ap. J. (Letters)*, **243**, L89.
Feigelson, E. D., Kriss, J. A. 1981, *Ap. J. (Letters)*, **248**, L35.
Feigelson, E., Nelson, P. L. 1985, *Ap. J.*, **293**, 192.
Gahm, G. F. 1980, *Ap. J. (Letters)*, **242**, L163.
Haisch, B. M., Simon, T. 1982, *Ap. J.*, **263**, 252.
Haisch, B. M. 1983, in *Activity in Red Dwarfs Stars*, ed. P. B. Byrne, M. Rodono, Dordrecht, Reidel, p. 255.
Haisch, B. M., Linsky, J., Harnden, F. R., Jr., Rosner, R., Seward, F. D., Vaiana, G. S. 1980, *Ap. J. (Letters)*, **242**, L99.
Harnden, F. R., Jr., *et al.* 1979, *Ap. J. (Letters)*, **234**, L51.
Hartmann, L., Dupree, A. K., Raymond, J. C. 1980, *Ap. J. (Letters)*, **236**, L143.
Helfand, D. J., and Caillault, J. P. 1982, *Ap. J.*, **253**, 766.
Hearn, A. G. 1975, *Astron. Astrophys.*, **40**, 355.
Hoffleit, D. 1982, The Bright Star Catalogue.
Holt, S. S., White, N. E., Becker, R. H., Boldt, E. A., Mushotzky, R. F., *et al.* 1979, *Ap. J. (Letters)*, **234**, L65.
Isobe, T., Feigelson, E. D., Nelson, P. I. 1986, *Ap. J.*, **306**, 490.
Kahler, S., Golub, L., Harnden, F. R., Jr., Liller, W., Seward, F., *et al.* 1982, *Ap. J.*, **252**, 239.
Ku, W., H-M., Chanan, G. A. 1979, *Ap. J. (Letters)*, **234**, L59.
Ku, W., H-M., Righini-Cohen, J., Simon, M. 1982, *Science*, **215**, 61.
Linsky, J. L., 1982 in *Advances in Ultraviolet Astronomy: Four Years of IUE Research*, NASA Conference Publ. 2238, p. 17.
Linsky, J. L. 1985, *Sol. Phys.*, **100**,.
Linsky, J. L., Haisch, B. M. 1979, *Ap. J. (Letters)*, **229**, L27.
Long, K., White, R. 1980, *Ap. J. (Letter)*, **239**, L65.
Lucy, L. B. 1982, *Ap. J.* **255**, 286.
Lucy, L. B., White, R. L. 1980, *Ap. J.* **241**, 300.
Maggio, A., *et al.* 1986, in preparation.

Maggio, A., Sciortino, S., Vaiana, G. S., Majer, P., Bookbinder, J., Golub, L., Harnden, F. R., Jr., and Rosner, R. 1987, *Ap. J.*, in press.
Majer, P., Schmitt, J. H. M. M., Golub, L., Harnden, F. R., Jr., Rosner, R. 1984, *Bull. Am. Astron. Soc.*, **16**, 514.
Majer, P., Schmitt, J. H. M. M., Golub, L., Harnden, F. R., Jr., Rosner, R. 1986, *Ap. J.*, **300**, 360.
Mewe, R., Heise, J., Groneshield, E. H. B. M., Brinkman, A. C., Schrijver, C. J., den Boggende, A. J. 1975, *Ap. J. (Letters)*, **202**, L67
Mewe, R., Groneshield, E. H. B. M., Westergard, N. J., Heise, J., Seward, F. D., *et al.* 1982, *Ap. J.*, **260**, 233.
Mewe, R. 1984, *Proc. Int. Colloq. EUV and X-ray Spectrosc. Astrophys. and Lab. Plasmas, 8th, Washington, DC.*
Micela, G., Sciortino, S., and Serio, S. 1984, in *X-Ray Astronomy*
Micela, G., Sciortino, S., Serio, S., Vaiana, G. S., Bookbinder, J., Golub, L., Harnden, Jr., F. R., and Rosner, R. 1985, *Ap. J.*, **292**, 172.
Micela, G., Sciortino, S., Vaiana, G. S., Stern, R., Harnden, Jr., F. R., and Rosner, R. 1986, in preparation.
Montmerle, T., Koch-Miramond, L., Falgarone, E., Grindlay, J. E. 1983, *Ap. J.*, **269**, 182.
Mundt, R. 1981, *Astron. Astrophys.*, **95**, 234.
Nugent, J., Garmire, G. 1978, *Ap. J. (Letters)*, **226**, L38.
Pallavicini, R. P., Golub, L., Rosner, R., Vaiana, G. S., Ayres, T., and Linsky, J. L. 1981, *Ap. J.*, **248**, 279.
Ramella, M., *et al.* 1986, in preparation.
Rosner, R. 1983, in *Solar and Stellar Magnetic Fields: Origins and Coronal Effects.*, ed Stenflo, J. O., Dordrecht: Reidel, p. 279.
Rosner, R., Golub, L., and Vaiana, G. S. 1985, in *Ann. Rev. Astron. Ap.*, **23**, 413.
Schmitt, J. H. M. M. 1984,in *Symp. X-ray Astron. 84*, Bologna, Italy, p. 17.
Schmitt, J. H. M. M. 1985, *Ap. J.*, **293**, 178.
Schmitt, J. H. M. M., Rosner, R. 1983, *Ap. J.*, **265**, 901.
Schmitt, J. H. M. M., Golub, L., Harnden, Jr., F. R., Maxson, C. W., Rosner, R., and Vaiana, G. S. 1985a, *Ap. J.*, **290**, 307.
Schmitt, J. H. M. M., Harnden, F. R., Jr., Peres, G., Rosner, R., Serio, S. 1985b,*Ap. J.*, **288**, 751.
Schrijver, C. J., Mewe, R., Walter, F. M. 1984, *Astron. Astrophys.*, **138**, 258.
Schussler, M. 1983, in *Solar and Stellar Magnetic Fields: Origins and Coronal Effects.*, ed Stenflo, J. O., Dordrecht: Reidel, p. 213.
Schwartz, R. D. 1978, *Ap. J.*, **223**, 884.
Serio, S. 1985, *Proc. ESA Workshop: "Cosmic X-ray Spectroscopy Mission, Lungby, Denmark*, (ESA SP-239), p. 59
Seward, F. D., Forman, W. R., Giacconi, R., Griffiths, R. E., Harnden, F. R., Jr., *et al.* 1979, *Ap. J. (Letters)*, **234**, L55
Simon, T., Cash, W., Snow, T. P., Jr. 1985, *Ap. J.*, **293**, 542.
Smith, M. A., Pravdo, S. H., Ku, W. H.-M. 1983, *Ap. J.*, **272**, 163.
Snow, T. P.,Jr., Cash, W., Grady, C. A. 1981, *Ap. J. (Letters)*, **244**, L19.
Stern, R. A. 1983, *Adv. Space. Res.*, **2**, 39.
Stern, R. A., Zolcinski, M. C., Antiochos, S. C., and Underwood, J. M. 1981, *Ap. J.*, **249**, 647.
Stern, R. A., Underwood, J. H., Antiochos, S. K., 1983, *Ap. J. (Letters)*, **264**, L55.
Stern, R. A., Antiochos, S. K., Harnden, F. R., Jr. *Ap. J.***305**, 417.
Swank, J. H., White, N. E., Holt, S. S., Becker, R. H. 1981, *Ap. J.*, **246**, 208.
Ulrich, R. K. 1976, *Ap. J.*, **210**, 377.
Vaiana, G. S., Rosner, R. 1978, *Ann. Rev. Astron. Astrophys.*, **16**, 393.
Vaiana, G. S., *et al.* 1981, *Ap. J.*, **245**, 163.
Vaiana, G. S. 1983, in *Solar and Stellar Magnetic Fields: Origins and Coronal Effects.*, ed Stenflo, J. O., Dordrecht: Reidel, p. 165
Waldron, W. L. 1984, *Ap. J.*, **282**, 256.
Walter, F. W. 1982, *Ap. J.*, **253**, 745.
Walter, F. W. 1987, this colloquium.
Walter, F. W. 1986, *Ap. J.*, **306**, 573.
Walter, F. W., Cash, W., Charles, P. A., Bowyer, C. S. 1980, *Ap. J.*, **236**, 212.
Walter, F. M., Kuhi, L. V. 1981, *Ap. J.*, **250**, 254.

THEORY OF STELLAR CORONAE

S. Serio
Astronomical Observatory and *IAIF-CNR*
Palazzo dei Normanni
90134 Palermo, Italy

ABSTRACT. X-ray observations have shown in recent years that the existence of high temperature plasmas in stellar atmospheres is far more widespread than extrapolated from preexisting theories of the solar and stellar coronae, forcing a radical change in our understanding of the mechanisms of coronal physics. This paper reviews our current ideas on stellar coronae, in particular on the role of magnetic confinement in the atmospheres of late type stars.

1. INTRODUCTION

The new field of stellar coronal physics enjoys already of a large and consistent set of observations (Vaiana and Sciortino 1987). Moreover, much of our understanding of the theoretical aspects of stellar coronae relies on the high level of detail and high spatial resolution of solar coronal observations. Since the extension of solar coronal theory to stellar coronae requires its extrapolation over a wide range of parameters, the comparison of theory and observations for stellar coronae provides also a feedback on our understanding of the solar corona.

The high degree of spatial structuring is among the most striking characteristics of the solar corona. It is easily interpreted as a manifestation of the existence of coronal magnetic fields, that shape the low β coronal plasma (Vaiana and Rosner 1978). The general idea is, therefore, that stars similar to the Sun, i. e. stars in which convective motions can interact with rotation to generate magnetic fields, can experience similar phenomena. This picture assumes that the magnetic fields produced by a dynamo mechanism inside the stars cause coronal and chromospheric activity, and feedback, through mass loss and magnetic spindown, on the rotation rate and differential rotation that are, together with convection, the essential ingredients of the dynamo. High energy particles produced during flares, and loss of mass through the wind, create a link of coronal physics to the physics of the circumstellar medium.

Since the picture is complicated by the feedback loops, and by the richness of the phenomena in magnetized plasmas, it is easily understandable that many problems are still open. In some of the basic areas of coronal physics, such as those relating to the heating of the coronal atmosphere, and to the source of the magnetic field, the theory does only provide a qualitative framework to understand the basic phenomena, rather than detailed predictions.

Although one might think of a role for some primordial magnetic fields (see for example Uchida, 1986) in early type stars, the picture that has attracted some consensus for the coronae of these stars, is one in which high temperatures plasmas are produced by instabilities and shocks in their strong winds (Lucy and White 1980; Lucy 1982).

I. Appenzeller and C. Jordan (eds.), Circumstellar Matter, 347–356.

Since the theory of these instabilitis is treated elsewhere in this volume (Hearn 1987), I shall limit my discussion only to late type coronae. In the progress of this paper I shall review some of the critical aspects of the theory of magnetically confined coronae, namely the dynamo mechanism and the formation and heating of the coronal structures (§ 2). I shall then concentrate on "loop models" of stellar coronae of late type stars. This last topic has been recently investigated by several authors (Zolcinski et al. 1982; Giampapa et al. 1985; Schmitt et al. 1985; Landini et al. 1985; Stern et al. 1986; Mewe 1986), and it constitutes, perhaps, the area of coronal physics most satisfactory for comparison of theory and observations.

2. CORONAE OF LATE TYPE STARS

Our understanding of the coronal mechanism in late type stars is somewhat guided by observations of the solar corona showing that it is dominated by magnetic fields; that these fields are highly structured forming what we call loops or complexes of loops; that the coronal plasma is highly dynamic, with characteristic times ranging from the few minutes of flares to the several months of long term evolution of active regions.

The questions that I will address in this context include: 1. the production of magnetic fields, 2. the heating of the magnetized plasma, and 3. the structuring of the magnetized plasma.

2.1 Production of magnetic fields - Dynamo

Magnetic fields in late type stars are supposed to be produced by a dynamo mechanism. To understand how this mechanism works, let's look at the diffusion equation for the magnetic field $\underline{H}$:

$$\frac{\partial \underline{H}}{\partial t} = \underline{\nabla} \times (\underline{v} \times \underline{H}) - \underline{\nabla} \times (\eta \, \underline{\nabla} \times \underline{H}) \quad , \tag{1}$$

where $\underline{v}$ is the fluid velocity and η is the diffusivity.

Owing to the presence of velocity eddies, this equation is intractable. However, if one averages over eddies, one gets the equation for the *mean field* $\underline{B}$:

$$\frac{\partial \underline{B}}{\partial t} = \underline{\nabla} \times [(\underline{\Omega} \times \underline{r}) \times \underline{B} + \alpha(r)\underline{B}] - \underline{\nabla} \times (\eta_t \, \underline{\nabla} \times \underline{B}) \quad , \tag{2}$$

where Ω is the rotation speed. The "α" term and the "Ω" term drive the dynamo, while the η (turbulent diffusivity) term acts as a damping term. If we now neglect this last term and assume that $\underline{B}$ consists of a main poloidal component derived by a vector potential A in the azimuthal direction, and by a seed azimuthal component B_ϕ, we find:

$$\frac{\partial^2 B_\phi}{\partial t^2} = r \left[\underline{\nabla}(\alpha B_\phi) \times \underline{\nabla}(\Omega \sin\theta) \right] + O(\alpha^2) \quad . \tag{3}$$

Therefore, we can easily see that B_ϕ grows with a rate

$$\tau^{-1} \approx \sqrt{\alpha |\underline{\nabla}\Omega|} \quad . \tag{4}$$

Obviously, for a strong amplification we expect that the newly created B_ϕ stays around at least for a time $\sim \tau$.

It was pointed out by Parker (1975), that a magnetic flux tube embedded in the convection zone experiences a buoyancy force, that lifts it to the surface with a rate of the order of the

Alfvèn speed. For example, if the flux tube is created at the depth of $\sim 10^{10}$ cm, the time of rise is of the order of one day -- too short for appreciable amplification -- and field amplification has to take place in a region where the field can be stored long enough. This region can be identified in the overshoot layer at the top of the radiative zone. There, the drag force caused by downward flows is not balanced by buoyancy, and the result is that flux tubes can leave enough for magnetic amplification.

As suggested by Schmitt and Rosner (1983), this configuration is eventually unstable. Vertical motions of thin flux tubes will tend to be amplified because of the difference in magnetic and thermal diffusivity (doubly diffusive instability).

The time scale of the instability can be estimated by simply taking the pressure scale height at the overshoot layer L, divided by the Alfvèn velocity. This results in the constraint $\tau \geq L/v_A$. A rough estimate of α in Eq. 2 can be given by $\alpha \propto L^2 \Omega/R$, where R is the stellar radius, implying:

$$B \sim \sqrt{L^2 (\Omega |\nabla \Omega| \rho/R)} \quad , \tag{5}$$

where ρ is the mass density.

Since simple models for magnetic heating of coronae predict that the power going into heat is $\propto B^2$ (e. g. Golub et al. 1980), a quadratic relationship among X-ray luminosity and angular velocity is predicted, in general accord with observations (Pallavicini et al. 1981a), whenever $\nabla \Omega$ scales as Ω/R.

For early F type stars, however, the dependence on Ω is masked by the strong dependence of the depth of the convection zone (or of the convective turnover time) on mass; hence one expects, as found in the survey of Schmitt et al. (1985), that the X-ray luminosity of the early F stars be correlated to Rossby number rather than to angular velocity. Moreover, when the gradient of angular velocity is not fully developed, simple correlations are lost. Example: the Pleiades G and K stars, which show little correlation between X-ray luminosity and rotational velocity, but also between X-ray luminosity and Rossby number (Micela et al. 1984, 1985).

In summary, dynamo theory is in rough agreement with observations, although a detailed comparison of theory and observations is not yet possible because of the complicated physics of convection in the magnetized and rotating stellar interior.

2.2 Structuring and Heating of the stellar coronae.

The next problem I want to address is that of structuring and heating of the coronae. The question is: how does the magnetic field emerging at the photosphere acquire the fine structuring in thin elongated loops?

One approach has been pursued by Ferrari, Rosner, and Vaiana et al. (1982) and by Bodo et al. (1985). They consider a vertical magnetic field embedded in the cool photosphere, with a current parallel to the field produced by photospheric motions. they show that such configuration is unstable against Joule heating. Increased local heating increases the current locally, in a runaway process, thus producing filamentation on a small scale.

Of course this result pertains only to the linear regime, but it gives, nonetheless, an indication that the coronal structures are highly filamented. This might help to solve an other problem, that about heating the corona. The essential starting point, and the difficulty of the topic, stays in the vastness of the zoo of instabilities in magnetized plasma that can contribute to the thermal balance of the corona. The basic picture is that of a loop anchored in the photosphere. Turbulent motions of the footpoints will induce both MHD waves propagating along the field, and currents parallel to the field. The dissipation of the waves (AC heating) and the possibly anomalous Joule heating by the currents (DC heating) both contribute to coronal energy balance (Kuperus, Ionson, and Spicer 1981; Serio 1983).

To determine which of the two mechanism is most important, and its detailed working, it is extremely difficult (if not impossible because of concurrence of different mechanisms). DC heating has generally the shortcoming that narrow current sheets are necessary, thus rising the problem of heat transport across the field lines. AC heating models using body waves or Alfvèn waves do not encounter this problem, but the heating rates are difficult to estimate.

As for current heating, it has been suggested that observations of microwave emission from dwarf M flare stars can be explained in terms of coronal currents (Holman 1986). The evidence comes because the microwave emission observed during flares in dMe stars can be attributed to gyrosyncrotron emission from high energy ($\sim$ 10 keV) electrons (Linsky and Gary 1983). On the other side, an important feature of current dissipation mechanisms is that, in a plasma at temperature T, in the presence of an electric field E, thermal electrons with velocity greater than the critical velocity defined by $v_c^2 = [m \nu_e (KT/m)^{3/2}]/eE$, where ν_e is the thermal collision frequency, will be accelerated freely. Holman (1985, 1986), finds that a current dissipation model with current drift speed of the order of the ion sound speed can accelerate enough electrons to explain both X-ray and microwave gyrosyncrotron emission in flares and in the stationary emission of dMe stars.

3. CORONAL LOOP MODELS

The effect of confinement by the magnetic field on a static coronal plasma is to force scaling relationships for base pressure p, maximum temperature in the loop T, loop length L, and volumetric heating rate Q (Rosner, Tucker, and Vaiana 1978).

$$T \quad \sim 1.4 \times 10^3 \, (pL)^{1/3} \; K \quad , \tag{6}$$

$$Q \quad \sim 10^5 \, p^{5/6} \, L^{-7/6} \; ergs \; cm^{-3} \, s^{-1} \quad . \tag{7}$$

Although individual loops may have a wide range of base pressures and length, those contributing most to X-ray emission for unresolved stellar observations, will have a length about equal to 1/2 the pressure scale height h. The emission measure above that height, in fact, will be negligible, while smaller loops are likely to fill a small fraction of the available coronal volume. Under these assumptions, it is easy to see that the X-ray luminosity is related to coronal temperature by:

$$L_x \quad \sim 5.5 \cdot 10^{11} \, T^{5/2} \, q(h/R) \, f \quad ergs \; s^{-1} \quad , \tag{8}$$

where $q(h/R)$ is a slowly varying function of the ratio of the pressure scale height to the stellar radius R, and f is a "filling factor", describing the fraction of photospheric surface covered by the footpoints of coronal loops (Rosner, Golub, and Vaiana 1983; Stern et al. 1981). Eq. 8, with $f = 1$, sets an upper limit on the X-ray luminosity of a stellar corona. Alternatively, it can be used, once the X-ray luminosity and coronal temperature are known, to estimate the surface filling factor.

The question naturally arising is whether our picture, i.e. that of a corona dominated by identical static loops of "convenient" height, is realistic or not. This question cannot be answered unless we have an independent way of estimating f in Eq. 8, or unless we use some more refined means, such as, for example, comparing the distribution of emission measure predicted by coronal models with that deduced from observations. Static and dynamic loop models shall be discussed in the following subsections.

3.1 Static Loop Models

Assuming uniform density and temperature across the loop's cross section, the model involves the (numerical) solution of the unidimensional energy and force equations for a plasma confined in a tube of assigned shape and dimensions,:

$$Q = n^2 P(T) - \frac{\partial F_c}{\partial s} \quad , \tag{9}$$

$$n \mu g_s = \frac{\partial p}{\partial s} \quad , \tag{10}$$

where n is the plasma density, $P(T)$ is the plasma emissivity (emitted power per unit emission measure), F_c is the conductive flux, μ is the average ion mass, and g_s is the component of gravity along the loop's coordinate s. The boundary conditions usually used for this model assume that the loop is symmetric at the top, and that the conductive flux F_c vanishes at the footpoints (e.g. Rosner, Tucker, and Vaiana 1978), although different conditions have also been discussed (e.g. Vesecky, Antiochos, and Underwood 1979).

Hydrostatic numerical models of the kind described in Eqs. 9 and 10 have been developed for the solar corona (Vesecky, Antiochos and Underwood 1979; Serio et al. 1981) and have been shown to be generally in good agreement with resolved observations of active region loop complexes (Pallavicini et al. 1981b).

The extrapolation to stellar coronae, however, is not straightforward: owing to the lack of spatial resolution, it involves the assumption, as discussed above, that only loops of one size and base pressure are important in determining the observational characteristics of the corona. Despite this unavoidable strong assumption, static models have been applied to a variety of stellar coronal observations (Giampapa et al. 1985; Schmitt et al. 1985; Landini et al. 1985; Stern et al. 1986). Two different methods have been used, one relying on the comparison of the predicted emission in the EUV region and in the X-ray region (Giampapa et al. 1985; Schmitt et al. 1985; Landini et al. 1985) with IUE and Einstein or Exosat observations, the other on the detailed fit of the emission predicted by the model and the observed pulse height spectral distribution in the Einstein Imaging Proportional Counter (IPC -- Schmitt et al. 1985; Stern et al. 1986).

To illustrate the relative merits of both approaches, and their fundamental limitations, I shall briefly review some of their applications. Testing the basic assumption, i.e. that a corona can be characterized by a single loop, is particularly desiderable to establish the validity of the approach. This can be done on the only star for which we have resolved X-ray and EUV observations: the Sun. Giampapa et al. (1985), have used full disc estimates of EUV and X-ray solar fluxes to constrain single static loop models of the solar corona both during solar maximum and solar minimum.

The procedure for the fitting is straightforward. One builds a series of static models having the maximum temperature suggested by the X-ray observations, but differing in base pressure (the constraint of confinement and the fixed maximum temperature do obviously limit the parameter space to a trajectory in the $p L$ plane -- Eq. 6). For each model one computes the predicted fluxes, assuming that the emission comes entirely from identical loops whose integrated footpoint area is a fraction of the stellar surface:

$$F_i = f_i \int_0^L n^2 G_i(T) ds \quad , \tag{11}$$

where F_i is the line flux in any of the EUV lines under consideration (whose line emissivity function is $G_i(T)$), or the X-ray flux (here $G_i(T)$ is the X-ray emissivity function folded through the instrument bandpass). If the rhs of Eq. 11 is evaluated for n different EUV lines and/or X-ray bandpasses, and compared to the corresponding observed fluxes, it gives n values of f_i, whose mean represents the "filling factor" for the model, and whose scatter is a measure of how good, or how bad, is the loop model we are testing. In this way it is generally possible to select the loop model for which the scatter of the different f_i's is minimum; the length of the model loop and the average filling factor will then be a representation of the data. In the Sun this can be compared to direct, spatially resolved observations, to assess the validity of the approach. The "best fit" models for solar minimum and solar maximum have loop lengths $7.0 \cdot 10^{10}$ and $6.9 \cdot 10^{9}$ cm, and filling factors 0.56 and 0.03. While deviation of individual filling factors are appreciable, this appears to be a fair representation of the coronal structures in both cases: large scale structures dominating most of the corona during solar minimum, and active regions covering a small fraction of the solar surface during solar maximum.

Although this approach appears to work fairly well for the Sun, one should be careful in interpreting its results for other systems, for at least the following reasons:

i)   EUV and X-ray data are usually not simultaneous, thus we have to rely on the additional hypothesis that no significant variations have occurred between the times of acquisition of the different sets of data;

ii)   the actual heating distribution along the loop is difficult to assess, while in the discussion above it has been assumed uniform; a different distribution makes little difference for the X-ray flux, but it may give significantly different values for the EUV fluxes;

iii)   single loop models, while realistic for the solar corona, may not be realistic in the wide stellar context (e.g. Jordan et al. 1986).

In the second approach, only the X-ray spectral information is used. Here one computes the deviations of the IPC pulse height spectra from the predictions of the model. One may think that, having more degrees of freedom (the number of IPC channels available for the fit is generally ≥ 10), one can break the "degeneracy" implied in Eq. 6, and find a minimum χ^2 in the $p\,L$ plane. This is not true in general (Serio 1985; Schmitt et al. 1985), but by using more parameters, such as, for example, a variable cross section of the loops, one can obtain some constraints on the model parameters from the data (Stern et al. 1986). However, since the surfaces of constant χ^2 have no sharp minima in the parameter space, the probable error in the determination of the parameters is inherently high. Thus, even in this case, loop models suffer some inadequacy, and we may think of their validity in a statistical rather than individual way, in the sense that the ensemble of a large number of best fit loop models is representative of the ensemble of stellar loop atmospheres. The problem, at this stage in the analysis of the *Einstein* data, is that we do have only a few fitted models, although we might have more in the near future.

3.2 Dynamic Loop Models.

The success of hydrodynamic loop models in describing the evolution of solar flares (Strong 1986; Peres et al. 1987) is certainly stimulating and pointing to the possibility that similar models can be applied to stellar flares. These models are based on the equations of mass, momentum, and energy conservation, and on the equations of state and of ionization balance for plasma confined in a rigid tube and subject both to steady-state and transient heating:

$$\frac{dn}{dt} = -n \frac{\partial v}{\partial s} \quad , \tag{12}$$

$$n\mu \frac{dv}{dt} = -\frac{\partial p}{\partial s} + n\mu g_s + \frac{\partial}{\partial s}\left(\lambda \frac{\partial v}{\partial s}\right) \quad , \tag{13}$$

$$\frac{dE}{dt} + (E+p)\frac{\partial v}{\partial s} = Q - n^2 \beta P(T) + \lambda\left(\frac{\partial v}{\partial s}\right)^2 + \frac{\partial F_c}{\partial s} \quad , \tag{14}$$

$$p \;=\; (1 + \beta)\, n\, K\, T \quad , \tag{15}$$

$$E \;=\; \frac{3}{2}\, p \;+\; n\, \beta\, \chi \quad , \tag{16}$$

where v is the plasma velocity, λ its viscosity coefficient, Q describes both the steady state and the transient volumetric power input, β is the ionization fraction, and χ the hydrogen ionization potential.

Peres *et al.* (1983) have shown that the static solutions of Eq. 10 and 11 are indeed stable under the dynamics described by Eqs. 12-16, if one takes into account also the response of the chromosphere to coronal transients, i.e. if the modeling is extended to the denser layers below the transition region, that can act as a plasma reservoir.

Fig. 1 shows the comparison of X-ray line fluxes derived by numerical solutions of Eqs. 12-16, and of X-ray observations of a compact solar flare (Peres *et al.* 1987).

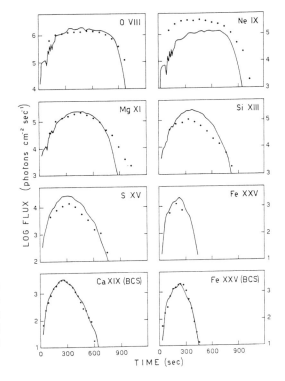

Fig.1 Comparison of computed (solid lines) light curves of X-ray lines, and of observations by the SMM Flat Crystal Spectrometer for the solar flare of 1980, Nov. 12, 17:00 UT. The computations assume that the flare impulsive energy is deposited near the top of a loop $2 \cdot 10^{9}$ cm high. The time profile of the impulsive heating, as well as its peak value, are optimized to best fit the observations. The flux is in units of photons $cm^{-2}\ s^{-1}$ at Earth (Peres *et al.* 1987).

It is apparent that the evolution of the flare is well described. Of course, for a solar flare observed with some spatial resolution, the dimensions of the loop in which the flare is supposedly exploding are known, and the numerical calculations, therefore, can be used to get some insight into the physics of the heating mechanism phenomenon, for example the site of energy release.

For stellar flares, however, the basic information we can hope to extract from a dynamic model is the very same spatial picture of the flaring region that we cannot perceive directly. Since the dynamics of plasma cooling is sensitive to the dimensions and shape of the region in which it is confined, we can hope to be able to obtain some information on the structure of stellar coronae just by comparing numerical calculations under different hypothesis with observed light curves of X-ray stellar flares. To do this in a simple way, we have to assume that the flare is confined during its evolution, which inevitably limits our considerations to small flares.

The approach is therefore the following: estimate the total impulsive energy related to the flare and the duration of the impulsive phase; assume starting values for the length and the cross section of the model loop, from the observed decay rate and of the integrated count rate, and compute the evolution of the plasma it confines, subject to the corresponding transient energy input; finally, compare the cooling rate of the computed light curve with the observed one, to estimate how the length of the loop, in a successive iteration, has to be changed to provide a better agreement.

In this way it is possible, in principle, to determine the best parameters of the loop (length and cross section) corresponding to a given pattern of impulsive energy input; successive refinements can be obtained by acting on the time evolution of the impulsive energy term and/or its total amount, and comparing computed with observed spectra (IPC spectra, for example, for observations with Einstein).

This approach is being followed by Reale *et al.* (1987) in a study of one flare observed by *Einstein* on Prox Cen. Some preliminary results of this study are illustrated in Fig. 2, showing the sensitivity of the flare cooling rate to the length of the model loop, and in Fig. 3, which compares the evolution of the computed average loop temperature with that measured by a isothermal fit to the IPC data.

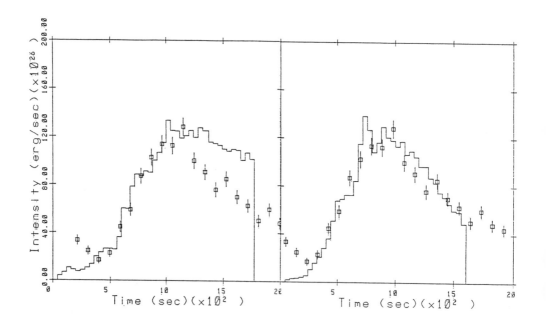

Fig. 2 Comparison of computed (solid lines) X-ray light curves and Einstein IPC observations for a Prox. Cen. flare: a) in a loop $1 \cdot 10^{10}$ cm high; b) in a loop $5 \cdot 10^{9}$ high, with the same energy input as in a).

As for the solar flare, the agreement between calculations and observations is good, and this is certainly comforting, considering the large amount of computing time that is necessary to run the numerical code. A few words of caution are of order, however. Some of the basic hypothesis underlying the simple physics described by Eq. 12-16 are difficult to verify for a

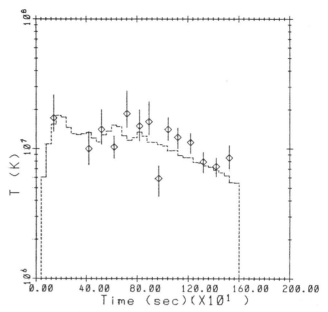

Fig.3 Comparison of the average plasma temperature derived by the hydrodynamic model for the Prox. Cen. flare and IPC single temperature fits to the observed count distribution.

stellar flare (for example that of confinement throughout the evolution), and therefore the agreement between calculations and observations can hardly be construed as a proof of the validity of the loop model itself, but rather as an independent confirmation of the wide body of evidence, from the solar analogy to the success of static loop models, pointing to a picture of stellar coronae as dominated by magnetic structures.

4. SUMMARY AND CONCLUSIONS

Although stellar coronae have been observed only recently, we have already a wide body of data and a consistent framework for their theoretical understanding, especially for late type stars where the *Solar analogy* can be used.

According to our picture, the coronae of late type stars, similarly to the solar corona, are confined and heated by the magnetic fields that also shape the coronal plasma in loop-like structures. Models trying to explain the observations by the emission of plasma confined in a *static* magnetic loop are reasonably successful, although they are not entirely constraining the physical parameters of the loop atmospheres, because of the high conductivity of the coronal plasma and of the poor spectral resolution of available observational data. Moreover, the assumption on which they are based, that the coronal emission is dominated by loops of the same size and base pressure is difficult to test, although for the Sun it works fairly well. Models describing the *dynamic* evolution of plasma flaring in a compact loop are more promising in this respect, because the flare event is presumably occurring in a localized coronal region whose emission can easily overwhelm the background coronal emission.

More theoretical insight and a post *Einstein* generation of stellar X-ray observations, however, will have to address several problems relating to stellar physics, including the origin of X-ray emission in early type stars, the dependence of late type stellar X-ray emission on parameters such as age, rotation rate, and the depth of the convection zone, the determination of the boundary of the region of the HR diagram in which dynamo activity is present, the contribution of flares and winds to the mass budget of the interstellar medium and to the cosmic ray flux, and a more detailed picture of the geometrical and thermal structure of late type coronae.

ACKNOWLEDGMENTS
This work was supported by Piano Spaziale Nazionale and Ministero Pubblica Istruzione.

BIBLIOGRAPHY
G. Bodo, A. Ferrari, S. Massaglia, R. Rosner, and G. S. Vaiana, 1985, *Ap. J.*, **291**, 798.
A. Ferrari, R. Rosner, G. S. Vaiana, 1982, *Ap. J.*, **263**, 944.
M. Giampapa, L. Golub, G. Peres, S. Serio, and G. S. Vaiana, 1985, *Ap. J.*, **289**, 203.
L. Golub, C. Maxson, R. Rosner, S. Serio, and G. S. Vaiana, 1980, *Ap. J.*, **238**, 343.
A. G. Hearn, 1987, *this volume*.
G. D. Holman, 1985, *Ap. J.*, **293**, 584.
G. D. Holman, 1986, preprint.
C. Jordan, A. Brown, F. M. Walter, and J. L. Linsky, 1986, *MNRAS*, **218**, 465.
M. Kuperus, J. A. Ionson, and D. S. Spicer, 1981, *Ann. Rev. Astron. Ap.*, 19, 7.
M. Landini, B. C. Monsignori Fossi, F. Paresce, and R. A. Stern, 1985, *Ap. J.*, **289**, 709.
J. L. Linsky, and D. E. Gary, 1983, *Ap. J.*, **274**, 776.
L. B. Lucy, 1982, *Ap. J.*, **255**, 286.
L. B. Lucy, and R. L. White, 1980, *Ap. J.*, **241**, 300.
R. Mewe, 1986, Proc. of the COSPAR Symposium on *Stellar and Solar Activity* (Toulouse), in
    press.
G. Micela, S. Sciortino, and S. Serio, 1984, in M. Oda and R. Giacconi (eds.): *X-Ray Astronomy
    '84*, Tokyo (Inst. of Sp. and Astronau. Sci.), 43.
G. Micela, S. Sciortino, S. Serio, G. S. Vaiana, J. Bookbinder, F. R. Harnden Jr., L. Golub, and R.
    Rosner, 1985, *Ap. J.*, **292**, 172.
R. Pallavicini, L. Golub, R. Rosner, G. S. Vaiana, T. Ayres, and J. L. Linsky, 1981a, *Ap. J.*, **248**,
    279.
R. Pallavicini, G. Peres, R. Rosner, S. Serio, G. S. Vaiana, 1981b, *Ap. J.*, **247**, 692.
E. N. Parker 1975, *Ap. J.*, **198**, 205.
G. Peres, F. Reale, S. Serio, and R. Pallavicini, 1987, *Ap. J.*, in press.
G. Peres, R. Rosner, S. Serio, and G. S. Vaiana, 1983, *Ap. J.*, **252**, 791.
F. Reale, *et al.*, 1987, in preparation.
R. Rosner, W. H. Tucker, and G. S. Vaiana, 1978 *Ap. J.*, **220**, 643.
R. Rosner, L. Golub, and G. S. Vaiana, 1983, *CfA Preprint* 1719.
J. H. M. M. Schmitt, L. Golub, F. R. Harnden Jr., C. W. Maxson, R. Rosner., and G. S. Vaiana,
    1985, *Ap. J.*, **290**, 307.
J. H. M. M. Schmitt, and R. Rosner, 1983, *Ap. J.*, **265**, 901.
S. Serio, G. Peres, G. S. Vaiana, L. Golub, R. Rosner, 1981, *Ap. J.*, **243**, 288.
S. Serio, 1983, *Adv. Space Research*, **2**, 271.
S. Serio, 1985, *Proc. ESA Workshop on a Cosmic Spectroscopy Mission*, ESA SP-239, 59.
R. A. Stern, M. C. Zolcinski, S. K. Antiochos, and J. H. Underwood, 1981, *Ap. J.*, **249**, 647.
R. A. Stern, S. K. Antiochos, and F. R. Harnden Jr., 1986, *Ap. J.*, **305**, 417.
K. T. Strong, 1986, Proc. of the COSPAR Symposium on *Synopsis of the Solar Maximum
    Analysis* (Toulouse), in press.
Y. Uchida, 1986, Proc. of the COSPAR Symposium on *Stellar and Solar Activity* (Toulouse), in
    press.
G. S. Vaiana, and R. Rosner, 1978, *Ann. Rev. Astron. Ap.*, 16, 393.
G. S. Vaiana, S. Sciortino, 1987, *this volume*.
J. F. Vesecky, S. K. Antiochos, and J. H. Underwood, 1979, *Ap. J.*, **233**, 987.
M. C. Zolcinski, S. K. Antiochos, A. B. C. Walter, and R. A. Stern, 1982, *Ap. J.*, **258**, 177.

EVIDENCE FOR STELLAR CHROMOSPHERES IN GLOBULAR CLUSTERS

William Liller and Gonzalo Alcaino
Instituto Isaac Newton
Ministerio de Educación
Casilla 8-9, Correo 9
Santiago
Chile

ABSTRACT. We suggest that the conspicuous gaps frequently seen in the color-magnitude diagrams of globular clusters are caused by a sudden increase in chromospheric activity shortly after stars evolve away from the main sequence.

Color-magnitude diagrams (CMDs) of globular clusters which include large numbers of randomly-selected stars belonging to the sub-giant branch (SGB) nearly always display conspicuous gaps at some point on the SGB (see, e.g., Alcaino and Liller 1980, 1984; Buonanno et al. 1984; Harris and Canterna 1980). The exact point where the gap occurs on the SGB seems to be weakly correlated with the metallicity of the cluster; a better correlation exists between gap location and Kukarkin's K index, a measure of the ratio of red to blue stars in the horizontal branch.

Standard stellar evolution theory does not predict a sudden change in the rate of evolution (Iben, private communication; Armandroff and Demarque 1984), and attempts to produce these gaps by hypothesizing discontinuities in the chemical composition of the stellar interior or changes in atmospheric structure have not been entirely successful (Armandroff and Demarque 1984).

In looking for a probable cause, it occurred to us that the gaps come just in that region of the CMD where chromospheric activity is always found, although it must be remembered that the metal-to-hydrogen ratio in globular clusters is one or two orders of magnitude smaller than in solar-type stars. For example, in NGC 288 there is a clearly defined gap at $M_V = +3.1 \pm 0.2$ and $B-V = 0.71 \pm 0.03$ (Buonanno et al. 1984), corresponding to $M(Bol) = +2.8$ and $T(eff) = 5100^0$ K. (VandenBerg 1983). A star with these characteristics almost definitely will possess a well-developed chromosphere (Linsky 1980).

The usual spectral manifestations of chromospheric

357

I. Appenzeller and C. Jordan (eds.), Circumstellar Matter, 357–358.

activity are emission lines seen in the deep cores of
saturated absorption lines, especially those of Ca II,
Mg II, and H-alpha. Indeed, these emission lines normally
play an important role in the radiative cooling of chromospheres (Linsky 1980). However, in the cooler regions of
the chromosphere and especially in low-metallicity stars,
the situation is radically changed, and another cooling
mechanism dominates, namely H⁻ emission (Athay 1976;
Osterbrock 1961).

We suggest, then, that the SGB gaps in globular clusters arise owing to a sudden increase in the amount of
chromospheric activity shortly after a star evolves away
from the main sequence. Given a sufficiently extended and
optically thick chromosphere, the H⁻ emission would affect
the total emergent flux of the star causing it to appear
slightly more luminous and cooler as is observed.

Because of its greater importance in the near-infrared, H⁻ emission should be more evident in globular cluster CMDs using a B-I color index instead of B-V, and
indeed this is exactly what is found. In the V, B-I CMD
for M4 (NGC 6121), the start of the SGB appears totally
detached from the main sequence lying some 0.6 magnitudes
to the right of the turnoff (Alcaino and Liller 1984).

What is needed now, of course, is a careful quantitative study of the expected conditions in the chromospheres
of low-metallicity stars, including possible reasons for
the sudden increase in nonradiative heating necessary to
create a chromosphere.

REFERENCES

Alcaino, G., Liller, W. 1980. Astron. J. 85, 1592.
Alcaino, G., Liller, W. 1984. Astrophys. J. Suppl. 56, 19.
Athay, R.G. 1976. The Solar Chromosphere and Corona:
    Quiet Sun. (Dordrecht:Reidel).
Armandroff, T.E., Demarque, P. 1984. Astron. Astrophys.
    139, 305.
Buonanno, T., Corsi, C., Pecci, F., Alcaino, G., Liller,
    W. 1984. Astron. Astrophys. Suppl. Ser. 57, 75.
Harris, W.E., Canterna, R. 1980. Astrophys. J. 239, 815.
Linsky, J.L. 1980. Ann.Rev.Astr.Astrophys. 18, 439.
Osterbrock, D. 1961. Astrophys. J. 134, 347.
VandenBerg, D. 1983. Astrophys. J. Suppl. 51, 29.

EMISSION MEASURES AND HEATING MECHANISMS FOR STELLAR TRANSITION REGIONS AND CORONAE

E. Böhm-Vitense
Department of Astronomy FM-20
University of Washington
Seattle, Washington 98195
U.S.A.

ABSTRACT. In order to determine the heating mechanisms for stellar transition regions and coronae we try to determine the damping lengths for the mechanical flux(es) responsible for the heating. For the lower part of the transition regions (30,000< T≤100,000 K) the damping lengths are consistent with shockwave damping. This appears to be also true for the upper part of the transition region in Procyon, while for the upper part of the solar transition region the damping length is much larger.

1. THE LOWER TRANSITION LAYER

In the Lower Transition Layer (L Tr) 30,000 K<T<100,000 K we find an equilibrium between the mechanical energy input and the radiative losses Erad, i.e.,

$$(1) \quad -\frac{d\,Fm\ell}{dh} = \frac{Fm\ell}{\lambda_\ell} = E_{rad} = n_e^2 \cdot f(T) = n_e^2 \cdot B \cdot T^\beta \quad \text{where } \beta \sim 2$$

Here Fmℓ is the mechanical energy flux in the L Tr and λ_ℓ its damping length. f(T) is the radiative loss function which in the L Tr increases approximately as T^2. B is a constant. Assuming $\lambda_\ell = \lambda_0 T^\alpha$ equation (1) leads to

$$(2) \quad T^{\beta+\alpha-2} = \frac{Fm\ell}{\lambda_0} \cdot \frac{1}{B} \cdot \frac{1}{P_e^2} \quad \text{with } P_e = n_e \cdot T$$

For the emission measures we find

$$(3) \quad Em = 0.35\,P_{eo}^2\,\frac{(\beta+\alpha-2)\cdot R}{\mu\,g_{eff}} \cdot \frac{1}{T}\left(\frac{T_o}{T}\right)^{\beta+\alpha-2}\frac{e^{-\int dh/\lambda_\ell}}{1-H/2\lambda_\ell}$$

I. Appenzeller and C. Jordan (eds.), Circumstellar Matter, 359–360.

with $H = \dfrac{RT}{\mu g_{eff}}$ and R=gas constant, g_{eff}=effective gravity

μ=atomic weight,

The observed Em(T) permit the determination of $P_e^2(T)$, which in turn permits the determination of $Fm\ell/\lambda_0$ from equation (2). The observed temperature dependence of the Em determines $\alpha=0.4 \pm 0.5$, in agreement with expectations for shockwave damping.

2. THE UPPER TRANSITION REGION

In the Upper Transition zone (U Tr) with $10^5K<T<10^6K$ the radiative loss function f(T) decreases for increasing T, a stable equilibrium between mechanical energy input and radiative losses is therefore not possible. The temperature stratification is governed by the conductive heat flux Fc(h). The energy equation tells us that the downward flowing conductive flux must equal the upward flowing mechanical flux Fmu(h) reduced by the amount of energy lost above the height h due to radiation and the stellar wind. For the emission measure in this layer we obtain

$$(4) \quad E_m(h)=(P_e^2(h)/F_c(h) \; T^{1.5}\cdot0.7 \; e^{-2\Delta h/\overline{H}}$$

For constant $P_e^2(h)/Fc(h)$ the observed increase of Em with $T^{1.5}$ is recovered (see also Jordan 1980). From the observed Em only the conductive flux Fc can be determined which relates to Fmu but not to λ_u.

3. THE CORONAL TEMPERATURES

Integration of the equation for the conductive flux from the base of the U Tr with $h=h_2$ and $T=T_2$ to the height h_c, where the conductive flux becomes zero and $T=T_c$, leads to the equation for the coronal temperature

$$(5) \quad T_c^{7/2}-T_2^{7/2} = -\frac{7}{2}\, \eta\cdot\lambda_u\cdot Fmu(h_2)\cdot[1-e^{-\Delta h_c/\lambda_u}(1+\Delta h_c/\lambda_u)]-E_r$$

where E_r describes the integral over the radiative losses in the U Tr. The coronal temperature T_c increases with increasing λ_u. The observed coronal temperatures thus permit a determination of the λ_u. For Procyon (Jordan et al. 1986) the derived value agrees with expectations for shockwave damping while for the sun the value is at least an order of magnitude too large for this heating mechanism.

REFERENCES

Jordan, C. 1980, *Astr. Astrophys.* 86, 355.
Jordan, C., Brown, A., Walter, F.M., Linsky, J.L. 1986, *MNRAS* 218, 465.
Pottasch, R. S. 1963, *Ap.J.* 137, 347.

ON THE EXISTENCE OF HOT CORONAE AROUND COOL STARS

R. Hammer
Kiepenheuer-Institut für Sonnenphysik
Schöneckstr. 6
D-78oo Freiburg, FRG

ABSTRACT. A star cannot have a solar-like corona if the available mechanical energy flux in the chromosphere is either too large or decreases outward more rapidly than the pressure. This result might be relevant for hybrid stars and cool giants.

The canonical explanation for the existence of the hot solar corona is based on a discussion of the local energy balance between radiation and heating in the chromosphere. The effectively optically thin emission can be approximated by pressure squared times an emissivity function f(T). Theoretical arguments and empirical models show that the heat input into the solar chromosphere, and thus also the available energy flux F itself, decreases outward less rapidly than linearly with p (case C in Fig. 1). Nevertheless, initially the chromosphere can achieve energy balance by means of a gentle outward temperature rise, since f(T) increases steeply with T for small temperatures.

Finally, however, a critical temperature is reached where the emissivity has a maximum. Beyond this point, which is marked by the **asterisk** on curve C in Fig. 1, energy balance at cool chromospheric temperatures is no longer possible. Therefore, the transition region to the solar corona, which is governed by a different type of energy balance since thermal conduction is important, must lie at or below this critical position. Its actual location is determined by the intersection of the curve F(p) with another curve that specifies the total coronal energy losses as a function of the coronal base pressure (cf. Hammer et al. 1982). Theoretical models of closed (e.g. Rosner et al. 1978, eq. (4.4)) and open (e.g. Hammer 1982, Fig. 2) coronal regions as well as semi-empirical studies (e.g. Jordan 198o, Fig. 4) show that the coronal energy losses increase with the base pressure to some power that is slightly larger than one.

It is interesting to apply this picture to stars near the dividing "line" that appears to separate the solar-like stars with hot coronae from the cool giants with massive winds and extended chromospheres. When we go from the Sun (case C in Fig. 1) towards these stars, it is well possible that the run of energy flux vs. pressure changes. Recently, 35hm-Vitense (1986) discussed the possibility that in the cool giants

I. Appenzeller and C. Jordan (eds.), Circumstellar Matter, 361–362.
© *1987 by the IAU.*

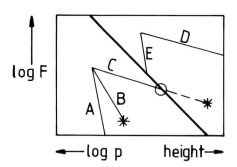

Figure 1. Available energy flux F in a stellar chromosphere as a function of height or pressure (thin curves) and energy requirements of a
hot corona (thick curve).

F decreases more rapidly than pressure squared (case A), so that the
chromospheric temperature decreases outward, and a corona is not formed.
If that is true, however, we should also find stars in which F varies
with p to some power between 1 and 2 (case B). In such a star, the
chromospheric temperature increases outward, and beyond a certain critical height the heat input can no longer be radiated away at cool
temperatures. On the other hand, a solar-like hot corona is also not
possible because at any height F is far too small to balance the coronal losses. Such a star would need other means of solving its energy
dilemma. Its outer atmosphere could, e.g., oscillate temporally between
the cool (overheated) and the hot (underheated) state. Or it could have
a warm envelope (with T near the maximum of f(T), beyond which energy is
transported outward by means of convection. Such stars, should they
exist, might exhibit some characteristics of hybrid stars.
        Fig. 1 suggests another possibility for a star to have no corona;
namely, if at some chromospheric level F is larger than the energy losses of a given type of corona. If now F drops slowly with p (case D),
no equilibrium solution exists. And if F drops rapidly (case E), the
equilibrium solution can be shown to be thermally unstable (Hammer et al.
1982).

REFERENCES

Böhm-Vitense, E.: 1986, Ap.J. 3o1, 297
Hammer, R.: 1982, Ap.J. 259, 779
Hammer, R., Linsky, J.L., and Endler, F.: 1982, in Advances in UV Astro-
    nomy, NASA-CP 2238, p.268
Jordan, C.: 198o, Astron. Ap. 86, 355
Posner, R., Tucker, W.H., and Vaiana, G.S.: Ap.J. 22o, 643

EVIDENCE FOR EXTENDED CHROMOSPHERES AND TRANSITION ZONES IN THE UV SPECTRA OF FK COMAE STARS

L.Bianchi
Osservatorio Astronomico
di Torino
I-10025 Pino Torinese
ITALY

M.Grewing
Astronomisches Institut
Waldhauserstrasse 64
D-7400 Tuebingen
GERMANY,Fed.Rep.

ABSTRACT. The chromospheres and transition zones of the fast rotating giants of the FK Comae type can be studied by analysing their ultraviolet emission line spectra. From relative line intensities, electron densities of the order of 10^{10} to 10^{11} cm^{-3} are found for the region where the Si IV emission arises. The sizes of the chromospheres and transition regions can be inferred from the emission measure distribution, and a temperature-height relation can be found on the assumption that hydrostatic equilibrium holds. We find the atmospheres of these stars to be clearly more extended than those of normal giant stars, and the flux in the higher excitation chromospheric and transition zone lines (e.g. C II, C IV, Si IV) is significantly stronger than in other stars of similar spectral type. Indeed, the location of these stars in the standard rotation-activity-correlation diagrams places them close to or even above the saturation limit for main sequence stars.

At present there are four stars known which are fast rotating, apparently single late type giants which have been named after the prototype object as FK Comae type stars (Bopp and Stencel 1981). The other members of the group are HD 32918, HD 36705, and HD 199178 (Bopp and Rucinski 1981, Bopp 1982, Collier 1982). Also, UZ Lib is counted as a member of this group (see e.g. Bopp et al. 1984), despite the fact that radial velocity variations have been discovered which reveal the presence of a low mass companion. All five stars show signs of strong chromospheric activity at optical wavelengths (CaII and H-alpha emission) and optical light curves somewhat similar to those of RS CVn stars or other stars with surface spots. At ultraviolet wavelengths they all show strong chromospheric and transition region lines (Bopp and Stencel 1981, Bopp et al. 1984, Bianchi et al. 1984,1985, Grewing et al. 1986), and with the exception of HD 199178 they all have been observed to emit soft x-rays.

Here we shall focus the discussion on the ultraviolet spectra of the three stars FK Comae, HD 32918, and UZ Lib. The observational data,

I. Appenzeller and C. Jordan (eds.), Circumstellar Matter, 363–366.

which were obtained with the International Ultraviolet Explorer (IUE) satellite, are displayed in Fig.1 and 2., covering the 1200-2000 A and the 2400-3200 A range, respectively.

In Table 1 we have compiled the absolutely calibrated UV line fluxes for the three stars discussed here. These data have been corrected for interstellar extinction when necessary and refer to the emission at the surface of the stars. Also included in the Table are the surface fluxes for β Cet (Engvold et al. 1984), a K 1 III star which shows no sign of rapid rotation.

Table 1

Chromospheric and TZ line fluxes
for three FK Comae stars and β Cet

	FK Comae G2 III	HD 32918 K1 III	UZ Lib KO III	β Cet K1 III
N V 1240	2.8(30)	2.8(30)	0.2(30)	-
O I 1304	9.2(')	2.7(')	0.8(')	16.1(28)
C II 1335	3.2(')	1.8(')	1.2(')	1.3(')
SiIV 1393	2.5(')	1.6(')	1.1(')	1.1(')
C IV 1548/50	8.1(')	5.6(')	3.0(')	1.0(')
HeII 1640	2.6(')	3.3(')	3.2(')	1.1(')
SiII 1808/17	4.7(')	3.2(')	1.3(')	8.5(')
MgII 2800	125.9(')	69.4(')	44.9(')	-

Note: the fluxes are given in units of ergs/ s.

Table 1 shows that the UV line intensities of the FK Comae stars are similar to each other and differ significantly from those of a normal giant : their absolute intensities are higher by typically a factor of 100, and the transition zone lines are relatively more intense than the chromospheric lines. This clearly demonstrates the fact that their atmospheres are much more active - very likely due to their fast rotation.

The absolute emission line fluxes as given in Table 1 can be used to derive the emission measure distribution by assuming a spherically symmetric atmosphere, an effectively optically thin collisionally excited plasma. Furthermore, by assuming also hydrostatic equilibrium, a temperature-height-relation can be obtained. Results for HD 32918 are given by e.g. Grewing et al. 1986.

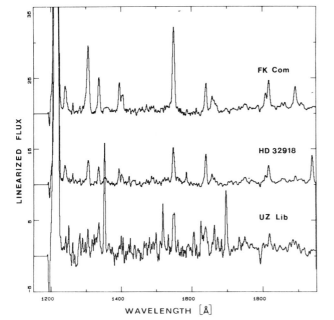

Figure 1:
The short-wavelengths
IUE-spectra of three
FK Comae stars showing
strong chromospheric and
transition zone lines.

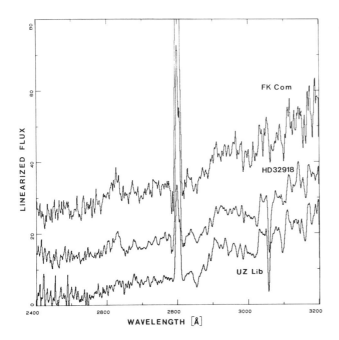

Figure 2:
The long-wavelengths
IUE-spectra of the same
three FK Comae stars.
Note the strong MgII-
emission which is found
to vary over a timescale
of months.

REFERENCES
Bianchi,L.,Grewing,M.,Kappelmann,N.,Cassatella,A.: 1984, ESA
        SP-218,p.355
Bianchi,L.,Grewing,M.,Kappelmann,N.: 1985,Astron.Astrophys. 149, 41
Bopp,B.,Stencel,R.: 1981,Astrophys.J. 347,L131
Bopp,B.,Rucinski,S.: 1981, IAU Symp. 93, p.177
Bopp,B.: 1982, SAO Special Report No.392, p.207
Bopp.B., et al.: 1984, Astrophys.J., 285, 202
Collier,A.: 1982, M.N.R.A.S.,200,489
Grewing,M.,Bianchi,L.,Cassatella,A.: 1986,Astron.Astrophys.,164,31
Engvold,O.,Kjeldseth Moe,O.,Jensen,E.,Jordan,C.,Stencel,R.,Linsky,J.:
        1984,Lecture Notes in Physics,Vol.193,p.359

DISCOVERY OF FK COMAE AND RS CVn SYSTEMS BY OBSERVATION OF THEIR
X-RAY EMISSION

Thomas A. Fleming
Steward Observatory, Tucson, AZ 85721 USA

Isabella Gioia* and Tommaso Maccacaro*
Harvard-Smithsonian Center for Astrophysics
Cambridge, MA   02138 USA

We are currently working with a statistically complete, unbiased sample
of 125 x-ray-bright stars which were serendipitously detected by the
Einstein Observatory Medium Sensitivity Survey (MSS). A program of
optical spectroscopy and photometry is currently underway to measure
radial velocities, distances, and such stellar parameters as rotation,
temperature, surface gravity, metallicity, chromospheric activity, and
age and to correlate them with absolute x-ray luminosity. So far, the
majority of the sample (which was defined at $|b^{II}| > 20^{\circ}$) appears to
be composed of either flare stars (e.g. dMe, dKe) or active binary
systems (e.g. cataclysmic variables, RS CVn, W UMa).

We have already identified six new RS CVn candidates. These stars
exhibit rapid rotation, strong Ca II H & K emission, and are binaries.
We also have two such stars which show no evidence of being binaries.
These are possible candidates for the class of FK Comae stars. This
class of star is rare because it represents a relatively short phase
in the evolution of a star: the moment at which the two cores of a
contact binary coalesce to form a single, rapidly rotating star. In
this paper, we discuss the x-ray characteristics of RS CVn and FK Comae
stars.

In searching for new candidates for the RS CVn and FK Comae
classes, the established method is to search objective prism plates
for Ca II H & K emission objects. As demonstrated by this sample of
stars from the Einstein MSS, one can easily find candidates for these
classes of stars by looking at stellar objects with high f_x to f_v
ratios. Unfortunately, at the moment it is more economical to take
objective prism plates than it is to put x-ray telescopes into orbit.

---

*on leave of absence from Istituto di Radioastronomia del CNR, 40126
Bologna, Italy

I. Appenzeller and C. Jordan (eds.), Circumstellar Matter, 367.
© *1987 by the IAU.*

RELATIONS BETWEEN CORONAL AND CHROMOSPHERIC ACTIVITY DIAGNOSTICS IN
T TAURI STARS

J. Bouvier
Institut d'Astrophysique de Paris
98 bis Boulevard Arago
75014 PARIS, FRANCE

ABSTRACT. The study of the relationships between various activity diagnostics in T Tauri stars (TTS) suggests that the CaII K, MgII k and
H_α lines are formed in a similar region of TTS' atmosphere. For the more
active TTS, an extended circumstellar region seems to be the major source of the emission, whereas a solar-type atmosphere alone may be able
to account for the emission spectrum of low-activity TTS.

1. INTRODUCTION

The atmosphere of TTS is the seat of a high degree of non-radiative heating which results in a number of emission lines (CaII, MgII, H_α) typical of the spectrum of these low-mass pre-main-sequence stars and in
a strong X-ray emission, up to 10^3 times larger than the X-ray flux observed in late-type dwarfs. Two broad classes of models have been proposed to account for TTS' emission line spectrum. The "deep chromosphere"
model assumes that TTS possess a solar-like chromosphere beginning however at higher optical depth than in the Sun /1/. The second class of
models assigns the origin of the emission line spectrum to an extended
circumstellar region of a few stellar radii /2/. It seems now widely accepted that both a chromosphere and an extended envelope are needed to
describe the various features of TTS' emission spectrum. However, the
detailed structure of the immediate circumstellar environment of TTS
remains unclear.

2. ACTIVITY DIAGNOSTICS

Informations about the structure of stellar atmospheres can be gained
from the analysis of relationships between activity diagnostics formed
at different atmospheric levels. Such relationships are known to exist
for late-type dwarfs where the intensities of various chromospheric diagnostics (CaII, MgII, H_α) are linearly correlated /3,4/ whereas coronal
X-ray emission varies with the intensity of chromospheric diagnostics
following a power-law with a slope of 2.6 /5/. The existence of these

369

I. Appenzeller and C. Jordan (eds.), Circumstellar Matter, 369–372.

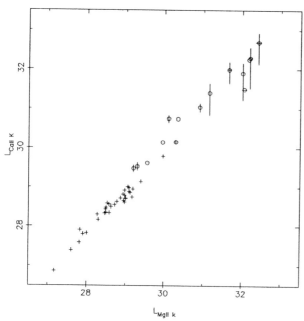

Fig. 1. CaII K-line luminosity versus
MgII k-line luminosity for dwarfs(+)
and T Tauri stars(o).

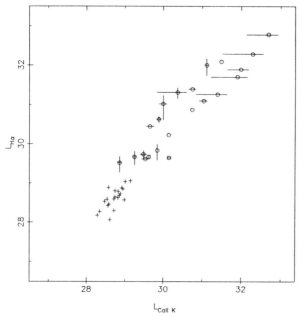

Fig. 2. H$_\alpha$-line luminosity versus
CaII K-line luminosity for dwarfs(+)
and T Tauri stars(o).

tight correlations between
various activity criteria
implies that the different
atmospheric layers are physically associated by a unifi-
ing mechanism which, in the
case of late-type dwarfs as
in the solar case, is believed to be the magnetic field /6/.
In Figures 1 to 4 we study
the relationships between several activity criteria in
TTS and compare them to those
found in dwarfs. In each figure, the crosses represent
late-type dwarfs and the open
circles represent TTS. The
bars associated with TTS reflect the range of observed
variability between consecutive measurements. The axis
are luminosities expressed
in erg/s on a logarithmic
scale. In figure 1 we have
plotted the stellar luminosity observed in the CaII K-
line versus that measured in
the MgII k-line. The one-to-
one correlation appears clea-
ly for dwarfs and seems also
to be fulfilled by TTS. Although this result doesn't
indicate a similar atmospheric structure between dwarfs
and TTS nor does it mean that
the heating mechanism is the
same in the two stellar
groups, it suggests that the
CaII K and MgII k lines are
formed in the same region of
TTS's atmosphere. This conclusion appears to be valid
also for the H$_\alpha$-line, the
luminosity of which is plotted versus the CaII K-line
luminosity in Figure 2. Although the scatter, both for
dwarfs and for TTS is much
higher than in Figure 1, these two diagnostics seem to be
linearly correlated. The in-

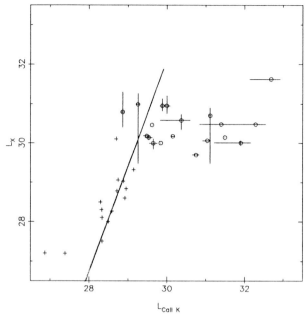

Fig. 3. X-ray luminosity versus CaII K line luminosity for dwarfs(+) and TTS (o). Solid line: $L_X \propto (L_{CaII\ K})^{2.6}$.

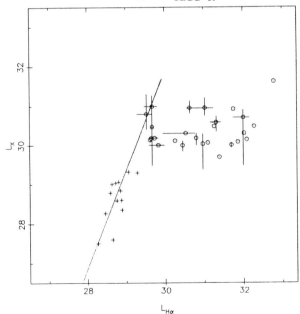

Fig.4. X-ray luminosity versus H_α-line luminosity for dwarfs(+) and TTS(o). Solid line: $L_X \propto (L_{H_\alpha})^{2.6}$.

creased scatter may arise from the fact that, in dwarfs, the H_α-line forms at higher chromospheric level than CaII K-line so that these two lines are more loosely connected than are the CaII K and MgII k lines. Remarkable is that the above relationships are valid over almost five decades and appear to remain the same for low-mass and high-mass TTS. In Figure 3 and 4 we plotted the stellar luminosity observed in the X-ray range versus the one observed in the CaII K and H_α-line respectively. In dwarfs the luminosities of these activity diagnostics are related by a power-law with a 2.6-slope which is represented in both figures by a solid line. Obviously this statistical relationship breaks down when dealing with TTS: whereas the intensity measured in the CaII K and H_α lines describes almost four decades, the X-ray luminosity varies only over one decade.

3. DISCUSSION

Clearly, most of the TTS in our sample namely the more active ones, do not appear to fit the solar-like atmosphere assumption. The failure appears in Figure 3 and 4 where the departure of a number of TTS from the correlation found in dwarfs goes in the direction of an excess of emission line intensity relative to X-ray emission. Moreover, emission line intensity and X-ray emission seem to a large extend uncorrelated in TTS contrary to what is expected in the case of a solar-

like atmosphere governed by magnetic fields. Thus it appears necessary
to call for an extended circumstellar envelope as the main contributor
of the emission line intensity observed in the more active stars of our
sample. And the conclusions drawn from the one-to-one correlations existing between CaII K, MgII k and H_α-line intensities in TTS seem to indicate that these three activity diagnostics form mainly in the circumstellar envelope. However, few TTS displaying lower emission characteristics lie on the extrapolation of the correlation between coronal and
chromospheric diagnostics verified by dwarfs, a result that suggests
that these low-active TTS do not possess large circumstellar envelopes
and that the emission arises mainly from a solar-like atmosphere.

4. CONCLUSION

The study of various activity diagnostics reinforces the growing evidence that two different circumstellar regions may play a leading role in
the emission characteristics of T Tauri stars. For low-active TTS a
solar-like atmospheric structure may account for the behaviour of the
different activity diagnostics although a larger non-radiative heating
input than in the Sun is necessary to reproduce the observed activity
level /7/. For more active TTS an hot, extended circumstellar region
seems to be the main contributor to the intense emission line spectrum,
keeping in mind that the additive contribution of an underlying solar-
type atmosphere cannot be dismissed.

REFERENCES

1. N. Calvet, G. Basri and L.V. Kuhi, Astrophys. J. 277, 725 (1984)
2. L. Hartmann, S. Edwards and A. Avrett, Astrophys. J. 261, 279 (1982)
3. C. Blanco, L. Bruca, S. Catalano and E. Marilli, Astron. Astrophys.
   115, 280 (1982)
4. G.H. Herbig, Astrophys. J. 289, 269 (1985)
5. E. Marilli and S. Catalano, Astron. Astrophys. 133, 57 (1984)
6. T.R. Ayres, N.C. Marstad and J.L. Linsky, Astrophys. J. 247, 545
   (1981)
7. J. Bouvier, Proc. of XXVI Cospar, Toulouse, July 1986, in press

WHAT CAN BE LEARNT FROM FULL DISK X-RAY OBSERVATIONS OF STELLAR FLARES?

J.H.M.M. Schmitt, H. Fink   F.R. Harnden Jr.
MPI f. extraterr. Physik   SAO
Karl-Schwarzschild-Str. 1   60 Garden Street
D-8046 Garching   Cambridge, MA 02139, U. S. A.

The **Einstein Observatory** demonstrated the existence of hot envelopes, i.e., stellar coronae, around most classes of normal stars (Vaiana et al. 1981). The coronae of late type stars of spectral type F through M are generally thought to be solar-like, i.e., structured and organised by the magnetic field topology and heated by some process(es) involving magnetic energy. Here the property "solar-like" does not refer to the optical appearance of a star, but rather to the role played by magnetic fields in the outer stellar envelope (Linsky 1985). Since it is difficult to measure magnetic fields on other stars directly, a number of indirect indicators is used in order to infer whether a corona should be considered "solar-like" or not.

Stellar flares are thought to be an excellent indicator of the magnetic nature of the underlying corona (Linsky 1985); in addition, the flare X-ray light curve and spectrum allow a determination of physical parameters such as density, temperature and scale size of the flaring plasma. As pointed out by Haisch (1983) most stellar flares observed so far in X-rays (with the possible exception of HD 27130) seem to be similar to solar flare events, except that the derived plasma densities in stellar flares are far higher than those of their solar counterparts.

The analysis procedures used to interpret stellar flares are rather crude, and further, only full disk observations with rather low spectral resolution and low signal to noise ratio (SNR) are available. Solar flares on the other hand are typically observed with rather high spatial, spectral and temporal resolution with good SNR, and we simply do not know what solar flares would look like if observed with the same instrumentation used on other stars.

Using the **Einstein Observatory** Imaging Proportional Counter (IPC) we have studied in detail solar X-ray light scattered in the upper atmosphere, i.e., data taken when the X-ray telescope was pointed at the Sun-lit Earth. By computing the propagation and scattering of solar X-rays in a realistic atmosphere model, we can determine the scattered X-ray flux as a function of viewing geometry and hence flight time; by comparing the expected and observed bright Earth X-ray light curves we can assess whether the incident solar X-ray flux was constant or not. In fig. 1 we show an example of a solar flare observed in scattered X-ray light; the medium panel shows the observed light curve as function of time, the lower one the hardness ratio increase during the flare. In fig. 2 we compare (upper panel) the observed and best fit light curve, and give in the lower panel the flare light curve of a fiducial observer, i.e., an observer in the subsolar point looking downwards.

Scattered solar X-ray light provides us with full disk low spectral resolu-

I. Appenzeller and C. Jordan (eds.), Circumstellar Matter, 373–374.

tion observations obtained with the same instrumentation used for stellar flare observations. Employing the same data analysis and data interpretation techniques, we find extremely good agreement between physical flare parameters derived from our IPC observations and "known" properties of compact solar loop flares, and hence full disk low resolution IPC observations can accurately reveal temperature and density of solar flare plasma. Thus we are confident that the interpretation of stellar X-ray flare observations is on a physically sound basis, and therefore stellar X-ray flares constitute an important diagnostic tool for the determination of physical conditions in stellar coronae. A detailed account of this work will be published elsewhere.

Haisch, B.M., 1983, in Activity in Red Dwarf Stars, ed. P.B. Byrne and M. Rodono, Reidel Publishing Company, Dordrecht.
Linsky, J.L., 1985, Solar Physics, 100, 333.
Vaiana, G.S. et al., 1981, Ap. J. 245, 163.

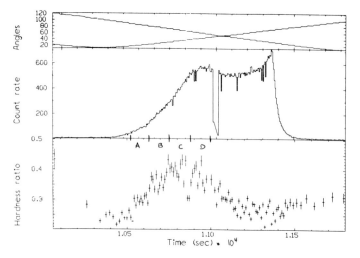

Fig. 1: X-ray light curve of bright Earth data segment on July 21, 1980 (medium panel) and hardness ratio (lower panel).

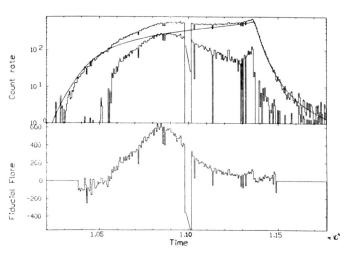

Fig. 2: X-ray light curve, best model fit and residual for the same data segment as in Fig. 1 (upper panel) and fiducial observer flare light curve (lower panel).

CIRCUMSTELLAR CaII LINES IN R Leo

M. Barbier
Observatoire de Marseille
2, place Le Verrier
F-13248 Marseille Cedex
France

M.O. Mennessier
Laboratoire d'Astronomie
Université Montpellier II
F-34060 Montpellier Cedex
France

ABSTRACT. Qualitative results on the H and K CaII lines in R Leo are presented from spectra throughout all the cycle of this mira star. A scenario of explaining the observations by an outward shockwave is proposed.

The figure shows tracings of the region of H and K CaII lines in spectra of the mira star R Leo that were taken at Haute Provence Observatory with a dispersion of 20 Å mm^{-1}. These spectra correspond to different phases of the light curve of R Leo.

A quantitative detailed study of these spectra is in progress, only qualitative results are presented here :
- Our observations confirm that an emitting hot region surmounted by an absorbing colder slab exists during the pre-minimun phases (Merrill, 1952 and Kraft, 1957).
- During the pre-maximum phases, largely blueshifted emission features seem to appear. They could be due to the front of a shockwave in the lowest region of the Ca circumstellar envelope.
- At the phase 0.48, the H and K CaII central features appear in emission.

So we propose the following scenario : the emission is due to an outward shockwave which reaches the Ca envelope during the premaximum phases. The emission lines are self-absorbed by the overlying cool ions. With increasing phase, the heated layer is higher and higher within the atmosphere and the absorbing slab diminishes, the apparent blueshift of the emission features decreases and tends to correspond to the velocity of the other emission lines due to the shock. Near the minimum, the hot layer is high in the atmosphere, so the absorbing slab disappears.

This agrees with the study of the infrared CaII triplet (Contadakis and Solf, 1981), the observations of H$_\alpha$ lines (Gillet et al., 1983) and the deduced model of shockwave (Gillet et al., 1985).

I. Appenzeller and C. Jordan (eds.), Circumstellar Matter, 375–376.

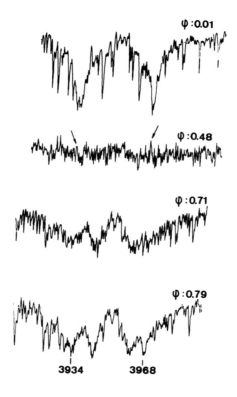

REFERENCES.

Contadakis, M.E., Solf, J., 1981, Astron. Astrophys. 101, 241.
Gillet, D., Maurice, E., Baade, D., 1983, Astron. Astrophys. 128, 384.
Gillet, D., Ferlet, R., Maurice, E., Bouchet, P., 1985, Astron. Astro-
phys. 150, 89.
Kraft, R., 1957, Astrophys. J. 125, 336.
Merrill, P.W., 1952, Astrophys. J. 116, 337.

CO MOLECULE IN TRANSITION REGION BETWEEN CHROMOSPHERE AND COOL STELLAR WIND: A NEW PROBE ON THE OUTER ATMOSPHERES OF COOL LUMINOUS STARS

T. Tsuji
Tokyo Astronomical Observatory, University of Tokyo
Mitaka, Tokyo 181, Japan

ABSTRACT. Presence of CO layer well separated from photosphere is con-firmed and this revealed a presence of quasi-static turbulent transition layer in normal red (super)giant stars. This layer may be related to an outer part of the extended chromosphere and/or a cool part of the chro-mospheric inhomogeneity, and will play major role in stellar mass-loss.

As is well known, the circumstellar matter in outer envelope of cool luminous stars is being lost from the stellar system(Deutsch,1956), but it is not clear where the mass-flow starts. There is a suggestion that the mass-loss already starts in chromosphere(Goldberg,1979), but the observed flow velocities are smaller than the local escape velocity in chromosphere and it is not clear if the chromospheric expansion could be a direct origin of stellar mass-loss. Furthermore, presence of a static layer, possibly situated above the chromosphere, is suggested not only in Mira variable stars(Hinkle et al.,1982) but also in non-Mira stars(Hall,1980). While little attention has been given to such a static layer in recent theories of stellar mass-loss, we have found some convincing evidences on the presence of such a static layer in normal red giant and supergiant stars during our analysis of high resolution infrared spectra of CO first overtone bands(Tsuji,1986a; to be referred to as Paper I).

Although CO lines originating from such a static layer show little Doppler shift against photospheric lines, they could be recognized by the following facts: 1) Equivalent widths of low excitation lines show systematic excess as compared with expected ones based on model atmos-phere(Fig.3 of Paper I), while higher excitation lines can quantitative-ly be well understood by the same model(Tsuji,1986b). 2) The low exci-tation lines show shifts and asymmetries that indicate excess absorption in blue wing in some stars and in red wing in other stars(Figs.4 & 5 in Paper I). 3) Radial velocities show differential variations between low excitation lines(remain almost stationary in the case of α Her shown in Fig.1) and high excitation lines(change is larger, possibly due to small amplitude pulsation of the photosphere). These observations suggest that at least a part of low excitation lines should be originating in a layer well separated from the photosphere. Further, comparison of the observ-ed spectrum with predicted photospheric spectrum revealed residual absorption for low excitation lines while there appeared no residual for

I. Appenzeller and C. Jordan (eds.), Circumstellar Matter, 377–378.
© 1987 by the IAU.

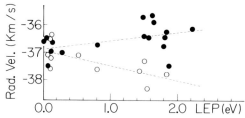

Fig.1 CO radial velocities in αHer plotted against lower excitation potential: Feb.19,1977(open circle) and June 24, 1977(filled circle).

TABLE 1 Physical properties of the quasi-static CO absorption layer

Star	Sp.Type	T_{ex}	v_{tur}	$\log N_{CO}$
α Ori	M2I$_{ab}$	1450K	>9Km/s	20.1
μ Cep	M2I$_a$	1100	>9	19.8
ρ Per	M4II	1940	>5	19.8
α Her	M5II	1670	>5	20.0
SW Vir	M7III	2160	>5	19.9

high excitation lines. In low excitation lines, the contribution by the CO layer has been separated by subtracting the photospheric contribution from the observed profile. A curve-of-growth analysis on equivalent widths of the separated CO profiles gave the results summarized in TABLE 1: note that the excitation temperature is surprisingly high(this is based on more consistent analysis than in Paper I that gave lower tempe-rature) and the turbulent velocity is rather large. Estimated total mass of the CO layer based on the deduced column density is as high as $10^{-4} M_\odot$.

As the excitation temperature is pretty high while the CO layer should be well separated from the photosphere as noted before, the CO layer may be an outer part of the extended chromosphere(which has been recognized only recently; see e.g., Linsky,1987) and/or a cool component of the chromospheric inhomogeneity. Anyhow, the CO absorption layer should represent a transition region between the chromosphere and the cool wind in luminous stars of non-coronal type. Probably, deposition of mass, momentum, and energy to the outer atmosphere from the photo-sphere may be sufficient to form the turbulent transition layer together with the extended chromosphere, but it may be not sufficient to be the direct driving force of stellar mass-loss. However, once the transition layer is formed, it provides an ideal environment for dust formation and radiation pressure on dust could drive mass-outflow. Such a hybrid model of mass-loss is well consistent with the known observations on the outer atmosphere of αOri,for example. Also, even if dust could not be formed, the Maxwellian tail of the turbulent motion could lead to mass-loss, since the local escape velocity in the transition layer may be al-ready small enough to be comparable with the observed flow velocities.

I am indebted to Drs.S.T.Ridgway and K.H.Hinkle for kind help in an observation at KPNO FTS, for archival data, and for useful discussions.

Deutsch,A.: 1956, Astrophys.J. 123, 210
Goldberg,L.: 1979, Quart.J.Roy.Astron.Soc. 20,361
Hall,D.N.B.: 1980, Interstellar Molecules ed. B.H.Andrew, Reidel, p.515
Hinkle,K.H., Hall,D.N.B., Ridgway,S.T.: 1982, Astrophys.J. 252, 697
Linsky,J.: 1987, This volume
Tsuji,T.: 1986a, Astrochemistry eds. M.S,Vardya & S.P.Tarafder,Reidel,p.
Tsuji,T.: 1986b, Astron.Astrophys. 156, 8

INFRARED AND RADIO EXCESSES OF LATE-TYPE STARS

C. J. Skinner
Dept. of Physics and Astronomy
University College London
Gower Street
London WC1E 6BT

The IRAS catalogues have been searched for cool (G,K,M) giant and supergiant stars to investigate the occurrence of circumstellar (C/S) silicate dust, revealed by its emission features at 9.7 and 18μm. Low Resolution Spectrograph (LRS) spectra covering the 7-23μm range were used, plus the 60 and 100μm photometric points. M Supergiants were found in White & Wing (1978), other stars by correlating the Bright Star Catalogue with the LRS catalogue: this discriminated against very cool stars reddened by dust; however it can be seen in Table I that there is a clear trend for cooler and more luminous stars to have a dust shell. M Supergiants almost all have dust shells, whilst only the cooler M bright-giants and giants do. Of the G and K stars, only a very few of the Supergiants have dust shells. The silicate features fell into two categories:

(1) narrow, sharply peaked 9.7 and 18μm features, typified by μ Cephei, observed in C/S shells of solitary cool giants/supergiants;

(2) much broader 9.7 and 18μm features (e.g. VX Sgr), reminiscent of the features seen in the Trapezium region, often seen in binary systems where the companion is a hot star. Few radio observations have been made of the continua of late-type stars; only the closest and brightest stars are above current observing thresholds. These observations are summarised in Drake & Linsky (1986), and it is apparent that the stars they observed have excesses at cm-wavelengths attributable to free-free emission from extended chromospheres. It is concluded that most stars in the present survey have radio excesses, while only the cooler and more luminous have infrared excesses (C/S dust shells).

An attempt has been made to model the prototype M-Supergiant α Ori (Skinner & Whitmore). Of the optical data available for amorphous silicate dust, that of Kratschmer & Huffman (1978) was found to give the best fit. With an optically thin C/S shell, and a photosphere approximated by a blackbody of 3600K appropriate to an M2Iab star, the observed spectrum cannot be satisfactorily reproduced. It was found necessary to change the spectral index of the IR continuum by invoking free-free emission from the extended chromosphere in order to fit the spectrum. The electron -density and -temperature distributions used were in keeping with those derived from chromospheric line-profile

379

I. Appenzeller and C. Jordan (eds.), Circumstellar Matter, 379–380.
© *1987 by the IAU.*

Table I

Stars with/without silicate dust shells.

	I	II	III
M	74/74	2/15	13/206
K	1/6	0/25	0/125
G	1/10	0/3	0/23

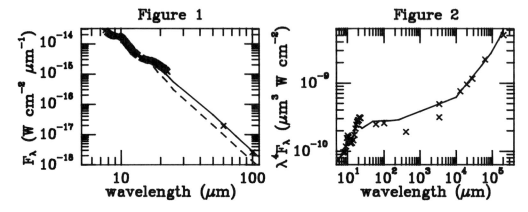

Figure 1   **Figure 2**

x observations   – – – dust–only model   ——dust+chromosphere model
Figure 2 : a black-body would be represented by a horizontal line

fitting, and the resulting spectrum fitted the IRAS observations and
all available radio observations. It appears that other M-Supergiants
can be fitted in the same way, the chromospheric contribution to the
continuum varying from star to star.
Free-free emission seems more important in M-Supergiants than other
cool, luminous stars, and mass loss is also more pronounced.
Schwarzschild (1975) suggested convective hot spots caused the
irregular variability of these stars, and speckle observations indicate
that mass is lost episodically in blobs, rather than continuously.
Mullan (1981) suggested that closed flux loops would be unstable above
the photosphere of cool giant/supergiants, and this author suggests
that convective cells may draw out bubbles of magnetic flux, which are
driven away from the star carrying plasma with them. In the hotter
stars, convection is less important and the magnetic field is more
stable.
This work was carried out whilst in receipt of an SERC studentship
which I gratefully acknowledge.

Drake,S.A., Linsky,J.L., 1986, Astr.J. 91, 602.
Kratschmer,W., Huffman,D.R., 1978, Astr.Spa.Sci. 61, 195.
Mullan,D.J. in I.A.U. Symposium No.102, D.Reidel.
Scwarzschild,M., 1975, Ap.J. 195, 137.
Skinner,C.J., Whitmore,B., M.N.R.A.S. (in press).
White,N.M., Wing,R.F., 1978, Ap.J. 222, 209.

CHEMICAL COMPOSITION AND CIRCUMSTELLAR SHELLS OF CARBON STARS – ANY OBVIOUS RELATIONS ?

Kjell Eriksson[1], Bengt Gustafsson[1,2] and Hans Olofsson[3]
[1] Uppsala Astronomical Observatory, Box 515,
  S-751 20 Uppsala, Sweden
[2] Stockholm Observatory, S-133 00 Saltsjöbaden, Sweden
[3] Onsala Space Observatory, S-439 00 Onsala, Sweden

Evidence for circumstellar absorption around the warm N-type carbon star TX Piscium was found in a high-resolution IUE spectrum by Eriksson et al. (1986). This investigation also included the search, with a positive result, for CO J=1-0 emission from the circumstellar shell. From the Mn I absorption and CO emission a column density of about $10^{20} - 10^{22}$ H atoms per cm^2 was estimated, as well as a mass loss rate around $10^{-7} - 10^{-6}$ $M_{\odot}$ per year.

Lambert et al. (1986) have recently determined CNO abundances and $^{12}C/^{13}C$ ratios for 30 bright, galactic N-type stars. From this sample we have selected twelve stars with different chemical profiles to survey

TABLE I Stellar parameters from Lambert et al. (1986) — Results from CO J=1-0 observations at Onsala Space Observatory

Star	T_{eff} (K)	$^{12}C/^{13}C$	$lg \frac{C-O}{O}$	[N/H]	[O/H]	T_{mb} (K)	v_{LSR} (km/s)	v_{exp} (km/s)	Notes
Z Psc	2870	55	-1.85	-.39	-.23	0.20	12.8	4.2	
U Cam	2530	97	- .52	-.42	-.42	~0.2	~10	~18	1,2
Y Tau	2600	58	-1.40	-.17	-.19	~0.25	~15	~13	1,3
BL Ori	2960	57	-1.41	+.05	-.29	?			1
UU Aur	2825	52	-1.20	+.15	-.18	0.46	6.7	12.4	4
VY UMa	2855	44	-1.22	-.31	-.29	≦0.2			
Y CVn	2730	3.5	-1.06	-.12	-.40	0.36	19.7	9.1	5
RY Dra	2500	3.6	- .74	-.05	-.38	0.15	- 4.9	10.9	
T Lyr	2380	3.2	- .54	-.83	-.50	≦0.08			
UX Dra	2900	32	-1.34	-.12	-.21	0.19	13.7	7.1	
V460 Cyg	2845	61	-1.21	-.06	-.32	0.27	26.3	13.1	
TX Psc	3030	43	-1.57	-.27	-.10	0.25	13.1	12.5	
WZ Cas	2850	4.5	-2.00	+.01	+.07	≦0.08			

Notes:

1. Interstellar lines

2. Previously detected:
   ZDC: 0.16 / 8.5 / 22.0 J=1-0

3. Previously detected:
   ZDC: 0.17 / 15.9 / 10.1 J=1-0

4. Previously detected:
   ZD: 0.66 / 3.6 / 11.5 J=2-1
   ZDC: 0.26 / 7.6 / 12.4 J=2-1
   K: 0.06 / 7.0 / 13.4 J=1-0

5. Previously detected:
   ZD: 0.37 / 23.7 / 6.3 J=2-1
   KM: (0.06 / 21.7 / 7.9) J=1-0
   WS: 0.35 / 21.1 / 7.3 J=2-1

In 2-5 the numbers given are T_{mb} / v_{LSR} / v_{exp}

K   Knapp(1986): Princeton Obs. preprint 167
KM Knapp & Morris(1985): Ap.J. 292, 640
WS Wannier & Sahai(1985): JPL preprint 106
ZD Zuckerman & Dyck(1986):Ap.J. 304, 394
ZDC Zuckerman,Dyck & Claussen(1986):ApJ 304,401

I. Appenzeller and C. Jordan (eds.), Circumstellar Matter, 381–382.

their possible CO emission in the J=1-0 transition with the Onsala 20 m
telescope. The observations were performed in December 1985 and April
1986. We detected CO emission from eight of the stars; the results are
presented in Table I and two examples are displayed in Figure 1. The
main beam brightness temperature, T_{mb}, is the antenna temperature
divided by the main beam efficiency ($\simeq 0.3$), and antenna and radome
transmission factors. Four of the stars have been detected in CO by
other groups independently.

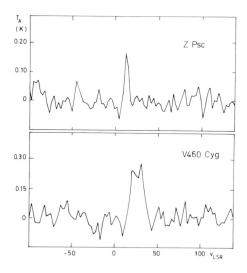

Figure 1: CO J=1-0 emission profiles for two N-type stars in
our sample.

Figure 2: CO expansion velocity
vs. carbon excess. Abscissa normalized such that $\lg \varepsilon_H = 12$.
Open symbols are from other
investigations.

We have investigated whether the shell emission and expansion velocities correlate with the chemical parameters, the effective temperatures
or the flux excess at 11 μm as measured on IRAS low resolution spectra.
No significant correlations were found. However, there may be a tentative
correlation between the expansion velocity and the carbon excess (relative to oxygen), Figure 2. This correlation, which has the right direction if radiation forces on dust grains is an important mass loss mechanism, is worth further study.

References:
Eriksson,K., Gustafsson,B., Johnson,H.R., Querci,F., Querci,M.,
    Baumert,J.H., Carlsson,M. & Olofsson,H. 1986: Astron.Astrophys.161,305
Lambert,D.L., Gustafsson,B., Eriksson,K. & Hinkle,K.H. 1986: Astrophys.
    J. Suppl., September issue

IMAGES OF THE ENVELOPE OF ALPHA ORIONIS

Jeremy C. Hebden, E. Keith Hege, and Andreas Eckart
Steward Observatory
University of Arizona
Tucson
AZ 85721, USA

ABSTRACT. Two images have been obtained, from observations made almost two years apart, of the H-alpha chromospheric envelope of Alpha Orionis at the diffraction limited resolution of the co-phased Multiple Mirror Telescope. Significant emission out to a distance of several stellar radii above the photosphere is observed.

1. INTRODUCTION

Several recent attempts have been made to design theoretical models for the extended atmosphere of the M type supergiant Alpha Orionis. However, the heights and thicknesses of the chromosphere within a few stellar radii of the star, in the region where the outward flow of matter is accelerated, has been relatively unknown. The H-alpha absorption line provides a valuable diagnostic since it is expected to be formed in this region. Using a narrow (1.2Å) H-alpha filter, and the fully-phased six-mirror Multiple Mirror Telescope (MMT)[1], images were obtained of the H-alpha chromosphere of Alpha Orionis at the greatest resolution available for imaging at optical wavelengths.

2. OBSERVATIONS AND DATA REDUCTION

Differential Speckle Interferometry (DSI) observations were made using the fully-phased MMT (Hege et al. 1985) on 1983 December 16/17, and again on 1985 November 2/4. The observational and data reduction procedures are discussed in detail by Hebden et al. (1986) and Hebden, Hege, and Beckers (1986). In order to extract images of the supergiant in the H-alpha line, the DSI imaging technique requires a reconstruction of the star's image in an adjacent continuum bandpass. A well defined photospheric radius, R_*, was found of 17 milli-arcseconds (mas) to 23 mas, dependent on the limb-darkening assumed.

[1] The Multiple Mirror Telescope Observatory is a joint facility of the Smithsonian Institution and the University of Arizona.

I. Appenzeller and C. Jordan (eds.), Circumstellar Matter, 383–384.
© 1987 by the IAU.

3. IMAGES OF THE H-ALPHA ENVELOPE

The images of the H-alpha envelope of Alpha Orionis for the 1983
December and 1985 November observations are shown in figure 1. The
contours are plotted at intervals of approximately five percent.

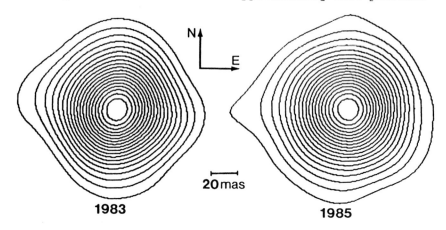

Figure 1. Images of the H-alpha envelope of Alpha Orionis

An intensity greater than one percent of maximum is detectable out to a
radius of about 95 mas, or 4.5 stellar radii. Both images exhibit a
small degree of asymmetry, corresponding to a position angle of about
280°. The absence of a distinct photospheric limb suggests that the
optical depth in H-alpha is probably very large. The radial profiles of
the images exhibit a remarkable agreement with a Gaussian-like
distribution, with intensity falling to I_0/e at $2R_\star$ and $I_0/10$ at $3R_\star$.
The size of the observed H-alpha envelope of Alpha Orionis appears to
conform to estimates of the chromospheric radius obtained from radio
observations (Altenhoff, Oster, and Wendker 1979; Newell and Hjellming
1982), and to the theoretical model of Hartmann and Avrett (1984). A
quantitative comparison of our results and this theoretical model is
described by Hebden, Eckart, and Hege (1987).

This work has been supported in part by the NSF (grant AST-8412206).

4. REFERENCES

Altenhoff, W. J., Oster, L., and Wendker, H. J. 1979, Astron.Astrophys.,
    73, L21.
Hartmann, L., and Avrett, E. H. 1984, Ap.J., 284, 238.
Hebden, J.C., et al. 1986, Ap.J., 309, In press.
Hebden, J. C., Eckart, A., and Hege, E. K. 1987, Ap.J., 312, In press.
Hebden, J. C., Hege, E. K., and Beckers, J. M. 1986, Opt.Eng., 25, 712.
Hege, E. K., Beckers, J. M., Strittmatter, P. A., and McCarthy, D. W.
    1985, Appl.Opt., 24, 2565.
Newell, R. T., and Hjellming, R. M. 1982, Ap.J., 263, L85.

MIRA MODEL PHOTOSPHERES

M. Scholz
Institut für Theoretische Astrophysik
Universität Heidelberg, F.R.G.

ABSTRACT. The temperature stratifications and the emitted fluxes of Mira model photospheres based upon the extended density distribution of a pulsation model differ substantially from those of conventional model photospheres based upon a hydrostatic density distribution. Hence, the interpretation of Mira spectra by means of hydrostatic models is inadequate, and the spectral characteristics of Miras may deviate significantly from those of non-Miras.

M type Mira model photospheres have been constructed in essentially the same way as the static models of Scholz (1985) with the technique of Schmid-Burgk (cf. Schmid-Burgk and Scholz 1984) for solving the spherical radiation transport and energy equations, and with opacities calculated from an improved version of Tsuji's (1978) program. The density distributions follow the isothermal limit of Wood's (1979) first overtone pulsation model, modified in order to account for recent spectroscopic results.

Fig. 1 shows the temperature and density stratifications in the model photosphere of a typical Mira at two different phases. The conventional definitions of a stellar radius, $R = r$ ($\bar{\tau}_{Ros}=1$), and of an effective temperature, $Teff^4 \propto L \cdot R^{-2}$, are adopted (r = distance from the stars's center; $\bar{\tau}_{Ros}$ = radial Rosseland optical depth) which, however, must here be considered with caution. Note the increase of R from maximum to minimum phase, leading to a lower effective temperature and to a general cooling of the photosphere near minimum even though the luminosity is kept constant in this exploratory computation.

First comparisons with observations show good agreement with the wavelength dependence of the radius of o Cet measured by Labeyrie et al. (1977) and Bonneau et al. (1982) and with typical differences between the spectra of Mira and non-Mira stars.

I. Appenzeller and C. Jordan (eds.), Circumstellar Matter, 385–386.

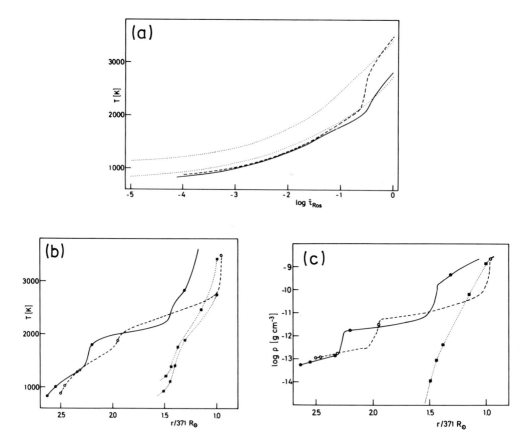

Fig. 1. Temperature and density stratifications in the model photosphere of a Mira variable of 1 M⊙ and 10000 L⊙ near
maximum (Teff = 3060 K, dashed line) and near minimum (Teff
= 2620 K, full) and in static model photospheres of the same
mass and luminosity (Teff = 3000 K {a,b,c} and 2500 K {a,b},
dotted). The squares, circles and dots in {b,c} mark the positions of log $\bar{\tau}_{Ros}$ = -4, -3, -2, -1 and 0.

REFERENCES

Bonneau, D., Foy, R., Blazit, A., Labeyrie, A. 1982, Astr.
    Ap. 106, 235
Labeyrie, A., Koechlin, L., Bonneau, D., Blazit, A., Foy, R.
    1977, Ap. J. Lett. 218, L75
Schmid-Burgk, J., Scholz, M. 1984 in Methods in Radiative
    Transfer, ed W. Kalkofen, Cambridge Univ. Press, p. 381
Scholz, M. 1985, Astr. Ap. 145, 251
Tsuji, T. 1978, Astr. Ap. 62, 29
Wood, P. R. 1979, Ap. J. 227, 220

A NEW TREATMENT OF WATER VAPOR OPACITY

D.R. Alexander[1], H.R. Johnson[2], G.C. Augason[3], R. Wehrse[4]
[1]The Wichita State University, Wichita, USA
[2]Indiana University, Bloomington, USA
[3]NASA-Ames Research Center, Moffett Field, USA
[4]University of Heidelberg, Heidelberg, FRG

ABSTRACT. Although the bands of H_2O are strong in cool stars, water vapor opacity has generally not been accurately treated due both to the inadequacy of laboratory data and the difficulty of treating the millions of lines involved. We report a new treatment based upon a statistical representation of the water vapor spectrum derived from available laboratory data. The statistical spectrum of water follows an exponential distribution of line strengths and random positions of lines to reproduce the line spacings and mean opacities observed in the laboratory. This statistical spectrum is then randomly sampled in the spirit of opacity sampling. Significant improvements are made in both the opacities and in the thermal structure and emergent fluxes of the model atmospheres.

1. INTRODUCTION
A principal difficulty in modeling the atmospheres of cool giant stars is the computation of the opacity, for the principal absorbers are molecules, which may have millions of individual lines. Much of the recent progress has been due to better molecular data and opacity treatments (cf. Johnson, 1986). Previous treatments of H_2O opacity include harmonic mean opacities (Auman, 1967); straight mean (SM) opacities (Johnson, Bernat and Krupp, 1980); and VAEBM opacities (Tsuji, 1976). We here present the results obtained by a new, accurate method of opacity sampling (a statistical representation of the water vapor spectrum). Additional details are provided in a more comprehensive paper (Alexander et al. 1986).

2. CALCULATIONS
To incorporate water vapor opacity correctly into an opacity sampling calculation requires that the monochromatic absorption coefficient be available at a few hundred frequencies. What is important is that profiles of the absorption coefficient of each absorber be appropriately sampled, including line centers, wings, and inter-line gaps (cf. Carbon 1984). Necessary input data are wavelength, line strength, and energy of the lower level of the transition. Unfortunately, this information is not available for water vapor. However, since the OS method randomly samples the spectrum, hypothetical lines yield the same effect as real lines so long as their distribution of line strengths and line positions closely

I. Appenzeller and C. Jordan (eds.), Circumstellar Matter, 387–388.

matches the real spectrum. Based upon an analysis of laboratory data, Ludwig, et al., (1973) have shown that an exponential distribution of line strengths adequately reproduce the opacity of hot water vapor. They also tabulate the average line spacing and mean opacity as functions of wavelength.

To utilize this data statistically, we consider a molecule with an exponential distribution of line strengths (S): $P(S) = \exp(-S/S_o)/S_o$. To determine S_o, the average line strength, we require the average absorption over a $25 \, cm^{-1}$ interval to equal that observed in the laboratory. At each chosen frequency, we then sample the number of lines expected from the mean line spacing at that frequency within ± 4 doppler half widths. For each line, a random displacement within this range and a random line strength, (but with an exponential distribution) are selected.

3. RESULTS

Four model atmospheres (solar abundances, $3200 \geq T_{eff} \geq 2750$ K, $\log g = 0$; $T_{eff} = 3200$, $\log g = 2$) were computed with the new H_2O opacities using the ATLAS6 model atmosphere program (Kurucz 1970). Plane-parallel, horizontally homogeneous media are assumed. The CN, CH, C_2, CO, OH, NH, and MgH were included through the OS method.

The thermal structure ($T(\tau)$) of the models is not radically altered from that of a SM opacity by the use of the OS treatment of water vapor, but the pressure in the OS model is significantly less than in the SM model, indicating that the OS opacity is actually higher than the SM. This is due to the fact that the opacities tabulated by Ludwig, et al., (1973) are often greater than those by Auman (1967), on which the SM opacities were based, and to the fact that the OS treatment of water vapor includes far larger numbers of weak lines.

We acknowledge support by NSF grant 8205800, a NASA consortium agreement with Indiana University, the computing center of Indiana University, Wichita State University, and the Advanced Computation Facility of NASA-Ames Research Center. DRA gratefully acknowledges the hospitality and support of the Astronomy Department of Indiana University.

REFERENCES

Alexander, D.R., Johnson, H.R., and Augason, G.C. 1986, Ap.J., (in preparation).
Auman, J.R., 1967, Ap.J. Suppl., 14, 171.
Eriksson, A., Gustafsson, B., Jorgensen, U.G., and Nordlund, A. 1984, Astr.Ap., 132, 37.
Johnson, H.R. 1986, in Atmospheres of M, S, and C Stars, ed. H.R. Johnson and F. Querci, NASA SP (in press).
Johnson, H.R., Bernat, A.P., and Krupp, B.M. 1980, Ap.J. Suppl., 42, 501.
Kurucz, R.L. 1970, Smithsonian Ap. Obs. Spec. Rept., No. 309.
Ludwig, C.B., Malkmus, W., Reardon, J.E., and Thomson, J.A.L. 1973. Handbook of Infrared Radiation from Combustion Gases, NASA SP-3080.
Tsuji, T. 1976, Publ. Astr. Soc. Japan, 28, 543.

NEUTRAL COMPONENT OF THE OUTERSTELLAR MEDIUM IN THE VICINITY OF THE SUN

Vladimir G. Kurt
Space Research Institute
USSR Academy of Sciences
Moscow, USSR.

Jean-Loup Bertaux
Service Aeronomie,
Verierre, France.

Optical observations of the solar radiation resonance scattering in the hydrogen and helium lines (λ1216Å and 584Å) provide a unique opportunity for determining numerous parameters of the local interstellar medium closest to the Sun (LISM). The distance exceeds 10 to 10^2 AU where the interstellar medium is no longer perturbed by the gravity field, and the hard and corpuscular radiation of the Sun.

Such observations took place in 1976/77y aboard the Soviet Prognoz-5 and -6 satellites and in 1978/79 aboard the Venera-11, -12 interplanetary stations. Both experiments were fulfilled by Soviet-French groups. The first experiment employed a photometer with narrow-band filters and a hydrogen absorption cell, the second, a diffraction spectrometer with 10 detectors to record bright lines of scattered radiation in the range 300÷1600Å.

The observational results were compared with the theoretical intensity value and, by way of optimization in terms of the "hot model", the following parameters of the unperturbed interstellar medium were determined: the density of hydrogen and helium atoms, their temperature, the velocity (its value and direction) of the Sun's motion relative to the interstellar gas, the mean ionization time of helium and hydrogen atoms. One more parameter was derived for hydrogen, which characterizes the ratio of the light pressure force (L_α-line) to the gravity force $-\mathcal{M}$. Assuming the normal abundance of helium relative to hydrogen (0.1 to 0.06) it was possible to estimate the degree of hydrogen ionization x_H from the measured quantity n_{He}/n_H, and, thus, the value of electron density n_e. The radius of the HI zone near the Sun (or near any cold

389

I. Appenzeller and C. Jordan (eds.), Circumstellar Matter, 389–391.

dwarf star) differs from the classical radius of the Strom-
gren zone, and the reasons for this is the absence of stability since the ionization time is by a factor of $5 \cdot 10^7$
shorter than the recombination time - over 1 A.U. During
the recombination time $5 \cdot 10^6$ yr the Sun shifts over 100pc
in the interstellar medium.

The measurements described made it possible to obtain
- Sun's motion velocity rela-
  tive to LISM 24 ± 2 km/s
- direction of this motion $\alpha = 254 \pm 3^{\circ}$
                                      $\delta = -17^{\circ} \pm 3^{\circ}$
- helium density, $n_{He} = 0.015 + 0.020$ cm^{-3}
- temperature $T_{He} = 12000^{\circ} + 16000^{\circ}$K
- ionization time $_{He} = (1 \div 2) \times 10^7$ sec
- hydrogen density $n_H = .06 + .03$ cm^{-3}
- temperature $T_H = 7000^{\circ} - 9000^{\circ}$K
- $\mathcal{M} = F_A/F_G = .75 \pm .1$

Data on the hydrogen distribution within the Solar system
are in better agreement with the anisotropy in the solar
wind flow which obeys the law $(1 - A \sin^2 \lambda)$, where $A = .4 \pm 0.1$, while λ is the heliographic latitude. The
hydrogen ionization degree is $X_H = .3 \pm .3$ which corres-
ponds to $n_e = 0.035 \pm 0.035$ cm^{-3}. The values derived in
measurements in hydrogen and helium lines agree well; except for temperature, where the difference reaches 5000
to 10000°K.

The data from the four experiments may be found in
the following references:

REFERENCES

Bertaux J.L, Blamont J.E., Mironova H.N., Kurt V.G.,Bour-
    gin M.S., 1977, Nature 270, 156.
Bertaux J.L., Lallement R., Kurt V.G., Mironova H.N.,1985,
    Astron. Astroph. 150, 1, 1.
Bertaux J.L., IAU Colloquium No. 81, 1983.
Bourgin M.S., 1981, Astroph. Comments 9, 157
Chassefiere H., Bertaux J.L, Lallement R., Kurt V.G.,1986,
    Astron. Astroph. 160, 229
Dalaudier F., Bertaux J.L., Kurt V.G., Mironova H.N. 1984,
    Astron. Astroph. 134, 171.
Lallement R., Bertaux J.L., Kurt V.G., Mironova H.N.,1985,
    Astron. Astroph. 140, 243.
Lallement R., Bertaux J.L., Kurt V.G. 1985, Journ. Geoph.
    Res. 90, 1413.
Kurt V.G., Mironova H.N., Bertaux J.L., Dalaudier F.,1983,
    Kosmicheskie Issledovania 21, 1, 83.

Kurt V.G., Mironova M.N., Bertaux J.L., Dalaudier F.,1984,
    Kosmicheskie Issledovania, 22, 1, 97.
Kurt V.G., Mironova H.N., Bertaux J.L,, Dalaudier F.,1984,
    Kosmicheskie Issledovania, 22, 2, 225.
Kurt V.G., Mironova H.N., Bertaux J.L., Dalaudier F.,1986,
    Preprint IKI, 1099.

MASS LOSS FROM HOT STARS

THEORY OF WINDS FROM HOT STARS

A.G. Hearn
Sterrewacht Sonnenborgh,
Utrecht
The Netherlands

1. INTRODUCTION.

Mass loss from hot stars was first established by Morton (1967). He observed 3 OB supergiants, δ, ε and ζ Orionis, with an ultraviolet spectrograph sent up with a rocket. In the wavelength range of 1200 A to 2000 A he observed 6 resonance lines of highly ionized atoms such as C III, C IV, N V and Si IV. These resonance lines showed a P Cygni type line profile with an absorption component displaced to the blue corresponding to a velocity away from the stars of 1400 km s^{-1}. Since the escape velocity from these stars is about 800 km s^{-1} these observations indicated the loss of mass from the stars. With rather simple assumptions he deduced a mass loss from the stars of 1 to 3 10^{-6} M$_\odot$ yr^{-1}. In spite of many more refined satellite observations and interpretations, the accepted value for mass loss from these stars has not changed by more than a factor of 2 or 3.

The basic mechanism for the mass loss from hot stars was explained by Lucy and Solomon (1970). When the photospheric radiation of the star is scattered isotropically by the resonance lines the momentum of the radiation is transferred to the gas. The resulting force can be considerable. With one unsaturated resonance line of an ion of carbon or nitrogen situated at the peak of the Planck function of an OB supergiant, the resulting radiative force outwards is typically 300 to 1000 times the force inwards due to gravity. But brute force is not sufficient to drive a supersonic continuous mass loss from a star as Marlborough and Roy (1970) showed. By modifying the basic equations of the Parker solution of the solar wind with an extra force outward in the equation of motion, they showed that if the extra force outward (be it a radiative force or some other force) is always greater than the inward force due to gravity the stationary subsonic supersonic solution with its critical point disappears and the flow remains subsonic. This is contrary to the observations of mass loss from hot stars since the stellar wind is highly supersonic. The reason for the flow remaining subsonic is that the extra force outward compresses the gas, thereby decelerating it instead of accelerating it. This is a good example of the contrariness of nature,

I. Appenzeller and C. Jordan (eds.), Circumstellar Matter, 395–408.
© *1987 by the IAU.*

the harder you push it the slower it goes.

There were two solutions proposed to overcome this problem. Hearn (1975) suggested on the basis of the interpretation of Balmer α line profiles that a corona of limited geometrical extent might exist at the base of the stellar wind of hot stars. Since the coronal temperature would be high, the ions responsible for accelerating the wind would not be present and the radiative force in the coronal region would be negligible apart from the force resulting from electron scattering. The ions that are present in the corona have their resonance lines at such short wavelengths that the star is not emitting there. Beyond the corona the temperature falls to photospheric temperatures and the ions of C III, C IV, N V etc. are present to yield the radiative forces necessary to sweep the stellar wind away to infinity with high velocity. In this model the initial acceleration of the stellar wind to supersonic velocities is done by a thermal mechanism, the Parker mechanism. The critical point exists in the corona and the mass loss is determined by the mechanical heating of the corona and not directly by the radiative forces.

There is no observational evidence of the existence of a small corona, either in the X-ray or infrared. Hearn (in preparation) has constructed fully self consistent theoretical models of a corona plus cool wind for an OB supergiant, but none of the models calculated has a mass loss greater than $3 \ 10^{-11} M_\odot$ yr^{-1}. Further work by Morse and Hearn (in preparation) leads to the conclusion that small corona models for stellar wind in OB supergiants do not exist for mass loss rates greater than $10^{-10} M_\odot$ yr^{-1}. It would appear from the existing observational and theoretical studies that small coronae do not exist in the winds of OB supergiants and that if they exist at all it can only be in some late B main sequence stars where the mass loss is less than $10^{-10} M_\odot$ yr^{-1}.

The second solution to the problem posed by Marlborough and Roy (1970) was proposed by Castor, Abbott and Klein (1975). They pointed out that the resonance lines are strong lines which saturate rapidly. This reduces the radiative force to less than that due to gravity. They developed a self-consistent theory in which the velocity gradient in the subsonic region is sufficiently small to give saturation of the resonance lines. In the supersonic region the velocity gradients and radiative forces are high. The theory gives a very rapidly accelerating stellar wind. The critical point lies just above the photosphere. In the Castor, Abbott and Klein theory the mass loss rate and the velocity distribution are determined together self-consistently and uniquely by the radiative forces alone.

In a review of this length it is not possible to discuss all the interesting developments in the theory of winds from hot stars which have taken place in the last few years, and the choice reflects personal taste and interest. I will discuss three developments:

a) Significant improvements in the Castor, Abbott and Klein theory which have substantially improved the agreement of the theory with the observations.
b) The problem of Wolf Rayet stars.

c) The theoretical problems raised by recent observations of Be stars.

This leaves unfortunately no room to discuss other interesting developments such as radiative driven instabilities in the winds of hot stars, the explanations of X-ray from OB supergiants or the problem of the narrow absorption components.

2. DEVELOPMENTS IN THE CASTOR, ABBOTT AND KLEIN THEORY.

The Castor, Abbott and Klein theory made a number of assumptions and simplifications.
a) The star is treated as a point source, so that all radiation streams radially.
b) A representative list of 900 lines of C III was used. It was assumed that the lines from all ions and elements would behave in a similar way.
c) The Sobolev approximation for the radiative transfer was used. This approximation is used if the thermal broadening velocity of the ions is negligible compared with the range of velocity occurring in the stellar wind. This gives a great simplification in the radiation transfer changing it from the solution of an integral equation over all space to a purely local solution. The result of the Sobolev approximation for strong resonance lines is that the radiative force is proportional to the velocity gradient $\frac{dv}{dr}$.
d) Optically thin lines give a radiative force which is independent of the velocity gradient $\frac{dv}{dr}$, and proportional to the number density of absorbing ions. Castor, Abbott and Klein assumed that the radiative force resulting from the mixture of optically thick and optically thin lines is proportional to $(\frac{dv}{dr})^{\alpha}$. From the representative line list of C III they found the radiative force was fitted by a coefficient α of 0.7.
e) They assume that continuous opacity in the wind is negligible, but of course important in the star. This is called the core-halo approximation.
f) Any modification of the stellar photosphere by the presence of the wind is ignored.
g) Mutiple line transfer is neglected. The momentum of a photon is only used once.

This theory gives simple results. The velocity distribution is

$$v^2 = v_o^2 + (v_\infty - v_o)^2 (1 - \frac{r_o}{r}) \tag{1}$$

where v is the velocity at distance r from the star's centre, v_o and v_∞ are the velocities at the stellar radius r_o and at infinity respectively. Since v_o is very small compared with v_∞ this is usually written

$$v = v_\infty (1 - \frac{r_o}{r})^{\frac{1}{2}} \tag{2}$$

In the interpretation of observations this velocity distribution is usually parameterized by setting the power of the function of distance equal to β. The Castor, Abbott and Klein theory therefore gives β equal to 0.5. The ratio of the final wind velocity v_∞ to the escape velocity v_{esc} from the surface of the star is a function only of the radiative force fitting parameter α

$$\frac{v_\infty^2}{v_{esc}^2} = \frac{\alpha}{1-\alpha} \tag{3}$$

For the value of α used in the original theory, 0.7, v_∞ equals 1.5 v_{esc}. The mass loss rate $\dot{M}$ is also uniquely specified by the radiative forces

$$\dot{M} \propto L \left(\frac{\Gamma}{1-\Gamma}\right) \tag{4}$$

where L is the total luminosity of the star and Γ is the ratio of the force due to electron scattering and that due to gravity. Γ is a function of the luminosity and also of the evolved state of the star, but Equation (4) gives an approximate relation that the mass loss is proportional to the luminosity to the power 1.5. The precise power depends on the details of the assumptions made.

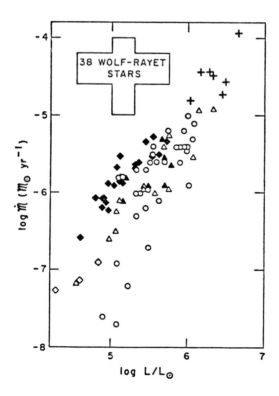

Figure 1.   The mass loss rates and luminosities of O type and Wolf Rayet stars. (From P.S. Conti, 1982)

Figure 1 shows the observed mass loss rate plotted logarithmic-
ally against the luminosity. Whether the spread in the points is
intrinsic or experimental is uncertain. There are uncertainties in the
derivation of the mass loss, particularly in the ionization balance
needed to convert a mass loss rate of ions to a total mass loss rate,
and in the bolometric correction needed to deduce the total luminosity. A least squares fit gives a mass loss rate proportional to $L^{-1.6}$.
The cross shows the region in which the Wolf Rayet stars come according to Willis (1982). It is clear from these results that the Wolf
Rayet stars do not fit the relationship derived from the other stars.
This discrepancy is discussed in the next section of this review.

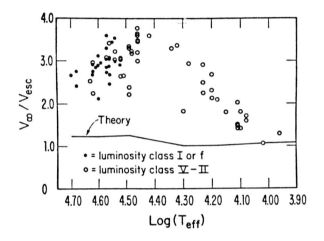

Figure 2.   The ratio v_∞/v_{esc} is plotted against effective temperature.
            The solid line is the constant of proportionality predicted
            by the Castor, Abbott and Klein theory. (From D.C. Abbott
            1982)

Figure 2 shows the ratio v_∞/v_{esc} derived for a number of stars
(Abbott 1982). According to the theory this ratio should be almost
constant, whereas the observations show a significant decrease towards
the hotter and cooler ends of the range. In terms of the original
Castor, Abbott and Klein theory there is no explanation for this
variation. A very illustrative method of plotting these results has
been done by Garmany and Conti (1984) and this is shown in figure 3.
Stars with a v_∞/v_{esc} ratio greater than 3.0 are plotted with an open
circle and stars with a v_∞/v_{esc} ratio less than 3.0 are plotted with a
filled circle. This diagram shows clearly that as stars evolve from
the main sequence the final velocity v_∞ of the stellar wind increases.
Since the mass loss also increases with evolution this is a surprising
result.
    The approximation that the star is treated as a point source has
been removed by Pauldrach, Puls and Kudritzki (1986) and by Friend and
Abbott (1986). The results of Pauldrach et al. (1986) are shown in
Table 1.

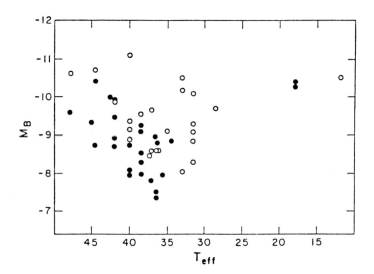

Figure 3.   An H-R diagram with bolometric magnitude plotted against
            effective temperature. Stars with a v_∞/v_{esc} ratio greater
            than 3.0 are plotted with an open circle and stars with a
            ratio less than 3.0 are shown with a fitted circle. (From
            Garmany and Conti 1985)

Table 1 Stellar wind solutions for an 05 V star

	$\dot{M}$ $(10^{-6} M_\odot \, yr^{-1})$	v_∞ $(km \, s^{-1})$
radial streaming	6.7	1559
finite core	3.52	5123
Abbott's line list		1270
Abbott's line list + finite core	6.76	2915

(From Pauldrach et al. 1986)

    Pauldrach et al. (1986) calculated first the Castor, Abbott and
Klein solution following the original paper for a 05 V star of 13.8
solar radii, log g 3.94 and an effective temperature of 49290 K. When
the star is assumed to be a point source the radiation streams
radially. If the star has a finite size, so that the radiation streams
in a finite cone, the mass loss is halved and the final wind velocity

is increased by more than a factor 3. This final velocity is now much
higher than the observed final velocity of about 3000 km s^{-1}. The
reason for these results is that near the star the radiative forces
are used less effectively because they are not directed completely
radially. Since the critical point is near the star this gives a lower
mass loss. This lower mass loss is then accelerated to higher final
velocities in the outer regions of the wind.

Abbott (1982) included a realistic line list in the theory
instead of the representative C III lines used in the original paper.
He included the ions from the 1st to 6th stages of ionization for the
elements from hydrogen to zinc. More than 200 000 lines were involved
in the calculation. The inclusion of Abbott's line list with radial
streaming reduces the final velocity to a value that is far lower than
the observed velocity. But the combination of the finite size of the
star with the Abbott line list gives results for the mass loss and
final velocity that are in very good agreement with the observations.

Pauldrach et al. (1986) have applied the modified theory to a
range of stars including 3 OV stars, 2 evolved O stars and 2 B
supergiants. A summary of their results is shown in Table 2.

Table 2. Comparison of calculated and observed stellar winds

	spectral	$(10^{-6} \dot{M}$ M$_\odot$ yr$^{-1})$		v_∞ (km s^{-1})		
star	type	obs	calc	obs	calc	β
P Cyg	B1 Ia	20 −30	29	400	395	0.98
ε Ori	B0 Ia	3.1	3.3	2010	1950	0.72
ζ Ori A	O9.5 I	2.3	1.9	2290	2274	0.72
9 Sgr	O4(f)V	4.0	4.0	3440	3480	0.81
HD48099	O6.5 V	0.63	0.64	3500	3540	0.81
HD42088	O6.5 V	0.13	0.20	2600	2600	0.79
λ Cep	O6 e f	4.0	5.1	2500	2500	0.79

$$v = v_\infty \left(1 - \frac{r_0}{r}\right)^\beta$$

(From Pauldrach et al. 1986)

The stars chosen for Table 2 all have not too uncertain stellar
parameters, but during the fitting procedure these parameters have
been adjusted within the allowed limits of error to give the best
agreement with the observations. The agreement between the theory and
observations is very good. Even for the highly evolved supergiant P
Cygni is the calculated mass loss and final velocity in good agreement
with the observations. However in this case the predicted velocity
distribution is far from the observed velocity distribution. β equal
to 1 in the parameterization of the velocity distribution gives a very
rapid rise in the velocity, reaching 200 km s^{-1} at 2 stellar radii.
This is contrary to the results obtained by Waters and Wesselius

(1986) from the IRAS infrared observations. They concluded that the velocity rises almost linearly from 30 km s^{-1} near the photosphere to 100 km s^{-1} at 5 stellar radii. Why this discrepancy should exist is not clear.

In a second paper Kudritzki, Pauldrach and Puls (1986) have studied the effect of reducing the metal abundance on the stellar wind models.

The stars in the two Magellanic clouds have significantly lower metallicities than the stars in our own galaxy. Since the metal ions are mainly responsible for driving the wind, differences are to be expected. Kudritzki et al. (1986) have calculated stellar wind solutions using the finite cone approximation and the Abbott line list for an O5 V star with an effective temperature of 45000 K, log g 4.0 and a radius of 12 solar radii. The solutions have been calculated for normal abundances, and for metal abundances reduced to 0.28 and 0.1 of the normal abundances. These models represent a star in our galaxy, the larger Magellanic cloud and the smaller Magellanic cloud respectively. The results are shown in Table 3.

Table 3.   Stellar wind models of an O5 V star with Galactic and Magellanic cloud metallicities

	Z	$\dot{M}$ $(10^{-6}$ $M_\odot$ yr^{-1})	v_∞ (km s^{-1})
Z_{gal}	Z_o	2.12	3350
Z_{lmc}	0.28 Z_o	1.35	2900
Z_{smc}	0.10 Z_o	0.72	2435

(From Kudritzki et al. 1986)

Garmany and Conti (1985) studied a sample of normal OB stars with well defined spectral types. They concluded that O stars of luminosity class III and V in the larger Magellanic cloud have a final wind velocity about 600 km s^{-1} slower than comparable stars in our galaxy, and stars in the smaller Magellanic cloud 1000 km s^{-1} slower. Further they concluded that the mass loss was about the same for comparable stars in the three systems.

Pauldrach et al. (1986) conclude that the smaller final velocity measured in stars in the Magellanic clouds can be reproduced by the calculations. The comparison of the mass loss rates is more complicated since, as they point out, if evolved objects of lower mass in the smaller Magellanic cloud are compared with more massive but less evolved stars in the galaxy because of a selection effect, then very similar mass loss rates can be obtained.

3. THE THEORY OF MASS LOSS FROM WOLF RAYET STARS.

Figure 1 shows that with the accepted parameters for Wolf Rayet stars, the mass loss is far higher than for comparable OB stars and that the Castor, Abbott and Klein theory cannot explain the observed mass loss rates. In fact the problem is more severe than this. If every photon emitted by the star is used once to drive the mass loss then from the momentum balance one obtains a theoretical maximum mass loss from the star given by L/cv_∞ where L is the total luminosity of the star, c is the velocity of light and v_∞ is the final velocity of the stellar wind. For Wolf Rayet stars the observed mass loss is far higher than this maximum. Barlow, Smith and Willis (1981) derived mass loss rates for WN 5, 6 and 7 and WC 5 stars which are 10 to 20 times this limit, and up to 50 times the maximum for WC 7 and 8 stars. Willis (1982) quotes even more extreme examples of mass loss 100 times greater than the theoretical maximum mass loss. For comparison, stars such as ζ Puppis have a mass loss equal to about half the theoretical maximum.

Panagia and Macchetto (1982) proposed that photons could be used more than once and estimated that the maximum mass loss could be increased by a factor of 10. More recently Abbott and Lucy (1985) have performed multiple scattering Monte Carlo calculations for ζ Puppis with a prescribed velocity distribution. They find that many scatterings occur within the same hemisphere as opposed to the scattering from one hemisphere to the other as envisaged by Panagia and Macchetto (1982). From their calculation the increase in mass loss through multiple scattering is only a factor 3. This would appear to be insufficient to solve the Wolf Rayet mass loss problem.

Cherapashchuk, Eaton and Khaliullin (1984) have taken photometric measurements from 2460A to 3.5μ of V444 Cygni. This is an eclipsing binary with a 4.2 day period composed of a WN5 star with an O6 star. From the observations they deduced the density and velocity structure of the WN5 star and concluded that the effective temperature of the star is 90000 K. The radius of the star, defined by the electron scattering optical depth being 2/3, is 2.9 solar radii. The velocity of the stellar wind at the stellar surface is supersonic, 400 km s^{-1}. The temperature in the wind drops gradually to 20000 or 30000 K at 3 stellar radii. The mass loss rate is 1.2 to 1.8 10^{-5} M$_\odot$ yr^{-1}.

Pauldrach, Puls, Hummer and Kudritzki (1985) have applied the Castor, Abbott and Klein theory with the Abbott line list and a finite star size to the result of Cherapashchuk et al. (1984) and obtain good agreement between the theory and the observation. The results of the comparison are shown in Table 4 and figure 4.

Table 4. Stellar parameters of the WN5 star

	PPHK	CEK
T_{eff}	91317 K	80–100 000 K
L/L_o	$5.29 \ 10^5$	$3.1 - 7.6 \ 10^5$
R_*/R_o	2.70	2.90
$\dot{M}$	$1.39 \ 10^{-5} \ M_o \ yr^{-1}$	$1.2 - 1.4 \ 10^{-5} \ M_o \ yr^{-1}$
V_∞	$1950 \ km \ s^{-1}$	$2000–2500 \ km \ s^{-1}$
Γ_e	0.7	

(From Pauldrach, Puls, Hummer and Kudritzki 1985)

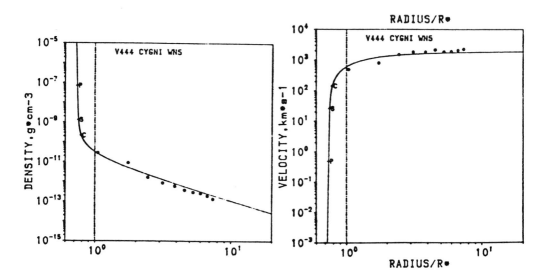

Figure 4.   The density and velocity is plotted against the distance in
            stellar radii for the WN5 component of V444 Cygni. The
            dots   are   the   results   of   Cherapashchuk,   Eaton   and
            Khaliullin (1984) and the lines from Pauldrach, Puls,
            Hummer and Kudritzki (1985)

    Figure 4 shows the density and velocity plotted against the
distance in stellar radii: the dots are the results of Cherapashchuk,
Eaton and Khaliullin (1984) and the lines are the best fit obtained by
Pauldrach, Puls, Hummer and Kudritzki (1985). The points labelled C, S
and P are the critical point, the sonic point and the location of
optical depth 10 measured in the electron opacity.
    If   the   interpretation   of   the   photometric   observations   by
Cherapashchuk et al. (1984) is correct and Wolf Rayet stars have a
much higher effective temperature than had been previously thought,
then it appears that their mass loss can be explained by the Castor,
Abbott and Klein theory. This conclusion has in fact been contested by
a poster paper at this meeting by Schmutz, Hamman and Wessolowski. On

the basis of comoving frame calculations of the helium lines they dispute that the effective temperature can be higher then 60000 K.

Cassinelli and v.d. Hucht (1986) have argued that evolutionary calculations of Wolf Rayet stars show large abundances of carbon and oxygen. If these abundances are included in the mass loss determination from the radio observations instead of the pure ionized helium wind used by other authors, then Cassinelli and v.d. Hucht conclude that the mass loss is 2 or 3 times greater still and the mass loss still exceeds the theoretical maximum by a factor of 20 to 60 for WC 7 to 9 stars. They use only the theoretical L-M relation and believe that still some other mechanism is necessary.

4. MASS LOSS FROM Be STARS.

There have been a number of developments in the mass loss from Be stars recently. These seem to show that at least for a part of the mass loss from Be stars some other mechanism is necessary. Although this is a review of the theory of mass loss from hot stars, this part of the review is a discussion of the work which shows that further theory is necessary.

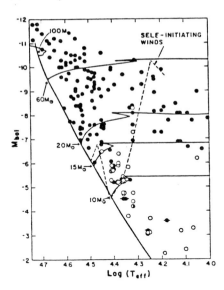

Figure 5.   An H-R diagram with the bolometric magnitude plotted against the effective temperature. Stars with a mass loss detected in the ultraviolet profiles are plotted by a filled circle. Stars for which no mass has been found are plotted as an open circle. If the mass loss from the star is probable, then the star is plotted with a half filled circle. Be stars are plotted with a line through the circle. The dotted line is the limit for self-initiating radiative driven wind (From Abbott 1984)

In the earlier discussions on the Castor, Abbott and Klein, one of the objections raised against the theory was that the radiative driven wind is not self-starting. In answer to this criticism, Abbott (1984) calculated the radiative forces for stars in hydrostatic equilibrium. If the radiative force in the outer layers of the star is greater than the force due to gravity, then Abbott argues that for these stars the radiative driven wind is self-starting.

Figure 5 shows in an H-R diagram the limit of self-initiating winds. In this diagram the bolometric magnitude is plotted against the effective temperature. The stars for which mass loss has been detected in the ultraviolet profiles are plotted as filled circles. Stars with no evidence of mass loss are plotted as open circles, and for stars with a probable mass loss the circle is half filled. The Be stars are represented by a line through the circle. Figure 5 shows that in the main sequence and stars just evolved off the main sequence the only stars having a detected mass loss that lie below the limit of self-initiating radiative driven winds are Be stars. This suggests that some other mechanism is at work in Be stars.

Infrared observations of Be stars made with IRAS have been interpreted by Waters (1986) (See the review by Lamers in this conference proceedings). Waters finds that the observed infrared excess of 4 Be stars, α Eri, ϕ Per, δ Cen and χ Oph can be explained by a disc extending round the equator of the star. The infrared observations show that the density of gas in this disc decreases as $r^{-2.4}$, and extends to 6 or 8 stellar radii. This density dependence means that the velocity of the wind in the disc is increasing very slowly with distance. By assuming a velocity for the wind in the disc near the surface of the star, Waters concludes that the mass loss is 50 to 100 times greater than that deduced from the ultraviolet P Cygni lines profiles, which is presumably measuring the mass loss from the polar regions of the Be star. The mass loss which is deduced from the infrared observations scales directly with the velocity that is assumed at the base of the disc. This interpretation of Waters (1986) also implies a different mechanism for mass loss from Be stars. Radiative driven winds of the Castor, Abbott and Klein type do not have very slow accelerations and the observed mass loss rate may be very much higher than can be explained by that theory.

Non-radial pulsations have been observed in Be stars. Absorption lines in these stars are found to have asymmetric features which cross the absorption profile steadily in about 5 hours. These features have been interpreted by Vogt and Penrod (1983) and Baade (1984) as non-radial pulsations in the Be stars. Figure 6 shows the distortions in the line profile and the pattern of non-radial pulsation deduced by Vogt and Penrod for ζ Oph 0.9 Ve. The contours are drawn every 5 km s^{-1}. The darkest shaded region represents material that is moving away from the observer, while the lightest regions represent material moving towards the observer. The non-radial pulsation shown here is an $l = 8$, $m = 8$ mode.

The disc model deduced by Waters (1986) for the Be stars is reminiscent of the old Struve model. But one sees from the work of Vogt and Penrod (1983) and Baade (1984) that it is not that the

equator of the star is rotating so rapidly that it is near break up
that is important, but that the non-radial pulsations are confined to
the region round the equator of the star. What the physical connection
is between the non-radial pulsation and the mass loss is not clear.

What is also very interesting is that there is some evidence that
Be episodes in ζ Oph coincide with changes in the mode of the non-
radial pulsation. Further there appears to be a systematic difference
between normal B stars and Be stars. Normal B stars appear to pulsate
with 1 or 2 short period high ℓ modes whereas Be stars appear to
pulsate in a long period ℓ equals 2 mode plus a short period high
ℓ mode.

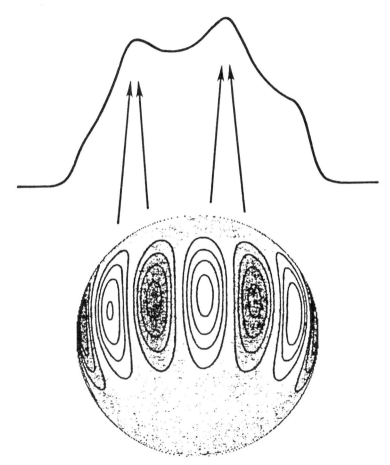

Figure 6.   The formation of distortions in the line profiles of a
rapidly rotating star undergoing non-radial pulsations. The
contours of the non-radial pulsation are drawn every 5 km
s^{-1}. The darkest region represents material moving away
from the observer, the lightest region represents material
moving towards the observer. (From Vogt and Penrod 1983)

ACKNOWLEDGEMENTS
    I am grateful to Dr. Kudritzki and Dr. Friend for sending me copies of papers before publication.

REFERENCES

Abbott, D.C.: 1982, Astrophys. J. 259, 282
Abbott, D.C.: 1984, Proceedings 3rd Trieste Workshop on "Relations between chromospheric-coronal heating and mass loss in stars" at Sacramento Peak Observatory, p. 265
Abbott, D.C., Lucy, L.B.: 1985, Astrophys. J. 288, 679
Barlow, M.J., Smith, L.J., Willis, A.J.: 1981, Mon. Not. R. Astr. Soc. 196, 101
Cassinelli, J.P., van der Hucht, K.A.: 1986, Astron. Astrophys., in press
Castor, J.I., Abbott, D.C., Klein, R.I.: 1975, Astrophys. J. 195, 157
Cherepashchuk, A.M., Eaton, J.A., Khaliullin, Kh. F.: 1984, Astrophys. J. 281, 774
Conti, P.S.: 1982, Proceedings I.A.U. Symposium no. 99 "Wolf Rayet stars: observations, physics, evolution" ed. de Loore, C.W.H. and Willis, A.J., p. 3
Friend, D.B., Abbott, D.C.: 1986, Astrophys. J. in press
Garmany, C.D., Conti, P.S.: 1984, Astrophys. J. 284, 705
Garmany, C.D., Conti, P.S.: 1985, Astrophys. J. 293, 407
Hearn, A.G.: 1975, Astron Astrophys. 40, 227
Kudritzki, R.P., Pauldrach, A., Puls, J.: 1986, Astron. Astrophys., in press
Lucy, L.B., Solomon, P.M.: 1970, Astrophys. J. 159, 879
Marlborough, J.M., Roy, J.-R.: 1970, Astrophys. J. 160, 221
Morton, D.C.: 1967, Astrophys. J. 150, 535
Panagia, N., Macchetto, F.: 1982, Astron. Astrophys. 106, 266
Pauldrach, A., Puls, J., Hummer, D.G., Kudritzki, R.P.: 1985, Astron. Astrophys. 148, L1
Pauldrach, A., Puls, J., Kudritzki, R.P.: 1986, Astron. Astrophys. in press
Vogt, S.G., Penrod, G.D.: 1983, Astrophys. J. 275, 661
Waters, L.B.F.M., Wesselius, P.R.: 1986, Astron. Astrophys. 155, 104
Waters, L.B.F.M.: 1986, Astron. Astrophys. in press
Willis, A.J.: 1982, Proceedings Workshop "Mass loss from astronomical objects", Rutherford Appleton Laboratory, p. 1

SOME OBSERVATIONS RELEVANT TO THE THEORY OF EXTENDED ENVELOPES

B. Wolf
Landessternwarte Königstuhl
6900 Heidelberg
West Germany

ABSTRACT. Recent observations of S Dor variables and B[e]-supergiants are reviewed. These objects belong to the visually and bolometrically brightest stars in the universe. They have gained considerable interest for their exceptional stellar wind properties and represent rare cases of hot blue stars which form dust envelopes. S Dor variables are characterized by major outbursts. The ejecta have been spatially resolved in some cases. These luminous emission line stars are supposed to represent a short-lived phase in the evolution of the very massive stars prior to becoming WR stars.

1. INTRODUCTION

A great variety of objects which are topics of this symposium are characterized by extended circumstellar envelopes. I shall concentrate on observations of luminous blue stars which have been studied by our group within the SFB 132 (Baschek, 1987, this volume) over the past few years. Among the luminous blue stars are two groups which are particularly distinguished due to their strong emission-line spectra: the S Dor variables and the B[e]-supergiants. These stars which are supposed to represent a short-lived phase in the evolution of the very massive ($M \gtrsim 50 \, M_\odot$) stars (cf. e. g. Maeder, 1983, Wolf, 1986, Zickgraf et al., 1986) form due to their particular wind characteristics cool extended shells. The main properties of these objects will be discussed in the following.

2. S DOR VARIABLES

The S Dor variables or Hubble-Sandage-variables are the visually (up to $M_V \approx -11$) and bolometrically brightest stars

409

I. Appenzeller and C. Jordan (eds.), Circumstellar Matter, 409–423.

in the universe and were among the first stellar objects
observed in other galaxies (Pickering, 1987, Duncan, 1922,
Wolf, 1923).

        S Dor variables are blue hypergiants with occasional
eruptions of more than one magnitude on timescales of years
to decades. From detailed spectroscopic observations, particularly of the LMC-S Dor variables R71 (Wolf et al.,
1981) and of R127 (Walborn, 1982, Stahl et al., 1983) these
luminous blue variables turned out to be OB supergiants.
During eruption they exhibit spectra equivalent of later
type (late B to A, F) formed in the expanding envelopes and
characterized by strong P Cygni type emission of the Balmer
lines and of lines of singly ionized metals. With an increase in brightness the stars become redder and their
equivalent spectral type later. This behaviour was particularly well studied in the case of R127 (Stahl et al., 1983).
This object had been classified as an Of/WN-transition type
star during minimum. R127 had an outburst in the early
1980's and has since then still increased in brightness. In
the beginning of 1986 it was the visually second brightest
star in the LMC (V = 9.5). Its recent light curve is given
in Fig. 1. Sections of spectra taken with CASPEC at the
3.6 m telescope of ESO, Chile, in 1984 and 1986, respectively, showing the spectral evolution of R127 are presen-
tend in Fig. 2. Note that during the very bright phase in
1986 R127 exhibits a spectrum which is almost a clone of
the spectrum of the prototype S Dor during outburst.

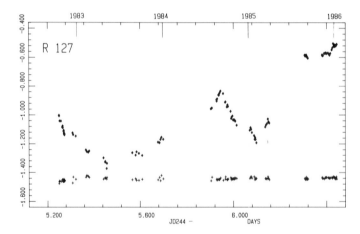

Figure 1. Visual (Strömgren y) differential light curve
($\diamond$) of R127. The y-magnitude of C_1 is 8.86. (+) denotes
differences between comparison and check star. The observations are from the "Long-term photometry of variables"
group, initiated by C. Sterken

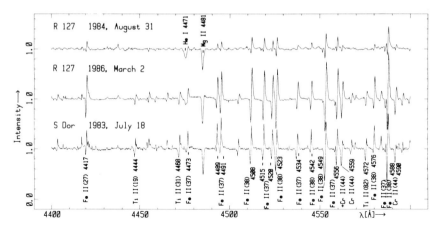

Figure 2.   The spectral evolution of R127 and comparison
with S Dor

A major finding by our group (cf. Wolf et al. 1981, Appenzeller and Wolf, 1982) has been that although the visual
brightness during outburst increases by more than a magnitude the bolometric luminosity remains essentially unchanged. This former hypothesis has been recently observationally established in the case of the galactic star AG
Car which was visually very bright in the early 1980's and
about two magnitudes fainter in 1985 (for the light curve
see Stahl, 1986). Its bright phase spectrum was shown by
Wolf and Stahl (1982) to be of S Dor type during outburst
whereas the spectrum in 1985 was shown by Stahl (1986) to
be typical for Of/WN-stars. Since the launch of IUE in
1978 AG Car was monitored in the satellite UV. From these
observations bolometric magnitudes were derived by direct
flux integration (Cassatella and Viotti, 1986) showing that
M_{bol} remained practically constant inspite of the above
mentioned visual brightness variations. As outlined in a
series of papers by our group (cf. e. g. Wolf, 1986, and
literature quoted therein) this behaviour can be understood
by stellar flux redistribution due to strong density variations in the highly variable wind of S Dor types. The mass-
-loss rate is typically 5.10^{-5} $M_{o}yr^{-1}$ and is about a factor
of ten lower than that during minimum.
     A detailed study of the wind characteristics of the
prototype S Dor during its bright phase has been recently
carried out by Leitherer et al. (1985) on the basis of co-
ordinated spectroscopic and photometric observations from
the satellite UV to the infrared.
     The high dispersion IUE spectrum is characterized by
copious absorption lines of singly ionized metals formed in
the expanding envelope. All lines are blueshifted against
the systemic velocity and in addition the edge velocity

v_{edge} was found to depend on the excitation energy χ; the edge velocity increases with decreasing χ from v_{edge} = 50 km s^{-1} (for χ = 5 eV) to v_{edge} 150 km s^{-1} (for χ = 0.0 eV). This dependence can be interpreted in terms of a depth-dependent temperature- and density-field where the velocity is increasing outwards. The edge velocity of the highly excited lines agrees resonably with the maximum outflow velocity as derived from the full width at zero intensity of the [Fe II]-lines which are best observed during fainter phases and which are formed at larger distances from the star. Modelling the observed IR-excess following the method discussed by Lamers and Waters (1984) yielded a smoothly increasing linear velocity law and a mass-loss rate of $\dot{M} \approx$ 10^{-4} $M_{\odot}$yr^{-1}. The velocity law derived in this way is very similar to the one found for P Cygni by Waters and Wesselius (1985) on the basis of infrared observations. This velocity law can not only account for the χ-dependence of v_{edge} of the UV lines but also for the prominent Balmer line P Cygni profiles. Particularly very sharp narrow emission peaks (about ten times continuum intensity in the case of H$_{\beta}$) indicate that a considerable fraction of the wind is at low velocities. Since even the typical high wind densities ($N_e \approx 10^{11}$ cm^{-1}) of S Dor variables during outburst require a large emitting volume to generate the observed high Balmer line flux, the velocity law has to be very gradual in order to produce the small Doppler velocities necessary for the narrow emission component. Both the extremely high value of $\dot{M}$ and the very gradual velocity law with a very low terminal velocity of only 150 km s^{-1} indicate that a wind mechanism different from that of normal OB stars must be active.

S Dor like other luminous emission line stars of the LMC (cf. Stahl and Wolf, 1986a,Appenzeller et al. 1987) show also narrow emission lines of [N II] ($\lambda\lambda$6548, 6583, 5755) corresponding to flow velocities of only about 60 km s^{-1}. Hence at the distance where these lines form ($N_e < 10^5$ cm^{-3}) the flow velocity is significantly lower than the maximum velocity in the inner parts of the envelope. Presumably [N II] is emitted in a region of interaction between the wind of an earlier epoch and ambient interstellar matter. Then we observe in the [N II] lines of S Dor an unresolved "ring nebula" of the kind seen around AG Car (Thackeray, 1977) and around R127 (see below).

S Dor (like R127, R71 and other S Dor variables) contains in the visual spectrum also a few pure absorption lines; namely collisionally populated high excitation lines like Mg II, He I and Si II. The radial velocities of these lines are slightly variable ($\Delta v \approx$ 20 km s^{-1}) around the systemic velocity. From these lines Leitherer et al. (1986) inferred the existence of an extended pseudo-photosphere which appears to be the base of the stellar wind of the

S Dor variables. The pseudo-photosphere where the continuum forms is in the case of S Dor of spectral type A and its extent is estimated to be about eight times larger during maximum than the stellar core. The radial velocity variations around the systemic velocity may indicate the presence of pulsation-like large-scale expanding and receding motions of the pseudo-photosphere. No theoretical explanations of how these pseudo-photospheres are formed during outburst have yet been provided. But an understanding of the pseudo-photosphere appears to be of fundamental importance for a realistic wind model of the eruptions of S Dor variables (cf. also Davidson, 1986). When the extended cool pseudo-photosphere is formed (i. e. hot stellar plasma is cooling down) the highly ionized metals are recombining to the second and first ionization state. From these ions the vast number of the before mentioned absorption lines in the UV originate which finally may produce a strong opacity driven wind as suggested by Lamers (1986) and Appenzeller (1986).

A consequence of the slow flow velocities of S Dor variables may be that dust can form in the winds of these intrinsically hot stars. In a recent paper Wolf and Zickgraf (1986) have shown that the LMC-S Dor variable R71 is an IRAS point source. The IRAS-measurements at 12, 25 and 60 microns are shown along with groundbased observations in the UBVRIJHKN bands, carried out during minimum phase, in Fig. 3. The very strong excess in the far infrared is ascribed to a very cool ($T_{Dust} \approx 140$ K) and very loosely bound dust shell ($R_{Dust} \approx 8000$ R_*) around R71. The amount of dust condensed in this shell has been estimated to be of the order of $M_{Dust} \approx 3 \cdot 10^{-4}$ $M_\odot$. Using the canonical value of the gas to dust ratio of 100 a total mass of the shell of 3.10^{-2} $M_\odot$ is estimated. Assuming a mass-loss rate of 5.10^{-5} $M_\odot yr^{-1}$ (Wolf et al., 1981) this amount of matter has been ejected within about 600 years. This agrees reasonably well with the kinematic age of 400 years derived from the radius R_{Dust} and from the very low expansion velocity ($v_{exp} \approx 20$ km s^{-1}) at this distance as derived from the splitted [N II]-lines $\lambda\lambda 6548$, 6583 (see Fig. 4).

To study the mass-loss history of S Dor variables and to get some observational basis for the estimate of the duration of the S Dor phase the spatial resolution of extended nebulosities is of crucial importance. Nebulosities have been detected around the galactic S Dor variables η Car (Davidson et al. 1986 and literature quoted therein) and AG Car (Thackeray, 1977). Quite recently the nebulosity around the LMC-S Dor variable R127 has been resolved on long-slit high dispersion spectra in [N II] 6548, 6583, [S II] 6717, 6731, and also in H_α. For contour plots see Appenzeller, Stahl and Wolf (1987, this volume). The nebula shows obviously deviations from total spherical symmetry.

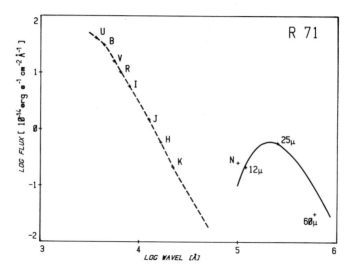

Figure 3. Broad-band fluxes of R71 during minimum phase
from ground-based observations (UBVRIJHK and N) and IRAS-
measurements at 12, 25 and 60 microns. Note the presence of
a very strong excess in the far infrared which is ascribed
to a very cool (T_{Dust} = 140 K), losely bound (R_{Dust} = 8000
R_*) dust shell around R71. The solid line is the black body
curve for T = 140 K.

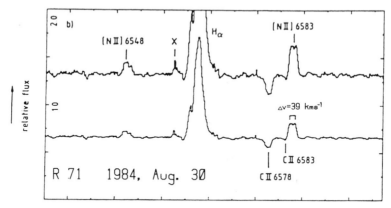

Figure 4. H_α and [N II] $\lambda\lambda 6548$, 6583 of R71. The [N II]
lines are double peaked. An expansion velocity of the nebu-
la of only 20 km s^{-1} is derived from the separation. Note
that [N II] $\lambda 6583$ is disturbed by absorption of C II 6583.

Such deviations provide evidence for a non spherical wind
during outburst and could e. g. cause the multi component
substructure of the UV Fe II-lines shown in Fig. 5 and
amply discussed by Stahl et al. (1983) and Stahl and Wolf
(1986b).A kinematic age of 2.10^4 yrs was derived for the

nebula. This corresponds closely to the expected life time
of the S Dor evolutionary phase.

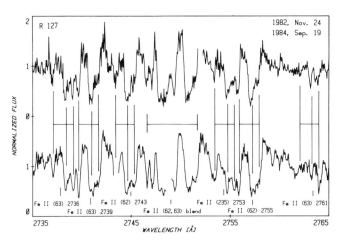

Figure 5. Fe II lines (mainly of multiplets Nos. 62 and
63) on IUE spectrograms taken in 1982 and 1984, respectively. Three components with heliocentric velocities 15, 157,
and 232 km s^{-1} are discernible. The radial velocities of
the three components have not changed during the two years
interval. Likewise no conspicuous variations of the relative intensities of the components have occurred.

The location of the S Dor variables (or Hubble-Sandage variables) of the LMC and of M31 and M33 in the HRD (Humphreys
et al. 1984) is shown below (see Fig. 10). Earlier suggestions (Sterken and Wolf, 1978, Humphreys and Davidson, 1979,
Wolf et al. 1980) according to which the S Dor variables
represent a short-lived phase of massive stars (M $\gtrsim$ 50 M$_O$)
as immediate progenitors of the massive WR stars got further support from Maeder's (1983) computation of evolutionary tracks of very massive stars including the S Dor phase
and are now widely accepted. According to Maeder's computations CNO processed material should appear at the surface
already during rather early phases of the evolution and
should be detectable in the atmospheres or ejecta of S Dor
variables. In fact Davidson et al. (1982) and Davidson et
al. (1986) found e. g. in the S condensation of η Car that
most of the CNO is indeed nitrogen.

3. B[e]-SUPERGIANTS

Another particularly interesting subgroup of the luminous
blue emission-line stars are the dusty B[e]-supergiants.
These objects are located in the HRD in the same region as

the S Dor variables (cf. Fig. 10) and have been recognized
as a group particularly from studies of members of the
MC's by Zickgraf et al. (1986).

B[e]-stars are characterized by the following typical
properties:
a) strong Balmer emission lines, frequently with P Cygni
profiles
b) permitted and forbidden emission lines of Fe II, [Fe II],
[O I] etc.
c) strong infrared excess due to thermal radiation of circumstellar dust.

The determination of the spectral type of B[e]-stars
is often very difficult due to the lack or scarcity of photospheric absorption lines. Hence only a few galactic B[e]-
stars like MWC 349 (Hartmann et al. 1980), CPD -52°9243
(Swings, 1981), and MWC 300 (Wolf and Stahl, 1985) have
been convincingly shown to be luminous supergiants.

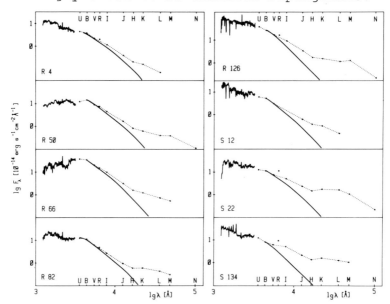

Figure 6.   Continuum energy distributions of the MC's B[e]-
supergiants from the satellite UV to the IR. The visual and
IR-fluxes were deduced from broad-band photometry. An excess in KLM(N) with colour temperatures of 900 K to 1200 K
is clearly discernible. In order to better judge the amount
of free-free envelope emission and dust emission estimated
visual and infrared fluxes for underlying Kurucz atmospheric models are also shown (full line). The conspicuous excess in the R-band of most objects is due to the very
strong H_α-emission.

Eight B[e]-supergiants are known to be members of the MC's. Their spectral energy distribution has been well studied from the satellite UV to the infrared and are shown in Fig. 6. The dust component with typical dust temperature of 900 to 1200 K is clearly evident. By carefully looking for photospheric absorption lines it was found that apart from R66 (B8Ia) all stars are early B supergiants (B0-B3).

The spectral appearance of B[e]-supergiants in the optical range is very similar to the S Dor variables. Fig. 7 shows a section of the optical spectrum of the prototype R126. However, unlike the S Dor variables the B[e]-supergiants of the LMC e. g. have only shown (thus far?) little or no photometric variations. Their mass loss appears to be much more stationary and stable.

High dispersion IUE observations of the prototype R126 revealed an intriguing result. The LWP spectrum is dominated by narrow emission lines as expected from the spectral appearance of the optical range. However, the SWP range is dominated by broad UV-resonance lines of N V, C IV, Si IV etc. (see Fig. 8). These resonance absorption lines of high

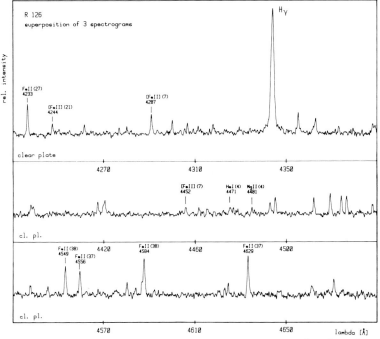

Figure 7.   Section of the optical spectrum of the dusty B[e]-supergiant R126 around H_γ. Numerous Fe II- and [Fe II]-lines are conspicuous. The Doppler velocity derived from the width of the Fe II-lines is only 20 km s^{-1}. The [Fe II]-lines are not resolved.

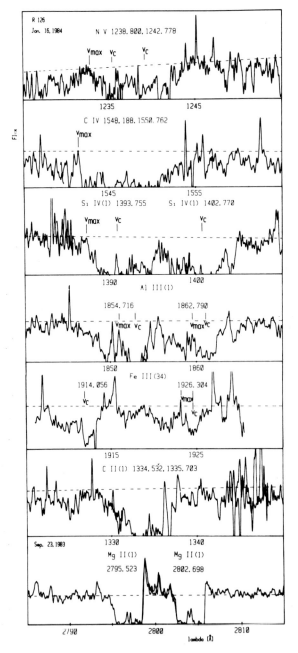

Figure 8. UV resonance absorption lines and Fe III(34)-lines of the dusty B[e]-supergiant R 126. Maximum expansion velocities v_{max} are ranging from 750 km s^{-1} (Fe III-lines) to 1760 km s^{-1} (N V(1)).

expansion velocities ($v_{max} \sim$ 1800 km s^{-1}) are typical for a line-driven "normal" wind for hot stars (e. g. Castor et al., hereafter CAK). The hybrid character of the line spectrum, i. e. narrow emission lines (FWHM $\approx$ 20 - 50 km s^{-1}) in the optical spectrum and in the IUE-LWP range and broad absorp-

tion resonance lines ($v_{max} \approx$ 1800 km s^{-1}) can be interpreted by a two component wind model as depicted in Fig. 9
(Zickgraf et al. 1985).
      According to this model the UV-resonance lines of
highly ionized species (N V, C IV, Si IV etc.) originate in
a high velocity line-driven CAK-type wind near the pole of
the hypergiant star. The disk is supposed to be formed by a
slow ($\approx$40 km s^{-1}), cool dense wind. The observed narrow
emission lines of Fe II, [Fe II] and of other singly ionized
metals in the IUE-LWP spectrum and in the optical wavelength range, as well as the thermal dust emission are produced in the disk. Stellar rotation at close to the break-
-up velocity is the driver of this two-component structure.

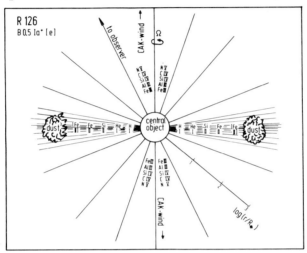

Figure 9.   Proposed bipolar flow model for the prototype
R126 of the B[e]-supergiants. The ions in the disk region
are ordered corresponding to their line widths and in the
polar region according to the expansion velocity. According
to the model dust formation occurs in the slow dense wind
in the equator region whereas a normal CAK wind is present
in the polar region.

Zickgraf et al. (1986) extended this model which has many
similarities to the model suggested for classical Be stars
(Poeckert and Marlborough, 1978) to all eight known B[e]-
supergiants of the MC's. Marked spectral differences observed between the individual B[e]- supergiants could be
explained by assuming different inclination angles. (However, I should like to note the cautionary example R4 of
the SMC which shows a very peculiar line spectrum being
possibly formed in a binary system (Zickgraf and Wolf, 1987,
this volume).)
      A bipolar flow pattern of an angular extension of 0".5

separated by a dark lane confirming the existence of a non-
-spherical mass loss (as e. g. suggested in the two-component wind model) was found by VLA observations of the luminous galactic B[e]-star MWC349 (White and Becker, 1985;
see also Cohen, 1987, this volume). In addition Leinert
(1986) resolved this star in the infrared, using speckle
interferometry, and found a gaussian FWHM of 0."085 in the
east west direction but only of 0."038 in the north south
direction. This last result could also indicate the existence of a disk, in this case for a dust distribution. In
the case of the LMC B[e]-supergiant Hen S22 evidence of deviations from spherical symmetry was also found from the
study of Fe II emission and absorption lines and of the
continuum energy distribution (Bensammar et al., 1983,
Muratorio and Friedjung, 1986).

As noted above, stellar rotation is supposed to be the
driver of the slow equatorial wind. At a first glance it
appears unlikely that rotational effects can play an important role in supergiant stars as - for the case of local
conservation of angular momentum - the ratio between centrifugal and gravitational acceleration at the surface decreases rapidly when the star evolves away from the main sequence. However, the B[e]-supergiants are in the same region in the HRD as the S Dor variables. Hence like in the
S Dor variables and as outlined above the combined opaci-

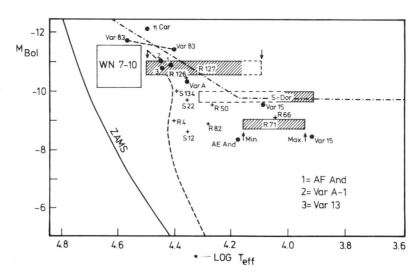

Figure 10. HRD showing the S Dor variables of the LMC
(bars) and of M31 and M33(.). The position of the MC's-
B[e]-supergiants are denoted by (+). Also given are the
boundaries of the main-sequence band (ZAMS = solid line,
TAMS = broken line according to Maeder, 1981). The dash-
-dotted line indicates the upper limit of stellar luminosities (cf. Humphreys and Davidson, 1979).

ties of many merging lines exert an additional acceleration.
In such a situation even a centrifugal acceleration which
is smaller than the gravitational term can lead to a negative acceleration at the equator, which can be compensated
only by an outward acceleration of matter, i. e. an enhanced
mass loss. In addition, due to convective angular momentum
transport from the contracting innermost layers and the
loss of the outermost stellar layers in the main sequence
phase the rotation of the surface layers of very massive
supergiant stars may be higher than expected from the local
angular momentum conservation hypothesis.

As mentioned above the B[e]-supergiants are located in
the same region of the HRD as the S Dor variables (Fig. 10).
As shown by the figure the B[e]-supergiants are located to
the right of the main sequence band. From a comparison with
evolutionary tracks it is clear that they represent evolutionary stages of massive ZAMS O stars. However, inspite of
their similar luminosity B[e]-supergiants have not shown
major outbursts. Possibly stellar rotation leading to an
enhanced mass loss from the equatorial regions prevents
them from becoming S Dor-type unstable.

4. CONCLUSIONS

The aim of the talk was to present "Some Observations Relevant to the Theory of Extended Envelopes" of luminous blue
stars, the S Dor variables and the B[e]-supergiants. Considerable technical advances have been made during the past
few years which allowed us to obtain of these objects e. g.
spectrograms of high resolution and high S/N-ratio (e. g.
wich CASPEC attached to the 3.6 m ESO telescope), to get
spectra in the satellite UV with IUE and infrared data both
from ground-based (JHKLMN) observations and from space observations with IRAS. A lot of interesting results have
considerably improved our knowledge on S Dor variables and
B[e]-supergiants. For a previously heterogenious group of
blue luminous stars in the uppermost corner of the HRD a
logical order is emerging in the form of an evolutionary
sequence.

Yet very basic questions such as the nature of the
wind mechanisms active in these objects, the nature of the
pseudo-photospheres and the cause of the outburst of S Dor
variables are essentially unanswered. However, we have
learnt to put these questions and what we do need most urgently is "Some Theory Relevant to the Observations ...".

ACKNOWLEDGEMENTS. I am obliged to the "Long-term photometry of variables" group, initiated by C. Sterken and to
Dr. L. B. Lucy for carefully reading the manuscript. This
work was supported by the DFG, SFB 132.

REFERENCES

Appenzeller, I., Wolf, B.: 1982, in ESO Workshop "The Most
     Massive Stars", eds. S. D'Odorico, D. Baade, K. Kjär,
     pg. 131
Appenzeller, I.: 1986, in Proc. IAU Symp. 116 (eds. C. de
     Loore, A. Willis, P. G. Laskarides), pg. 139
Appenzeller, I., Stahl, O., Wolf, B.: 1987 (this volume)
Baschek, B.: 1987 (this volume)
Bensammar, S., Friedjung, M., Muratorio, G., Viotti, R.:
     1983, Astron. Astrophys. 126, 427
Cassatella, A., Viotti, R.: 1986 (private communication)
Castor, J. I., Abbott, D. C., Klein, R. I.: 1975, Astrophys.
     J. 195, 157
Cohen, M.: 1987 (this volume)
Davidson, K., Walborn, N. R., Gull, T. R.: 1982, ApJ (Let-
     ters) 254, L47
Davidson, K., Dufour, R. J., Walborn, N. R., Gull, T. R.:
     1986, ApJ 305, 867
Davidson, K.: 1986, ApJ (submitted; preprint)
Duncan, J. C.: 1922, Publ. Astron. Soc. Pacific 34, 290
Hartmann, L., Jaffe, D., Huchra, J. P.: 1980, Astrophys. J.
     239, 905
Humphreys, R. M., Davidson, K.: 1979, Astrophys. J. 232,
     409
Humphreys, R. M., Blaha, C., D'Odorico, S., Gull, T. R.,
     Benvenuti, P.: 1984, Astrophys. J. 278, 124
Lamers, H. J. G. L. M., Waters, L. B. F. M.: 1984, Astron.
     Astrophys. 136, 37
Lamers, H. J. G. L. M.: 1986, in Proc. IAU Symp. 116 (eds.
     C. de Loore, A. Willis, P. G. Laskarides), pg. 157
Leinert, Ch.: 1986, Astron. Astrophys. 155, L6
Leitherer, C., Appenzeller, I., Klare, G., Lamers, H. J. G.
     L. M., Stahl, O., Waters, L. B. F. M., Wolf, B.: 1985,
     Astron. Astrophys. 153, 168
Maeder, A.: 1981, Astron. Astrophys. 102, 401
Maeder, A.: 1983, Astron. Astrophys. 120, 113
Muratorio, G., Friedjung, M.: 1986, Astron. Astrophys. (in
     press)
Pickering, E. C.: 1897, Harvard Circ. No. 19
Poeckert, R., Marlborough, J. M.: 1978, Astrophys. J. Suppl.
     38, 229
Stahl, O., Wolf, B., Klare, G., Cassatella, A., Krautter,
     J., Persi, P., Ferrari-Toniolo, M.: 1983, Astron.
     Astrophys. 127, 49
Stahl, O., Waters, L. B. F. M., Wolf, B.: 1985, Astron.
     Astrophys. 153, 168
Stahl, O., Wolf, B.: 1986a, Astron. Astrophys. 158, 371
Stahl, O., Wolf, B.: 1986b, Astron. Astrophys. 154, 243
Stahl, O.: 1986, Astron. Astrophys. 164, 321
Sterken, C., Wolf, B.: 1978, Astron. Astrophys. 70, 641

Swings, J. P.: 1981, Astron. Astrophys. 98, 112
Thackeray, A. D.: 1977, Monthly Notices Roy. Astron. Soc.
    180, 95
Walborn, N. R.: 1982, Astrophys. J. 256, 452
Waters, L. B. F. M., Wesselius, P. R.: 1986, Astron.
    Astrophys. 155, 104
White, R. L., Becker, R. H.: 1985, Astrophys. J. 297, 677
Wolf, M.: 1923, Astron. Nachr. 217, 475
Wolf, B., Appenzeller, I., Stahl, O.: 1981, Astron.
    Astrophys. 103, 94
Wolf, B., Stahl, O.: 1982, Astron. Astrophys. 112, 111
Wolf, B., Stahl, O.: 1985, Astron. Astrophys. 148, 412
Wolf, B., Zickgraf, F.-J.: 1986, Astron. Astrophys. 164,
    435
Wolf, B.: 1986, "Luminous Stars and Associations in Gala-
    xies". IAU Symp. No. 116 (eds. C. de Loore, A. Willis,
    P. G. Laskarides) pg. 151
Zickgraf, F.-J., Wolf, B., Stahl, O., Leitherer, C., Klare,
    G.: 1985, Astron. Astrophys. 143, 421
Zickgraf, F.-J., Wolf, B., Stahl, O., Leitherer, C.,
    Appenzeller, I.: 1986, Astron. Astrophys. 163, 119
Zickgraf, F.-J., Wolf, B.: 1987 (this volume)

LUMINOUS BLUE VARIABLES - AN EVOLUTIONARY PICTURE

S.R. Sreenivasan and W.J.F. Wilson
Department of Physics
The University of Calgary
Calgary, Alberta   T2N 1N4
Canada

1. INTRODUCTION

A considerable amount of observational information has been gathered on the so-called Hubble-Sandage Variables or S Doradus variable stars, largely due to the efforts of the Heidelberg group (see Wolf, this volume for a review). It is believed that the star P Cygni also belongs to this category. This group of objects is now being referred to as Luminous Blue Variables because they are massive (>10 $M_\odot$), evolved (spectral type B or later) objects showing variability of more than one kind, and evidence of nuclear processed material at the surface.

From an evolutionary point of view they are thought to be in a state just prior to those of Wolf-Rayet Stars (Maeder 1983). Stothers (1983) has presented a detailed first look that these objects and examined several scenarios, without identifying any preferences. We present, in the following pages, a possible evolutionary picture, although alternatives have been presented at this symposium.

2. THE EVOLUTIONARY PICTURE

We have studied recently the evolutionary history of pop I massive stars in the mass range 15 $M_\odot \geq M_* \geq$ 120 $M_\odot$, including the effects of mass-loss, spin and enlargement of the core due to differential rotation (Sreenivasan and Wilson 1978a,b; 1982, 1985a,b; 1986; Narasimha and Sreenivasan 1986a,b), to the point of helium exhaustion in the core. One evolutionary track so far unpublished is shown below. We find, as reported earlier, at other mass ranges, that mass-loss increases not only due to rotation but with evolution due to enhanced differential rotation with age. Surface rotation is reduced to zero well before hydrogen is exhausted in the core, angular momentum is continuously carried away by stellar winds.

The models are overstable to nonradial pulsation for all tested prograde modes ($\ell=2k$ m=-2; $\ell=4$, m=-4, etc...). Typical periods are between a few hours to days (Narasimha and Sreenivasan 1986b) with a slow modulation of amplitude over a period of years. As the core

I. Appenzeller and C. Jordan (eds.), Circumstellar Matter, 425–427.

expands and the outer layers are being peeled off at an increasing rate
with age, efficient meridional circulation is brought into action not
only aiding angular momentum transfer outwards from the core but
mixing the nuclear processed core material with the outer layers.

PARAMETERS FOR 120 $M_\odot$ MODEL

EVOLUTIONARY POINT	AGE (10^6 yr)	MASS ($M_\odot$)	$\dot{M}_{AC}$ ($M_\odot$/yr)	$\dot{M}_{RAD}$ ($M_\odot$/yr)	$\dot{M}_{TOT}$ ($M_\odot$/yr)	$V_{surface}$ (km/s)	V_{core} (km/s)	X_s	Y_c
ZAMS	0	120	0	9×10^{-6}	9×10^{-6}	100	202	0.70	0.27
$X_c = 0$	2.636	77.9	0	1×10^{-5}	3×10^{-4}	0	319	0.48	0.97
BASE RGB	2.641	77.7	7×10^{-5}	0	7×10^{-5}	0	0	0.47	0.96
TOP RGB	2.670	75.1	9×10^{-5}	0	9×10^{-5}	0	0	0.35	0.81

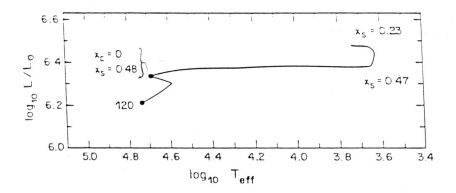

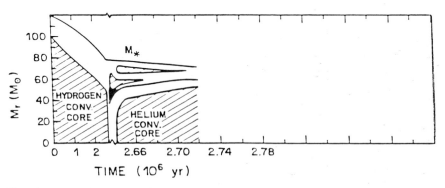

The increasing mass-loss rate with age enables the winds ensuing
at later stages to sweep up material carried by those at earlier epochs
to form a shell. Such mechanisms have been invoked earlier in
connection with planetary nebulae (Kahn 1983, Kwok 1983). The pulsations could couple with the winds not only to excite acoustic modes as
has been suggested by Narasimha and Chitre (1986) but also to result
in large episodic mass ejections as the differential rotation exceeds

critical levels (see also Stothers 1983). We had suggested a similar mechanism for the shell stars earlier (Sreenivasan and Wilson 1980). It is possible that the Luminous Blue Variables are dynamically similar to the Be and Shell stars, being their massive cousins.

We are thus provided with a single unifying mechanism to understand Be and Shell stars and the Luminous Blue Variables through understanding the role of rotation and mixing in massive stars. A more detailed study and a fuller account of it will be presented elsewhere.

Our work is partially supported by an NSERC research grant to SRS.

REFERENCES

Kahn, F.D., (1983). in Planetary Nebulae, ed. D.R. Flowers (D. Reidel).
Kwok, S., (1983). in Planetary Nebulae, ed. D.R. Flowers (D. Reidel).
Maeder, A., (1983). Astron. Ap. 120, 113-129.
Narasimha, D. and Chitre, S.M., (1986). Ap. J. (submitted).
Narasimha, D. and Sreenivasan, S.R., (1986a). BAAS 17, No. 4 (Houston Mtg AAS).
Narasimha, D. and Sreenivasan, S.R., (1986b). in Stellar Pulsation, a memorial to J.P. Cox (in press) ed. A. Cox.
Sreenivasan, S.R. and Wilson, W.J.F., (1978a). A & A 70, 755.
Sreenivasan, S.R. and Wilson, W.J.F., (1980). in Stellar Pulsation Instabilities, ed. D. Fischel et al: NASA TM: 80625, 363-380.
Sreenivasan, S.R. and Wilson, W.J.F., (1982). Ap. J. 254, 287-296.
Sreenivasan, S.R. and Wilson, W.J.F., (1985a). Ap. J. 290, 653.
Sreenivasan, S.R. and Wilson, W.J.F., (1985b). Ap. J. 292, 506-510.
Sreenivasan, S.R. and Wilson, W.J.F., (1986a). Ap. J. (submitted).
Sreenivasan, S.R. and Wilson, W.J.F., (1986b). in Internal Angular Velocity of the Sun: Theory and Observations. ed. B. Durney and S. Sofia (in press).
Stothers, R., (1983). Ap. J. 264, 583-593.
Wolf, B., (1987). these proceedings.

AN EXTENDED NEBULOSITY SURROUNDING THE S DOR VARIABLE R 127

I. Appenzeller, B. Wolf
Landessternwarte, Königstuhl
D-6900 Heidelberg
Federal Republic of Germany

O. Stahl
European Southern Observatory
Karl-Schwarzschild-Str. 2
8046 Garching bei München
Federal Republic of Germany

ABSTRACT. Using the CASPEC echelle spectrograph of the European Southern Observatory, La Silla, Chile, we obtained new high resolution spectrograms of the LMC S Dor variable R 127 in the blue and red spectral range.

The red spectrogram, which contains the [N II] 6548 and 6583 and the [S II] 6717 and 6731 lines shows the presence of a well resolved extended gaseous nebula around R 127 (see Figures 1 and 2). The nebula (which is also detected at the Balmer lines) shows blueshifted and redshifted emission (projected) on the position of the stellar continuum, and no wavelength-shift at the maximum (East-West) distance from the star. Hence, the nebulosity appears to be an expanding shell, reminiscent of the nebula around the galactic extreme supergiant AG Car. The angular diameter (or East-West extension) of the nebula around R 127 is of the order 4", corresponding to ≈ 1 pc at the distance of the LMC. The expansion velocity of the R 127 nebula is found to be 28 km s^{-1} from our spectrograms. Hence, assuming a constant expansion velocity we derive a kinematic age of the R 127 nebula of $\approx 2 \cdot 10^4$ years. This corresponds closely to the expected lifetime of the S Dor evolutionary phase.

A more detailed description of our results will be published in the proceedings of the Workshop on "Instabilities in Luminous Early Type Stars" (C. de Loore, H. Lamers, eds.) Lunteren 1986.

I. Appenzeller and C. Jordan (eds.), Circumstellar Matter, 429–430.
© *1987 by the IAU.*

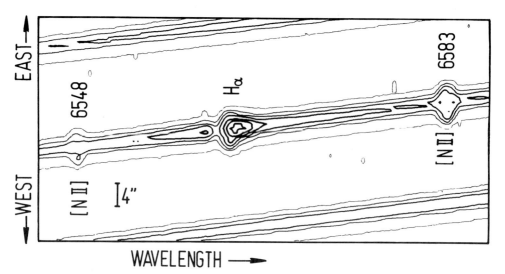

Figure 1. Contour plot of a section of the "red" echello-
gram of R 127, showing the angular extent of the [N II] and
Hα lines. The intensity difference between two contours corresponds to a factor of two. The lowest contour corresponds
to about 1 % of the maximum intensity.

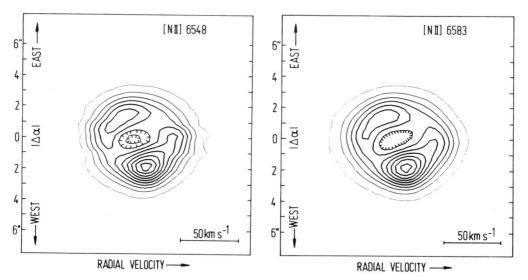

Figure 2. Contour plots of the velocity distribution of
the [N II] emission along the projected spectrograph slit
east and west of the star. The lowest intensity contour corresponds to 5 % of the maximum emission. Between the contours the intensity increases by 10 %. These plots have
been derived from Figure 1 by subtracting the stellar continuum and rebinning to obtain equal ordinate values for
points of identical angular distance from the star.

A METHOD FOR CALCULATION OF LINE PROFILES IN EXPANDING
ATMOSPHERES: APPLICATION TO WINDS FROM CENTRAL STARS OF
PLANETARY NEBULAE

M. Cerruti-Sola[1], H.J.G.L.M. Lamers[2], M. Perinotto[3]
1) Osservatorio Astrofisico di Arcetri
   Largo Enrico Fermi 5 - 50125 Firenze, Italy
2) SRON Laboratory for Space Research
   Beneluxlaan 21, 3527 HS, Utrecht, The Netherlands
3) Istituto di Astronomia, Università di Firenze
   Largo Enrico Fermi 5, 50125 Firenze, Italy

1. INTRODUCTION

The structure of a stellar wind and its associated mass loss can
be derived from the analysis of P Cygni-like line profiles. These do
better occur in the UV range and have been extensively observed with the
Copernicus (900 - 1450 A) and IUE Satellites(1150 - 3200 A).
Two main theoretical approaches have been developed so far to interpret these lines:

1)   The Sobolev approximation with no turbulence, used e.g. in the
     Castor and Lamers (1979) atlas of theoretical profiles.
2)   The comoving-frame method (Mihalas et al., 1975; Hamann, 1980).

The first method is not always able to reproduce the observed P
Cygni profiles in early type stars. That is in the theoretical profiles:

a)   The violet edge of strong lines is often too steep in comparison
     with the observed profile.
b)   For strong lines the theoretical profiles are not able to reprodu-
     ce the whole saturation of the absorption component.
c)   The strength of the emission component is often overestimated.

The presence of turbulence in the wind has been shown to improve
the fitting of theoretical profiles with observations (Hamann, 1980,
1981).
The second method is quite more accurate, but requires substantial
CPU time on large computers.
A method more accurate than 1) and more practical than 2) would be
then worthly.

I. Appenzeller and C. Jordan (eds.), Circumstellar Matter, 431–432.

2.   THE "SEI" METHOD

We have developed a method which uses the Sobolev approximation and solves exactly the equation of radiative transfer. ("SEI" = Sobolev with Exact Integration).

In particular:

a)   it calculates line transfer even for doublets;

b)   it takes into account the presence of underlying photospheric absorption lines;

c)   it is accurate up to large optical depths;

d)   it takes into account the presence of turbulence in the wind;

e)   it allows to represent a wind having a wide range of physical conditions in temperature and in density;

f)   it is fast enough to be of practical use with a medium size computer.

The profiles computed with the SEI method agree fairly well with the profiles computed with the comoving-frame method.

As an application, IUE high resolution profiles of the central star of the planetary nebula NGC 6826 have been fitted with theoretical profiles computed with the SEI method. The figures show the fits for an optically thick line (λ 1548.2, 1550.8 CIV) and for an optically thin line (λ 1718.6 NIV).

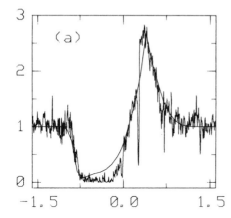 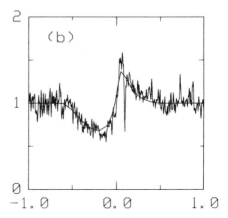

Figure 1.   Theoretical profiles computed with the "SEI" method superimposed to the normalized observed profiles of CIV (a) and NIV (b) of the planetary nebula NGC 6826.

REFERENCES

Castor, J.I. and Lamers, H.J.G.L.M.: 1979, *Astrophys. J. Suppl.* **39**, 481.
Hamann, W.-R.: 1980, *Astron. Astrophys.* **84**, 342.
Hamann, W.-R.: 1981, *Astron. Astrophys.* **93**, 353.
Mihalas, D., Kunasz, P.B. Hummer, D.G.: 1975, *Astrophys. J.* **202**, 465.

THE VELOCITY LAW OF P CYGNI

Rudolf Duemmler
Astronomical Institute
Domagkstr.75, 4400 Muenster
Fed. Republic of Germany

Nevena Markova
National Astronomical Observatory
Lenin Boul.72, Sofia-1184
Bulgaria

ABSTRACT. The multiple absorption components in the Balmer lines of P Cygni indicate that shells are ejected regularly with a period of about 200 days. The newly found multiple, flat-topped emission components are attributed to thin, spherical, expanding shells. Together with the absorptions they allow the derivation of a velocity law, illustrated in the following by a sequence of figures.

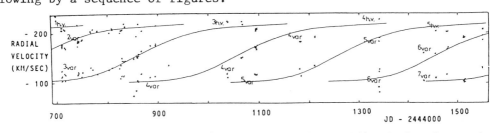

Figure 1. The radial velocities (averages of H9 to H14) of the absorption components (Markova, 1986) of 1981–83 are shown. The identifications of components at different epochs are based on the assumption that their velocities always increase up to the terminal velocity. The results are similar to those of Lamers et al. (1985) for UV-lines. The mean curve connecting the measurements (repeated every 200 days) shows that new components appear with this period and follow roughly the same velocity law.

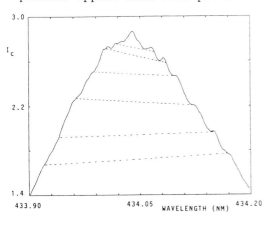

Figure 2. The representative Hγ emission profile was obtained from 13 spectra observed on Nov. 10 and 13, 1984. Averaging was necessary because (sometimes drastic, but largely chaotic) changes occur on time scales of hours or less. The flat-topped components, identified by dotted lines connecting related steps on both wings of the emission, are consistent with the profiles which Beals (1931) derived for thin, spherically expanding shells.

433

I. Appenzeller and C. Jordan (eds.), Circumstellar Matter, 433–434.

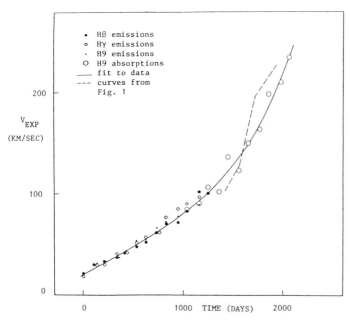

Figure 3. The expansion velocity of a shell can be determined from the width of the plateau. Velocities for the Hβ, Hγ and H9 mean profiles of Nov. 1984 and Sept. 1985, respectively, are shown. The velocities for different lines of the same shell agree well. The time interval between two subsequent shells is assumed to be 200 days. This period is used to shift the observations of 1985 relative to those of 1984, such that they form each second group of points. The excellent fit supports the 200-day period.

DISCUSSION. So far, we have studied in detail the emission structure of only three hydrogen lines. Comparable features are present in He I and N II lines. For a given spectrum their radial velocities agree with those of the hydrogen components. Furthermore, the velocities of some emission components in lower Balmer lines correspond to those of absorptions in higher Balmer lines and of other ions, such as O II, N II, Si III, up to hydrogen velocities of ~160 km/sec. The final increase in velocity of the heavier elements is noticeably slower (Markova, Kolka, 1987). The above results lead to the model of thin, spherical shells of accelerated expansion, ejected with a 200-day period. Integration of the full line in Fig.3 gives a preliminary velocity law for the hydrogen shells in the envelope of P Cygni. Following a steep rise near the stellar surface, an almost linear increase in velocity is observed between 50 and 350 R_* (V~60...240 km/sec). Assuming that the high velocity absorptions in Fig. 3 are not subject to systematic effects, the weakly indicated sudden increase in acceleration at ~200 R_* could be real.

ACKNOWLEDGEMENTS. This work is based on material collected at the National Astronomical Observatory, Rojen, Bulgaria, and was supported by the Bulgarian Academy of Sciences and the Deutsche Forschungsgemeinschaft.

REFERENCES.

Beals, C.S.: 1931, *Mon. Not. R. Astron. Soc.* **91**, 966
Lamers, H.J.G.L.M.; Korevaar, P.; Cassatella, A.: 1985,
            Astron. Astrophys. **149**, 29
Markova, N.: 1986, *Astron. Astrophys.* **162**, L3
Markova, N.; Kolka, I.: 1987, submitted to *Astrophys. Space Sci.*

THE DETECTION OF A CIRCUMSTELLAR SHELL AROUND P CYGNI BY DIRECT CCD IMAGING

Claus Leitherer and Franz-Josef Zickgraf
Landessternwarte Königstuhl
D-6900 Heidelberg
Germany

The star P Cygni (=HD 193237) is the proto-type of a class of stars with peculiar spectra, the so-called P-Cygni-stars (or Luminous Blue Variables). They are characterized by extreme mass-loss characteristics and wind densities surpassed only by the winds of Wolf-Rayet stars.

The high wind densities in these objects and the resulting high surface brightness in certain nebular emission lines even at large ($>10^3$ R_*) stellar distances led to the direct discovery of circumstellar shells on deep H_α-images around several objects (e. g. AG Carinae and R 127). We report on direct narrow-band H_α-images of P Cygni obtained in an attempt to find evidence for circumstellar matter in the immediate surroundings of P Cygni.

The observations were performed at the 1.23-m telescope of the Spanish-German Astronomy Center on Calar Alto (Spain) using a CCD camera on October 12, 1985. The image scale was 0".46 pixel^{-1}. The CCD camera was equipped with a set of three interference filters: H_α (λ_0 = 6553 Å, $\Delta\lambda$ = 36 Å), [N II] (λ_0 = 6587 Å, $\Delta\lambda$ = 22 Å), continuum (λ_0 = 4784 Å, $\Delta\lambda$ = 118 Å). The observational technique was as follows: We chose a nearby comparison star of similar magnitude and spectral type (55 Cyg; V = 4^{m}84, B3Ia) and exposed P Cyg and 55 Cyg alternatively in each filter. The exposure time was 0.75 sec in each filter and for the two stars. The zenith distance was less than ≈ 50 and seeing was better than 0".7 in a photometric night. The observations were organized in several exposure sequences: We obtained CCD frames in H_α, [N II], and the continuum of 55 Cyg and immediately afterwards of P Cyg. The optimum focus had been determined for each filter at the beginning of the first sequence and was adjusted for each exposure. The sequence for 55 Cyg and P Cyg was repeated three times resulting in a total of 24 CCD frames for the two stars.

We analyzed the brightness profiles of P Cyg and 55 Cyg in H_α, [N II], and the continuum: In order to determine the

I. Appenzeller and C. Jordan (eds.), Circumstellar Matter, 435–437.

centers of light we fitted 2-dimensional Gaussian profiles
to the observed light distribution. The radial brightness
profile was then derived by averaging the counts per pixel
in concentric annular segments around the light center. The
resulting profiles generally have a full width at half ma-
ximum of $0\overset{\prime\prime}{.}6$. Subsequently we multiplied the four best-fit
profiles of each filter for 55 Cyg and P Cyg by constant
factors in order to normalize the peak intensity to a common value. The normalization factor was chosen as to give
best agreement for the count rates of the (radially) three
innermost pixels of the profiles. After normalization the
four respective profiles of H_α, [N II], and the continuum
were superposed leading to seeing-disk profiles for P Cyg
and 55 Cyg in H_α, [N II], and the continuum, respectively.
The light distribution of P Cyg is more extended in H_α and
[N II] than the seeing disk of 55 Cyg. In contrast, P Cyg
and 55 Cyg have identical seeing disk in continuum light.
The agreement of the radial brightness profiles of P Cyg
and 55 Cyg in continuum light and the extended structure of
P Cyg in H_α and [N II] can be interpreted in terms of a
shell surrounding P Cyg. This shell can be most readily seen
in nebular emission lines such as H_α and [N II]. On the
other hand, in continuum light the shell is essentially
transparent and P Cyg's and 55 Cyg's light distributions
are identical.

After conversion to absolute intensity units the intensity profile in H_α and [N II] of the circumstellar shell
around P Cyg is obtained from the difference of P Cyg's
brightness profile and a point-spread function represented
by 55 Cyg. In both cases the intensity as a function of
central angular distance behaves as $\sim\alpha^{-3}$. An α^{-3} behavior
is theoretically expected for the intensity of an expanding,
spherically symmetric shell. The intensity in H_α is given by

$$I_O(H_\alpha) = \int \frac{1}{4\pi} N_P N_E a_{32} (T_E, N_E) h\nu_{32} \, dl \quad \left[erg \; cm^{-2} \, sec^{-1} \atop sterad^{-1} \right]$$

with

N_P, N_E: proton and electron number density, respectively
               (here: $N_P = N_E$)
a_{32}:   recombination coefficient for H_α
$h\nu_{32}$:   energy of Balmer photon
dl: path length along line of sight.

Using the equation of continuity together with the stellar
parameters of P Cygni from Lamers (Astron. Astrophys. 159,
90 [1986]) the intensity can be expressed as a function of
P Cyg's mass-loss rate $\dot{M}$ and α (after correction for red-
denning):

$$I(H_\alpha) = 1.66 \cdot 10^{-20} \frac{(\dot{M} [M_\odot yr^{-1}])^2}{(\tan \alpha [''])^3} \left[erg \; cm^{-2} \; sec^{-1} \; arc \; sec^{-2} \right]$$

If $\dot{M} \sim 4.2 \cdot 10^{-4}$ $M_\odot yr^{-1}$ is inserted in this equation the theoretical intensity profile is in good agreement with the observed one. A mass-loss rate of $\dot{M} \sim 4 \cdot 10^{-4}$ $M_\odot yr^{-1}$ is not unreasonable for P Cygni. Its present-day mass-loss rate is $2 \cdot 10^{-5}$ $M_\odot yr^{-1}$ (Lamers 1986) but Luminous Blue Variables are known to show outbursts exceeding $\dot{M} = 10^{-4}$ $M_\odot yr^{-1}$. Interestingly, P Cygni has shown a major outburst in the 17th century. The flow time-scale $t \sim a/v_\infty$ (accidentally?!) turns out to be ~ 300 yr. One may speculate that the extended structure found in P Cyg's light distribution is associated with the above outburst.

ON THE PUZZLING LINE SPECTRUM OF THE B[e]-SUPERGIANT R 4 OF THE SMC

F.-J. Zickgraf, B. Wolf
Landessternwarte, Königstuhl
D-6900 Heidelberg
FRG

The early-type supergiants of the Magellanic Clouds (MC's) include a class of objects, the B[e]-supergiants, which are described in detail in Zickgraf et al. (1985, 1986) who suggested a two-component wind model consisting of a cool, dense and slowly expanding disk-like equatorial wind component and a hot line-driven fast polar wind. This model could explain many properties of most of the B[e]-supergiants as e. g. the occurrence of "shell" absorption lines. R 4 of the SMC, however, although well fitting to the general scheme of B[e]-supergiants exhibits several peculari- ties of its line spectrum, which make this particular star an exceptional case among the B[e]-supergiants of the MC's.

The spectrum is dominated by very rich and complex emission-lines of H, FeII, [FeII], [FeIII], [SII], [SIII], [NII], and [OI]. The Balmer-lines have P Cygni profiles with a composite emission consisting of a sharp (FWHM = 75 km s^{-1}) strong central emission and a shallow and broad component.

In Fig. 1a the emission-line profiles of several ions are shown on a velocity scale. Obviously large differences of the radial velocities and line widths of the various ions are present with the very narrow FeII-lines being sig- nificantly less red-shifted than the broad [OI]-lines. Whereas the emission-lines generally are symmetric the line-profiles of [SII] and [SIII] are exceptional, being clearly asymmetric.

In addition to the emission lines several photospheric absorption lines of SiIII, SiIV, OII, and HeI (spectral type BO-BO.5) and some very sharp "shell" absorption lines of TiII and CrII are present. Surprisingly these "shell"- lines are red-shifted with respect to the systemic velocity (as given by the SiIV-lines (see Fig. 1b)) and with respect to all emission lines apart [OI]. A binary system with spa- tially separated absorption and emission regions could pos- sibly provide an explanation for the complex line-spectrum of R 4.

439

I. Appenzeller and C. Jordan (eds.), Circumstellar Matter, 439–440.

A more detailed discussion of the spectrum of R 4 is forthcoming.

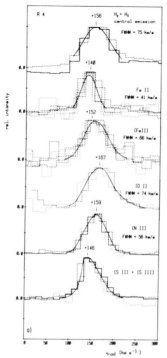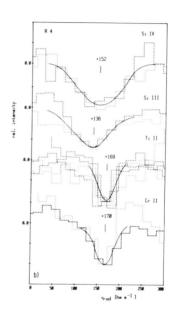

Figure 1. a) Emission-line profiles of various ions. After
subtraction of the continuum the line fluxes were scaled to
equal height of the emission peaks. Gaussian profiles were
fitted to the mean line profile of each ion (solid lines)
except the asymmetric lines of [SII] and [SIII] , for which
only the mean profile is indicated. Large differences of
the radial velocities and line widths are present.
     b) same as a) for absorption lines. The "shell" absorp-
tion lines of TiII and CrII are much narrower than the photospheric lines of SiIII and SiIV. Surprisingly they are
red-shifted with respect to the systemic velocity (given by
SiIV) and with respect to all emission lines apart [OI] .
The spectra (λλ 3920 - 4950, λλ 5740 - 6800 Å) were taken
in August 1984 with the CASPEC spectrograph at the 3.6 m
telescope at ESO/La Silla.

ACKNOWLEDGEMENT. This work was supported by the Deutsche
Forschungsgemeinschaft (Wo 269, 2-1)

REFERENCES
Zickgraf, F.-J., Wolf, B., Stahl, O., Leitherer, C., Klare,
     G.: 1985, Astron. Astrophys. 143, 421
Zickgraf, F.-J., Wolf, B., Stahl, O., Leitherer, C.,
     Appenzeller, I.: 1986, Astron. Astrophys. 163, 119

A COMPACT NEBULOSITY SURROUNDING THE PECULIAR BLUE EMISSION-LINE SUPERGIANT HD 37836 OF THE LMC

O. Stahl
European Southern Observatory,
Karl-Schwarzschild-Str. 2, D-8046 Garching b. München, FRG

B. Wolf
Landessternwarte Heidelberg
D-6900 Heidelberg-Königstuhl, FRG

HD 37836 (= R 123; Feast et al., 1960) is one of the brightest slightly variable (V ≈ 10.5-10.7; Stahl et al., 1984) emission-line stars of the LMC. Since it is also likely to be a hot star (see below) it is also bolometrically very bright (M_{bol} ≼ -11).

We have observed HD 37836 in the ultraviolet with IUE (1200 Å ≼ λ ≼ 3100 Å) and in the optical with CASPEC (3900 Å ≼ λ ≼ 4950 Å and 5750 Å ≼ λ ≼ 6800 Å) with high dispersion. In addition we derived the continuum energy distribution from low dispersion IUE spectra and ground-based broad-band photometry (UBVRIJHKL). A strong IR excess was found. The most prominent absorptions in the IUE spectrum are stellar-wind lines of NV, SiIV and CIV with a terminal velocity of 2400 km sec^{-1}. This is typical for an O star.

By far the strongest emission lines in the visual spectrum of HD 37836 are the Balmer lines and the HeI lines (see Stahl et al., 1985). From the CASPEC spectra we found in addition lines of FeII, [FeII], [FeIII], SiII, SiIII, [SIII], MgI, MgII, NaI and [NII]. Three different groups of lines can be distinguished.

a) The forbidden lines [NII] λλ5755,6548,6583, [SIII] λ6312 and [FeIII] λλ4658,4702 are narrow and approximately gaussian shaped with a FWHM of ≈ 40 km sec^{-1}.

b) The low-excitation lines of allowed transitions (FeII, SiII λλ6347,6371, MgI λ6318, NaID) are much wider and double-peaked. The separation of the peaks is about 70 km sec^{-1}.

I. Appenzeller and C. Jordan (eds.), Circumstellar Matter, 441–442.

c) Emission lines of higher excitation, i.e. the Balmer lines, the HeI
   lines and SiIII]λ1892 are also wider than the forbidden lines but
   do not show a double-peak structure.

   The different line-widths of the various emission lines indicate
that the velocity is decreasing outwards since the forbidden lines are
particularly narrow. It appears that the blue-shifted absorption lines
in the UV are formed in a normal stellar wind whereas the emission
lines and the IR excess form in a region of considerably lower
velocity, which could be a disk.

   From the line intensity ratio [NII] λ5755 / (λ6548 + λ6583) we
derived n_e ≳ 4·10^6 cm^{-3}. Assuming the nebula to be photoionized by the
star, the linear size cannot be larger than ≈ 10^{-2} pc and the mass of
the ionized gas is ≲ 10^{-1} M$_\odot$. The small size of the nebula indicates
that it is very different from the nebula surrounding R 127 which has
a linear size of ≈ 1 pc (Appenzeller et al., 1987). The ring nebulae
around Wolf-Rayet and Of stars are even much larger. In the LMC they
have a diameter of 20 pc or more (Chu and Lasker, 1980). Thus, low-
velocity gas can be present around luminous hot stars at a large range
of distances.

   A more detailed paper is forthcoming in Astronomy and Astrophysics.

References

Appenzeller, I., Stahl, O., Wolf, B.: 1987, in Instabilities in
     luminous early-type stars, eds. de Loore and Lamers, Reidel,
     Dordrecht.
Chu, Y.-H., Lasker, B.M.: 1980, Publ. Astron. Soc. Pacific 92, 730.
Feast, M.W., Thackeray, A.D., Wesselink, A.J.: 1960, Monthly Notices
     Roy. Astron. Soc. 121, 337.
Stahl, O., Wolf, B., Leitherer, C., Zickgraf, F.-J., Krautter, J., de
     Groot, M.J.H.: 1984, Astron. Astrophys. 140, 459.
Stahl, O., Wolf, B., de Groot, M.J.H., Leitherer, C.: 1985, Astron.
     Astrophys. Suppl. 61, 237.
Stahl, O., Wolf, B.: 1986, Astron. Astrophys. 158, 371
Walborn, N.R., Panek, R.J.: 1984, Astrophys. J. Letters 280, L27.

EVIDENCE FOR DISKS AROUND CERTAIN LUMINOUS MAGELLANIC CLOUD STARS FROM THE STUDY OF FeII

G. Muratorio
Observatoire de Marseille
2, place Le Verrier
F-13248 Marseille Cedex 04

M. Friedjung
Institut d'Astrophysique
98bis, Boulevard Arago
F-75014 Paris

ABSTRACT. Study of FeII emission lines in emission and absorption using both emission line self absorption curves and ultraviolet spectral synthetis, shows that line emission is produced in the case of certain luminous Magellanic Cloud stars, in a region not in front of the photosphere. This region is most easily understood as being a disk. Absorption lines of FeII are either produced by a wind, or if the disk inclination is small with respect to the line of sight, in layers associated with the disk. The presence of disks also helps one to explain the form of the continuous energy distribution.

Certain highly luminous stars with a fairly high effective temperature have optical spectra which are very rich in FeII emission lines, while low resolution UV spectra suggest that blends of FeII absorption lines are more often dominant than emission in that spectral region. We have studied stars of this kind in the Magellanic Clouds.

We analyze emission lines using the self absorption curve method, i.e. by plotting graphs of $\log(F_\lambda W_\lambda \lambda^3/gf)$ against $\log(gf\lambda)$. Here F_λ is the continuum flux corrected for interstellar absorption, W_λ the line equivalent width, g the lower level statistical weight, and f the oscillator strength. λ is the wavelength. It can be shown that :

$$\log\left(\frac{F_\lambda W_\lambda \lambda^3}{gf}\right) = \log(2k\pi hc) + 2\log\left(\frac{R_c}{d}\right) + \log(\phi_c V_c) + Q(\tau_c) \qquad (1)$$

In this equation k is a constant equal to $0.02654\,\mathrm{cm}\;\mathrm{s}^{-1}$, h and c have their usual meanings, d is the distance of the star, while R_c, ϕ_c, V_c and τ_c are characteristic values of the radius, the upper level population of a unit column per unit velocity range/statistical weight, velocity and optical thickness of the medium emitting the line. $Q(\tau_c)$ is a function of τ_c, which depends on the nature of the medium. Noting that τ_c is a multiple of k, $gf\lambda$ and of the characteristic value of the lower level population of a unit column per unit

I. Appenzeller and C. Jordan (eds.), Circumstellar Matter, 443–444.
© *1987 by the AU.*

velocity range/statistical weight, the shape of a self absorption curve for lines of the same multiplet gives the shape of the variation of $Q(\tau_c)$ with log τ_c, as long as the levels of each term have relative populations proportional to their statistical weights. Population ratios of different terms can be found by shifting self absorption curves of different multiplets relative to each other, relative horizontal shifts for multiplets having the same upper term give population ratios for the lower terms, while relative vertical shifts for multiplets having the same lower term give population ratios for the upper terms.

The function $Q(\tau_c)$, giving the shape of the self absorption curve, has been calculated for various simple models, including some of stellar winds. This enabled observations of emission lines to be fitted, and various parameters such as R_c and the column density to be determined.

Spectral synthesis of low resolution IUE spectra has also been performed. The parts of a wind in front of the photosphere produce absorption components, which have been taken into account after calculation of the curve of growth. The form of this for a high velocity wind assuming only Doppler broadening of the lines, was specially calculated.

We conclude that firstly the FeII lines are not photospheric, and that the stars studied appear to have line emission formed in a region not in front of the photosphere, most easily understood as a disk, in addition to the indications found for the presence of a wind in at least some cases. Analysis of the continuum energy distribution also supports the presence of disks, which reprocess radiation from the central star.

THE O6.5IIIf STAR BD +60°2522 AND ITS INTERACTION WITH THE SURROUNDING INTERSTELLAR MEDIUM

C. Chavarría-K., C. Jäger, and C. Leitherer
Landessternwarte Königstuhl
D-6900 Heidelberg
Germany

The luminous O-type star BD +60°2522 is embedded in the extended H II region S 162. Part of S 162 is NGC 7635, the striking spherically symmetric bubble nebula surrounding BD +60°2522. This star itself is unique in that it is the only known O star apparently associated with warm dust.

In an attempt to study the interaction of the central O star with the surrounding H II region we obtained flux calibrated narrow-band CCD frames (H_α, H_β, O III, N II) and highly resolved coudé spectrograms.

Fig. 1 shows the northern part of NGC 7653. This figure underlines the sharp boundary between the bubble nebula and the nearby H II region. This boundary represents a drop in density, since the output of stellar Lyman photons suffices to ionize the whole S 162 complex (i. e. NGC 7653 is density bounded). A marked peculiarity within NGC 7635 is the comet-like condensation west of BD +60°2522.

The H_α/H_β ratio is found to be very uniform all over the bubble nebula, $E(B-V) \approx 0.5$, except for the bright westward condensations, $E(B-V) \approx 1.0$.

The density structure is derived from the S II 6716/6731 ratio obtained from the long-slit spectrograms. The maximum density in the knots is about 10^5 cm^{-3}, whereas the average density outside the knots is $10^2 - 10^3$ cm^{-3}.

The dereddened H_β flux integrated over the bubble nebula without immersed stars is about 5×10^{-10} erg sec^{-1} cm^{-2}. The corresponding total mass of NGC 7653 including the knots turns out to be ~ 3 $M_\odot$. An estimate of the nitrogen abundance (relative to sulfur) can be obtained from the $I(6548 + 6584)/I(6717 + 6731)$ ratio. From this ratio we derive $N(N^+)/N(S^+) \approx 7.3$ for slit position 0 to 35 and $N(N^+)/N(S^+) \approx 15$ for slit position 55. These values can be converted to total abundances using the ionization correcting factors. We find $(N(N)/N(S))/N(N^+)/N(S^+)) \approx 0.3$ for the average bubble region and approximately 0.18 for slit position 55. Thence within the limits of uncertainty the N/S abun-

I. Appenzeller and C. Jordan (eds.), Circumstellar Matter, 445–446.
© 1987 by the IAU.

dance is the same, 0.4.

The velocity field can be investigated from our coudé long-slit spectrograms. We did not find significant radial--velocity variations from line to line eastward of the star (slit position 0 to 35). We derived a mean LSR velocity of (-39 ± 7) km sec^{-1} for NGC 7635. On the other hand the bright knots show a significant radial-velocity difference of $(+12 \pm 1)$ km sec^{-1} (slit position 50 to 75) relative to slit position 0 to 25.

Moreover, a significant difference in the line width (FWHM) between the bubble nebula and the bright knots is observed: At slit positions 0 to 25 we find for the forbidden lines as well as for H_α line widths of 30 km sec^{-1}. In contrast, the line widths of the forbidden lines at slit positions 50 to 75 are 20 km sec^{-1}, whereas H_α shows a line width of 50 km sec^{-1}, indicating an origin of H_α and the forbidden lines in different regions in the knots. This interpretation is supported by the different radial velocity behavior of H_α and the forbidden lines.

The significant line-width and radial-velocity differences cast serious doubts on the physical association between the bubble nebula and the bright knots. The bubble nebula itself can be interpreted as a stellar-wind blown shell caused by the stellar wind of BD $+60^\circ 2522$ interacting with the ambient ISM. The mechanical power of the stellar wind, $L_W \sim 3.4 \times 10^{36}$ erg sec^{-1}, is transferred to the ISM leading to an expansion of the swept-up matter by $v_{EXP} \approx$ 20 km sec^{-1}, then the age of the bubble is of the order $10^4 - 10^5$ yr.

Our derived nitrogen supports the wind-blown shell model, since if the bubble nebula were due to an explosive stellar event, an overabundance of processed matter should be expected.

We conclude that the molecular cloud and the dust detected at the knots and around BD $+60^\circ 2522$ is not physically associated with the bubble itself but is (together with BD $+60^\circ 2522$) part of the large star-forming region S 162: The existence of the molecular cloud and the dust ($M_{dust} \approx$ 0.2 M_O) close to BD $+60^\circ 2522$ within the bubble is highly improbable due to (i) the destructive stellar UV radiation and (ii) the gas-to-dust ratio which would be of the order ~ 0.13.

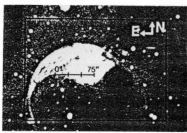

Fig. 1: The H II region S 162 with the bubble nebula NGC 7635. CCD orientations and slit positions are shown.

THE UNSTABLE O6.5f?p STAR HD 148937 AND ITS INTERSTELLAR ENVIRONMENT

Claus Leitherer and Carlos Chavarría-K.
Landessternwarte Königstuhl
D-6900 Heidelberg
Germany

The massive, early-type star HD 148937 (spectral type O6.5f?p) is surrounded by a unique set of nebulosities. A spherically symmetric Strömgren sphere (Radius $\simeq$ 25 pc), an ellipsoidal filamentary nebulosity (semimajor axis $\simeq$ 5 pc) interpreted as a stellar-wind-blown shell, and a bipolar nebular complex (semimajor axis $\simeq$ 1 pc) with HD 148937 located in the apparent center of symmetry. Here we report on observations of these nebulosities (narrow-band CCD imaging, IDS spectrophotometry, high resolution ($\Delta V \simeq$ 7 km sec^{-1}) spectroscopy) in an attempt to establish a consistent model of HD 148937 and its nebulosities.

The two bipolar nebulosities (known as NGC 6164/5) show striking resemblance. They are of similar extent and distance from HD 148937 implying a simultaneous origin due to an explosive event in HD 148937. A comparison of Hα- and Hβ-images and IDS fluxes gives a rather homogeneous dust distribution all over the bipolar nebula with an extinction in agreement with E(B-V) of HD 148937. We conclude that the apparent morphology of NGC 6164/5 is not simulated by variable extinction but rather reflects the actual distribution of (ionized) gas. Moreover, it is safe to assume that the distribution of Hα emission is identical with the total amount of gas in the inner nebulosities (i. e. NGC 6164/5 is density-bounded). The output of Lyman quanta of HD 148937 (log $N_L \simeq$ 49.15 sec^{-1}) even suffices to keep ionized the stellar-wind-blown shell and the Strömgren sphere. The ratio [O III]/Hβ drastically decreases with increasing distance from HD 148937. This behavior is paralleled by a corresponding increase of [N II]/Hα indicating the transition from the high-ionization zone to the low-ionization zone. We stress the similarity to the structure of ejecta from novae (see Gallagher and Anderson 1976).

In establishing a geometrical model for the bipolar nebula we are led by its striking resemblance to ejecta from novae or symbiotic stars (e. g. Solf 1983). The high

447

I. Appenzeller and C. Jordan (eds.), Circumstellar Matter, 447–448.
© 1987 by the IAU.

rotational velocity of HD 148937 (v · sin i = 200 km sec^{-1},
Conti and Ebbets 1977) strongly implies axial symmetry for
the system and thus giving rise to the ejection of matter
preferentially along the axis of symmetry. The detailed kinematic structure is derived from our highly-resolved line
profiles. We find a double-cone structure nearly perpendicular to the line of sight with a kinematic age of $10^3 - 10^4$
yr. The ejection of the nebulosities may be due to instabilities in this star close to the Of-WR transition phase. The
high rotational velocity may play a crucial role in these
instabilities.

The influence of rotation on the stellar wind of HD
148937 and the resulting non-isotropic mass flow could also
account for the oval-shaped structure of the wind-blown
shell. We underline that the spectral appearance of this
shell is typical of wind-blown shells from e. g. WR-stars
and clearly excludes an origin as due to a SN explosion. The
age of the bubble as determined from the luminosity of the
wind is a few times 10^5 yr.

Since NGC 6164/5 is clearly ejected by the star and expands into space with virtually no contamination by ISM (the
density in the stellar-wind cavity is $\sim 10^{-2}$ cm^{-3}) it provides the unique possibility of studying the chemical abundances of an evolved Of star. Our physical analysis gives
an electron density of $\sim 10^4$ cm^{-3}. The electron temperature
is 6700 K. In these respects the nebulosities closely resemble normal H II regions. We find no evidence for shock
excitation in the lines. The abundance analysis is treated
in the usual way following Peimbert and Costero (1969). We
find an overabundance of nitrogen relative to hydrogen by a
factor of 6 relative to the sun for NGC 6164/5. On the other
hand, the outer border of the Strömgren sphere (NGC 6188)
shows abundances typical of normal H II regions (cf. Orion).
An overabundance of N by a factor of 6 is in good agreement
with theoretical evolutionary modelling of stellar abundances by Maeder (1983). This provides strong support for the
evolutionary state of Of stars as being intermediate between O- and WR-stars.

References:
Conti, P. S., Ebbets, D.: 1977, Astrophys. J. 213, 438
Gallagher, J. S., Anderson, C. M.: 1976, Astrophys. J. 203,
    625
Maeder, A.: 1983, Astron. Astrophys. 120, 113
Peimbert, M., Costero, R.: 1969, Bol. Obs. Ton. Tac. 5, 3
Solf, J.: 1983, Astrophys. J. Lett. 266, L 113

Ionization fractions and mass-loss in O stars

Raman K. Prinja & Ian D. Howarth
Dept. of Physics and Astronomy, University College London

Introduction — The most sensitive indicators of mass-loss for stars in the upper left part of the HR diagram are the UV P Cygni profiles observed in the resonance lines of common ions such as N V, Si IV, and C IV. We present here some results from a study of these lines in the high resolution *IUE* spectra of 197 O stars. Profile fits were carried out in the manner described by Prinja & Howarth (1986) for all unsaturated P Cygni resonance doublets. The parameterisations adopted enable the product of mass-loss rate $(\dot{M})$ and ion fraction (q_i) to be determined at a given velocity, such that $\dot{M} \, q_i \propto N_i \, R_* \, v_\infty$, where N_i is the column density of the observed ion i, v_∞ is the terminal velocity, and R_* is the stellar radius. The accompanying figures illustrate the behaviour of $\dot{M} \, q_i$ (evaluated at 0.5 v_∞) for N V and C IV.

Figures 1 and 2 — The product $\dot{M} \, q_i$ is plotted as a function of luminosity for N V and C IV in Fig. 1 and 2 respectively. Almost the same linear relation of the form $\mathrm{Log} \, (\dot{M} \, q_i) \propto \mathrm{Log} \, ((L_*/L_\odot)^{1.6})$ is observed for both ions. This dependence on luminosity of $\dot{M} \, q_i$ is in turn almost exactly the same as that found between $\dot{M}$ and L_* from radio observations of thermal OB stellar sources (which give an almost model independent estimate of $\dot{M}$; see *e.g.* Abbott *et al.*, 1984). These figures suggest, therefore, that the ionization fractions of N V and C IV are, in a statistical sense, constant as a function of luminosity.

Figures 3 and 4 — The residuals from a linear fit of $\mathrm{Log}(\dot{M} \, q_i)$ *vs.* $\mathrm{Log}(L_*)$ are plotted against effective temperature in Figures 3 and 4, for N V and C IV respectively. If we assume that the scatter observed in Figures 1 and 2 is primarily due to differences in the ionization fractions, then surprisingly, the ion fractions are not a simple function of T_{eff}.

This result — and the conspicuous luminosity class dependence of the Si IV doublet — reveals the inadequacy of current models to accurately predict the ion fractions. The determination of mass-loss rates from UV resonance lines (which do not normally represent the dominant stages of ionization) is therefore severely restricted by the uncertainty in the ionization fractions.

449

I. Appenzeller and C. Jordan (eds.), Circumstellar Matter, 449–450.

Abbott, D. C., Telesco, C. M., and Wolff, S. C. 1984, *Ap. J.*, **279**, 225.

Prinja, R. K., and Howarth, I. D. 1986, *Ap. J. Suppl.*, **61**, 357.

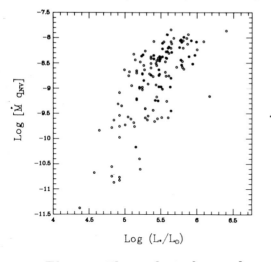

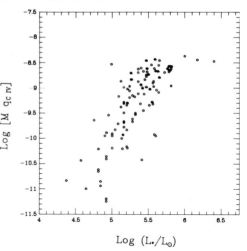

Fig. 1 — *The product of mass-loss rate ($\dot{M}$) and N V ionization fraction (q_{NV}) versus luminosity. $\dot{M}$ is in $M_\odot \, yr^{-1}$.*

Fig. 2 — *The product of mass-loss rate and C IV ionization fraction versus luminosity.*

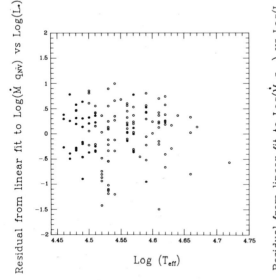

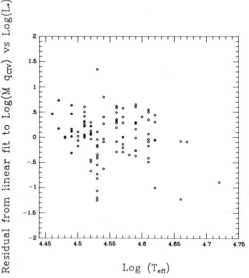

Fig. 3 — *Residuals from the linear fit to Log ($\dot{M} \, q_{NV}$) vs. Log (L_*) (Fig. 1) plotted against effective temperature.*

Fig. 4 — *Same as Fig. 3, except for C IV. In neither case is the dispersion about a straight-line fit substantially reduced.*

DYNAMICAL EFFECTS OF STELLAR WINDS AND ASSOCIATED HII REGIONS ON THE INTERSTELLAR MEDIUM

Dieter Breitschwerdt
Max-Planck-Institut fuer Kernphysik
P.O. Box 10 39 80
D–6900 Heidelberg
West Germany

ABSTRACT. A dynamical model has been developed, describing the joint evolution of a stellar wind and an associated HII region during the main sequence life time of an OB-star. It is proposed that in case of spherical symmetry all dynamical features are obtained by dividing the flow into spatially homogenous regions separated by discontinuities (shock fronts, contact discontinuity and ionization front). Different phases in the dynamical evolution can be distinguished. The results show that the extension and structure of the HII region depends sensitively on the initial density of the ambient medium, the Lyman continuum photon rate and the mechanical wind power of the central star. This will be important for future high resolution observations.

1. INTRODUCTION

The state of the Interstellar Gas is strongly determined by energy input of early type stars in the form of a) radiation, b) stellar wind (SW) and c) supernovae (not discussed here). Already a decade ago Copernicus observations have shown that a SW exists for all stars of spectral type B5 and earlier [1]. On the other hand these stars also produce sizeable HII regions. Furthermore the conversion of the stellar energy output by a) and b) into *kinetic energy* of the Interstellar Gas is comparable for both processes [2]. Consequently, SW and HII regions are **associated** phenomena which must be adequately taken into account in a dynamical description. However, most elaborate theoretical models so far have been dealing either with pure photoionized HII regions or SW bubbles alone, each subject to a variety of different boundary conditions. The objective of this work has been to analyze how basic physical processes modify the combined dynamics.

2. MODEL

Consider a single star of spectral type OB, emitting an isotropic steady wind and Lyman continuum photon flux into a homogenous and extended surrounding medium. The flow is then described in spherical geometry, "switching on" the SW and radiation field simultaneously.

It turns out that the flow can be divided into spatially homogenous regions (fig. 1) connected by conservation of mass, momentum and energy across the discontinuities (details s. [3]). In region (1), SW gas is moving with uniform velocity V_w, most of its kinetic energy being transformed into heat thus leading to an **isobaric** and **adiabatic** hot "bubble" (region (2)). It is in pressure equilibrium with the dense, thin and **isothermal** shell of shocked HII region (region (3)). When the unshocked HII region (region (4)) expands gasdynamically it creates a strong shock giving rise to a dense shell

451

I. Appenzeller and C. Jordan (eds.), Circumstellar Matter, 451–452.

of neutral gas (region (5)). We start numerical integration using similarity solutions for the SW [4][5] and follow the time-dependent evolution of the dynamical variables.

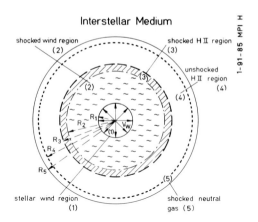

Figure 1. Model for the gas flow in the vicinity of a central OB-star; the various regions and their outer boundaries are: unshocked SW (1) and terminal shock (R_1), shocked SW (2) and contact discontinuity(R_2), shocked HII region (3) and outer wind shock (R_3), unshocked HII region (4) and ionization front (IF) (R_4), shocked HI gas (5) and IF-precursor shock (R_5).

3. RESULTS AND DISCUSSION

Model calculations have been performed for an O9 star (mass loss rate $\dot{M}_w = 10^{-7} M_\odot$/yr, wind velocity $V_w = 2000$km/s and Lyc photon rate $S_* = 2.1 \cdot 10^{48}$s^{-1}) in media with number densities 10 and 10^4cm^{-3}. Three phases of time evolution can be distinguished. In the early phase the ionization front(IF) and SW regions evolve independently (very unlikekly to observe). In the intermediate phase they are coupled by the gasdynamical expansion of the unshocked HII region and the shock (R_2) behind the IF can catch up with it (interactions). The late phase strongly depends on the ambient density. For low densities (10cm^{-3}) the evolution of the HII region is determined by its gasdynamical expansion leading to extended HII regions whereas for high densities (10^4cm^{-3}) the SW dominates the dynamics and large bubbles with "skinlike" HII regions form. The "thickness" and the dimensions of the HII region are also strongly correlated with the ratio of wind power ($L_W = (1/2)\dot{M}_w V_w^2$) to photon rate. For high values of L_W/S_* we get thin ionized regions and huge bubbles and for low values rather thick and extended HII regions. The model is able to reproduce these time-dependent results quantitatively for given initial parameters of the star and its ambient medium. It is strongly suggested that VLA observations should be carried out in the future to reveal the detailed structure predicted here. Finally, the dimensions and dynamics of supernova remnants would also be affected by that.

References.

[1] Snow, T.P. and Morton, D.C., 1976 *Ap. J. Suppl.* **32,** 429
[2] Dyson, J.E., 1981 *Investigating the Universe*, F.D. Kahn (ed.), D. Reidel Publ, Comp.
[3] Breitschwerdt, D., 1985 *Dissertation*, Univ. Heidelberg
[4] Dyson, J.E. and deVries, J., 1972 *Astr. Astrophys.* **20,** 223
[5] Weaver, R., McCray, R., Castor, J., Shapiro, P., Moore, R., 1977 *Ap. J.* **218,** 377

EPISODIC DUST FORMATION IN THE WIND OF HD 193793

P.M. Williams,[1] K.A. van der Hucht,[2] D.R. Florkowski,[3]
A.M.T. Pollock,[4,5] & W.M.Wamsteker [6]

1. Royal Observatory, Blackford Hill, Edinburgh, Scotland.
2. SRON Laboratory for Space Research, Beneluxlaan 21, Utrecht,
   The Netherlands.
3. U.S. Naval Observatory, Washington, D.C. 20309, U.S.A.
4. Dept Space Reasearch, University of Birmingham, Birmingham,
   B15 2TT, England.
5. ESA-EXOSAT Observatory, ESOC, Robert Bosch Strasse 5,
   6100 Darmstadt, F.R.G.
6. ESA-IUE Tracking Station, Villafranca del Castillo, Madrid, Spain

ABSTRACT.   In 1985 April, the WC7+abs star HD 193793 was observed, using
UKIRT, to have brightened significantly in the infrared owing to the formation
of a new dust shell. Examination of infrared photometry of this star since 1979
and previously published data indicates that the dust formation occurs at intervals
of 7.9 years.   Phasing the published radial velocities of the absorption line
component with this period confirms that it is a member of an eccentric (e =
0.7-0.8) binary system having periastron passage shortly before dust formation.
The X-ray spectrum also changed between 1984 and 1985 in becoming
significantly "harder" while the non-thermal radio source disappeared, both changes
indicating greater extinction. This suggests a model wherein the source of the
non-thermal radio and X-ray emission moves deep into the Wolf-Rayet wind.

1.   INFRARED, RADIO AND X-RAY VARIATIONS

During 1985, the infrared emission from HD 193793 (= WR 140) was observed to
rise by over 2.5 mag. at 3.8 μm over an interval of about ten weeks, reaching a
maximum in 1985.4. This is attributed to the formation of 5 x 10^{-9} $M_\odot$ of
amorphous carbon grains. Comparison with the 1977 dust formation event studied
by Williams et al. (1978)   and Hackwell, Gehrz & Grasdalen (1979) confirmed
that grain formation is a recurrent phenomenon and indicates a "period" of
7.9±0.1 years.   In 1975.8, Florkowski & Gottesman (1977) observed HD 193793 to
be a strong radio source having a spectrum   quite different from that observed
from γ Vel or expected from free-free radiation by a steady stellar wind. During
and after the 1977 dust event, the radio souce had faded significantly (Florkowski
1982). In 1984.4, Pollock (1985) observed a strong X-ray source associated with
HD 193793,   confirming   the presence of relativistic electrons in its wind. We
re-observed HD 193793 with EXOSAT in 1985.5 and 1985.8 and found that the
X-ray source was significantly "harder" than in 1984.4 (Williams et al. 1987).

I. Appenzeller and C. Jordan (eds.), Circumstellar Matter, 453–454.

2. A COLLIDING WIND MODEL

Although HD 193793 has a composite spectrum, WC7+O5, its status as a physical binary has been controversial (Lamontagne, Moffat & Seggewiss 1984 and Conti *et al.* 1984). Re-examination of the 60 years of radial velocities presented by these authors using the 7.9 year period determined from the infrared maxima indicated that the system *is* a binary of high eccentricity having periastron passage shortly before infrared maximum (Williams *et al.* 1987). Other WC+O binaries, including γVel, show X-ray emission probably originating in standing shocks where the WC and O star winds collide. Because the mass-loss rate of the WC stars are 1-2 orders of magnitude greater than those of the O stars, the standing shocks are much closer to the latter than to the WC components. In a system with a highly eccentric orbit, like HD 193793, the shocked region will move in and out of the dense Wolf-Rayet wind along with the O star. Near periastron, the intrinsic X-ray emission will be greatest; but the low energy flux will suffer more absorption in the wind, resulting in the "harder" spectrum we observed with EXOSAT. The same region must be responsible for the non-thermal radio emission. The extinction of this from HD 193793 near infrared maximum (and periastron passage) and the fact that non-thermal radio emission is not observed from systems like γ Vel are a consequence of free-free absorption in the dense Wolf-Rayet winds. The radius of the 5GHz radio "photosphere" (Wright & Barlow 1975) around HD 193793 or γ Vel is about 5×10^{14} cm. This is 30 times the semi-major axis of γ Vel's orbit so we would never expect to see non-thermal emission from this system at 5 GHz. For reasonable masses, the semi major axis of WR 140's orbit is 2.6×10^{14} cm so that the separation of the WC and O components varies between 4.6×10^{14} cm and one-seventh this value. Apart from times around periastron, non-thermal radio emission will escape from the wind. The phases at which we can observe it depend on the orientation of the orbit. The dust formation itself must the consequence of the compression of part of the stellar wind by the standing shock, greatest at periastron, and the advection of this material by the stellar wind to a region sufficiently far from the star that it can make dust. The infrared data indicate that dust formation first occurred at a radius of 2.5×10^{15} cm. Given a wind velocity of 3000 km/s, this implies that the critical compression occurred ≈3 months before dust formation began.

REFERENCES

Conti, P.S., Roussel-Duprè, D., Massey, P. & Rensing, M., 1984. *Astrophys.J.*, 282, 693.

Florkowski, D.R., 1982. In: *Wolf-Rayet Stars: Observations, Physics, Evolution, (I.A.U. Symp. 99)* eds: C.W.H. de Loore & A.J. Willis, p. 63.

Florkowski, D.R. & Gottesman, S.T., 1977. *M.N.R.A.S.*, 179, 105.

Hackwell, J.A., Gehrz, R.D. & Grasdalen, G.L., 1979. *Astrophys.J.*, 234, 133.

Lamontagne, R., Moffat, A.F.J. & Seggewiss, W., 1984. *Astrophys.J.*, 272, 258.

Pollock, A.M.T., 1985. *Space Science Reviews*, 40, 63.

Williams, P.M., Beattie, D.H., Lee, T.J., Stewart, J.M. & Antonopoulou, E., 1978. *M.N.R.A.S.*, 185, 467.

Williams, P.M., van der Hucht, K.A., van der Woerd, H., Wamsteker, W.M., Geballe, T.R., Garmany, C.D. & Pollock, A.M.T., 1987. In: *Instabilities in Luminous Early-Type Stars*, eds: H. Lamers & C. de Loore, (Reidel, in press)

Wright, A.E. & Barlow, M.J., 1975. *M.N.R.A.S.*, 170, 41.

DUST FORMATION IN, AND THE STRUCTURE OF WOLF-RAYET STELLAR WINDS

K.A. van der Hucht, SRON Space Research Utrecht
P.M. Williams, Royal Observatory Edinburgh
P.S. Thé, Astronomical Institute "Anton Pannekoek"

ABSTRACT. An infrared photometric survey of all 40 known galactic WC7-10 stars shows that around most of them hot amorphous carbon is condensing continuously at distances of about 8500 $R_\odot$, well within the ionized stellar winds around these stars. Typical dust production rates are of the order of 10^{-7} $M_\odot/yr$.

1. INTRODUCTION

Evolved massive stars in their Wolf-Rayet phase are a particularly appropriate subject for this Symposium on Circumstellar Matter, because this is all we can see of the WR stars! UV and optical continua arise in the lower regions of the dense steller winds, IR and radio continua are formed in the outer regions, while X-rays are apparently observed where winds of binaries collide.

IR photometric observations of WR stars since the early 1970's have shown two kinds of IR excesses: free-free radiation caused by dense stellar winds ($\dot{M}_{WN} \approx 1$-12×10^{-5} $M_\odot/yr$, $\dot{M}_{WC} \approx 2.5$-15×10^{-5} $M_\odot/yr$, van der Hucht et al., 1986) and thermal emission by hot ($T \approx 1000$ K) circumstellar dust. This dust radiation is observed from the latest subtypes of the WC sequence only and known for some cases since the work of Allen et al. (1972). In order to study origin, composition and mass of this hot circumstellar dust around WR stars, we carried out an IR ($JHKLMN_1N_2N_3Q_0$) photometric survey at ESO and UKIRT since 1982 of all the 40 known galactic WC7-10 stars (van der Hucht et al., 1981, and updates), supplemented with IRAS data.

2. WC STAR DUST

We find hot dust around 5 of the 10 known WC8 stars, around 15 of the 17 known WC9 stars and around the one and only known WC10 star. In addition, episodic presence of hot dust is found around 2 of the 12 known WC7 stars: the WC7+a star WR137 = HD192641 (Williams et al., 1985) and the WC7+O4 system WR140 = HD193793, recently recognized as a spectroscopic binary (P = 7.9 yr, Williams et al., 1987b; this Symposium). Typical energy distributions can be found in van der Hucht et al. (1985). Heated dust emission from two very late WN stars was

I. Appenzeller and C. Jordan (eds.), Circumstellar Matter, 455–456.

reported by van der Hucht et al. (1984): the WN10 star WR122 and the suspected WN11 star LSS4005. High resolution spectroscopy of the latter shows that it is better classified as Ofpe/WN9 or B[e] (van der Hucht et al., 1987).

The hot dust shells around WC stars can be characterized by the following aspects: (a) the IR energy distributions are featureless, except for the interstellar 9.7μ absorption feature, for which we find a relation $A_V/\tau_{9.7} = 19.8\pm1.7$; (b) the dust shells are optically thin in most cases; and (c) the dust is being formed continuously at a fixed distance from the star *within the ionized stellar wind*. The fact that the IR energy distributions are featureless rules out dielectric grains, and the $1/\lambda$ emissivity found for the episodic dust producer WR140 (Williams et al., 1987) points to amorphous carbon. Table 1 gives some typical WC dust shell parameters. The dust production by the 10 late WC stars within 3 kpc from the Sun amounts to 6×10^{-7} $M_\odot$/yr, or, projected on the galactic plane, 1.4×10^{18} $g/kpc^2.s$.

Van der Hucht et al. (1986) argue that WC7-10 stars have mass loss rates 2 to 3 times greater than the WC4-6 stars and that, because of their lower terminal wind velocities, WC7-10 stars consequently have 4 to 7 times larger stellar wind densities. The late WC subtypes apparently provide the proper circumstances for dust to form: at those radii

Table 1. WC dust shell parameters

average values	WC8 (5)	WC9 (15)	WC10 (1)
L_{IR}/L_*	.02	.10	.59
T (K)	1550	1350	1330
R ($R_\odot$)	7300	8900	6600
ρ (g/cm^3)	5.8E-21	3.4E-20	5.7E-18
ρ/ρ_{gas}	0.0008	0.008	
M ($M_\odot$)	3.1E-8	2.9E-6	1.0E-5
$\dot{M}$ ($M_\odot$/yr)	5.4E-8	3.7E-7	2.5E-5

where the radiation temperature for the dust grains is about 1000 to 1500 K, the gas density in the ionized carbon-rich (25 % by number, Prantzos et al., 1986) WC stellar wind is sufficiently high to allow dust formation. WC stars hotter than WC7 apparently have winds of insufficient density at radii where dust could survive in the stellar radiation fields. This gas density limit above which dust can form is 4.3×10^{-18} g/cm^3, i.e. $n = 3.4\times10^5$ cm^{-3}.

Particulars of this study are published in Astronomy and Astrophysics (Williams et al., 1987a).

REFERENCES

Allen, D., Swings, J.P., Harvey, P.M.: 1972, Astron. Astrophys. **20**, 333
van der Hucht, K.A., Conti, P.S., Lundström, I., Stenholm, B.: 1981, Space Science Reviews **28**, 227
van der Hucht, K.A., Williams, P.M., Thé, P.S.: 1984: in: A. Maeder & A. Renzini (eds), *Observa-tional Tests of the Stellar Evolution Theory*, Proc. IAU Symp. **105** (Dordrecht: Reidel), p. 273
van der Hucht, K.A., Jurriens, T.A., Olnon, F.M., Thé, P.S., Wesselius, P.R., Williams, P.M.: 1985, in: W. Boland & H. van Woerden (eds), *Birth and Evolution of Massive Stars and Stellar Groups*, Proc. of a Colloquium in honour of Adriaan Blaauw (Dordrecht: Reidel), p. 167
van der Hucht, K.A., Cassinelli, J.P., Williams, P.M.: 1986, Astron. Astrophys, in press
van der Hucht, K.A., Williams, P.M., de Loore, C.W.H., Mulder, P.M.: 1987, Astron. Astrophys., in preparation
Prantzos, N., Doom, C., Arnould, M., de Loore, C.: 1986, Astrophys. J. **304**, 695
Williams, P.M., Longmore, A.J., van der Hucht, K.A., Wamsteker, W.M., Talavera, A., Abbott, D.C., Telesco, C.M.: 1985, Monthly Notices Roy. Astron. Soc. **215**, 23P
Williams, P.M., van der Hucht, K.A., Thé, P.S.: 1987a, Astron. Astrophys., submitted
Williams, P.M., van der Hucht, K.A., van der Woerd, H., Wamsteker, W.M., Geballe, T.R., Garmany, C.D., Pollock, A.M.T.: 1987b, in H. Lamers & C. de Loore (eds) *Instabilities in Luminous Early Type Stars*, Proc. of a Workshop in honour of Cornelis de Jager (Dordrecht: Reidel), in press.

WOLF-RAYET NEBULAE - ENRICHMENT IN He AND N AND EFFECTIVE TEMPERATURES OF WOLF-RAYET STARS *)

Michael R. Rosa[1]
The Space Telescope-European Coordinating Facility
European Southern Observatory, Karl-Schwarzschild-Str. 2
D-8046 Garching, Federal Republic of Germany
[1] Affiliated to the Astrophysics Division, Space Science
Department, European Space Agency

SUMMARY. Nebulae surrounding isolated Pop I WR stars provide observational constraints on as yet poorly determined chemical surface abundances and FUV energy distributions of their central stars. An integral of the mass loss history and the chemical evolution is stored in those parts of the nebulae that have suffered only very little mixing with the ISM (cf. Kwitter 1984). Effective temperatures in the Lyman continuum region are reflected in the ionization structure of oxygen and sulfur (cf. Mathis 1982). For intrinsic problems involved refer to the papers cited above.

Long slit spectrophotometry of the nebulae RCW 58, RCW 104 and MR 26 (cf. Chu et al. 1983) was obtained with a B & Ch. Cassegrain spectrograph and CCDs at the 2.2m telescope at ESO, La Silla. A total of 40 hours was spent on 6 positions in the 3 nebulae. The present data represent averages over 1.5 arcmin slits, reduced in a standard way.

The abundances in table 1 are based on N and S electron temperatures and on ionization correction factors determined from (O,S) diagrams calculated by Mathis (1982, 1985). All 3 nebulae are of low ionization, making a correction for unseen neutral helium difficult. Total He abun-

Object	RCW 58	RCW 104	MR 26	Orion
T(e;N,S)	8000	7500	8600	7800
n(e)	190	70	40	–
log(O/H)	8.55	8.52	8.52	8.54
log(He/H)	11.48	11.32	11.15	11.02
log(N/O)	-0.14	-0.30	-0.77	-1.15
log(S/O)	-1.5	-1.2	-1.6	-1.5
log(Ne/O)	-1.0		-1.2	-0.8
log(Ar/O)	-1.8	-1.7	-1.9	-1.9
log(Cl/O)	-3.4	-3.6	-3.6	-3.3

*) Based on observations collected at the European Southern Observatory, La Silla, Chile

I. Appenzeller and C. Jordan (eds.), Circumstellar Matter, 457–458.

dances quoted are likely lower limits. He and N are overabundant (relative to the Orion nebula values) by factors between 1 and 10. The heavy
elements O,Ne,S,Ar,Cl are normal within the uncertainties. These results
pertain to only a minute fraction of the nebulae and may not be representative for the global values, albeit significant differences between
the various slit positions have not been found.

The most important new result is visualized in Figure 1. The ionic
fractions (O+/O) and log(S+/S++) of the 3 nebulae are compared with the
expected values for different line of sights in model nebulae, characterized by (T(eff), Kurucz or Mihalas), from Mathis (1982) and with observations of Orion and 30 Dor.

The WN 4 star in RCW 104 ($\approx$ 40 000 K) is intrinsically hotter than
the WN 6 star in RCW 58 ($\approx$ 30 000 K). MR 26 occupies the degenerate part
of the diagram were widely different atmospheres produce the similar low
ionization values. Since MR 26 is a thin shell, other positions along its
periphery will likely yield similar results. However, unless the conditions in this shell are exceptional an upper limit of 40 000 K may be appropriate. Very high values of T(eff), above say 50 000 K are not compatible with the present observations. It is particularly rewarding, that
the NLTE models by Schmutz Hamann and Wessolowski presented at this conference indicate the same temperature domain.

REFERENCES

Chu,Y.-H.,Treffers,C.,Kwitter,K.B.: 1983, Astrophys.J.Supp., 53, 937
Kwitter,K.B.: 1984, Astrophys.J., 287, 840
Mathis,J.S.: 1982, Astrophys.J., 261, 195
Mathis,J.S.: 1985, Astrophys.J., 291, 247

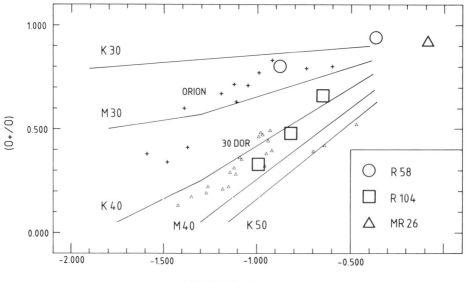

THE BEHAVIOUR OF λ4686 LINE OF He II IN WOLF-RAYET BINARIES

B. S. Shylaja
Indian Institute of Astrophysics
Koramangala, Bangalore 560034, India

SUMMARY   Being one of the brightest lines in emission in the optical region of Wolf-Rayet (WR) stars, the   λ4686 line of He II has attracted the attention of many investigators (Sahade, 1958; Ganesh and Bappu, 1967). Here, the binaries with subgroup WN components have been chosen, with periods ranging from 1.64d to 21.63d, for understanding the effect of the companion on the formation of this line.
         Spectrophotometric data used for the total flux measures in this line in two short period systems CQ Cep and HD 50896, clearly show the enhancements of fluxes at phases corresponding to the eclipse of the components. Such an increase in flux may be infered from differences in line profiles in many other binaries as well. Sudden sporadic brightening, general positive shift of the γ value in the radial velocity (RV) curves (with reference to the N IV line of λ 4058, chosen to represent the true motion of the WN component) and eccentric orbital solutions from the RV curves are other peculiar features of this line.   Another interesting point that results from a comparison of various RV curves is the amplitude K, which is smaller than that of N IV line at λ 4058 in short period systems and appears to be larger in long period systems.   Systems with periods around 4d show little difference in K between these two RV curves. From the line profile studies of the eclipsing system V444 Cyg (Sahade, 1958), it was possible to infer the concentration of line emitting material towards the inner Lagrangian point.   This may be taken to mean that the line emitting region is extended.   The red shift of the γ velocity has been explained by electron scattering processes (Auer & van Blerkom, 1972).   The amplitude difference may be due to the contribution from the companion, or due to the line emitting material distributed nearer to the companion, or due to the multiple structure of the line profile. The likelihood of line emitting material nearer to the companion is possible in a wind dominent Roche surface as has been calculated by Niemela (c.f. Sahade & Wood, 1980), which can explain the increase in flux at eclipses as well (Shylaja, 1986).   The same effect can explain the decrease of eclipse depth in case of V444 Cyg, dip in the λ 4680 light curve of HD 90657 and the asymmetric distribution in HD 5980.
         Thus it appears that the peculiarity is mainly due to the large

I. Appenzeller and C. Jordan (eds.), Circumstellar Matter, 459–460.

extended line emitting region, multiple component line profile and the distortion caused by strong stellar winds.   Since the proximity of the companion is responsible for the distortion of the line emitting region, which in turn may result in the other two factors mentioned, it may lead to the inequality K(4686) < K(4058) for short period binaries only.

A detailed investigation will be published elsewhere.

References

Auer, L.H., van Blerkom, D., 1972, Astrophys.J., 178, 175.
Ganesh, K.S. Bappu, M.K.V., 1967, Kodaikanal Obs.Bull., Ser A, 185, 104.
Sahade, J., 1958, Mem.Soc.Roy.Sci., Liege, Ser 4e, 20, p.46.
Sahade, J., Wood, F.B., 1980, in Interacting Binary Stars, Pergamon Press,p.101.
Shylaja, B.S., 1986, J.Astrophys.Astron (in press).

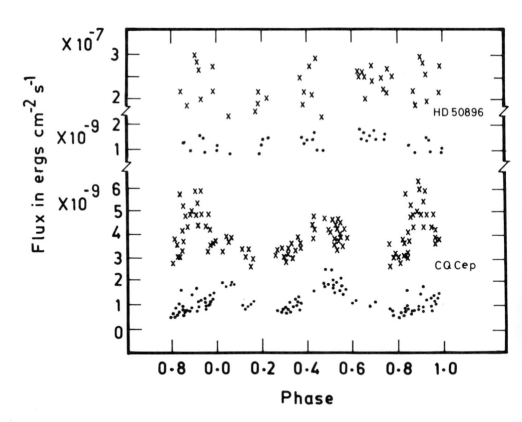

Figure 1. Variation of the fluxes of   λ 4686 (crosses) and
           λ 4058 (dots) with orbital phase in case of two
           binaries C Q Cep and H D 50896.

UPPER LIMITS FOR THE EFFECTIVE TEMPERATURE OF WOLF-RAYET STARS FROM THE PRESENCE OF HE I

W. Schmutz, W.-R. Hamann, U. Wessolowski
Institut für theoretische Physik
und Sternwarte der Universität
Olshausenstrasse, D-2300 Kiel,
Federal Republic of Germany

ABSTRACT. A recently developed non-LTE code for realistic semi-empirical models of Wolf-Rayet atmospheres is used to calculate synthetic helium lines. From the resulting line strenghts it can be concluded that if He I lines are present, the effective temperatures of these stars have to be less than an upper limit. This limit depends on the stellar radius and is approximately 40kK for $R_* = 20 \, R_\odot$ to 60kK for $R_* = 5 \, R_\odot$.

1. Introduction

The basic stellar parameters of the Wolf-Rayet stars are still essentially unknown though much effort has been invested in this subject. In particular, the published values of the effective temperature strongly disagree: for the subclass WN5, e.g., the temperatures range from 29kK (Underhill, 1983) over 41kK (Nussbaumer et al., 1982) to 90kK (Cherepashchuk et al., 1984).

Our new approach to the temperature problem is based on semi-empirical model calculations. It is demonstrated that the He I / He II line ratio may be used as a sensitive temperature indicator.

2. Non-LTE model calculations for WR atmospheres

The calculations are performed as described by Hamann (1985, 1986) and Hamann and Schmutz (1986). In order to avoid high electron temperatures in the envelope the temperature law was forced to approach to 8kK at great distances from the star, in contrast to the temperature structure adopted by Hamann and Schmutz (1986).

In Fig. 1 line profiles of He I $\lambda 5876$ and He II $\lambda 4686$ are presented for several star temperatures, while the remaining parameters characterizing the model are kept fixed at $R_* = 20 \, R_\odot$, $M = 4 \, 10^{-5} \, M_\odot/\mathrm{yr}$, $V_{max} = 2100$ km/sec, $ß = 1$, and $\log(g_{eff}) = 3.5$. The line strengths as well as the profiles depend also on the other model parameters: the stellar radius, the mass-loss rate, the velocity law and the temperature stratification in the wind. Next to the effective temperature, the stellar radius has the largest influence on the line strength.

461

I. Appenzeller and C. Jordan (eds.), Circumstellar Matter, 461–462.

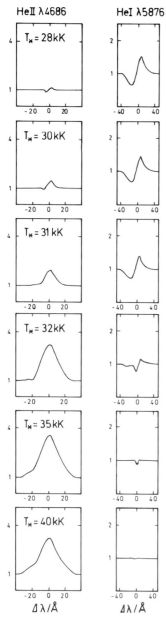

In the temperature sequence shown in Fig. 1, He I λ5876 would not be observable for temperatures higher than about 35kK. It is obvious that the presence of He I lines in a Wolf-Rayet spectrum gives an upper limit for the effective temperature of the star. Varying parameters other than temperature and radius may influence the line strengths up to a factor of two or three. Taking into account other parameter combinations, it can be stated that the He I lines vanish for temperatures higher than 40kK for $R_* = 20\ R_\odot$ to 60kK for $R_* = 5\ R_\odot$.

3. Conclusions

He I lines are observed in the spectra of the Wolf-Rayet subtypes WN5 or later and WC7 or later. From the upper temperature limit derived above these Wolf-Rayet stars are clearly below the temperature threshold, for which He^{++} recombines in the outer envelope (Schmutz and Hamann, 1986). It is therefore certain that these stars have He^+ and not He^{++} as the dominating ion in the radio-emitting region. Therefore, the mass-loss rates hitherto published have to be corrected as given by Schmutz and Hamann (1986). On the whole the mean mass-loss rates of the Wolf-Rayet stars are of the order of $4\ 10^{-5}\ M_\odot$/yr, rather than $2\ 10^{-5}\ M_\odot$/yr as given by Abbott et al. (1986).

References

Abbott,D.C., Bieging,J.H., Churchwell,E., Torres,A.V.: 1986, Astrophys. J. **303**, 239
Cherepashchuk,A.M., Eaton,J.A., Khaliullin, Kh.F.: 1984, Astrophys. J. **281**, 774
Hamann,W.-R.: 1985, Astron. Astrophys. **145**, 443
Hamann,W.-R.: 1986, Astron. Astrophys. **160**, 347
Hamann,W.-R., Schmutz,W.: 1986, in press
Nussbaumer,H., Schmutz,W., Smith,L.J., Willis, A.J.: 1982, Astron. Astrophys. Suppl. **47**, 257
Schmutz,W., Hamann,W.-R.: 1986, Astron. Astrophys. Letters, in press
Underhill,A.B.: 1983, Astrophys. J. **266**, 718

Figure 1. Synthetic line profiles of He I λ5876 and He II λ4686 calculated for different star temperatures.

Acknowledgements: W.S. and U.W. were supported by the Deutsche Forschungs-Gemeinschaft. W.S. thanks the Swiss Science Foundation for travel support and the IAU for the registration fee.

THE EFFECTS OF BOUNDARY CONDITIONS ON STELLAR EVOLUTION

Amos Harpaz[1], Attay Kovetz[2], Giora Shaviv[1]
1. Department of Physics and Space Research Institute
    The Technion, Haifa, Israel
2. Tel Aviv University, Ramat Aviv, Israel

ABSTRACT.   The effects of using different treatments of the surface boundary conditions are investigated in the context of the mass of He White Dwarfs. We find that since the White Dwarf progenitor is a star with a very extended atmosphere, the results are sensitive to the degree of accuracy implemented in the handling of the boundary conditions.

INTRODUCTION

The influence of mass-loss on the evolution of stars with extended atmospheres has become recognized in the past decade. Several authors investigated mass-loww effects during the RGB phase of the evolution and performed detailed stellar evolution computation.

Practically all mass-loss formulae are empiric and based on some fit to the observed data and expressed in terms which are combinations of L, R and M (the luminosity, radius and mass of the star). The expression given for this purpose by Reimers (1975) for example, is LR/M, which has the dimensions of $\dot{M}$.

The mass-loss rate is proportional to the photospheric radius. At this stage, the luminosity depends mainly on the core mass. The radius, on the other hand, depends on the structure of the very outer envelope. Small changes in the radius lead to small changes in the mass-loss rate. However, its effect is cumulative and the final result may be quite different.

We compare in this paper a simplified treatment with an elaborate and quite accurate method.

THE SURFACE BOUNDARY CONDITIONS

The fast changes in the thermodynamic quantities and the opacities near the stellar surface cause problems in stellar modeling because they demand the introduction of many thin mass shells, which in turn lead to prohibitively small time steps.

The simplest method to implement the boundary conditions is (Sweigart et al., 1974, Aizenman et al., 1969, and Demarque et al., 1968) to apply 'radiative surface conditions', namely the Eddington approximation

463

I. Appenzeller and C. Jordan (eds.), Circumstellar Matter, 463–464.

in one form or another and write:

$$L = 4\pi\sigma \; a \; R^2 \; T_b^4 \tag{1}$$

where a is some function of τ, the optical depth of the grey atmosphere and T_b is the temperature at the base of the atmosphere.

We compared this method (method 1), with a detailed integration of the grey stellar atmosphere equations proposed by Mihalas (1978). By this method (method 2) the boundary conditions are not imposed at $\tau=2/3$ (where their diffusion approximation breaks down), but at rather sufficiently inward mass. At this point the optical depth is much larger than unity. The range from $\tau=2/3$ down to the place where the boundary conditions is imposed is not included in the interior calculations, but integrated independently. The results of this integration are called here the boundary conditions for the evolution computation. The principles of the method are not new, however, the computational scheme, the detailed treatment of the equation of state and the opacities, and the mode of integration render a very flexible method that operates over a very wide range of stellar models and enveloped. A similar method was recently implemented by Van den Berg et al. (1983).

The results of the grey atmosphere calculations were compared with those obtained from model atmospheres and the agreement was excellent.

NUMERICAL RESULTS

Two sequences of evolutionary computation were carried out, for a star of initial mass of 0.7 $M_\odot$. One with the simplified radiative boundary conditions (method 1), where we used for a (in Eq. 1) a=2./(1+3τ/2), and the second with the detailed integration of the grey atmosphere (method 2). We find that the effective temperatures obtained by method 1 are systematically higher, and the model radii are respectively smaller than those obtained by method 2. In both methods Reimer's formula for mass loss rate was used. The larger radius produces a correspondingly larger mass-loss rate. While the gross feature of the evolution by both methods are essentially similar, the accumulated difference in mass-loss leads eventually to quite different WD mass, when the entire envelope is lost. The effective difference in mass-loss rate is about 10%, and this is also the order of the effect in the remnant mass.

ACKNOWLEDGEMENT: This research was supported in part by the Eppley Foundation of Research .

REFERENCES:
Aizenman, M., Demarque, P., Miller, R.H., 1969, Ap. J. 155, 973.
Demarque, P., Hartwick, F.D.A., Naylor, M.D.T., 1968, Ap. J. 154, 1143.
Ibcn, I., 1963, Ap. J. 138, 452.
Mihalas, D., 1978, in "Stellar Atmospheres", Freeman, San Francisco.
Sweigart, A.V., Mengel, J.G., Demarque, P., 1971, Ap. J. 164, 317.
Reimers, D., 1975, "Problems in Stellar Atmospheres and Envelopes",
    p.229, Ed. B. Baschek, W.H. Keggel, G. Tremaine, Springer, Berlin,
    Heidelberg, New York.
Weidman, V., Koester, D., 1983, Astran. Astrophys. 121, 77.
Van den Berg, D.A., Hartwick, F.D.A, Dawson, P., Alexander, D.R., 1983,
    Ap. J. 266, 747.

A PRELIMINARY MODEL FOR HR 8752

E. Zsoldos
Konkoly Observatory, Budapest
Hungary

HR 8752 is a G5 hypergiant in the Cep OB1 association. It is an SRd variable according to the GCVS. The V and B-V curves of the star are plotted in Figure 1. One point in the figure is the mean of all measurements made in a ten-day interval. The circles contain the observations of Arellano Ferro (1985) which show large deviation in two cases. Neglecting these circles, it is clear from Figure 1 that HR 8752 has alternating brighter and fainter maxima, the difference between them being some hundredth of magnitude. A possible explanation will be outlined for this phenomenon.

Willson and Bowen (1984) proposed a close link between pulsation and mass loss in post-main sequence stars. This connection was investigated further in the case of nonradial pulsation (Abbott et al., 1986). In the case of P Cygni Lamers et al. (1985) and Markova (1986) observed periodic shell ejection which is triggered by nonradial pulsation (van Gent and Lamers, 1986).

HR 8752 pulsates (Lambert et al., 1981) with a rather large amplitude ($A\sim0.2$ mag) which suggests radial pulsation. Shock waves, generated by pulsation, can enhance mass loss (Willson and Bowen, 1984). The proposed model for HR 8752 is the following. In every second cycle there is a sudden increase in mass loss arising from the above mechanism. It means, that a part of the circumstellar shell will have larger density and thus larger optical depth, which can cause the fainter maxima observed. This irregularity in the density structure of the shell may be smoothed out during the following minimum. There is, however, no spectroscopic evidence for this model. A long-term detailed spectroscopic study of HR 8752 would be highly desirable.

This scenario is similar to that of P Cygni. Van Gent and Lamers (1986) found a characteristic timescale of about 30 days for the pulsation of this star, while a shell ejection occurs approximately in every 200 days (Markova, 1986). There are, of course, differences between the two stars, but it is, however, an interesting similarity.

I wish to thank Drs. J. R. Percy and E. N. Walker for sending me their unpublished observations of HR 8752, and Drs. K. Olah and B. Szeidl for stimulating discussions.

I. Appenzeller and C. Jordan (eds.), Circumstellar Matter, 465–466.

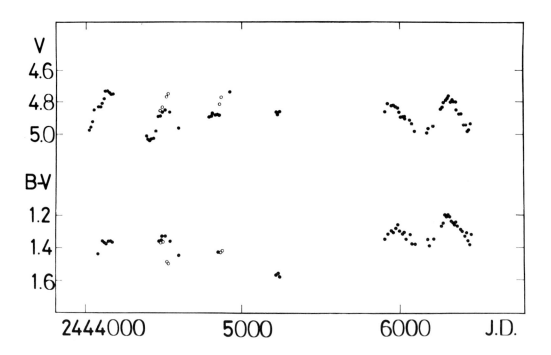

Fig. 1. V and B-V curves of HR 8752. Sources: Arellano Ferro, M.N. <u>216</u>,
         571; Halbedel, IBVS Nos. 2718, 2876; Parsons and Montemayor,
         Ap.J.Suppl. <u>49</u>,175; Percy, unpubl.; Percy and Welch, P.A.S.P.
         <u>93</u>,367; Walker, unpubl.; Zsoldos, unpubl.; Zsoldos and Olah,
         IBVS No. 2715.

REFERENCES

Abbott, D.C., Garmany, C.D., Hansen, C.J., Henrichs, H.F. and Pesnell,
     W.D. (eds.) 1986, P.A.S.P. <u>98</u>,29
Arellano Ferro, A. 1985, Mon. Not. R. astr. Soc. <u>216</u>,571
Lambert, D.L., Hinkle, K.H. and Hall, D.N.B. 1981, Astrophys. J. <u>248</u>,638
Lamers, H.J.G.L.M., Korevaar, P. and Cassatella, A. 1985, Astron.
     Astrophys. <u>149</u>,29
Markova, N. 1986, Astron. Astrophys. <u>162</u>,L3
Van Gent, R.H. and Lamers, H.J.G.L.M. 1986, Astron. Astrophys. <u>158</u>,335
Willson, L.A. and Bowen, G.H. 1984, Nature <u>312</u>,429

SYMBIOTIC STARS

THE R AQUARII JET

A. Cassatella[1], M. Kafatos[2], A.G. Michalitsianos[3],
L. Piro[4], R. Viotti[5]

1. IUE Observatory, European Space Agency, Madrid, Spain
2. George Mason University, Dept. Physics, Fairfax, USA
3. Laboratory for Astronomy and Solar Physics, NASA Goddard
   Space Flight Center, Greenbelt, USA
4. Istituto TESRE, CNR, Bologna, Italy
5. Istituto Astrofisica Spaziale, CNR, Frascati, Italy

ABSTRACT. The X-Ray (EXOSAT) and ultraviolet (IUE) observations of
R Aqr and its jet are discussed in the light of a proposed model.

Ultraviolet spectra in the 1200-3200 A wavelength range were obtained of
the symbiotic Mira R Aqr and associated jet-like feature over the course
of four years with IUE (Kafatos et al. 1986). The spatial extent of the
jet has enabled us to isolate it from the central compact HII region sur
rounding the unresolved binary, with the large 10x20" IUE aperture. The
appearance of HeII 1640 A, and particularly NV 1240 A in the jet indic
tes that excitation in this region has recetly increased, and is now
higher than in the central source. A modulation of the intensity of the
high ionization lines in the jet with a time scale of about 550 days
could be related to the Mira pulsation and the receding velocity of the
R Aqr components after the passage at the periastron (Figure 1).
In contrast, the UV spectrum of the central HII region remained almost
constant over four years, with NV and HeII very weak or absent.

A weak X-Ray flux have been detected with EXOSAT, in the low energy Thin
Lexan mode, in June 1985 at the Mira light maximum (Viotti et al. 1985).
This is the first positive detection of R Aqr, because the marginal Einstein Observatory detection (Jura and Helfand 1984) is consistent with
background fluctuation in the IPC image (Viotti et al. 1986). We have
observed R Aqr again with EXOSAT in December 1985, and found no significant change of the X-Ray flux (Viotti et al. 1986), in spite of the large luminosity difference of the Mira.

We can explain the high ionization of the jet in terms of a cone of in-

I. Appenzeller and C. Jordan (eds.), Circumstellar Matter, 469-470.

tense ionizing radiation which escapes mainly perpendicular from a thick
accretion disk (Kafatos and Michalitsianos 1982). During episodes of enhanced mass accretion onto the hot subdwarf, X-Ray and UV radiation in-
tensifies at the inner layers of the accretion disk. Ionizing photons
emerge primarily normal to the disk plane, in two oppositely directed
radiation cones. Parcels of gas that were expelled during previous outbursts, upon being illuminated by the intense ionizing cone of radiation,
thermalize and scatter ionizing photons. If the accretion disk is orient
ed nearly edge-on with respect to our line-of-sight, X-Rays will not
escape from the central HII region, because of the high column densities
in the disk plane. Thus the weak X-Ray flux recently detected and the
high ionization lines are probably emitted by the jet or are produced in
the invisible central region and partly scattered in the jet towards us.
This latter hypothesis is in agreement with the high near-UV polarization
of R Aqr (cf. Serkowski 1970). Future UV and X-Ray polarimetry will certainly be fundamental for the modelling of R Aqr.

REFERENCES

Jura, M., Helfand, D.J.: 1984, Astrophys.J. 287, 785.
Kafatos, M., Michalitsianos, A.G.: 1982, Nature 298, 540.
Kafatos, M., Michalitsianos, A.G., Hollis, J.M.: 1986, Astrophys.J.
    Supplement Series, in press.
Serkowski, K.: 1970, Astrophys.J. 160, 1083.
Viotti, R., Piro, L., Friedjung, M., Cassatella, A.: 1985, IAU Circ.4083.
Viotti, R., Rossi, L., Cassatella, A., Piro, L.: 1986, IAU Circular 4168.

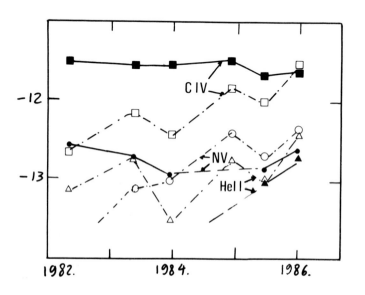

Figure 1. The UV line
fluxes of the high io
nization lines in R
Aqr (filled symbols)
and its jet (open sym
bols). Ordinates are
log fluxes in erg s^{-1}
cm^{-2}.

SPECTROSCOPY AND POLARIMETRY OF THE R AQUARII SYSTEM

Hugo E. Schwarz
Mullard Space Science Laboratory, Holmbury St. Mary, Dorking, Surrey, RH5 6NT, UK
Colin Aspin
Royal Observatory, Blackford Hill, Edinburgh EH9 3HJ, UK

ABSTRACT. Optical spectroscopy is used to derive the electron temperature and density of the jet of R Aqr. We present polarimetric data which shows that the degree of polarization in R Aqr can vary by up to two orders of magnitude and that this variation is correlated with the phase of the Mira. The polarization angle also varies with the phase of observation.

1. INTRODUCTION

R Aqr is a peculiar symbiotic with a 387 day Mira as cool component. The system has a binary period of 44 years and (perhaps) nova-like mass ejections at irregular intervals. The binary is embedded in a complex nebula consisting of three inner knots forming a curved jet out to ~7 arcseconds and a much larger nebula extending over ~2 arcminutes in the shape of two intersecting arcs.

Observations of R Aqr have been made at radio (Kafatos and Michalitsianos, 1982), infra-red (Stein et al., 1969), optical (Mauron et al., 1985; Wallerstein, 1986), ultra-violet (Kafatos et al., 1986) and X-ray wavelengths (Baratta et al., 1985). Broad-band polarimetric observations have been made by several authors (e.g. Serkowski, 1971) and recently a high resolution polarization spectrum has been obtained by Aspin et al. (1985).

2. SPECTROSCOPY

Optical spectra of the jet were obtained using the FOS on the Isaac Newton telescope at the Roque de los Muchachos Observatory, La Palma in January 1986.

From the [O III] and [N II] line ratios we calculate an electron density of $\sim 10^6$ cm^{-3} at a temperature of ~8000 K. These densities are above the critical density for the $\lambda 5008$ line. The [O III] 4363 to Hγ ratio gave 10^6 cm^{-3} at 8300 K. Using the [S II] line ratios we obtain either 10^6 cm^{-3} with 4600 K or 3.10^5 cm^{-3} at 8000 K. This might indicate that these lines are formed in regions with differing densities or temperatures. Our results compare well with those of Wallerstein and Greenstein (1980) (WG) who quote a density of $\sim 10^6$ cm^{-3} and T$_e \sim$ 10,000 K using forbidden line ratios. Using only the [O III] lines our density at a given temperature is somewhat lower than that of WG which might indicate expansion of the feature over the 9 year period between the observations. The results of Kafatos et al. (1986) based on UV line ratios give much lower densities at $\sim 10^4$ cm^{-3} at roughly the same temperature. This discrepancy is as yet unexplained. By obtaining emission measures from our optical data and comparing these with the emission measures of Kafatos et al. this question might be resolved. This work is now in progress.

I. Appenzeller and C. Jordan (eds.), Circumstellar Matter, 471–473.
© *1987 by the IAU.*

3. POLARIMETRY

It has been known for several years that R Aqr is intrinsically polarized and that its polarization is variable. Here we present evidence for a correlation between the degree and angle of polarization and the Mira phase.

Firstly, Figure 1 shows the dramatic variation in the degree of polarization with the phase of observation. The variation is up to two orders of magnitude in the blue but very small in the red. All data runs show a dip in polarization across the [O III] emission line at λ5008 A. This might indicate that the central star is the source of polarization: when a large forbidden line contribution is present the relative polarization drops. The ratio of [O III] to nearby continuum is about 13; in the more heavily smoothed polarization data the depolarization is a factor of about 5. This is at least qualitatively correct.

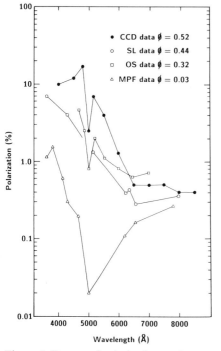

Figure 1. Degree of polarization against wavelength for four phases of R Aqr. CCD (Aspin et al., 1985) , SL (Schulte-Ladbeck, 1985), OS (Oestreicher, un-published) and MPF (Schwarz and Aspin, new data).

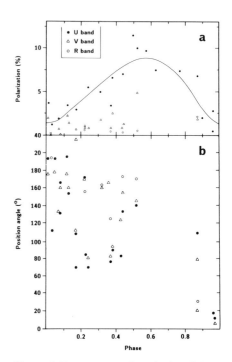

Figure 2. Degree (a) and angle (b) of U, V and R polarization versus phase of R Aqr.

Figure 2 shows the variation with phase of the degree (a) and angle (b) of polarization. The U band measures are clearly correlated with the phase of the Mira, in the sense that maximum polarization coincides with minimum light. This behaviour is opposite to that of o Ceti (Shawl, 1974) where maximum polarization occurs near maximum light. The angle variation for R Aqr is also much larger than for o Cet but similar angular modulation is found for α Ori, a semi-regular M-type supergiant (Hayes, 1984; Schwarz, 1986).

Clearly, the modelling of phase-related polarization phenomena in Miras is not a straightforward task as also exemplified by the recently discovered, extremely complicated spectral features in the polarization of R Aqr (Aspin et al., 1985).

4. SUMMARY

We have shown that the electron density and temperature of the jet in R Aqr have not changed significantly since 1977. A slight decrease in density is perhaps indicated. There is a significant difference between electron density as determined by optical and UV methods. This discrepancy is as yet unexplained. Collation of 17 years of polarization data shows Mira phase related behaviour of the U band polarization and U, V and R band angle changes. This is possibly because angle changes are generally position-dependent while degree of polarization changes tend to be strongly wavelength-dependent. The complex structure in polarization spectra cannot as yet be explained by any polarization models.

ACKNOWLEDGEMENTS. This work was supported by the SERC. HES would like to thank Prof. J. L. Culhane for his support and encouragement and Dr. J. R. Allington-Smith for donating some observing time. We acknowledge PATT for allocation of observing time at La Palma.

REFERENCES
Aspin, C., Schwarz, H.E., McLean, I.S., Boyle, R.P. (1985) Astron. Astrophys. 149, L21.
Baratta, G.B. et al. (1985) ESA SP239, pp95.
Hayes, D. P. (1984) Ap. J. Suppl. Ser. 55, 179.
Kafatos, M., Michalitsianos, A.G. (1982) Nature 298, 540.
Kafatos, M., Michalitsianos, A.G. Hollis, J.M. (1986), preprint.
Mauron, N. et al. (1985) Astron. Astrophys. 142, 413.
Schulte-Ladbeck, R. (1985) Astron. Astrophys. 142, 333.
Schwarz, H. E. (1986) Vistas in Astronomy, in press.
Serkowski, S. (1971) Kitt Peak Contr. No.554, 107.
Shawl, S. J. (1974) in : Planets, Stars and Nebulae studied with photopolarimetry, Arizona, UAP.
Stein, W.A. et al. (1969) Ap. J. 155, L3.
Wallerstein, G. (1986) Publ. Astron. Soc. Pac., 98, 118.
Wallerstein, G., Greenstein, J.L. (1980) Publ. Astron. Soc. Pac., 92, 275.

VARIATION OF LINEAR POLARIZATION IN R AQUARII SYSTEM

M. R. Deshpande, U. C. Joshi and A. K. Kulshrestha
Physical Research Laboratory
Navrangpura
Ahmedabad-380009
India

R Aquarii is a very interesting symbiotic system. There is an emission nebulosity close to the star which extends in North-South direction and a "jet" feature of about 6" with a position angle of 29° has also been observed (Sopka et al. 1982; Mauron et al. 1985; Kafatos et al. 1983). Polarization measurements are important to understand the peculiar geometry of the circumstellar material around the central objects. Wavelength and time dependence of polarization can be used to put constraints on the geometry of the object and to identify the mechanism(s) responsible for polarization. In view of this linear polarization measurements of R Aquarii were carried out by us in UBVRI bands. Observations were made during November-December 1984 on 1 meter telescope of Indian Institute of Astrophysics, Bangalore, with a dual channel photopolarimeter discussed elsewhere (Deshpande et al. 1985). Measured values of percent polarization and position angle at different phases alongwith the earlier observations of Serkowski (1974) and Ladbeck (1985) are plotted in Figure 1.

Figure 1 shows several interesting features -

a) In ultraviolet polarization is large and varies between 5 and 19 percent. The polarization in VRI band is small ($\sim$ 0.5 to 4 percent)

b) The position angle in U band shows small variation about the mean value ($\sim 120^\circ$) with time. The position angle in VRI bands, unlike the U band, show large variation ranging 20° to 170°.

All these observations support and may be explained through the binary model for R Aquarii system in which a hot white dwarf companion orbits around a Mira variable. Detailed discussion is given elsewhere (Deshpande et al. 1986). Angle between the position angle of "jet" and mean polarization vector is $\sim 90^\circ$. The small variation in position angle in U band around expected value of 119° may be due to the precession of jet around an axis. Emission knots found at different position angle (Kafatos et al. 1983) near the star may perhaps be explained with this model.

475

I. Appenzeller and C. Jordan (eds.), Circumstellar Matter, 475–476.

References

Deshpande, M. R. , Joshi, U.C. , Kulshrestha, A.K. , Banshidhar,
Vadher, N. M. , Mazumdar, H.S. , Pradhan, N.S. and Shah, C.R.1985.
 Bull. Astron. Soc. India, 13, 157.
Deshpande, M.R. , Joshi, U.C. , Kulshrestha, A.K. 1986. Publ.
    Astron. Soc. Pacific (In press).
Kafatos, M. , Hollis, J.M. and Michalitsianos, A.G, 1983, Astrophys.
    J. (Letters) 267, L103.
Ladbeck, R.S. 1985, Astron. Astrophys. 142, 333.
Mauron, N. , Nieto, J.L., Picat, J.P. Lelivre,G. and Sol, H. 1985,
    Astron. Astrophys. 142, L13.
Serkowski, K. 1974, I.A.U. Circular No. 2712.
Sopka, R.J. , Herbig, G. , Kafatos, M. and Michalistianos, A.G.
    1982, Astrophys. J. (Letter) 258, L35.

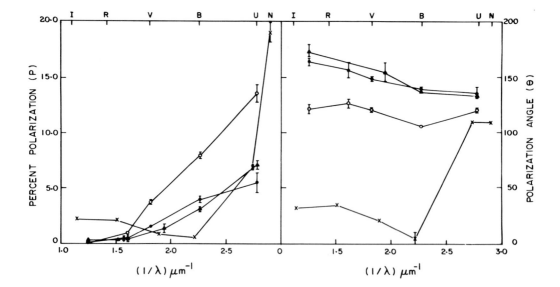

Figure 1.   Plots showing wavelength dependence of polarization
            and position angle. Symbols represent the data taken
            from -
            x Serkowski (1974);   ▲ Ladbeck (1983)
            ● Present data (Nov.1984); 0 Present data (Dec.1984).

FAR INFRARED OBSERVATIONS ON THE PECULIAR VARIABLE STAR R AQUARII[*]

B.G. Anandarao[1] and S.R. Pottasch[2]
1. Physical Research Laboratory, Ahmedabad-380009, India
2. Kapteyn Laboratorium, Post Box 800, 9700 AV Groningen, The Netherlands.

SUMMARY. First results on the peculiar Mira variable star R Aquarii obtained by the Infrared Astronomical Satellite (IRAS) in the far-infrared region (12-100 microns) are presented. A simple radiative transfer model for plane, isothermal and homogeneous layer is given to interpret the far-infrared excess radiation in terms of circumstellar dust emission. A two dust shell model with silicate grains is proposed to explain the observed fluxes. This model explains as well the low-resolution spectra (LRS) in the 8-20 micron region from IRAS. The equivalent sizes of the emitting regions (shells) in R Aquarii are found to be 0.1 arc sec and 3 arc sec at temperatures of 800 ± 80 K and 87 ± 8 K respectively. Silicate grains of size 1 micron seem to be compatible with the observations. There is no clear indication for the presence of an isolated dust cloud in R Aquarii to support the eclipsing-cloud hypothesis for explaining the minima observed at 44 yr interval.

* Full paper appeared in Astronomy and Astrophysics, 1986, 162, 167-170.

I. Appenzeller and C. Jordan (eds.), Circumstellar Matter, 477.

THE ENIGMA OF RX PUPPIS

Alan E. Wright* and David A. Allen†
*Division of Radiophysics, CSIRO, PO Box 76, Epping, NSW 2121,
   Australia
†Anglo-Australian Observatory, PO Box 296, Epping, NSW 2121,
   Australia

RX Puppis is a southern symbiotic star (R.A. $08^h12^m28^s.2$, Dec. -41°33'18" (B1950). For almost a century it has been seen to have had a violent history at optical wavelengths. In 1974 it was discovered, at 5 GHz, to be a weak radio source. Shortly after, its radio spectrum was found to be of the optically thin type, having approximately the same flux density at all frequencies between 2.7 and 22 GHz. This implied that the RX Puppis system contained dispersed, ionized gas having a linear size $>10^{15}$ cm.

From 1974 we have monitored RX Puppis regularly at different frequencies with the Parkes radio telescope. We paid it particular attention following the optical outburst reported by several authors in the early 1980s. During 1984 we found that its radio spectrum had dramatically changed. At 22 GHz its flux density had increased from ~20 to 120 mJy whilst its 5 GHz flux had increased from 20 to only 30 mJy. The radio spectral index had changed from near 0 to ~+0.8.

The radio spectrum of thermally emitting radio stars provides direct evidence about the distribution of ionized gas around the stars. For example, a star losing ionized gas at a uniform rate has a density distribution that drops off as the inverse square of the distance from the star. This produces a radio spectral index of +0.6. The value of +0.8 now observed for RX Puppis is not easy to understand. It is, however, similar to the value that we find for many other symbiotic stars.

Models to explain the radio emission from these stars have been proposed by Allen (1984) and Taylor and Seaquist (1984). In essence, a Mira giant star is losing (neutral) gas at a rate of ~10^{-5} solar masses per year and at a velocity of ~20 km s^{-1}. A distant, hot, compact companion star - probably a white dwarf - ionizes part of the circumstellar gas. But both models suffer from the disadvantage that the parameters needed to produce the +0.8 index are highly specific: we would not expect so many stars to have indices in the range +0.8 to

I. Appenzeller and C. Jordan (eds.), Circumstellar Matter, 479–480.

+1.0, as we in fact observe. Furthermore, these models cannot explain how the index of RX Puppis rose from ~0 to +0.8.

What, then, is the distribution of ionized gas around RX Puppis? It seems likely to us that the models mentioned above are correct in their essentials. However, there is observational evidence for mass-loss from the hot, compact star, and we feel that successful modelling of the RX Puppis sytem must take this into account. The distribution of ionized gas would then be decided by the momentum balance of the two stellar winds as well as the ionizing flux from the hot star.

Valuable new evidence from observations with the Very Large Array has recently been published by Hollis et al. (1986). They confirm that the +0.8 index applies to the radio spectrum of RX Puppis even up to frequencies around 100 GHz. This result sets very tight constraints on any modelling attempts. In addition, they report an angular size of around 1 arcsec for the radio emitting region. At the distance of RX Puppis (2 kpc) this corresponds to a linear size of ~3 x 10^{16} cm, comparable to, or greater than, the separation of the two stars.

This latter finding presents us with the enigma. On at least three different occasions RX Puppis has been seen to flare at 5 GHz over a period of ~10 min. But the size of 3 x 10^{16} cm from the VLA measurements corresponds to a light-travel time of ~10 days. How can a fast flare be produced by a region that is thousands of times bigger?

One possibility is that there are TWO radio emitting regions, the ionized gas and a compact non-thermal source located on, or near, the compact star. Even so, the extended ionized gas must not fill the circumstellar volume if the fast flares are to be seen. This follows, since a fast variable source cannot be seen through surrounding, ionized gas which is optically thick at the frequency of the variations. Another possibility is that the radio flares come from the Mira star and that we are viewing the symbiotic system from such a direction that only neutral gas lies between us and the flaring star. However, we have no convincing evidence that other Miras can emit radio flares.

RX Puppis is one of the strongest symbiotic stars at radio wavelengths: it is also one of the most active. At present we have no plausible model that accounts for all the changes at radio, optical and other wavelengths. Thus we believe that it merits close and continuing study in the future: it may hold the key to a better understanding of all symbiotic stars.

REFERENCES

Allen, D.A. (1983). Proc. Astron. Soc. Aust., 5, 211-213.

Hollis, J.M., Oliversen, R.J., Kafatos, M., Michalitsianos, A.G. (1986). Astrophys. J., 301, 877-880.

Taylor, A.R. and Seaquist, E.R. (1984). Astrophys. J., 286, 263-268.

A NON-ECLIPSING BINARY MODEL OF THE SYMBIOTIC STAR AG DRA

T. Iijima
Asiago Astrophysical Observatory
I-36012 Asiago (Vicenza)   ITALY

ABSTRACT. The light variation of AG Dra is considered as a result of a variation of free-free emission from a gaseous envelope around the hot component. The emission measure of the envelope at a light maximum is estimated to be about $N_e^2 V = 6.5 \times 10^{59}$ cm^{-3}. The amplitude of the light variation in the U band should be less than 1.4 mag. which is consistent with the observed results. The emission measure of the envelope might vary according to a variation of mass transfer rate in an elliptical binary system.

1. Introduction

AG Dra (BD +67°922) is sometimes called as a yellow symbiotic star, because it has a relatively early type (K 3 ∿ 5 Ⅲ) cool component (Doroshenko and Nikolov, 1967; Boyarchuk, 1969; Belyakina, 1969; Viotti et al., 1983; etc.). Many works made in the recent years covering the regions from X ray to infrared seem to support a binary model consisting of a K type giant and a hot compact star (Belyakina, 1969; Anderson et al., 1982; Taranova and Yudin, 1982; Kenyon and Webbink, 1984; Viotti et al., 1983; 1984). Meinunger (1979) found out that the luminosity in the U band periodically varies with the following elements,

$$\text{JD (Max)} = 2438900 + 554 \text{ (days)} \times E.$$

The light curve is different from those of usual eclipsing binaries, namely the light minima are wide and there are significant fluctuations in their depth and phase (Meinunger, 1979; Oliversen and Anderson, 1982). After the large outburst in 1980 – 81 an anticipated light minimum on JD 2445326 and a maximum on JD 2445549 were not detected (Luthardt, 1985), which means that the light variation was strongly disturbed by the outburst. On the other hand, the light variation according to the elements of Meinunger (1979) reappeared in the new quiescent stage in 1983 – 84, that is a light minimum was observed on JD 2445850 (Phase 0.55) and it was again wide (Luthardt, 1985). In this paper a possible model to explain these phenomena is discussed.

I. Appenzeller and C. Jordan (eds.), Circumstellar Matter, 481–483.

2. Models

Viotti et al. (1984) proposed an eclipsing binary model and suggested
that the wide light minima may be due to a reflection effect of the radiation from the hot component by the cool component. It seems difficult,
however, to explain the complete disappearance of the periodical light
variation with such models, because eclipsing phenomena are usually detectable also during outburst stages (e.g. those of CI Cyg: Belyakina,
1983). Taranova and Yudin (1982) proposed another model in which the
light variation is due to a variation of free-free (f-f) emission from a
pulsating gaseous envelope around the hot component. Since the envelope
may be disturbed by outbursts, the light variation soon after the outburst is in agreement with this model. The reappearance of the same light
variation, however, seems to be inconsistent with this model, because
after the large disturbance very probably a new pulsation will start with
a different mode. It might be possible to overcome these difficulties
with an elliptical binary model, in which the light variation is due to a
variation of f-f emission from a gaseous envelope. The envelope, however,
does not pulsate by itself, but varies according to a variation of mass
transfer in the elliptical binary system. The envelope can be strongly
disturbed by outbursts, but since the physical parameters of the binary
are not changed, the same light variation will start again in the new
quiescent stage.

   Viotti et al. (1984) showed that the continuum radiation of AG Dra in
the IUE ultraviolet region consists of two components. The one, which is
dominant in the region $\lambda < 2000$ A, has a steep energy distribution corresponding to a black-body curve whereas the other one shows a rather flat
energy distribution. They suggested that these components may correspond
to the hot compact star and the accretion disk, respectively. It is also
possible, however, to construct a model of AG Dra assuming that the latter
component is due to f-f emission from a gaseous envelope, as follows. The
intensity of the flat component is 1.3×10^{31} erg sec^{-1} A^{-1} at 2860 A
(Viotti et al., 1984), where the distance to AG Dra is 700 pc (Anderson
et al., 1981). The contribution from the hot compact star ($R = 0.02$ R$_\odot$,
$T = 150000$ K : Iijima, 1987) to the continuum is about 10% of the observed
flux and that from the cool component (K 5 III) is nearly the same. Assum-
ing that 1.1×10^{31} erg sec^{-1}A^{-1} is due to the f-f emission, an emission
measure $N_e^2 V = 6.5 \times 10^{59}$ cm^{-3} is derived (Allen, 1973), where N_e is the
electron number density and V is the volume of the envelope. Since He II
4686 emission line was very strong (Blair et al., 1983), an electron temperature 20000 K is used. The observation of Viotti et al. (1984) was made
at phase 0.96 (June 27, 1980), that is nearly on the phase of the maximum
U luminosity. Lower values will be found in other phases. In the U band
the f-f emission is about 2.6 times as bright as the cool component
(Allen, 1973). Infrared photometries showed that the luminosity of the
cool component is stable (Viotti et al., 1983), therefore the maximum amplitude of the light variation in the U band is given as $2.5 \times \log((2.6+1)/1)$
$= 1.4$ mag., which is consistent with the observed results (Meinunger, 1979;
Oliversen and Anderson, 1982). Further sources of the light variation,
for example eclipsing phenomena, are not necessary. The variation in the
V band should be less that 0.2 mag., which may correspond to the nearly

stable V magnitude (Meinunger, 1979; Luthardt, 1985).

Since we assumed that the intensity of the f-f emission varies according to a variation of mass transfer, the envelope should be included in the Roche lobe of the hot component: $\sim$130 $R_\odot$. This radius is derived using the formula of Paczyński (1971) under the assumption that the masses of the primary and the secondary are 4 $M_\odot$ and 1 $M_\odot$, and the orbital period is 554 days. If the electron density in the envelope is $10^{10} \sim 10^{11}$ cm^{-3}, which is suggested from the lack of the forbidden lines (Blair et al., 1983), this requested condition is satisfied. In the case of $N_e = 5\times10^{10}$ cm^{-3} we have a radius 60 $R_\odot$ and a mass 1×10^{-8} $M_\odot$. The mass accretion rate onto the hot component is estimated to be $\sim 10^{-7}$ $M_\odot$/year (Iijima, 1986). The mass transfer rate in the binary system seems to be high enough to maintain the envelope.

3. Conclusion

The light variations of AG Dra may be due to a variation of f-f emission from a gaseous envelope. The envelope, however, may not pulsate by itself, but probably varies according to a variation of mass transfer rate in the binary system. The emission measure of the envelope at a light maximum is about 6.5×10^{59} cm^{-3} and a probable radius and a mass are 60 $R_\odot$ and 1×10^{-8} $M_\odot$. The maximum amplitudes of the light variation in the U and the V bands are 1.4 and 0.2 mags., which are consistent with the observed results (Meinunger, 1979; Luthardt, 1985).

I wish to thank Profs. L. Rosino, R. Barbon and Dr. F. Sabbadin for careful reading of the manuscript and for useful suggestions.

References :

Allen, C.W.: 1973, *Astrophysical Quantities*, The Athlone Press, London
Anderson, C.M., et al.: 1981, *Astrophys. J. (Lett.)* 247, L 127
Anderson, C.M., et al.: 1982, in *The Nature of Symbiotic Stars*, IAU Coll.
    No. 70, eds. M. Friedjung, R. Viotti, D. Reidel Publ. Co., p. 117
Belyakina, T.S.: 1969, *Izv. Krymskoj Astrofiz. Obs.* 40, 39 (in Russian)
Belyakina, T.S.: 1983, *Izv. Krymskoj Astrofiz. Obs.* 68, 108 (in Russian)
Blair, W.P., et al.: 1983, *Astrophys. J. Suppl.* 53, 573
Boyarchuk, A.A.: 1969, in *Non-periodic Phenomena in Variable Stars*, ed.
    L. Detre, Academic Press, Budapest, p. 395
Doroshenko, V.T., Nikolov, N.S.: 1967, *Soviet Astron.* 11, 453
Iijima, T.: 1987, in Proceedings of IAU Coll. No. 93 on Cataclysmic
    Variables, ed. H. Drechsel.
Kenyon, S.J., Webbink, R.F.: 1984, *Astrophys. J.* 279, 252
Luthardt, R.: 1985, *Inf. Bull. Variable Stars* No. 2789
Meinunger, L.: 1979, *Inf. Bull. Variable Stars* No. 1611
Oliversen, N.A., Anderson, C.M.: 1982, in *The Nature of Symbiotic Stars*,
    IAU Coll. No. 70, p. 177
Paczyński, B.: 1971, *Ann. Rev. Astron. Astrophys.* 9, 183
Taranova, O.G., Yudin, B.F.: 1982, *Soviet Astron.* 26, 57
Viotti, R., et al.: 1983, *Astron. Astrophys.* 119, 285
Viotti, R., et al.: 1984, *Astrophys. J.* 283, 226

POLARIMETRY OF SOUTHERN SYMBIOTIC STARS

Regina E. Schulte-Ladbeck[1] and Antonio Mario Magalhães[2]
1) Landessternwarte, Königstuhl, D-6900 Heidelberg, F.R.G.
2) Instituto Astronômico e Geofísico, Universidade de São Paulo, Caixa Postal 30627, São Paulo 01051, Brazil

ABSTRACT. We present the first results of an ongoing survey of the linear polarization properties of symbiotic stars. With these multifilter optical measurements, we aim to obtain a statistically significant sample from which to study the nature and geometrical arrangement of their circumstellar matter and its relationship to the stellar components. Up to now, the observations show:

1. Among the five objects for which we have obtained a complete filter coverage, we find one new intrinsically polarized dusty object, BI Cru.

2. In total, intrinsic linear polarization is now detected in 9 out of 23 symbiotic stars.

3. Polarized symbiotic stars are observed at a slightly higher frequency when the associated cool component is of spectral type later than M4.

4. Intrinsic polarization is found to be more common among systems which belong to the D (warm dust) infrared type than among the S (stellar IR continua) type objects.

5. The three symbiotic subgroups S, D and D' seem to be distinguishable as well from the IRAS infrared observations. Intrinsically polarized objects occur in both type S and D objects of the IRAS sources.

We conclude, from our limited sample, that dust scattering in the asymmetric circumstellar environment of symbiotic systems does play a role in producing the polarization, at least in D-type objects. In S-type systems, photospheric scattering might originate part of the observed intrinsic polarization. The full results will be published elsewhere.

I. Appenzeller and C. Jordan (eds.), Circumstellar Matter, 485–486.

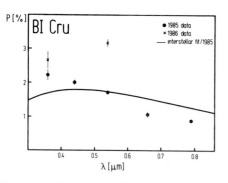

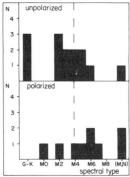

Table 1 : Southern Symbiotic Stars newly observed for Polarization

OBJECT	IR TYPE	SPECTRAL TYPE	INTRINSIC POL.
SY Mus	S	M4	no
BI Cru	D	M0 - M1	yes
RW Hya	S	M2	no
HD 330036	D'	F - G	no
RR Tel	D	> M5	no

Figure 1. The degree of po-
larization P versus wavelength for the D-type symbiotic
BI Cru. The filled circles show the measurements taken in
1985. The data are not well represented by interstellar polarization alone. When measured again in 1986, as shown by
the crosses, the polarization had varied significantly in
the V-filter. We consider the observed polarization of BI
Cru to possess an intrinsic component.

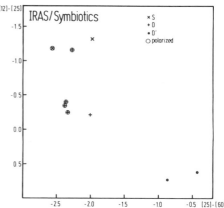

Figure 2. Histograms showing
the distribution of unpolarized and polarized symbiotic
stars in our sample by spectral type. There is a slight
tendency for polarized objects to appear at later
spectral types of the associated cool companions.

Figure 3. The (12μm)-(25μm)
versus (25μm)-(60μm) IRAS colors separate well the three
symbiotic types S, D and D'.
R Aqr groups here with the
S-type systems. More than
half of the intrinsically polarized objects are present
in this diagram.

Acknowledgements. R.E.S.-L. was supported by the Deutsche
Forschungsgemeinschaft (SFB 132A). A.M.M. wishes to acknowledge the brazilian Conselho Nacional de Desenvolvimento
Cientifico e Tecnologico (CNPq) and the Deutscher Akademi-
scher Austauschdienst (DAAD) for financing his working trip
to Germany.

UV AND OPTICAL SPECTROSCOPY OF CH CYGNI IN 1980-86

J. Mikołajewska[1], M. Mikołajewski[1], R. Biernikowicz[1], P.L. Selvelli[2] and Z. Turło[3]
1 Institute of Astronomy, UMK, Torun, Poland
2 Astronomical Observatory, Trieste, Italy
3 Astrophysical Laboratory, CAMK-PAN, Torun, Poland

CH Cyg is a binary (P~5750 days) consisting of a normal M6-7 giant and an unseen companion. During active phase its spectrum is similar to that of a symbiotic star - the strong B-A continuum and numerous low-excitation emission lines dominate the visual and UV spectrum. The last outburst, started in 1977, is conspicuous by the highest brightness level observed since monitoring begun in 1935. In mid 1984, a drop in brightness was accompanied by large continuum and emission line changes and correlated with a radio outburst and two expanding jets appearance (Taylor et al. 1985).

In the following, physical conditions in CH Cyg are analyzed on the basis of low (150 Å/mm) and high (18 Å/mm) resolution spectra obtained at Torun Observatory and low resolution IUE spectra.

Fig.1 presents several examples of energy distributions in the optical spectrum, normalized to continuum at $\lambda 4275$ Å.

Table 1. Observational data and parameters for HII region.

Period	Hα/Hβ	Hγ/Hβ	Hδ/Hβ	H8/Hβ	$n_e (cm^{-3})$	$R(R_\odot)$
1980	8.7*	.25	−	−	$10^{10}-10^{-1}$	25-100
1982/4	−	.28 /.34	.12 /.19	.07 /.1	$3\times10^{10}-10^{11}$	40- 85
1984/5	−	.36	.25	−	$5\times10^9-10^{10}$	57- 90
1985	2.7	.47	.20	−	$< 10^8$	4×10^3
1985/6	−	.39	.20	−	$\lesssim 10^{10}$	$\gtrsim60$

* according to Blair et al. 1983, Ap.J. Suppl., 53, 573.

The electron density range was estimated by comparison of observed Balmer decrement with the teoretical decrements given by Drake and Ulrich (1980), assuming T_e =10000 K and E(B-V)=0. Then, the radii of the HII region were derived using emission measures for the Balmer continuum, assuming a distance d=330 pc (Mikołajewska et al. 1986). The values are given for five periods: 1980 - before maximum of activity;

I. Appenzeller and C. Jordan (eds.), Circumstellar Matter, 487–489.
© 1987 by the IAU.

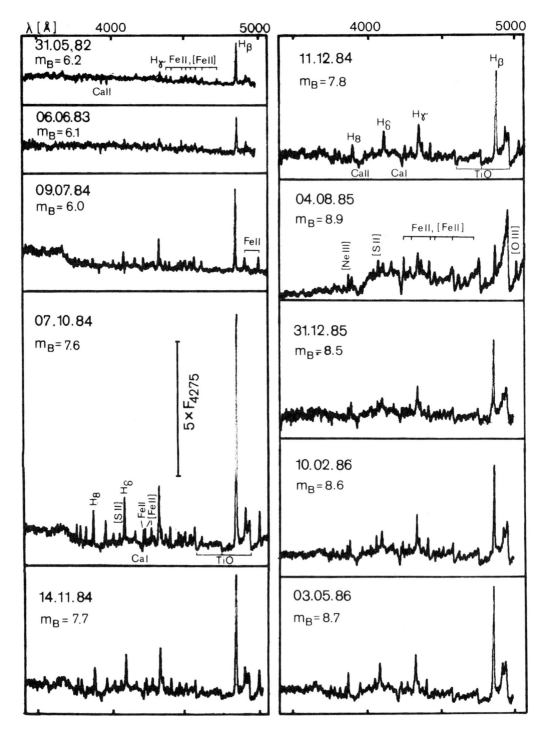

Figure 1. Low resolution optical spectra of CH Cyg: energy distribution F_λ / F_{4275} .

1982/mid 1984 - maximum; mid 1984/Feb 1985 - after drop in brightness; Mar-Oct 1985 - minimum (see also Mikołajewski et al. 1986); 1985/6 - after minimum.

In Sept 1980 Balmer lines were visible up to H38 (Farragiana and Hack 1980), while in 1982-83 they were observed up to H30. This corresponds to $n_e = 4 \times 10^{10}$ cm^{-3} and R(HII)=42 $R_\odot$ in 1980, and $n_e \lesssim 10^{11}$ cm^{-3}, R(HII)$\gtrsim$40 $R_\odot$ in 1982-83

After the drop in brightness in mid 1984, the "nebular phase" with strong forbidden lines in optical and very rich and high-excitation emission line spectrum in UV range was observed. The ratio of CIII] lines implies $n_e = 5 \times 10^6$ cm^{-3} (Selvelli and Hack 1985), much lower than the values suggested by the observed Balmer decrement.

In Mar-Oct 1985, a deep minimum in U band and IUE integrated flux was observed and interpreted as an eclipse of a hot component (Mikołajewski et al. 1986). Weak single profile Balmer lines were observed instead of the double peaked lines present in the spectrum before and after the minimum, while forbidden lines remained strong. The average ratio of [OIII] 5007/4363=3.6 implies $T_e > 8000$ K and $n_e > 10^6$ cm^{-3}. Assuming the cosmic abundance of O/Ne and considering [OIII] 5007/[NeIII] 3869=1.4 we have $T_e = 10000$ K and $n_e = 8 \times 10^6$ cm^{-3}. All these n_e values are close to the derived from the CIII] lines and in agreement with the observed Balmer decrement. The value of $n_e = 10^7$ cm^{-3} was adopted for the R(HII) estimation.

The Balmer jump measured on our spectra was always in very good agreement with the values predicted by the fits for IUE spectra (Mikołajewska et al. 1986).

If the 1985 minimum is indeed due to an eclipse, than:
- The double peaked Balmer lines are formed close to the hot component possibly in a disk or a rotating envelope.
- Variations of Balmer decrement suggest a presence of two different regions: relatively small (R$\sim$ 50-100 $R_\odot$) and dense ($n_e \sim 10^{10}$-10^{11} cm^{-3}) disk or envelope occulted during the minimum and extended (R$\sim$4000 $R_\odot$) low density ($n_e \sim 10^6$-10^7) region, possibly coinciding with jets, in which single profile HI lines and most of forbidden lines are emitted.
- X-ray detection in May 1985 (Leahy and Taylor 1986) suggests that X-ray emission arises from a region which is not close to the hot component (e.g. from jets).

REFERENCES

Drake S.A., Ulrich R.K., 1980, Ap. J. Suppl., 42, 351.
Farragiana R., Hack M., 1980, IBVS No. 1861.
Leahy D., Taylor A.R., 1986, preprint.
Mikołajewska J.,Selvelli P.L.,Hack M., 1987, IAU Coll.No.93.
Mikołajewski M., Tomov T., Mikołajewska J., 1986, ibid.
Selvelli P.L., Hack M., 1985, Astronomy Express, 1, 115.
Taylor A.R.,Seaquist E.R.,Mattei J.A., 1986, Nature, 359,38.

JETS FROM SYMBIOTIC STARS

M. Kafatos[1], A. Cassatella[2], A.G. Michalitsianos[3],
L. Piro[4] and R. Viotti[5]

1. George Mason University, Dept. Physics, Fairfax, VA, USA
2. IUE Observatory, ESA, Madrid, Spain
3. Laboratory for Astronomy and Solar Physics, NASA-GSFC,
   Greenbelt, MD, USA
4. Instituto TESRE, CNR, Bologna, Italy
5. Instituto Astrofisica Spaziale, CNR, Frascati, Italy

ABSTRACT. R Aquarii is the closest symbiotic variable that shows extended emission with multiple jet components. A number of other symbiotics
also show jet activity and this phenomenon may be common, particularly
among D-type symbiotics.

R Aquarii may be the prototype of jet activity in symbiotic stars.
Extended UV, radio, optical and X-ray observations are described
in Kafatos, Michalitsianos and Hollis (1986) and Cassatella et. al.
(1987). The emission from the jet of R Aqr is variable and the high
excitation lines such as He II, N V and C IV have been increasing in
intensity in the recent past, surpassing the emission of those lines
from the compact H II region surrounding the binary star system. The
jet is also a weak X-ray emitter (Cassatella et al. 1987) and it may
well be that both X-ray and UV line emission of the jet components comes
from component A and possibly B (Kafatos, Michalitsianos and Hollis 1986).
The nature of the alligned radio-optical features that comprise the R
Aqr jet indicates that directional mass outflow is taking place. The
most likely mechanism of ejection is radiation pressure on grains formed
in the outer, cool regions of a spatially thick accretion disk that forms
around the hot secondary star. The inner regions of this disk are hot,
with $T \gtrsim 10^5$ K and emit high energy radiation. The plane of the orbit
of the two stars is, however, along the line of sight and the cone of
radiation that escapes from the disk cannot be seen directly but goes
out perpendicular to the disk and ionizes the gas parcels which form
the jet components. This model explains the predominance of high excitation lines in the jet, the secular increase of UV line emission, the
semi-periodic variability of the UV line emission from the jet and
is consistent with the large amount of dust being present in R Aqr and
maybe even with the eclipses of R Aqr itself by a cooler object in the
late 70's. R Aqr is a D-type symbiotic (dust-type) in contrast to the

I. Appenzeller and C. Jordan (eds.), Circumstellar Matter, 491–492.

S-type (stellar-type) which show strong, hot star, stellar continuum. Recent observations of R Aqr with the Hat Creek interferometer (Hollis et al. 1986) indicate that the jet also emits in SiO masing line, $v = 1$, $J = 2 - 1$. This emission does not arise in the Mira primary but must come from an extended region south of the binary, most likely a cool jet component. Finally, optical observations of R Aqr resulting in slit spectra (Solf and Ulrich 1985) indicate that the expansion velocities of the gas components in the R Aqr nebula are low, ~ 50 km s^{-1}, consistent with ejection from low escape velocity regions.

Other symbiotic stars show jet emission or extended nebular emission such as: CH Cyg (Taylor, Seaquist and Mattei 1986); AG Peg (Hjellming 1985). Extended emission has been detected from V1016 Cyg (Hjellming and Bignell 1982); RX Pup (Hollis et al. 1987). CH Cyg and AG Peg are S-type symbiotics whereas V1016 Cyg and RX Pup are D-type symbiotics. CH Cyg is the only symbiotic where high velocity ejection ($\gtrsim 1000$ km s^{-1}) is present, although this interpretation depends on a somewhat contro- versial identification of distinct jet features. The sizes of these jets or extended emissions are all about a few x 10^{15} cm, the largest one being the size of the jet structure in R Aqr where a distance of the outermost component B from the central source is 10^{16} cm. Luminosi- ties seem to be well below the Eddington limit for all symbiotics where jet or extended emission has been detected and mass outflow rates are generally around 10^{-7} M$_\odot$ yr^{-1}. These results are in agreement with the results for R Aqr indicating that dust may be responsible for the ejec- tion mechanism in most symbiotics. Further studies of symbiotic jets would provide important clues for the important topic of astrophysical jets.

REFERENCES

Cassatella, A., Kafatos, M., Michalitsianos, A.G., Piro, L., and Viotti, R.: 1987, present volume.
Hjellming, R.M.: 1985, Radio Stars, edit. R.M. Hjellming and D.M. Gibson, D. Reidel Publishing Co., Dordrecht.
Hjellming, R.M., and Bignell, R.C.: 1982, Science, 216, 1279.
Hollis, J.M., Michalitsianos, A.G., Kafatos, M., Wright, M.C.H., and Welch, W.J.: 1986, Ap. J. (Lett.), in press.
Hollis, J.M., et al.: 1987, in preparation.
Kafatos, M., Michalitsianos, A.G., and Hollis, J.M.: 1986, Ap. J. Suppl., in press.
Solf, J., and Ulrich, H.: 1985, A&A, 148.
Taylor, A.R., Seaquist, E.R., and Mattei, J.A.: 1986, Nature, 319, 38.

THE CIRCUMSTELLAR MATTER IN THE BE + K BINARY KX AND SEEN IN THE UV SPECTRA *

S. Štefl
Astronomical Institute
251 65 Ondřejov, Czechoslovakia

KX And (HD 218393) belongs to well-known and often observed Be stars. The periodicity of spectroscopic characteristics as well as other observations led to the conclusion that the star is a peculiar inter-acting binary (Harmanec et al. 1980). The aim of this paper is to present the phase-dependent behaviour of the UV spectrum of the star.

Almost all (i.e. 12 SWP, 8 LWR and 1 LWP) available high-re-solution IUE spectra of the star were analysed. The radial velocities of the optical shell lines - presented for comparison with the UV spectra - are based on the measurements of 60 high-dispersion blue spectra obtained with the Ondřejov 2m telescope. The UV spectra were secured 1979 to 1983, the optical spectra in the period 1972 - 1981. The identification of the absorption lines in the UV region, rectification of the spectra and determination of T_{eff} , log g and v sin i were carried out with the help of computed synthetic spectra. A computer code, kindly provided by Dr. I. Hubený, was used in the LTE regime and using LTE model atmospheres and solar abundances (Kurucz 1979). The problems concerning a simulation of the subionized Be envelope by the synthetic spectra under the above assumptions were discussed in detail by Hubený et al. (1985). The phases of the observations were computed according to the ephemeris derived from the radial velocities of optical shell lines (Štefl et al. in preparation). Phase zero corresponds to the velocity maximum.

It is known that optical shell lines reach their intensity maximum near the phase 0.0 and minimum near 0.5. The ultraviolet line spectrum, which shows much weaker intensity variations, may be roughly simulated by the synthetic spectrum computed for the model atmosphere with T_{eff} = 9000 K, log g = 2.0 and v sin i $\simeq$ 40 km s^{-1} in any phase. The largest deviations from the computed spec-trum can be observed near phase 0.0, because of the more complicated shell-line profiles. The considerable departures from the computed spectrum appear in the region C IV (UV 1) doublet, but the presence of the true C IV lines in the KX And spectrum remains uncertain. The strong resonance lines with P-Cyg profiles are the most striking features in the UV spectrum;

* Based on data from International Ultraviolet Explorer, de-archived from the Villafranca Data Archiv of the European Space Agency.

I. Appenzeller and C. Jordan (eds.), Circumstellar Matter, 493–495.

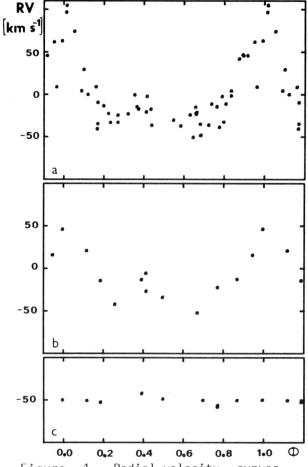

RV
[km s⁻¹]

Figure 1. Radial-velocity curves of a) the optical metallic lines, b) of the UV subordinate shell lines of Fe II, Ni II, Ti II, Ca II Mn II, Mg II and Al II in the region of SWP IUE camera, c) of Fe II (UV 1,2,3) resonance lines.

the emission components of Mg II (UV 1) and C II (UV 1)being the most intensive ones. The asymmetry of the short wavelength wings of absorption components varies from ion to ion but it seems to be more pronounced in phases 0.25-0.70 in general.

Radial velocities were determined for all identified UV absorption lines. The mean values were computed for selected sets of lines using the methods of robust statistics (with an error of about 1%). Figure 1 shows the radial-velocity dependence for the UV subordinate metallic lines in the region of SWP camera, for the resonance lines Fe II (UV 1,2,3) and - for comparison - for the optical metallic shell lines. The plots illustrate the following conclusions valid for all regions of SWP and LWR cameras: a) The UV subordinate lines follow the double-wave radial-velocity curve of the visual shell lines (though with a lower amplitude). b) Radial velocities of the UV resonance lines are almost phase-independent. In case of Fe II (UV 1,2,3), Fe III (UV 34), Si II (UV 1,2,3) and Si IV (UV 1) lines the largest deviations from the constant velocity can be seen near phase 0.4. Interestingly enough the most striking feature at the radial velocity curve of the UV resonance lines thus coincides with inconspicuous secondary maximum in the radial velocity curve of optical and UV shell lines. The amplitude of the radial velocity of the resonance lines is - with the exception of Si IV (UV 1) lines - smaller than 15 km s⁻¹. Neglecting this possible phase-dependent variations the radial velocity probably increases with the ionization potential - see Fig.2.

Most of features seen in the ultraviolet line spetrum between about 1290 and 3100 A probably originate in the subionized circum-

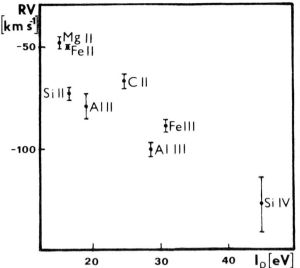

Figure 2. Radial velocity of the UV resoance lines versus ionization potential. The sum of the ionizat- ion and the excitation potential is used for the subordinate Fe III (UV 34) lines. These lines originate from the metastable energy levels and are thus formed similarly as resonance lines.

stellar envelope. The photo- sphere of the hot primary seems to be almost complete- ly obscured by this envelo- pe. Yet, considering a rather low number of strong non-reso- nance lines present in an early B-type UV spectrum, one cannot safely exclude the presence of some possibly dumped photospheric lines. The case of KX And demon- strates that circumstellar matter around hot stars, which is optically too thin to be observed in the optical spectra, can be detected in the ultraviolet.

In conclusion; the observed data seem to suggest that the B-type star in the binary system is surround- ed by a circumstellar envelo- pe or disc, which is optical- ly thick in the ultraviolet. The optical shell lines can be seen in phases when we can see the accretion stream, projected on the primary component. The ob- served radial-velocity curve reflects the binary motion but it differs from that of the primary component, especially near the velocity maximum (Barr effect). The structure of the accretion disc and the stream seem to be responsible for the secondary minimum and maximum on the radial-velocity curves of visual and ultraviolet shell lines. The UV resonance lines may originate inside the primary Roche lobe or in an envelope around the whole binary system.

The author thanks Drs. I. Hubený and P. Harmanec for the help- ful comments to the manuscript.

REFERENCES

Harmanec P., Horn J., Koubský P., Žďárský F., Kříž S., Pavlovski K.: 1980, Bull. Astron. Int. Czechosl. 31, 144.
Hubený I. Štefl S. Harmanec P.: 1985, Bull. Astron. Inst. Czechosl. 34, 214.
Kurucz R. L.: 1979, Astrophys. J. Suppl. Ser. 40, 1.

PLANETARY NEBULAE

NEAR IR OBSERVATIONS OF IRAS PLANETARY NEBULAE

P. PERSI, A. PREITE-MARTINEZ, M. FERRARI-TONIOLO,
L. SPINOGLIO
Istituto di Astrofisica Spaziale, CNR
C.P. 67
00044 Frascati
Italy

We have observed 117 faint planetary nebulae in the J, H, K, and sometimes L and M bands. The objects are all listed in the Perek-Kouteck
(PK) catalogue and were observed by IRAS.

The observations were carried out in three observing runs at the
1.0m and 3.6m E.S.O. telescopes and at the italian infrared telescope
(TIRGO), all equipped with InSb photometers / spectrophotometers. More
than 70% of the sources were observed for the first time in the near IR.

We selected the sources according to the following criteria:
(i) all the objects are listed in the PK catalogue, and have been detected by the IRAS satellite; (ii) they are preferentially optically
faint and compact sources. Two third of the selected sources have been
observed in the radio continuum at 6cm, and show small angular sizes
($\lesssim 15''$).

With these criteria, our sample may contain emission line objects,
symbiotic stars, peculiar emission line objects, other than classical
planetary nebulae. Ten sources are in common with the sample of faint
PNe observed by Whitelock (1985). Other 27 objects have been also previously observed by different authors, though most of them only in one
or two bands.

The results are shown in the colour-colour diagram J-H vs. H-K of
Figure 1. According to Fig.1 the observed objects may be divided in at
least three different classes:

- N (nebular): The near IR emission is dominated by ff+fb emission,
with the addition of ionic emission lines (H, He,...). N-objects are located in the region $(J-H) \lesssim 0.2$. 32 objects in our sample can be classified as N-type.

- S (stellar): These objects are located on the main sequence, and
in the region occupied by Mira variables, carbon stars, and symbiotic
stars (see Allen, 1973). The near IR emission is dominated by a stellar

I. Appenzeller and C. Jordan (eds.), Circumstellar Matter, 499–500.

continuum. 67 objects fall into this class.

— D (dust): In this class a combination of hot dust and either nebular or stellar emission(from a hot star) is present. The objects lie
in the region of the diagram to the left of the main sequence and blackbody curves, above the nebular emission curve (dashed line).

The presence of hot dust in N- or S-type objects can be inferred
from the analysis of the L-photometry. A study of the energy distribution for the objects of the sample from 1 to 100μm is necessary, adding
ground based observations to IRAS data. To complete the analysis of the
sources, a programme of spectrophotometry with CVF around the Brackett
lines and between 8–13μm is in progress.

From this additional information we can preliminary conclude that
S-type objects are most probably symbiotic stars or other kind of emission line stars.

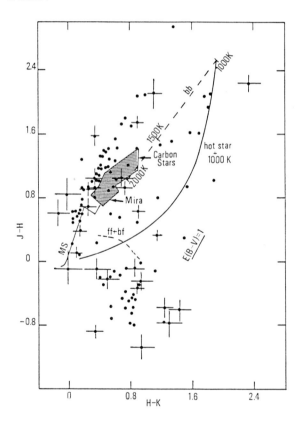

Fig.1: (J–H) vs.
(H–K) colours
of the observed
PK sources.

References

Allen,D.A.:1973,Mon.Not.R.astr.Soc.,161,145
Whitelock,P.A.:1985,Mon.Not.R.astr.Soc.,213,59

ATOMIC HYDROGEN IN THE PLANETARY NEBULA IC 4997

C. Giovanardi[1], D.R. Altschuler[2], S.E. Schneider[3],
P.R. Silverglate[4]

[1]Osservatorio Astrofisico di Arcetri
Largo E. Fermi 5, I-50125 Firenze, Italy

[2]Dept. of Physics, University of Puerto Rico
Rio Piedras, 00931 Puerto Rico
and
Max-Planck-Institut für Radioastronomie
Auf dem Hügel 69, D-5300 Bonn 1, F.R.G.

[3]Dept. of Astronomy, University of Virginia
Charlottesville, VA 22903, U.S.A.

[4]Perkin-Elmer Corporation, Danbury, CT 06810, U.S.A.

In the course of a sensitive search for atomic hydrogen emission associated with planetary nebulae having high velocities relative to Galactic HI (Schneider et al. 1986), we detected absorption in the spectrum of IC 4997 (PK 58-10.1). This is only the second definite detection of HI associated with a PN. The velocity of the feature coincides precisely with that expected if the gas is expanding with the measured optical expansion velocity of 14 km s^{-1} (Sabbadin 1984), strongly suggesting an association.

We observed IC 4997 with the Arecibo 305 m radiotelescope using the 21 cm dual-circular polarization feed and the 1008 channel autocorrelation spectrometer. Details of the observing procedure are given in Schneider et al. (1986).

Our results are presented in the figure. The upper panel shows the Galactic emission, the central panel the difference between the on-source spectrum and the average of the four off-source spectra, and the bottom panel the difference between the <N+S> and <E+W> averaged scans. The central panel represents our effort to remove Galactic HI emission from the spectrum of IC 4997. Cancellation is imperfect for the strongest Galactic emission, but beyond this region and over the entire velocity range of IC 4997 (centered around the velocity marked by an arrow), the <N+S> - <E+W> spectrum shows a flat baseline. This indicates that variation in the Galactic HI emission is linear outside of the region around V_{LSR}=0. However, in the central panel there is a dip at V_{LSR}=-64 km s^{-1} which was independently confirmed in follow-up measurements.

After deconvolution with our 8 km s^{-1} resolution the line width for

I. Appenzeller and C. Jordan (eds.), Circumstellar Matter, 501–502.

IC 4997 is 18 km s^{-1}. This value
is consistent with that expected
from a spherical structure of
neutral hydrogen which surrounds
the continuum source and expands
with the nebula.

The neutral hydrogen mass
$M_{HI} = 1.9 \pm 1.2 \ 10^{-3} M_{\odot}$ was computed
assuming a spin-excitation
temperature $T_{ex} = 100$ K and is
twice the HI actually detected to
allow, by symmetry, for gas on
the far side of the nebula. We
note that this value does not
significantly increase the total
computed mass of IC 4997.

Aside from other questions
such as geometric corrections
(discussed by Altschuler et al.
1986) a factor of two range in
the distance scaling yields a
factor of four uncertainty in M_{HI}.
We therefore find it useful to
compute the ratio of HI to ionized
mass in the nebula ($M_{HI}/M_i = 0.39 \pm$
0.08) which is only weakly dependent on the distance and angular
size. We used the formulation for
M_i of Milne and Aller (1975),
assuming a temperature of 18,000 K
for the ionized gas, which appears
appropriate for this very young
nebula, and a standard helium
abundance. Details of the computations, and a comparison of our
results with other planetary nebula
can be found in Altschuler et al. (1986).

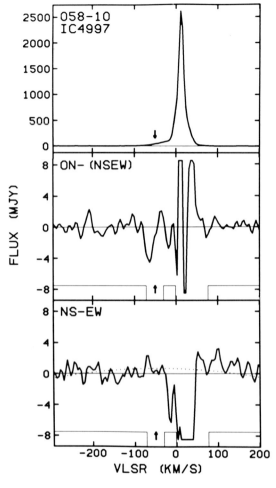

Spectra towards IC 4997

The Arecibo Observatory is part of the National Astronomy and
Ionosphere Center which is operated by Cornell University for the
National Science Foundation.

REFERENCES

Altschuler, D.R., Schneider, S.E., Giovanardi, C., Silverglate, P.R.:
   1986, Astrophys. J. Lett. 306, (in press)
Milne, D.K., Aller, L.H.: 1975, Astron. Astrophys. 38, 183
Sabbadin, F.: 1984, Astron. Astrophys. Suppl. 58, 273
Schneider, S.E., Silverglate, P.R., Altschuler, D.R., Giovanardi, C.:
   1986, (in preparation).

EVIDENCE OF A CIRCUMSTELLAR DUST CLOUDLET ORBITING AROUND THE CENTRAL
STAR OF NGC 2346

R. Costero[1], M. Tapia[2], J. Echevarría[2], A. Quintero[1],
J.F. Barral[1] and M. Roth[2]

[1] Instituto de Astronomía, Universidad Nacional Autónoma de
   México, Apartado Postal 70-264, 04510 México, D.F., México.
[2] Instituto de Astronomía, Universidad Nacional Autónoma de
   México, Apartado Postal 877, 22830 Ensenada, B.C., México.

ABSTRACT. The photometric behaviour of AGK3-0°965, the central star
of the bipolar planetary nebula NGC 2346, has been monitored photometrically for several months at the Observatorio Astronómico Nacional
at Tonantzintla and San Pedro Mártir, Mexico. A model is proposed in
which the eclipses were caused by the passage of an elongated cool dust
cloudlet of size $\sim$ 2-5 $\times$ 10^{12} cm and total mass $\sim$ 10^{-12} $M_\odot$. This model
can explain most of the observations. The velocity of the cloud in the
direction of the major axis of the projected central binary orbit is
v_p = 0.14 km s^{-1}. Another warmer (T $\lesssim$ 1000 K) circumstellar cloud is
responsible for the infrared excess at wavelengths from 3 to 12 μm.
Its emission, as seen from the Earth, has not changed significantly at
λ > 3 μm during the past twelve years, as shown by new infrared observations also reported. Its most relevant physical properties are still
to be determined. The present results provide the first evidence of a
dense circumstellar cloudlet of mass similar to that of a minor planet
which is probably the result of the fragmentation of a disk or toroid
around the central star of NGC 2346. Although the presence of many
other similar cloudlets in its vicinity is expected, the probability of
similar events occurring in the next few hundred years is very small.

The details of the present work will be published in the *Revista
Mexicana de Astronomía y Astrofísica*.

I. Appenzeller and C. Jordan (eds.), Circumstellar Matter, 503.

CO IN PLANETARY NEBULAE

P. J. Huggins and A. P. Healy
Physics Department
New York University
4, Washington Place
New York, NY 10003
U.S.A.

ABSTRACT. We report the first detection of millimeter CO emission in two highly evolved planetary nebulae: NGC 6720 and NGC 7293. The CO is a useful probe of the structure and kinematics of the molecular gas in the nebulae, and provides an estimate of the mass of material which remains un-ionized by the central star.

Millimeter emission from CO has been the single most useful probe of the kinematics and masses of the dense neutral circumstellar envelopes around the coolest giant stars. Since these objects are generally believed to be the direct precursors of planetary nebulae, observations of CO in the remnant neutral shells of the nebulae are of considerable diagnostic potential for understanding the evolutionary sequence. However, despite repeated searches, very few have so far proved to have detectable CO emission (cf. Knapp 1985). Here we report on the first results of an observational program to re-evaluate the question of CO in planetary nebulae.

The observations were made with the NRAO 12-m telescope at Kitt Peak in the CO (2-1) line at 230 GHz, where the beamsize is 30 arc seconds (FWHM). This telescope-frequency combination offers a considerable improvement in sensitivity to small scale, optically thin emission than was available for previous searches. Among the nebulae in the program are the Ring (NGC 6720, PK 63 +13.1) and the Helix (NGC 7293, PK 36 -57.1), and CO has been detected in both.

The CO emission from the Ring nebula has a peak temperature of 0.2 K and the line is centered at a radial velocity of 1 km/s (with respect to the local standard of rest), with a total velocity extent of 48 km/s. The kinematics are similar to those found from studies of optical lines. Very limited mapping indicates a CO source size which is comparable to that of the optical nebula, whose diameter is about 70 arc seconds. The CO molecules probably reside in the dense filaments seen, e.g., in OI images of the nebula.

The angular size of the Helix nebula (about 12 arc minutes in diameter) is much larger than the telescope beamsize and allows the CO distribution to be examined in more detail. Limited mapping

I. Appenzeller and C. Jordan (eds.), Circumstellar Matter, 505–506.
© *1987 by the IAU.*

indicates that the CO emission closely corresponds to the two ring
structures seen in optical images, and several positions have CO line
temperatures in excess of 1 K. The line widths at each position are
typically a few km/s, and the velocities vary systematically with
position angle in the nebula. Two distinct kinematic components are
seen, which correspond to the two optical rings, and the systematic
velocity variations can be reproduced quite closely with a simple model
of two expanding rings; complex helical models which have previously
been proposed are not required to explain the CO kinematics. The
radial velocity of the nebula is -23 km/s.

In addition to the structure and kinematics of the neutral
gas in the nebulae, the observations can be used to estimate the mass
of molecular material, using simplifying assumptions on the opacity,
excitation, distribution, and abundance of the CO. The results
indicate that the mass ratio of molecular to ionized matter is
approximately 0.1 for both the Ring and the Helix.

The few previous detections of CO in bona fide planetaries include
the young objects NGC 7027 and CRL 618 (cf. Knapp 1985 and references
therein). Both have very small ionized masses and thick neutral
envelopes, with mass ratios of molecular to ionized matter greater than
100, much larger than found here for the evolved nebulae. Thus for the
nebulae in which CO has been detected, the observed trend is for lower
molecular to ionized masses for the more evolved nebulae. This is in
accord with the work of Pottasch (1980), who has shown statistically
that the ionized masses of planetary nebulae increase with radius, and
interprets this as indicating a reservoir of neutral material which
becomes ionized as the nebulae expand. The results described here
provide support for the picture in which the formation and evolution of
planetary nebulae are directly controlled by the ionization of
circumstellar material commonly seen around their red giant precursors.

Further details of this work can be found in Huggins and Healy
(1986a, and 1986b).

This work has been supported by the National Science Foundation.
NRAO is operated by Associated Universities, Inc., under contract with
the NSF.

REFERENCES

Huggins, P. J., and Healy, A. P. 1986a, Ap. J.(Letters), 305, L29.
Huggins, P. J., and Healy, A. P. 1986b, M.N.R.A.S., 220, 33p.
Knapp, G. R. 1985, in "Mass Loss from Red Giants", eds. M. Morris and
    B. Zuckerman (Dordrecht: Reidel), p171.
Pottasch, S. R. 1980, Astr. Ap., 89, 336.

ANOMALOUS [NII]-EMISSION FROM Mz-3

G.F.O. Schnur[1] and W.H. Kegel[2]
[1]Astronomisches Institut, Ruhr-Universität
D-4630 Bochum, Federal Republic of Germany
[2]Institut für Theoretische Physik der Universität
D-6000 Frankfurt/M., Federal Republic of Germany

We obtained high resolution spectra of the red [NII]-lines in the bipolar nebula Mz-3. This object probably is a proto-planetary nebula (see e.g. Lopez and Meaburn, 1983 and Meaburn and Walsh, 1985). The spectra were taken in July and September 1982 at La Silla using the Coudé Auxiliary Telescope (CAT) combined with the Coudé Echelle Spectrograph (CES). The resolution depending on the slit width was 30000 and 100000, respectively. Fig. 1 shows the [NII]-lines at 6583.6 and 6548.3 Å together with H_α. The structure of the 6583.6 Å line, which is quite complicated in detail, is dominated by two narrow peaks, similar to the profiles of

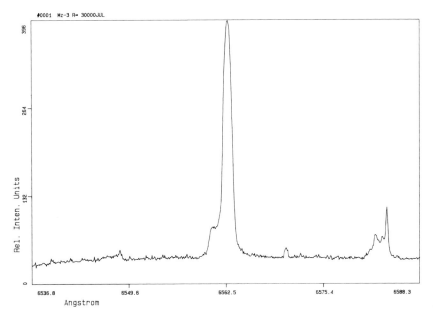

Figure 1. Spectrum showing [NII] $\lambda\lambda$ 6548, 6584 and H_α.

507

I. Appenzeller and C. Jordan (eds.), Circumstellar Matter, 507–508.

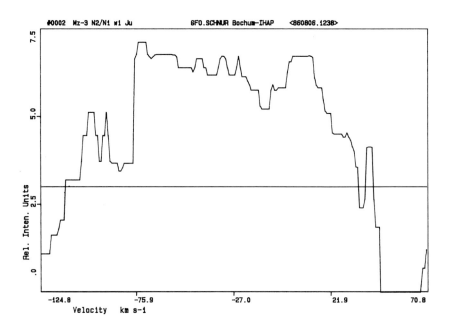

Figure 2. Intensity ratio I(6584)/(6548) as function of V_r.

circumstellar OH-masers. The profile varies strongly with
the slit width and shows substantial differences between
the observations in July and September. - The most striking
result of our observations is that the intensity ratio
6584/6548 deviates strongly from the canonical value of 3
as shown in Fig. 2.

We interpret the anomalous intensity ratio as being due
to optical pumping in an anisotropic radiation field. This
leads to a polarization of the NII-ions and renders also
the emission to be anisotropic, the angular dependence for
the two lines being different. A prediction of this model
is that the two lines should be linearly polarized with
the polarization vectors being perpendicular to each other.
Therefore, the deviation of the intensity ratio from the
canonical value will be even larger when one observes light
of only one polarization direction. This may have happened
in our observations, since the CAT/CES combination is
likely to be sensitive to polarized light due to some
grazing incidence of the light beam in the telescope and
spectrograph. - Further predictions of our model are an
intensity of the λ5755 Å-line comparable to that of the
λ6584 Å-line, and population inversion for the FIR lines
at 122 μ and 204 μ.

References
Lopez, J.A., Meaburn, J.: 1983, M.N.R.A.S. *204*, 203
Meaburn, J., Walsh, J.R.: 1985, M.N.R.A.S. *215*, 761

HOW UNIQUE IS THE PROTOPLANETARY NEBULA STAR HR 4049? (*)

C.L. Waelkens (1)   and L.B.F.M. Waters (2)

(1)   Research Associate of the Belgian National Science
      Foundation, University of Leuven, Belgium
(2)   Space Research Laboratory, Utrecht, The Netherlands

1.   INTRODUCTION

The late B-supergiant HR 4049 is peculiar in different respects: (1) It
is located far from the galactic plane (b = 23°); (2) It is a variable
with a large amplitude and on a long time scale (Waelkens and Rufener,
1983); (3) It has a spectacular infrared excess (Lamers et al., 1986).
Two models were proposed: (i) HR 4049 is a runaway hypergiant embedded
in a dust cloud, or (ii) HR 4049 is a low-mass star in a post-AGB stage
of evolution. In this paper we present evidence that favours the second
hypothesis. This evidence consists of new observational data on HR 4049
itself and of the discovery of a second very similar object, that is
located still farther from the galactic plane.

2.   SPECTROSCOPIC DATA ON HR 4049.

We have secured two 12 A/mm blue spectra of HR 4049 with the Coudé
spectrograph attached at the $1^{m}52$ telescope at ESO, at different phases
of the light curve. The low gravity of the object is confirmed by the
presence of 28 Balmer lines. A reversed and variable Balmer progression
is observed.
     Apart from the Balmer lines, the only conspicuous lines in the
spectra are the (essentially interstellar) Ca K line and several lines
of CI and one line of OI. Such lines are also conspicuous for the high-
latitude supergiant HD 46703, which is thought to be a post-AGB object
(Luck and Bond, 1984). The presence of CI and OI lines in HR 4049 and
the observation of CO emission in some other high latitude supergiants
(Omont, this conference) lend support to the leftward post-AGB evolution
hypothesis for these stars. One would then have a picture in which the
CO-envelope is still prominent in the coolest stars (like HD 161796), is
partly dissociated in early F-supergiants (like 89 Herculis) and is
dissociated completely in HR 4049.

(*) Based on observations made at the European Southern Observatory

I. Appenzeller and C. Jordan (eds.), Circumstellar Matter, 509–510.

3. A TWIN TO HR 4049

The southern late B-supergiant HD 213985 is very similar to HR 4049. It presents the same kind of variability (Figure 1): amplitude, time scale and color behavior are similar in both objects. IRAS data reveal that also HD 213985 shows a spectacular IR excess (Figure 2). As yet the star has not been observed in the near infrared. From the IRAS data it is however clear that the dust around HD 213985 is cooler than that around HR 4049 (T_{BB} = 350 K to be compared with 1250 K) and that the size of the emitting volume is larger for HD 213985 (R_{dust} = 260 R_* to be compared with 28 R_*). If HD 213985 were a truly high mass star, its apparent faintness (m_V = 9), spectral class (B9Ib) and high latitude (b = -57°) would put it at more than 5 kpc from the plane, so that it could not have been formed in the plane. It is therefore probable that both HR 4049 and HD 213985 are low mass stars.

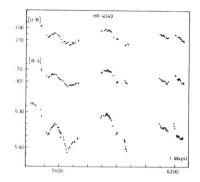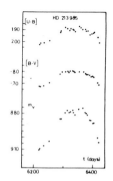

Fig. 1:   Photometric variability of HR 4049 and HD 213985.

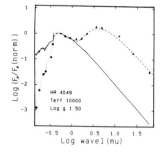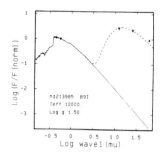

Fig. 2:   Energy distribution for HR 4049 and HD 213985

REFERENCES

Lamers, H.J.G.L.M., et al. (1986), Astron. Astrophys. 154, L20-L22.
Luck, R.E., Bond, H.E. (1984), Astrophys. J. 279, 729-737.
Waelkens, C., Rufener, F. (1983), Hvar Obs. Bull. vol.7, no.1, p. 29.

MILLIMETER WAVE OBSERVATIONS OF CIRCUMSTELLAR ENVELOPES
WITH THE IRAM TELESCOPE

A. Omont, T. Forveille, S. Guilloteau, R. Lucas
Observatoire de Grenoble - B.P. 68
38402 Saint Martin d'Hères Cedex - France

Nguyen-Q-Rieu
Observatoire de Meudon - 92195 Meudon Principal Cedex - France

L. Likkel, M. Morris
UCLA - Los Angeles - CA 90024 - USA

We report various recent observations of molecules in circumstellar envelopes of late type stars, mainly possible proto-planetary nebulae at various stages of evolution, and supergiants, including :

- First observation of SO_2 (Lucas et al. 1986) in five stars and SO in one. SO_2 is particularly strong in OH 231.8+4.2 (OH 0739-14) and OH 26.5+0.6, where one of its lines is the strongest non-maser line observed in these envelopes (Guilloteau et al. 1986).

- Search for HCO^+ with negative results in IRC +10216, but positive ones in OH 231.8+4.2.

- Observation of carbon rich molecules in oxygen rich envelopes. HCN observed in 8 stars is particularly strong in supergiants with massive shells (VYCMa, NMLCyg, IRC+10420, VXSgr) and in OH 231.8+4.2. In the latter, CS and HNC have a magnitude comparable to HCN ; CS is marginally detected in two others stars.

- Observation of various molecules in C-rich envelopes and in particular in CL 2688 and IRC +10216.

- Detailed studies of the particular bipolar nebula OH 231.8+4.2 including the detection of SO_2, SO, $H^{13}CN$, HNC, CS, HCO^+, OCS and thermal SiO, and a mapping of the 2.6 mm line of ^{12}CO. The ^{12}CO profile exhibits very broad wings (total width $\approx$ 200 km/s) and the spatial structure of a strong bipolar flow in agreement with the observation of Herbig-Haro like objects by Cohen et al. (1985).

I. Appenzeller and C. Jordan (eds.), Circumstellar Matter, 511–512.

- First millimeter wave observations of two high galactic latitude stars (HD 161796 and 89 Her) of F-supergiant spectral type with infra-red excess, which could be proto-planetary nebulae, and of a few other very cold and peculiar envelopes of IRAS stars, including the high galactic latitude stars SAO 163075 and IRC-20101.

These results have various implications on the structure and the kinematics of these objects, the chemical processes and the elemental abundances, and their evolution and nucleosynthesis.

References

COHEN, M., DOPITA, M.A., SCHWARTZ, R.D. and TIELENS, A.G.G.M. 1985, Astrophys. J. 297, 702

GUILLOTEAU,S., LUCAS, R., NGUYEN-Q-RIEU and OMONT, A., 1986 Astron. Astrophys. Lett. in press.

LUCAS, R., OMONT, A., GUILLOTEAU, S. and NGUYEN-Q-RIEU 1986, Astron. Astrophys. Lett. 154, L12

CIRCUMSTELLAR DUST AND CHEMISTRY

FORMATION AND DESTRUCTION OF DUST GRAINS IN CIRCUMSTELLAR REGIONS

Yu. A. Fadeyev
Astronomical Council
USSR Academy of Sciences
Pyatnitskaya Str. 48
Moscow 109017, USSR

ABSTRACT. The existence of circumstellar dust in late-type stars is connected with stellar pulsations since periodic shocks accompanying pulsations seem to be the most probable mechanism of the gas density increase needed for condensation of dust particles in outer layers of the stellar atmosphere. Most abundant solid materials formed in O-rich stars are forsterite (or enstatite), silicon monoxide and iron, whereas in C-rich stars these are carbon, silicon carbide, magnesium sulphur and iron. Application of the homogeneous nucleation theory shows that condensation proceeds at extremely high departures from thermal equilibrium due to a very long time between collisions of condensable monomers. The most efficient mechanism of destruction of dust grains is thermal evaporation caused by variations of the stellar effective temperature.

1. INTRODUCTION

Circumstellar dust exists not only around cold late-type stars, but also around supergiants with higher effective temperatures. These are, for instance, R CrB stars, RV Tau (Gehrz and Woolf, 1970) and W Vir (Lloyd Evans, 1985) pulsating variables. Recent IRAS observations allow to suspect that even long-period Classical Cepheids are surrounded by dust shells (Deasy and Butler, 1985).

The main property of pulsating stars surrounded by dust shells is an extremely high luminosity to mass ratio. In Classical Cepheids and W Vir variables this ratio is of about 10^3 $L_\odot/M_\odot$, whereas in R CrB stars, RV Tau variables and pulsating red supergiants the luminosity to mass ratio is $L/M \sim 10^4$ $L_\odot/M_\odot$. A simple linear analysis shows that radial pulsations in the atmospheres of such stars can exist only in the form of running waves (Unno, 1965). This conclusion is confirmed by nonlinear calculations for the

515

I. Appenzeller and C. Jordan (eds.), Circumstellar Matter, 515–527.

models of pulsating asymptotic-giant-branch stars (Wood,
1974), R CrB variables (King, 1980), and low-mass yellow
supergiants (Fadeyev, 1982). Moreover, nonlinear calculations show that periodic shock waves accompanying oscillations of such stars cause the mass ejection and seem to be
the main mechanism of mass loss. On the other hand, dust
grains cannot condense if the stellar atmosphere is in
hydrodynamical equilibrium. This is due to the fact that
gas density drops very fast with increasing radial distance, so that molecular compounds cannot be supersaturated.
Thus, it seems to be most possible that it is the periodic
shocks that cause increase of gas density needed for grain
formation in outer layers of the stellar atmosphere.

Dust formation in outer layers of the stellar atmosphere results in another mechanism of mass loss connected
with radiation pressure acting onto grains. So, mass loss
from pulsating stars can proceed as it is shown in Fig. 1.

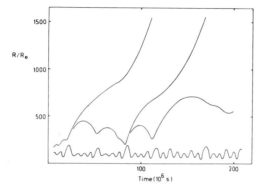

Figure 1. Temporal dependences of the outermost mass zones
of the pulsating supergiant.

Each pulsation period the shock wave propagating through
the stellar atmosphere causes expansion of gas with velocity comparable with the local escape velocity. As a result
vapors of some compounds becomes supersaturated and further
dynamical evolution of outer layers is governed by radiation pressure acting onto grains and momentum transfer from
grains to gas molecules.

2.   COMPOSITION OF DUST GRAINS

The main quantity determining the composition of dust is
the relative abundancy (by atoms) of carbon to oxygen
[C]/[O]. The most stable molecular compound formed in gas
cooling below of about 3000K is carbon monoxide CO. So, in
the O-rich stars almost all carbon atoms are tied in molecules of CO, so that carbon cannot take part in phase

transition. And vice versa, in C-rich stars all oxygen atoms are tied in CO molecules and formation of solid oxydes becomes impossible. The most abundant elements and their most abundant compounds existing in O-rich and C-rich stars are listed in Table I. Other elements such as natrium, aluminium, calcium etc. can be neglected due to their lower relative abundancy.

TABLE I

Element	Molecular compound	
	O-rich stars	C-rich stars
O	CO H_2O $(SiO)_s$	CO
C	CO	$(C)_s$ C_2H_2 CO $(SiC)_s$
Mg	Mg $(MgS)_s$	Mg $(MgS)_s$
Si	$(SiO)_s$	$(SiC)_s$
S	$(MgS)_s$	$(MgS)_s$
Fe	$(Fe)_s$	$(Fe)_s$

Some of the compounds listed in Table I can directly transform gas phase into solid state. In O-rich stars these are SiO, MgS and Fe. However, according to the thermochemical calculations done during last two decades (Wood, 1963; Lord, 1965; Gilman, 1969; Lewis et al., 1979) the most abundant solid materials formed in the atmospheres of O-rich stars are forsterite Mg_2SiO_4 and enstatite $MgSiO_3$. These compounds do not exist in gas phase and form due to reaction between gaseous cpecies. For instance, forsterite forms in the reaction

$$2Mg + SiO + 3H_2O = (Mg_2SiO_4)_s + 3H_2 . \qquad (1)$$

At lower temperatures solid forsterite reacts to solid enstatite according to the reaction

$$SiO + H_2O + (Mg_2SiO_4)_s = (2MgSiO_3)_s + H_2 . \qquad (2)$$

This conclusion is supported by the experimental studies of vapor-phase nucleation in the Mg-SiO-H_2 system (Day and Donn, 1978; Nuth and Donn, 1982; 1983a; 1983b). Abundant molecular compound which can directly transform from vapor into solid state is silicon monoxide. However, the experi-

mental studies of avalanche nucleation in the SiO-H$_2$ system
(Nuth and Donn, 1982) showed that SiO vapor condences
rather into Si$_2$O$_3$ than into more stable oxides SiO, SiO$_2$.
Unfortunately, these experiments were done at total pressures by 6-8 orders of magnitude higher than those typical
for grain-forming regions in the stellar atmospheres. The
presence of MgS particles in O-rich stars seems to be questionable since this solid material is extremely hygroscopic
and can be rapidly destroyed via reaction with water molecules (Nuth et al., 1985).

The most abundant compounds which can transform into
solid state in C-rich stars are carbon, silicon carbide,
magnesium sulphur and iron. Because the principal bearers
of gas phase carbon in typical C-rich stars are carbon
monoxide CO and acetylene C$_2$H$_2$, only a very small fraction
of gas phase carbon being in atomic form can take part in
nucleation process (Hoyle and Wickramasinghe, 1962). Further growth of carbon particles can proceed due to the
following exchange reaction:

$$C_n + C_2H_2 = C_{n+2} + H_2 \; , \tag{3}$$

where n is the number of carbon atoms in the particle. The
existence of MgS particles in C-rich stars was proposed
for explanation of the 30 μm emission feature observed in
some objects (Nuth et al., 1985; Goebel and Moseley, 1985).

3. INNER BOUNDARY OF CONDENSATION REGION

The main quantity governing the direction of the phase
transition is the supersaturation ratio

$$S = P_v/P_s(T) \; , \tag{4}$$

where P_v is vapor partial pressure, $P_s(T)$ is the equilibrium vapor pressure. The condition S=1 means that the
flux of molecules evaporating from the solid surface is
exactly the same as the flux of molecules incident onto
this surface, that is solid material is in equilibrium
with vapor. For S>1 vapor transforms into solid state since
the flux of incident molecules exceeds that of evaporating
molecules. And vice versa, solid material evaporates and
condensation is impossible for S<1.

The approximate temperature dependence of equilibrium
vapor pressure over a flat solid surface is determined by
Clapeyron-Clausius equation:

$$\frac{dP_s}{dT} \equiv \frac{h}{T(\Omega_g - \Omega)} \; , \tag{5}$$

where h is the latent heat of the phase transition. Let us assume that the latent heat does not depend of temperature and the bulk molecular volume Ω is neglegible in comparison with that of gaseous phase Ω_g. Then the solution of Clapeyron-Clasius equation (5) can be written in the following form:

$$\ln P_s(T) = -h/(kT) + C ,\qquad (6)$$

where the integration constant C is determined from laboratory experiments. It should be emphasized that equilibrium vapor pressure is determined for the flat solid surface. This is a very strong shortcoming of the condensation theory since usually the grain radii do not sufficiently exceed the molecular sizes.

The problem of grain formation in circumstellar envelopes is complicated due to the existence of the two different temperatures: gas temperature T and temperature of dust particles T_p. In the first approximation dust temperature can be estimated from the radiation energy balance equation

$$4\pi r^2 Q_p(r,T_e) W \sigma T_e^4 = 4\pi r^2 Q_p(r,T_p) \sigma T_p^4 ,\qquad (7)$$

where W is a dilution factor, r is a dust particle radias, T_e is the stellar effective temperature, Q_p is the Planck-mean absorption efficiency factor. As it was shown by Lefevre (1979), the flat solid surface of temperature T_p is in equilibrium with vapor of temperature T at vapor partial pressure

$$P' = P_s(T_p) (T/T_p)^{1/2} .\qquad (8)$$

Thus, the supersaturation ratio is determined from the following relation:

$$S = \frac{P_v}{P_s(T_p)} \left(\frac{T_p}{T}\right)^{1/2}\qquad (9)$$

Let us consider the spherically-symmetric steady expansion of gas. In this case gas density ρ in the layer with radius R and expansion velocity v is determined from the continuity equation:

$$\dot{M} = 1/(4\pi R^2 \rho v) ,\qquad (10)$$

where $\dot{M}$ is the matter outflow rate. Because temperature of expanding gas gradually decreases with increasing radial distance R, it is important to find the radius of the saturation level where S=1. For the fixed stellar effective

temperature T_e and matter outflow rate $\dot{M}$ this level represents an inner boundary of the grain-forming region because
the condition $S > 1$ is fulfilled at larger radii. Fig. 2
shows the radius of the saturation level as a function of
the effective ytemperature calculated for forsterite and
silicon monoxide at matter outflow rates $\dot{M} = 10^{-10}$ and 10^{-4}
$M_\odot yr^{-1}$. The bolometric magnitude of the model is -5 mag.

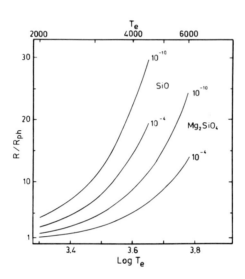

Figure 2. The radial distances of the saturation levels of
forsterite and silicon monoxide as a function of the effective temperature at matter outflow rates 10^{-10} and 10^{-4}
$M_\odot yr^{-1}$.

As it is seen from Fig. 2, the radius of the saturation
level exponentially increases both with increasing effective temperature and decreasing matter outflow rate. Such
a behaviour of the inner boundary of the condensation
region is connected with exponential temperature dependence
of equilibrium vapor pressure (6).

4.   HOMOGENEOUS NUCLEATION AND GROWTH OF DUST GRAINS

The saturation level gives only a lower limit of the
radius of the condensation region, whereas vapor has to
reach the critical supersaturation when the perceptable
fraction of vapor commences to transform into solid state.
This critical supersaturation can be found from equations
describing mass exchange between gaseous and solid states.
     Supersaturated vapor is in metastable state, so that
under equilibrium conditions vapor inevitably transforms
into stable solid state. At the beginning of the phase

transition the small droplets of the new phase appear in vapor. The probability of droplet formation due to thermodynamic fluctuations is

$$w \propto \exp(-\Delta G/kT) \; , \tag{11}$$

where ΔG is the change of the free Gibbs energy connected with formation of a droplet. Formation of droplets is accompanied by the competition of the two mutually opposite effects: creation of a spherical boundary of new phase and lowering of the chemical potential due to phase transition. Let us assume that droplets are spherical macroscopic clusters of molecules, so that macroscopic thermodynamics can be applied. In this case the change of the free Gibbs energy is

$$\Delta G = 4\pi r^2 \sigma - \frac{4}{3}\pi \frac{r^3}{\Omega} \, kT \ln S \; , \tag{12}$$

where σ is the surface tension energy, r is the radius of the droplet. The change of the free Gibbs energy as a function of the number of molecules in the cluster is shown in Fig. 3. The maximum of the free Gibbs energy is

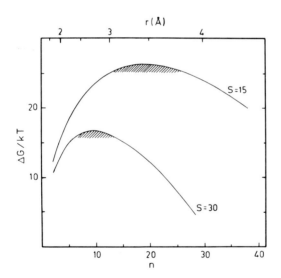

Figure 3.   The free Gibbs energy as a function of the number of carbon monomers contained in the cluster. Dashed areas represent the critical regions.

reached at the radius of the cluster

$$r_* = 2\sigma \Omega /(kT \ln S) \; . \tag{13}$$

For radii larger than the critical one r_* addition of a new

molecule onto the surface of the cluster causes decrease of
the free Gibbs energy, so that growth of such clusters is
more favourable than destruction due to evaporation. However, more detailed analysis shows that the number of molecules contained in the critical cluster stochastically
changes, so that there is the possibility that the critical
cluster may become the subcritical cluster due to fluctuations (Zel'dovich, 1942). So, within the critical region
where ΔG changes by of about kT the motion of the clusters
in the space of their sizes is ibeyed to Focker-Planck type
equation.

The nucleation rate, that is the number of critical
clusters formed per unit time per unit volume is written as

$$J = \alpha_s (4\pi r_*^2) \beta \, ZN_* \, , \tag{14}$$

where α_s is a sticking coefficient,

$$\beta = P_v / (2\pi m \, kT)^{1/2} \tag{15}$$

is impingement flux of monomers, m is the mass of the
monomer. The factor Z takes into account diffusion of the
clusters through the critical region (Feder et al., 1966).
The concentration of critical clusters is determined by

$$N_* = N_1 \exp(-\Delta G_*/kT) \, , \tag{16}$$

where N_1 is the concentration of monomers. According to
Draine and Salpeter (1977) it is convenient to introduce
the quantity

$$\Theta = 2(4\pi/3)^{1/3} \, \Omega^{2/3} \, \sigma/k \, , \tag{17}$$

whereby the expression for the free Gibbs energy may be
written as

$$\Delta G_*/kT = 1/2 \, (\Theta/T)^3 \, (\ln S)^{-2} \, . \tag{18}$$

The most uncertain quantity of the nucleation theory
is the surface tension energy σ. The review of the experimental data presented by Tabak et al. (1975) shows that
the surface tension energy of solid carbon is in the range
from 1000 to 3000 erg cm^{-2}. According to Draine (1979), the
surface tension energy of small graphite particles is $\sigma =$
1400 erg cm^{-2}. The estimates of σ for silicates are also
uncertain. Blander and Katz (1967) estimated the surface
tension to be $\sigma = 850$ erg cm^{-2}. Nuth and Donn (1982)
concluded that the surface tension energy of solid silicates is in the range from 500 to 650 erg cm^{-2}.

The phase transition in expanding circumstellar gas
is connected with competition of the two mutually opposite

effects: lowering of gas temperature and increase of the
mean collision time of monomers. So, at the beginning the
nucleation rate exponentially increases with decreasing
gas temperature. Simultaneously with formation of new
critical clusters the dust grains formed before commence
to grow due to accretion of vapor molecules onto their
surface. As a result, the nucleation rate reaches the
maximum and begins to diminish when growth of the dust
grains causes the perceptable depletion of the vapor and,
hence, decrease of the supersaturation. The calculations
show that the maximum nucleation rate is reached when from
1 to 3 percent of vapor transforms into solid state.

The quantities describing phase transition of forsterite in the O-rich star with T_e=2000K, M_{bol}=-5 mag and
M=3.16$\cdot$10^{-4} $M_\odot$ yr^{-1} are shown in Fig. 4. These calculations show that maximum of the nucleation rate is reached at
extremely high supersaturation ($S\sim 10^5$) corresponding to
the very small number of molecules in the critical cluster
($n_* \approx 3$), so that the applicability of the liquid droplet
approximation becomes doubtful. Such a high departure from
thermal equilibrium is due to the fact that the mean collision time of condensable monomers ($t_c \approx 10^4$ s) exceeds the
time interval of the nucleation burst by two orders of
magnitude only. The final radius of the forsterite grains
does not exceed 5$\cdot$10^{-7} cm.

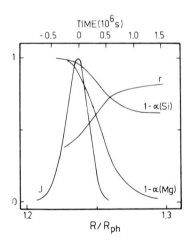

Figure 4.   Nucleation rate J, mean grain radius r, relative fractions of gas-phase magnesium and silicon as a
function of the radial distance from the stellar center.

Condensation of SiO proceeds at less favourable conditions since the nucleation peak occurs at $S \sim 10^6$ when
the mean collision time of SiO molecules is $t_c \approx 6\cdot 10^4$ s.
The final radius of SiO grains is of about 4$\cdot$10^{-7} cm. It

should be noted that in experimental studies the avalanche
nucleation of vapor-phase SiO was observed also at very
high supersaturations ranging from 10^3 to 10^5 (Nuth and
Donn, 1982).

5.   DESTRUCTION OF DUST GRAINS

Almost all late-type stars surrounded by dust shells are
pulsating variables, so that the periodic variations of
the effective temperature caused by stellar pulsations
seem to be the most important mechanism of destruction of
dust grains. Let us assume that the dust grains are spherical particles. In this case the rate of the change of
the radius of grains is determined by the following relation:

$$dr/dt = \Omega \, \alpha_s \, \beta \, (1 - S^{-1}) \; . \tag{19}$$

So, when the supersaturation ratio increases above unit
the growth rate of dust grains tends to its upper limit
depending of the impinging flux of monomers β . On the
other hand, decrease of S below unit is accompanied by
increase of the evaporation rate without any limit. This
means that the evaporation rate depends only on the rate
of increase of the stellar effective temperature.
    The simple criterion of the survival of forsterite
grains in the expanding circumstellar envelope can be
obtained with use of the diagram shown in Fig. 5. The
constant supersaturation of forsterite is maintained at

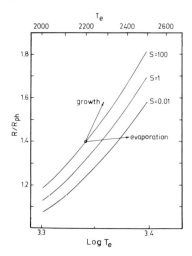

Figure 5.   The radial distances of the layers with constant supersaturation of forsterite as a function of the
effective temperature.

the radial distance obeying to the relation

$$\frac{d \log x}{d \log T_e} = 1.8 , \tag{20}$$

where $x=R/R_{ph}$, R_{ph} is the radius of the photosphere. The motion of the dust grains on this diagram can be represented by the following relation

$$\frac{d \log x}{d \log T_e} = \frac{T_e v}{R} / \frac{d T_e}{dt} . \tag{21}$$

Let us approximately assume that $dT_e/dt = 2 \Delta T_e/P$, where ΔT_e is the amplitude of the change of the effective temperature, P is the pulsation period. Then the condition of the survival of forsterite grains may be written as

$$T_e v P/(2R \Delta T_e) > 1.8 . \tag{22}$$

For instance, for the red supergiant with T_e=2200K, R_{ph}= $622R_\odot$ and dust velocity v=20 km s^{-1} the relation (22) may be rewritten as

$$P/ \Delta T_e > 4 \cdot 10^4 .$$

So, if the amplitude of effective temperature variation is ΔT_e=800K, the dust grains can be destroyed at pulsation periods less than 370 days.

Another mechanism of grain destruction which in principle may occur in circumstellar envelopes is sputtering caused by the drift of dust grains through gas. According to Wickramasinghe (1972) there is a few regimes of sputtering. Firstly, the energy of incident particles (atoms or molecules) E has to exceed the threshold energy E_t since at lower energies the collision does not cause escaping the atoms of lattice. If the energy of incident particles is less than a few tenth of keV, sputtering proceeds in the hard sphere collision regime and the sputtering yield (i.e. the number of atoms escaping the solid surface per one incident particle) is determined by the following empirical relation (Wehner, 1958):

$$Y = K (E - E_t) .$$

According to numerous studies reviewed by Barlow (1978) the threshold energy is approximately related to the sublimation energy of lattice H_s as $E_t=4H_s$, whereas the sputtering yield slope is $K=S_i/H_s$. The quantity S_i is constant for fixed solid material and incident particles. The values of S_i are given by Barlow (1978).

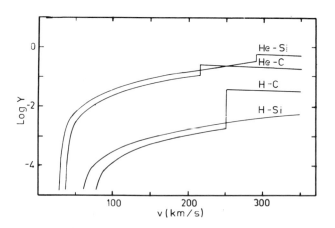

Figure 6. The sputtering yield Y calculated for solid
carbon and solid silicates colliding with atoms of hydrogen and helium as a function of the velocity of the
incident particle.

As it is seen from Fig. 6 sputtering occurs at too
high drift velocities which seem to be improbable in late-
type stars. For instance, destruction of silicate grains
due to sputtering by incident helium atoms commences at
the drift velocity of about 30 km s^{-1}.

REFERENCES

Barlow, M.J.: 1978, Monthly Not. Roy. Astron. Soc. 183, 367
Blander, M., and Katz, J.: 1967, Geochim. Cosmochim. Acta
         31, 1025
Day, K.L., and Donn, B.: 1978, Astrophys.J. (Letters) 222,
         L45
Deasy, H., and Butler, C.J.: 1985, preprint
Draine, B.T., and Salpeter, E.E.: 1977, J.Chem.Phys. 67,
         2230
Draine, B.T.: 1979, Astrophys. Space Sci. 65, 313
Fadeyev, Yu.A.: 1982, Astrophys. Space Sci. 86, 143
Feder, J., Russel, K.C., Lothe, J., and Pound, G.M.: 1966,
         Adv.Phys. 15, 111
Gehrz, R.D., and Woolf, N.J.: 1970, Astrophys.J. (Letters)
         161, L213
Gilman, R.C.: 1969, Astrophys.J. (Letters) 155, L185
Goebel, J.H., and Moseley, S.H.: 1985, Astrophys.J. (Let-
         ters) 290, L35
Hoyle, F., and Wickramasinghe, N.C.: 1962, Monthly Not.
         Roy.Astron.Soc. 124, 417
Jones, T.W., Ney, E.P., Stein, W.A.: 1981, Astrophys.J.
         250, 324

King, D.S.: 1980, Space Sci. Rev. 27, 519
Lefevre, J.: 1979, Astron. and Astrophys. 72, 61
Lewis, J.S., Barshay, S.S., and Noyes, B.: 1979, Icarus
        37, 190
Lloyd Evans, T.: 1985, Monthly Not. Roy. Astron. Soc. 217,
        493
Lord, H.C. III: 1965, Icarus 4, 279
Nuth, J.A., and Donn, B.: 1982a, Astrophys.J. (Letters)
        257, L103
Nuth, J.A., and Donn, B.: 1982b, J.Chem.Phys. 77, 2639
Nuth, J.A., and Donn, B.: 1983a, J.Chem.Phys. 78, 1618
Nuth, J.A., and Donn, B.: 1983b, J.Geophys.Res. 88,
        Supplement, A847
Nuth, J.A., Moseley, S.H., Silverberg, R.F., Goebel, J.H.,
        and Moore, W.J.: 1985, Astrophys.J. (Letters)
        290, L41
Tabak, R.G., Hirth, J.P., Meyrick, G., and Roark, T.P.:
        1975, Astrophys.J. 196, 457
Unno, W.: 1965, Publ. Astron. Soc. Japan 17, 205
Wehner, G.K.: 1958, Phys.Rev. 112, 1120
Wickramasinghe, N.C.: 1972, Monthly Not. Roy. Astron. Soc.
        159, 269
Wood, J.A.: 1963, Icarus, 2, 152
Wood, P.R.: 1974, Astrophys.J. 190, 609
Zel'dovich, Ya.B.: 1942, J. Exp. Theor. Phys. 11/12, 525

MOLECULAR CATASTROPHES AND THE FORMATION OF CIRCUMSTELLAR DUST

Robert E. Stencel
Center for Astrophysics and Space Astronomy
University of Colorado
Campus Box 391
Boulder, CO USA 80309-0391

ABSTRACT. Radiative instabilities due to simple molecules may convert chromospheric material into masering molecules and dust capable of being removed from the star by radiation pressure.

Among the four leading explanations for mass loss from red giant and supergiant stars (thermally driven winds, magnetically driven winds, pulsation and radiation pressure on dust grains), there is ample evidence that once dust grains form near such objects, radiation pressure is sufficient to drive them to infinity (Jura 1984). The problem of the formation of such dust is a classic one (Salpeter 1974) which requires understanding the combined roles of radiative transfer, gas dynamics and chemical reactions. Virtually all published studies have only attacked portions of the combined problem.

One key deficiency in previous efforts to understand the formation of circumstellar dust has been in using radiative equilibrium assumptions concerning the nature of the atmosphere underlying the circumstellar envelope (CSE). Jennings and Dyck (1972) asserted that chromospheres are 'quenched' in the presence of dust, based solely on optical Ca II K line emission. New ultraviolet and microwave analyses (Carpenter et al. 1985; Stencel et al. 1986; Hjellming and Newell 1983; Judge 1986) show the chromospheres of dusty red supergiant stars are persistent, and, unlike the solar chromosphere, may fill the entire volume out toward the base of the CSE (up to several stellar radii). I suggest that this extended chromosphere is prone to instabilities which ultimately result in the formation of dust grains. Such "thermo-chemical" instabilities are analogous to that discussed for the warm and cold phase of the ISM (Glassgold and Langer 1976; Lepp et al. 1985) and in the solar temperature minimum region (Kneer 1983). Kneer described this as "a dynamical situation far from radiative equilibrium, caused by molecules and the temperature dependence of their formation."

Compared to ISM conditions, the atmospheres of stars are denser and warmer (log density: 6-12; log T: 3-4). Simple molecules like CO, SiO, H_2O and OH are unique in that they have relatively high binding energies, absorb well in the UV and radiate efficiently in

I. Appenzeller and C. Jordan (eds.), Circumstellar Matter, 529–531.

the IR, and thus act as effective coolants. In high gravity stars like the Sun, the conditions in the upper photosphere tend to associate C and O. When this happens, the radiative cooling due to CO strongly cools the surroundings, leading to the formation of additional CO molecules which further enhance the cooling until complete CO saturation is achieved (a runaway process dubbed the "molecular catastrophe" by Kneer 1983 and Muchmore 1986. In the Sun, there is a striking difference between the brightness temperatures in the 2.3 micron CO band, and atomic features of the upper photosphere. It is this strong temperature sensitivity of molecular opacity which I propose can operate to ultimately lead to the formation of dust at the base of CSE in red supergiant stars.

Joseph Nuth (private communication) has computed the location of thermodynamic equilibrium in the $SiO = Si + \frac{1}{2}O_2$ reaction in density and temperature space. By comparing this locus with the temperature density run of model atmospheres for the Sun (VAL-C, 1981) and a low gravity star, Betelgeuse (Hartmann and Avrett 1984), we discover that the transition line lies just below the solar temperature minimum, but above the corresponding minimum in the low gravity atmosphere. We infer that the process leading to thermal bistability in the solar atmosphere involving CO is capable of operating with the analogous SiO molecule in red supergiants like Betelgeuse. Given that SiO is the basis for the formation of silicate dust grains, this molecular catastrophe may provide the process by which dust is formed and mass loss caused.

VLBI observations of the M4Ie supergiant VX Sgr by Chapman and Cohen (1986) and Lane (1984) are instructive in this context: the SiO masers lie closest to the stellar photosphere (at 1-2 radii), the OH and H_2O masers occur farther out (tens and hundreds of radii). Localized CO catastrophes in the stellar photosphere give rise to pressure perturbations which result in SiO formation catastrophes in the extended chromosphere of the star. The formation of SiO in excited states prompts the observed maser emission, and subsequent chemistry anneals the SiO into clusters and associations like olivine $(Mg, Fe)_2 SiO_4$ (Woolf and Nye 1969; Donn and Nuth 1985), which is removed from the star by radiation pressure. The OH and H_2O masers result from their formation catastrophes at lower temperatures and densities in the outer chromosphere/CSE where conditions associated with their lower binding energy phase change take place.

The molecular catastrophe description for the conversion of chromospheric gas into molecular masers and circumstellar dust holds promise for a coherent expanation of the formation of these entities and the process of mass loss from cool, high luminosity objects. We will report elsewhere on quantitative simulations of this scenario, in collaboration with David Muchmore and Joseph Nuth, incorporating a full treatment of gas dynamics, radiative transfer and chemical reactions. This work has been supported by CASA at the University of Colorado, for which the author is grateful. I also acknowledge the skillful assistance of Terry Armitage, Susan Barnes, and Michael Van Steenberg in manuscript preparation.

REFERENCES

Carpenter, K., Brown, A. and Stencel, R. 1985, *Ap.J.*, **289**, 676.

Chapman, J.M. and Cohen, R.J. 1986, *M.N.R.A.S.*, **220**, 513.

Clayton, D. 1985 in *Interrelationships Among Circumstellar, Interstellar and Interplanetary Dust*, NASA Conf. Publ. 2403.

Donn, B. and Nuth, J. 1985, *Ap. J.*, **288**, 187.

Glassgold, A. and Langer, W. 1976, *Ap. J.*, **204**, 403.

Hartmann, L. and Avrett, E. 1984, *Ap. J.*, **284**, 238.

Hjellming, R. and Newell, R. 1983, *Ap. J.*, **263**, L85.

Jennings, M. and Dyck, H. 1972, *Ap. J.*, **177**, 427.

Judge, P. 1986, *M.N.R.A.S.*, **221**, 119.

Jura, M. 1984, *Ap. J.*, **286**, 630.

Kneer, F. 1983, *Astron. Astrophys.*, **128**, 311.

Lane, A. 1984 in *VLBI and Compact Radio Sources*, eds. Fanti et al. (Dordrecht; Reidel), p. 329.

Lepp, S., McCray, R., Shull, J.M., Woods, T. and Kallman, T. 1985, *Ap. J.*, **288**, 58.

Muchmore, D. 1986, *Astron. Astrophys.*, **155**, 172.

Salpeter, E. 1974, *Ap. J.*, **193**, 585.

Stencel, R., Carpenter, K. and Hagen, W. 1986, *Ap. J.*, **308** in press.

Vernazza, J., Avrett, E. and Loeser, R. 1981, *Ap. J. Suppl.*, **45**, 635.

Woolf, N. and Ney, E. 1969, *Ap. J.*, **155**, L183.

DUST FORMATION IN M-STARS

H.-P. Gail
Institut für Theoretische Astrophysik
Universität Heidelberg
Im Neuenheimer Feld 561
D-6900 Heidelberg

ABSTRACT. A mechanism is proposed for the initiation of dust formation in stellar winds of M-type giants and supergiants. If Mg and Fe are ionized (M0...M4) dust formation is initiated by homogeneous nucleation of SiO, otherwise (later ≈ M2) by homogeneous nucleation of MgS. The condensation temperatures for these mechanisms agree well with observations.

1. INTRODUCTION

The temperature of the hottest grains at the inner edge of circumstellar dust shells according to model calculations for the radiative transfer problem range from ≈ 1000K down to ≈ 500K (R.-R. & H., 1982, 1983). High condensation temperatures are observed for stars later than ≈M2 while low temperatures are observed for spectral types between M0 and ≈ M4.
　　From experiment and theory it is known that nucleation suddenly occurs above a critical supercooling of the gas below the stability limit against vapourization of the bulk condensate. The observed large temperature range for condensation in stellar winds is incompatible with a single mechanism for all types of M-stars.

2. CHEMICAL COMPOSITION OF THE GAS

Crucial for the chemistry of dust formation is the composition of the gas phase. This determines which atoms and molecules are available. One has to consider two cases:
　　(i) In early type M-stars the chromosphere ionizes elements of low ionization potential. The chemistry is determined by ion-molecule reactions in this case (Clegg et al., 1983). The most abundant particles are H, H^-, H_2, O, Si^+, SiO, Si, OH, O_2 and H_2O.
　　(ii) Late type M-stars show no indication for the presence of a chromosphere and ion-molecules. The chemistry is determined by neutral radicals in this case. The most abundant particles are H, H_2, H_2O, Fe, SiO, MgS and Mg.
　　C and N are blocked in CO and N_2 in both cases.

I. Appenzeller and C. Jordan (eds.), Circumstellar Matter, 533–534.

3. THE INITIAL DUST FORMATION PROCESS

From observations it is known that dust forms rapidly once the temperature in the wind has dropped to sufficiently low values. This requires
that (i) only particles with high number density and (ii) only chemical
reactions with a high rate coefficient are involved in the condensation
process.

The first condition can be met only with particles formed from elements of high abundance (H, O, C, N, Fe, Si, Mg, S). C and N are blocked
in the extremely stable molecules CO and N_2. No H-O-compound exists
which condenses at high temperatures. Hence, in the first case (Fe, Mg,
....ionized) only SiO and in the second case (Fe, Mg not ionized) Fe,
SiO, MgS and Mg are available for the nucleation process. All these molecule tend to form condensates at high temperatures.

The second requirement cannot be met by
Fe and Mg since diatomic assoziation reactions
are slow. Clustering of SiO and MgS is fast,
however, since it involves no activation energy barrier and radiative assoziation of the collision complex is easy since already the dimer
contains four atoms. This favours high reaction rates.

Hence we have to conclude that in early
type M-stars dust formation is initiated by
clustering of SiO (at 500 .....600K) while in
late type M-stars clustering of MgS (at 800
.....900K) initiates dust formation (Gail et al.,
1986).

This does not mean that the observed dust
material is solid SiO or MgS since once stable clusters are formed they
will collect the remaining condensable elements and the material is subject to oxidation by water vapour.

4. CONCLUSION

By simple arguments based on the chemical composition of the gas and the
velocity of chemical reactions it is possible to isolate two chemical
reaction paths (clustering of SiO and MgS) which are responsible for the
onset of dust formation in early and late type M-stars, respectively.
The two mechanisms explain the observational facts.

This work is sponsored by the DFG (SFB) 132.

REFERENCES:

Clegg, van Ijsendorn, Allamandola: 1983, Month.Not.203,125.
Gail, Sedlmayr: 1986, Astronomy Astr., in press.
Rowan-Robinson, Harris: 1982, Month.Not. 200, 197,
                          1983, Month.Not. 202, 767.

THE DUST ENVELOPE OF IRC +10216

Stephen T. Ridgway [1]
Kitt Peak
National Observatory [2]
Tucson AZ, 85726

John J. Keady
Los Alamos
National Laboratory
Los Alamos, NM 87545

ABSTRACT. Infrared speckle measurements and photometry are used to assess the shape and thickness of the dust shell of the carbon rich Mira variable IRC +10216. A spherically symmetric radiative transfer model has been developed to approximate the physical parameters.

The carbon rich Mira variable IRC +10216 has been the subject of numerous observational studies. This work has mapped in some detail the characteristics of the spectrum and the spatial intensity distribution. However, the characteristics of the star and surrounding dust shell may not for the most part be inferred directly from the observational material. A modeling process must be applied to deduce the stellar parameters of astrophysical interest. With the aid of numerical radiative models for the dust envelope, the observational material suffices to specify within narrow ranges the principal model parameters.

IRC +10216 is a Mira type variable with a period of ≈ 640 days. The bolometric flux varies by a factor of ≈ 4 during the Mira cycle. The color temperature (based on K-L color) varies concurrently over the approximate range 530-575K (R. Joyce, private communication). The Mira periodicity is superimposed on a longer term secular decrease in mean luminosity of approximately 0.02 mag/year.

Spatial measurements by IR interferometry and lunar occultation show that the apparent angular size of IRC +10216 depends on position angle, wavelength, and phase. The time-averaged angular size in the E-W direction, defined by the half-intensity radius of the one-dimensional image, increases from ≈ 0.15 arcsec at $2\mu m$ to ≈ 0.4 arcsec at $11\mu m$. The variation with phase is not well determined observationally, but is probably $\approx \pm 20\%$.

The N-S dimension is greater than the E-W dimension by a factor ranging from $\approx 1.5x$ at $2\mu m$ to $\approx 1.2x$ at $5\mu m$. However, the N-S shape is more complex at some epochs and a single parameter description of the shape is inadequate.

The apparent bipolar geometry suggests that the E-W spatial data is more

[2] Research supported in part by NATO grant RG86/0080

[2] National Optical Astronomy Observatories, operated by the Association of Universities for Research in Astronomy, Inc., under contract to the National Science Foundation

I. Appenzeller and C. Jordan (eds.), Circumstellar Matter, 535–536.

suitable to describe the typical source dimension than either the N-S data or a mean, since the N-S extension may be the result of rather narrow polar 'windows' in the shell structure. For purposes of a model analysis, we attempt to characterize the situation near a typical recent maximum, since the most complete data are available for this phase. The situation at other phases can then be estimated by scaling the model parameters.

The models are spherically symmetric, with a sharp inner boundary where dust formation is assumed to occur, and an outer boundary sufficiently large to play no role in the analysis. The major grain constitituent is amorphous carbon. SiC grains are included as a secondary component to reproduce the $11\mu m$ SiC emission feature, but do not play a significant role in the analysis. The radiation transport and radiative equilibrium equations are solved by a partial linearization procedure to determine the dust temperature. The emergent intensity distribution is compared with observational data.

As is well known, it is possible to reproduce a given shell spectrum and size with a large range of model parameters. In the case or IRC +10216, however, the data strongly constrain the choice of parameters because of two additional pieces of information. The variation of diameter with wavelength constrains both the stellar temperature and the radial density law. And the observed stellar flux fraction at mid-IR wavelengths fixes the integrated dust extinction.

It was not possible to match all observed characteristics within observational uncertainty with any combination of parameters. This suggests that the assumption of spherical symmetry is not adequate to fully describe the source. However, it was possible to fit the observations well enough to conclude that the model parameters are probably a good approximation to the actual source characteristics. The central result is that the inner boundary of the dust shell is at approximately $5R_*$.

Interestingly, the dust density distribution in the inner regions of the shell is strongly constrained by the wavelength dependence of the angular diameter. If we assume that the dust density distribution is $\propto r^{-2.0}$ for $r \geq 12R_*$, then the distribution at smaller radii is $\propto r^{-2.3}$. Two arguments would suggest that the density distribution should vary as a higher power of r in the inner shell. First, the dust is thought to accelerate after formation as a result of radiation pressure. Second, molecular spectra show that the gas reaches a terminal velocity of approximately $14\frac{km}{sec}$.

The relatively shallow observed dependence of density on radius suggests that the acceleration of the dust occurs within a very narrow radial distance at the inner boundary of the dust shell, and that outside this 'impulsive' acceleration region, the acceleration continues at a relatively mild rate, $v \propto r^{0.3}$. This is consistent with the analysis of high resolution molecular line profiles, which have a 'plateau' in the velocity curve at approximately $9\frac{km}{sec}$ in the inner shell (Keady et al 1986). Additional details of the results of the shell modeling will be published elsewhere (Ridgway and Keady 1981, 1986).

Keady, J.J., Hall, D.N.B., and Ridgway, S.T. 1986, *Ap.J.*, in press.

Ridgway, S.T., and Keady, J.J. 1981, *Phil. Trans. Roy. Soc. Lond. A.*, **303**,497.

Ridgway, S.T., and Keady, J.J. 1986 (in preparation).

OPTICAL/INFRARED OBSERVATIONS OF RV TAURI STARS

M. J. Goldsmith[1], A. Evans[1], J. S. Albinson[1], M. F. Bode[2]

1. Physis Department, Keele University, ST5 5BG, UK.
2. Astronomy Department, Lancashire Polytechnic, PR1 2TQ, U.K.

ABSTRACT.   Optical/infrared observations of RV Tauri stars obtained at SAAO have allowed the natures of the dust shells around stars with infrared excess to be investigated.   The data suggest that dust formation occurs sporadically and that some stars have multiple shells. There is no photometrically discernible difference between carbon- and oxygen-rich stars or their dust shells. There is some evidence that stars with higher metallicity have more dust.

1. INTRODUCTION

Near simultaneous optical (UBVRI) and infrared (JHKLMN) broad band photometric observations of RV Tauri (RVT) stars were obtained at SAAO in June/July 1985.   The data were corrected for interstellar and circumstellar reddening; black body curves were fitted to the stellar fluxes and, where present, the infrared excesses (see Fig 1).   From these curves and the photometric colours, star and (where appropriate) dust shell parameters were inferred.

2. STELLAR PROPERTIES

A   number of our programme stars were identified in Kukarkin (1969) either as RVT?, or as variables other than RVT type; W Cen, for example, was classed as a Mira variable. However, while the present data cannot confirm RVT status, Mira classification is unlikely: there is no warm dust present and the spectral type is too early.   Similarly, EI Peg cannot be an A-type semiregular (SRA) variable since such stars are by definition class M or later, while EI Peg's intrinsic colours imply an early K type.

Infrared colour-colour plots revealed that the majority of the stars lay close to the G-M giant locus (see Fig. 2).   A small group (UY Ara, AI Sco, SX Cen, R Sge) had much redder colours, mainly due to the presence of circumstellar dust.

Comparing the colours and temperatures of the stars with Kurucz's (1979) models shows that, in general, their metallicity is similar to,

537

I. Appenzeller and C. Jordan (eds.), Circumstellar Matter, 537–540.

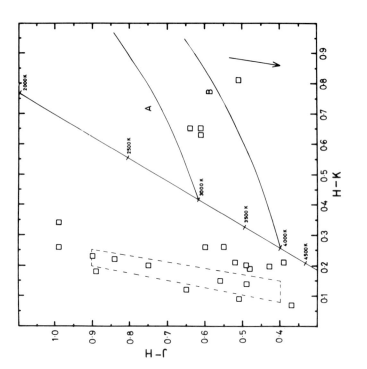

Fig. 2: (J-H)-(H-K) two colour diagram for RVT stars; data not dereddened. Dotted lines enclose region occupied by G-M giants; arrow denotes dereddening appropriate for visual extinction of 1 mag. Full lines denote blackbody locus, and combinations with 1000 K blackbody of 3000 K and 4000 K blackbodies.

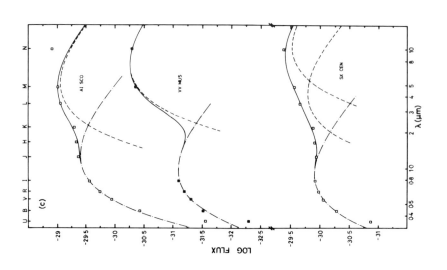

Fig. 1: Flux distributions of RVT stars with infrared excess; fluxes in W cm^{-2} Hz^{-1}.

or somewhat greater than, that of the sun. However V453 Oph seems to have extreme metal deficiency. This was the only C-type observed. Some of the globular cluster RVT stars are C-type, which would be consistent with the low metallicity of V453 Oph. No dust was found around this star.

3. DUST SHELLS

Of the 25 stars observed, infrared excesses were found in 11. The values of dust shell radii determined from their solid angles were in good agreement (~20%) with those calculated from dust temperatures, assuming the dust to be silicate based. Assuming carbon grains led to discrepancies by factors > 2. However, it is clear that some non-dielectric component is present and a dirty silicate seems the most probable grain material. The masses of dust present ranged from 10^{-} $M_\odot$ to 10^{-} $M_\odot$; mass loss rates, implied by the dust shell parameters, were typically ~ 10^{-} $M_\odot$ y^{-}. Such high values may reflect the sporadic nature of dust production in these stars, as they refer to the current state of their dust shells. A likely value for mean mass loss rate, on the basis of the stellar parameters, is 10^{-} $M_\odot$ y^{-}. The two velocity groups of RVTs (Joy 1952) may have different mean dust shell opacities, the low velocity, population I, metallic group having more massive dust shells than the high velocity objects.

While in some cases (e.g. AR Sgr and AC Her), the dust shells had their inner edges roughly at the condensation radius in others (e.g. VV Mus and RY Ara), the dust shells were situated much further from the star. The implication is that while dust is currently, or was recently, forming around some RVT stars, others have not formed dust for some considerable time: it is in any case clear that RVT stars do not produce dust at each pulsation, nor at each deep minimum but over some longer time scale, perhaps due to cepheid-like motion about the HR diagram (Deasy & Butler 1986).

Some RVT stars (e.g. SX Cen) had infrared excesses which must apparently be ascribed to multiple dust shells. In the case of SX Cen, optical and infrared variability on a timescale < 1 week was observed (see Fig. 3); the inner and probably the outer shell cooled significantly on this timescale. This behaviour cannot be ascribed to changes in radiative input to the dust from the underlying star, but may well be due to outward motion of the dust shells; such an interpretation may be supported by corresponding changes in the solid angles of the dust shells.

4. CONCLUSIONS

Simultaneous optical and infrared observations of RVT stars have enabled us to determine dust shell parameters for 11 stars. Dust production in RVT stars is apparently a sporadic process, with long periods of quiescence. When dust formation does occur, it is rapid, and shells may be observed to expand over timescales of days or less. There was no photometrically determinable difference between carbon and oxygen

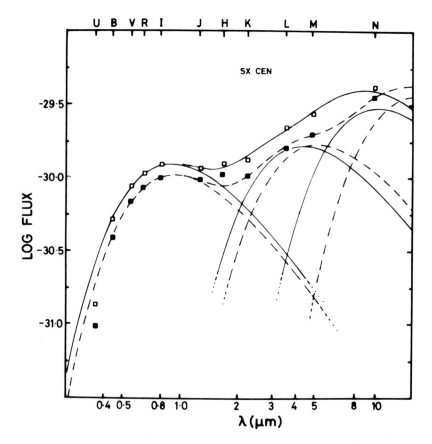

Fig. 3: Variations in the flux distribution of SX Cen over a 4 day period. Open squares (and full curves) denote data obtained prior to data represented by full squares (dotted curves). Flux in W cm^{-2} Hz^{-1}.

rich stars or their dust shells, although metallicity differences may be significant.

ACKNOWLEDGEMENTS

MJG is supported by the University of Keele, JSA and MFB by the SRC.

REFERENCES

Deasy, H. & Butler, C. J., 1986, Nature, 320, 726.
Joy, H. J., 1952, Ap. J, 115, 25.
Khukarkin, B. V. et al, 1969 "General Catalogue of Variable Stars".
Kurucz, R. C., 1979, Ap. J. Supplement, 40, 1.

SLOW VARIABILITY AND CIRCUMSTELLAR SHELLS OF RED VARIABLE STARS

T. Lloyd Evans
South African Astronomical Observatory
P.O. Box 9   Observatory 7935   Cape
South Africa

Infrared photometry shows that while all RV Tauri stars have circumstellar dust shells, the RVb stars with slow cyclic variations in mean light as well as the 30–100 day variations common to all RV stars have more hot dust close to the star (Lloyd Evans 1985).   Many M giant stars which are variables of semiregular type also show long-period variations in the mean light (O'Connell 1933; Payne–Gaposchkin 1954), with a roughly constant ratio between the two periods.   Payne–Gaposchkin (1954) found P_2/P_1 ~9.4 for red variables of type M and P_2/P_1 ~19.4 for stars of type F–K, most of which are RV Tauri stars. Re-analysis using the more extensive data available now indicates P_2/P_1 ~10 for the M giants and P_2/P_1 ~15 for the RV Tauri stars.   The nature of the long-period variability is unknown (Wood 1975).

K – [12] and [12] – [25] have values close to 0.65 and 0.10, respectively, throughout the spectral range M0–6 for stars with visual light amplitude Δm < 0.3 mag.   These colours become redder rapidly as a function of amplitude when Δm exceeds 0.3 mag.   Part at least of the increase in colour must result from the contribution of a circumstellar shell, as the IRAS Low Resolution Spectra show that silicate emission becomes common among the stars with larger K – [12] and [12] – [25]. Variability of small amplitude evidently promotes the formation of a dust shell just as occurs for the large amplitude Mira variables (Whitelock, Feast & Pottasch 1986; Jones, Ney & Stein 1981).   The stars with variable mean light have redder colours than average at a given amplitude, especially in the case of [12] – [25].

The distribution of the infrared spectral types of SR variables as a function of visual light amplitude has been studied using the classi-fication system described by Beichman et al. (1985) for the IRAS Low Resolution Spectra.   Stars with P_1 > 150 days were excluded to eliminate luminous stars which tend to have stronger silicate emission. The doubly periodic stars have a higher proportion of silicate emission spectra in four of the five amplitude subdivisions and show silicate emission more frequently than SRb, SRa or even Mira variables.

It is concluded that the occurrence of a long period variation in the mean light is conducive to the formation of circumstellar dust shells in semiregular variable stars of spectral type M, as it is in the case of the RV Tauri stars.

I. Appenzeller and C. Jordan (eds.), Circumstellar Matter, 541–542.
© 1987 by the IAU.

References

Beichman, C.A., Neugebauer, G., Habing, H.J., Clegg, P.E. &
    Chester, T.J., 1985.   IRAS Point Source Catalog Explanatory
    Supplement, Jet Propulsion Laboratory.
Jones, T.W., Ney, E.P. & Stein, W.A., 1981.   Astrophys. J., **250,**   324.
Lloyd Evans, T., 1985.   Mon. Not. R. astr. Soc., **217,**   493.
O'Connell, D.J., 1933.   Harvard Bull., **893,**   19.
Payne–Gaposchkin, C., 1954.   Harvard Annals, **113,**   191.
Whitelock, P.A., Feast, M.W. & Pottasch, S.R., 1986.   Calgary Workshop
    on Late Stages of Stellar Evolution.
Wood, P.R., 1975.   Multiple Periodic Variable Stars, p69, ed
    W.S. Fitch, D. Reidel, Dordrecht, Holland.

DUST FORMATION IN C-STAR SHELLS

E. Sedlmayr
Institut für Astronomie und Astrophysik
Technische Universität Berlin
Hardenbergstraße 36
D-1000 Berlin 12

ABSTRACT. The formation of carbon grains is described by a chemical pathway from acetylene via polyaromatic hydrogens (PAHs). The proposed mechanism is in excellent agreement with the observations and provides in particular the observed low condensation temperature which cannot be explained by classical nucleation theory.

1. INTRODUCTION

The cool extended envelopes of C-stars are well known to be places of copious dust formation. The observational evidence for this fact is manifested by the broad band extinction and reddening of the star light and by the occurrence of particular spectral features attributed to transitions of special functional groups within carbon compounds like the famous 2200 A band which is characteristic for electronic transitions of graphite-like structures or the IR transitions at 3.4, 6.2, 7.7, 8.6 and 11.3 μ which are due to particular vibrations of C-H or C-C groups of hydrogenated carbon or polyaromatic compounds, respectively, (e.g. Weast, 1976; Duley and Williams, 1979, 1981, 1983; Léger and Puget, 1984) but allow no definite conclusion about the real chemical and physical structure of the grains. Because of this rather vague observational and theoretical picture, any reliable theory on dust formation has to be based on the study of the elementary chemical processes (like soot formation in flames).

2. THE CONDENSATION TEMPERATURE AND THE BASIC NUCLEATING MOLECULE

Observations of the inner edge of dust shells confine effective dust formation to temperatures which are definitely lower than 1000 K (e.g. Rowan-Robinson and Harris, 1983). This low condensation temperature raises objections against explanation of carbon formation in the frame of classical nucleation theory which yields condensation temperatures around T $\simeq$ 1200 K, values which are definitely too large for being compatible with the observations (Gail and Sedlmayr, 1985).

I. Appenzeller and C. Jordan (eds.), Circumstellar Matter, 543–544.

In the relevant p-T-regime, the most abundant carbon bearing molecule able to condensate is acetylene (C_2H_2) which, therefore, has to be considered as the basic nucleating species. These two facts, the low condensation temperature and C_2H_2 as condensating molecule, provide the basis upon which any nucleation theory has to rely.

3. CARBON DUST FORMATION

A detailed study of hydrogen chemistry in order to construct a reliable pathway from acethylene to macroscopic "carbon grains" has been essentially performed in our group by R. Keller in his thesis (Keller, 1986a, b). In the following, we list the main results of our investigation:

- In the temperature-pressure regime where dust formation is observed PAHs are by far the most abundant large molecules. Therefore, the nucleation path has to proceed via a chain of PAHs.
- In order to obtain sufficiently large particle densities, i.e. a sufficient nucleation rate, the gas kinetic temperature has to be about 700...950 K.
- At this temperature the critical cluster is acephenanthrylene, a molecule containing 16 C-atoms.
- The growth of the supercritical molecules occurs by radical reactions, i.e. by a three step mechanism: i) formation of a free radical site by hydrogen abstraction, ii) attachment of an acethylene molecule as a side chain, and iii) ring closure. By the first two steps PAH growth and H_2 formation are coupled by a fixed proportion.
- In order to obtain sufficiently large growth rates, the H/H_2-ratio has to depart strongly from chemical equilibrium.

4. CONCLUSION

Along these guiding lines, a pathway to carbon grains can be constructed which is compatible both with fundamental chemistry and astrophysical observations.

REFERENCES

Duley, W.W.; Williams, D.A.: 1979, Nature, **277**, 40
Duley, W.W.; Williams, D.A.: 1981, Mon.Not.R.astr.Soc., **196**, 269
Duley, W.W.; Williams, D.A.: 1983, Mon.Not.R.astr.Soc., **205**, 67p
Gail, H.-P.; Sedlmayr, E.: 1985, Astron.Astrophys., **148**, 183
Keller, R.: 1986a, Thesis, Technische Universität Berlin, in prep.
Keller, R.: 1986b, Les Houches Workshop on Polycyclic Hydrocarbons and
     Astrophysics, in press
Léger, A.; Puget, J.L.: 1984, Astron.Astrophys., **137**, L5
Rowan-Robinson, M.; Harris, S.: 1983, Mon.Not.R.astr.Soc., **202**, 797
Weast, C.R.: 1976, CRC Handbook of Chemistry and Physics, CRC Press,
     Cleveland, Ohio

POSSIBLE RÔLE OF THE WHITE DWARF IN GRAIN FORMATION IN CV SYSTEMS

J. S. Albinson, A. Evans,
Department of Physics, University of Keele,
Keele, Staffordshire, ST5 5BG,
United Kingdom.

ABSTRACT. Zhilyaev & Zubko[3] described white dwarf model atmospheres in which carbon might condense. Whittaker[2] presented a version of the phase diagram for carbon on which there is a region where carbyne is thermodynamically favoured over graphite. Because the ratio of cross-section to mass is much higher for a thin cylinder, a carbyne cylinder has a much better chance of being blown out of a white dwarf atmosphere by radiation pressure. However even for small grains, gravity overcomes radiation pressure for cylinders, as it does for spherical grains. In the case of a non-polar cataclysmic variable (CV) system, the white dwarf is surrounded by an accretion disk and the luminosity of the disk may provide sufficient additional radiation pressure to drive a grain out of the system. If a grain can initially be transported out of the white dwarf atmosphere there may be several CV systems in which the combined radiation pressure of the disk and white dwarf can blow grains out of the system. For small grains and white dwarf masses > 0.5 M$_0$ radiation pressure can overcome gravity. These remarks also apply to grains that originate in the cooler regions of the disk, where the density may be high enough to sustain grain formation. Thus some non-polar CV systems may possess circumstellar dust shells, whose composition may reflect the nature of the white dwarf or secondary. For cases in which grains originate in the disk any circumstellar dust may be transient, as the mass transfer varies. Furthermore, in classical nova systems, this process may provide grain precursors, on which larger grains might grow during a nova outburst. A more complete discussion may be found in Albinson & Evans[1].

References.

1. Albinson, J. S. & Evans, A., 1986, in "Cataclysmic Variables", IAU Colloquium 92, D. Reidel, in press.
2. Whittaker, A. G., 1978, Science, **200**, 763.
3. Zhilayev, B. E. & Zubko, V. G., 1983, Sov. Ast. Lett., **9**, 122.

I. Appenzeller and C. Jordan (eds.), Circumstellar Matter, 545.

NUCLEATION IN NOVAE

C. M. Callus, J. S. Albinson & A. Evans
Department of Physics,
Keele University, Keele,
Staffs., ST5 5BG, U. K.

ABSTRACT. The observation of a deep minimum in the light curves of some novae, accompanied by a simultaneous rise in the infrared some weeks after outburst, is attributed to the rapid formation and growth of dust grains in the ejecta (Clayton & Wickramasinghe 1976). The observed nucleation rate $J(obs) \sim 10^{-10}$ cm^{-3} s^{-1} for typical dusty novae, whereas the expected homogeneous nucleation rate $J(hom) \sim 6 \times 10^{-26}$ cm^{-3} s^{-1}. We suggest that heterogeneous nucleation on ions could be a possible grain forming mechanism. In this case $J(het) \sim 3.3 \times 10^{-12}$ cm^{-3} s^{-1}, which is within an order of magnitude of the observed nucleation rate.

We have calculated ionization times for various species (e.g. Mg, Fe, Si) using the method described in Mitchell & Evans (1984). We find that, for typical nova abundances, several likely species are ionized before carbon, regardless of ejected mass. In fact for all elemental abundances (with the exception of very low hydrogen abundances) carbon is always the last of these elements to be ionized. Thus CI-ion reactions in suitable regions of the ejecta may initiate grain formation.

Observationally only novae of intermediate speed class produce large quantities of dust; this model provides some explanation for this observational phenomenon. Firstly in fast novae no dust is produced because carbon is ionized well within the radius of condensation. At the slowest end of the speed class range (e.g. HR Del), nucleation on ions becomes extremely inefficient as the density is too low by the time the nucleating species are significantly ionized. A more complete discussion may be found in Callus et al. (1987).

REFERENCES

Callus C.M., Albinson J.S., & Evans A., 1987 in "Cataclysmic Variables", IAU Colloquim 93, D. Reidel, in press.
Clayton D. D., & Wickramasinghe N. C.,1976, Ast. Sp. Sci., **42**, 451.
Mitchell R. M., & Evans A., 1984, M.N.R.A.S., **209**, 945.

I. Appenzeller and C. Jordan (eds.), Circumstellar Matter, 547.

CIRCUMSTELLAR CHEMISTRY OF COOL EVOLVED STARS

A.E. Glassgold and G. Mamon
New York University
New York NY, 10003
USA

ABSTRACT. A status report is given of the continuing development of the photochemical model for circumstellar envelopes around cool evolved stars, with emphasis on molecular ions in both O-rich and C-rich envelopes.

1. THEORY

In the photochemical model, molecules that enter the circumstellar envelopes (CSEs) of giant stars beyond the region of dust formation are broken down by interstellar UV photons that penetrate the envelope from the outside (Huggins and Glassgold 1982a; see also the reviews by Omont 1985 and Glassgold and Huggins 1986). The initial focus was on the distribution of the molecules and their photo products. For example, the dependence of the location of the OH masers on mass loss rate can be understood in terms of the shielding by dust of the radiation which dissociates H_2O (Huggins and Glassgold 1982b). More recently it has been possible to confirm that the photo destruction products generate strong ion molecule chemistry at intermediate regions in the CSE, typically 10^{16} to 10^{17} cm from the central star (Glassgold, Lucas, and Omont 1986, henceforth GLO). In C-rich CSEs, GLO showed that a variety of molecular ions are produced by interstellar UV and cosmic rays, the most abundant being $C_2H_2^+$ and C^+ produced by photoionization of acetylene. GLO also suggested that HCO^+ ought to be marginally detectable in IRC +10216.

During the last year we have improved the chemical model by including new laboratory information on the temperature dependence of ion polar molecule reactions and on recombination, and by using a new treatment of CO line self shielding (Mamon, Glassgold, and Huggins 1987) based on the Sobelev approximation. We present below our refinements to the C-rich model of GLO and our first results on O-rich CSEs.

I. Appenzeller and C. Jordan (eds.), Circumstellar Matter, 549–550.

2. C-RICH CSEs

A sensitive search for the 89 Ghz J=1-0 transition of HCO^+ in IRC +10216 (Lucas et al. 1986) yielded an upper limit to the antenna temperature of 20 mK, well below the GLO prediction of 95 mK. The improved model with standard parameters (e.g. $\dot{M} = 4\times10^{-5}$ M yr^{-1}) gives 24 mk, which is consistent with the observations. Further details, including a investigation of the abundance of HNC, will be presented in a paper in press (Glassgold, Mamon, and Omont 1986).

3. O-RICH CSEs

The spatial distribution of the ionization is similar for O-rich and C-rich CSEs: Just outside the region of dust formation, the ions are produced by cosmic rays; at intermediate distances, by photoionization of H bearing molecules (i.e. H_2O or C_2H_2); and at large distances by photoionization of C (from CO and, in the C-rich case, from C_2H_2). The two most abundant molecular ions in O-rich CSEs are HCO^+ and H_3O^+; H_3O^+ is produced by photoionization of water followed by rapid hydogenation. The surprisingly large abundance of HCO^+, almost 10^{-7}, is orders of magnitude larger than in C-rich CSEs. The main source of HCO^+ is the reaction $C^+ + H_2O \longrightarrow HCO^+ + H$, where C^+ is generated by the CO photochain. The quantitative aspects of the abundance of C^+ are critically dependent on the theory of CO line self shielding (Mamon, Glassgold, and Huggins 1987). The main conclusion here is that HCO^+ and H_3O^+ ought to be detectable in O-rich CSEs. Omont has in fact reported the detection of HCO^+ in OH 231.8 + 4.5 at this meeting. A more complete discussion of the ionization of O-rich CSEs will be presented by Mamon, Glassgold, and Omont (1987).

REFERENCES
Glassgold, A.E. and Huggins, P.J. 1986, in M, S, and C
Stars, Eds. H.R. Johnson and F. Querci (NASA and CNRS).
Glassgold, A.E., Lucas, R., and Omont, A. 1986,
        Astr. Ap. 157, 35.
Glassgold, A.E., Mamon, G. and Omont, A. 1986,
        Astr. Ap. in press.
Huggins P.J. and Glassgold, A.E. 1982a, Ap. J. 252, 201.
Huggins P.J. and Glassgold, A.E. 1982b, A.J. 87, 1828.
Lucas, R., Omont, A., Guilloteau, S., and NguyenQuang Rieu,
        1986, Astr. Ap. Lett. 154, 12.
Mamon, G., Glassgold, A.E., and Huggins, P.J. 1987, Ap.J.
        in press.
Mamon, G., Glassgold, A.E., and Omont, A. 1987, Ap.J.
        in press.
Omont, A. 1985, in Mass Loss in Red Giants, Eds. M. Morris
        and Ben Zuckerman (Reidel, Dordrecht), p. 269.

THE CHEMISTRY OF COOL CIRCUMSTELLAR ENVELOPES

L.A.M. Nejad and T.J. Millar
Department of Mathematics
UMIST
P.O. Box 88
Manchester M60 1QD

ABSTRACT. We have developed a time-dependent chemical kinetic model to describe the chemistry in the circumstellar envelopes of cool stars, with particular reference to IRC + 10216. Our detailed calculations show that ion-molecule reactions are important in the formation of many of the species observed in IRC + 10216.

1. INTRODUCTION

The envelope of IRC + 10216 is observed to contain many complex molecules nearly all of which have also been detected in the dark dust cloud, TMC-1. In this source, cosmic-ray ionisation drives the chemistry which, at least for the smaller molecules, is fairly well understood (Millar and Freeman 1984, Leung, Herbst and Huebner 1984). Ion-molecule chemistry can also occur in IRC + 10216 driven by both cosmic-rays and the external interstellar ultraviolet radiation field (Huggins and Glassgold 1982, Nejad, Millar and Freeman 1984). Parent species injected into the envelope at $r_0 = 10^{16}$ cm are CO, C_2H_2, HCN, CH_4, NH_3 and N_2, though this latter species has only a minor effect on the chemistry. These parents give rise to a system which totals 80 species whose radial abundances are calculated by integrating the resulting system of stiff differential chemical kinetic equations through use of the GEAR method. In Table 1 we present a summary of one of the many calculations performed by us.

2. DISCUSSION

The absorption of the external UV field by dust grains in the envelope ensures that at $r = 10^{16}$ cm the chemistry is driven by cosmic-ray ionisation of H_2. The H_3^+ ion so produced has a low proton affinity and undergoes proton transfer reactions with CO, C_2H_2, HCN and NH_3. The ion HCO^+ does not have a large abundance however because it transfers protons to C_2H_2, HCN and NH_3. The formation of $HCNH^+$ can lead to the production of HNC through dissociative recombination while

I. Appenzeller and C. Jordan (eds.), Circumstellar Matter, 551–552.

TABLE 1. Calculated column densities, $N(cm^{-2})$, for IRC + 10216 using $\dot{M} = 5 \times 10^{-5}$ $M_\odot$ yr^{-1} and V_e = 16 km s^{-1}.

Species	N	Species	N	Species	N
CN	7.4(14)	C_2H	1.8(15)	HCO^+	1.7(12)
HCN	2.1(13)	C_3H	5.5(13)	$HCNH^+$	1.8(12)
CCN	4.4(11)	C_3H_2	5.7(13)	CH_2CO	6.0(11)
C_3N	5.2(12)	C_4H	2.1(14)	CH_3CN	1.0(13)
HC_3N	4.6(12)	C_3O	3.7(12)	C_2H_3	1.1(14)

$C_2H_3^+$ reacts with C_2H_2, CH_4 and CO to form more complex species. As one proceeds outward in the envelope, the external UV dominates cosmic-rays and provides a source of ionisation as well as the creation of daughter species. In this region the chemistry is particularly interesting. The most important ions in driving the chemistry become $C_2H_2^+$, CH_3^+ and C^+. The ion $C_2H_2^+$ reacts with many species and leads to C_3O, HC_3N, C_3N, C_4H, C_3H_2, C_3H and CCN. CH_3^+ reacts slowly with H_2 but has an extremely rapid radiative association with HCN and an association with CO which produce CH_3CN and CH_2CO respectively.

Our detailed calculations enable us to make the following conclusions about ion-molecule chemistry in IRC + 10216.

1. Cosmic-rays and UV radiation provide a source of ionisation which drives an extensive ion-molecule chemistry in the envelope.

2. Given the uncertainties in both the observationally derived column densities and those calculated theoretically, we find a good agreement between observation and theory for many species including CN, C_2H, HNC, C_3H, C_3H_2, C_4H and CH_3CN.

3. The calculated abundances of C_3N and HC_3N are too low by more than an order of magnitude which may imply that they, or at least HC_3N, are formed in the warm, denser gas interior to 10^{16} cm.

4. Certain oxygen-bearing species such as C_3O, CH_2CO and HCO^+ may be detectable in IRC + 10216.

5. Similarities between the chemical composition of TMC-1 and IRC + 10216 can be explained by the occurance of a similar ion-molecule chemistry in both objects.

6. Differences in the absolute abundances of certain species in these two objects relate to the fact that in IRC + 10216, the chemistry involves the breakdown of stable parent molecules into atoms and atomic ions, while in TMC-1 it involves the build-up of these parents from atoms.

3. REFERENCES

Huggins, P.J. & Glassgold, A.E. 1982. Ap. J., 252, 201.
Leung, C.M., Herbst, E. & Huebner, W.F. 1984. Ap. J. Suppl., 56, 231.
Millar, T.J. & Freeman, A. 1984. MNRAS, 207, 405.
Nejad, L.A.M., Millar, T.J. & Freeman, A. 1984. Astron. Astrophys.,
    134, 129.

INVESTIGATION OF CIRCUMSTELLAR SHELLS BY MID-INFRARED
HETERODYNE SPECTROSCOPY

U. Schrey, S. Drapatz, H.U. Käufl, H. Rothermel,
and S.K. Ghosh
Max-Planck-Institut für extraterrestrische Physik
8046 Garching
W. Germany

Heterodyne spectroscopy at 11 µm combines high spectral re-
solution $(\lambda/\Delta\lambda \sim 10^6)$, high spatial resolution (< 1 arcsec
at 3 m telescopes) and high penetration depth. Therefore,
it seems promising to use it also for the investigation of
bright circumstellar atmospheres.
We have used our heterodyne spectrometer (Rothermel et al.
1983) for preliminary observations of a number of sources
(Schrey 1986). In the following the observation of IRC
10216 will be discussed, where a few absorption lines of
silane (SiH_4) have been observed (Fig. 1).
Fitting the experimental data to the relation:

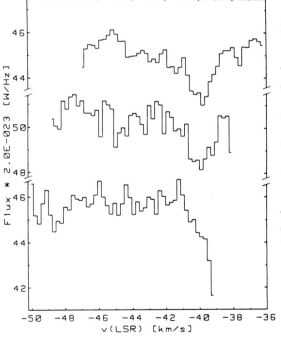

Fig. 1: Silane absorption
lines in IRC 10216

Transition, frequency ν,
equivalent width W:

$-P^+(4)F_2 \leftarrow F_1$:
  26939885±3 MHz, 15±3 MHz
$-Q^\circ(8)F_2 \leftarrow F_1$:
  27043262±10 MHz, 7±2 MHz
$-P^+(5)F_2 \leftarrow F_1$:
  26835800±200 MHz, 8±4 MHz

The spectral resolution is
0.2 km/sec, the integration
time per spectrum 2000 sec.

I. Appenzeller and C. Jordan (eds.), Circumstellar Matter, 553–554.

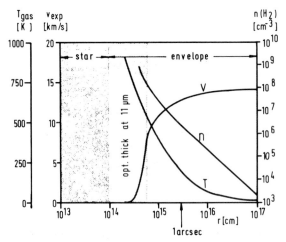

Fig. 2: H_2 density n, gas temperature T, and expansion velocity v as a function of the distance r from the star. The model of Lafont et al. 1982) is used with the asymptotic expansion velocity (14 km sec^{-1}) from Huggins and Healy (1986).

$$W/g\nu \sim N \exp(-E/kT_{rot})$$

with E, g the energy and statistical weight of the absorbing level, one obtains the rotational temperature T_{rot} and column density N:

$T_{rot} = 140 \pm 30$ K, $N = (8.5 \pm 1.3) \cdot 10^{14}$ cm^{-2}

A short interpretation of our measurement follows on the basis of Fig. 2.

The location (i.e. shell radius r) of the absorbing silane can be obtained:

- from the measured centroid velocity(13.5km/sec): $r = 8.10^{15}$cm
- from T_{rot}, which equals the kinetic temperature: $r = 5.10^{15}$cm
- from the range of expansion velocities over the length of the absorbing column, if given by the line width: $4.10^{15} < r < 10^{16}$cm. Since no absorption at lower expansion velocities has been found, SiH_4 has been formed at r, probably by radical reactions on silicate grains. Similar molecules are formed in that region, e.g. CH_4 (Clegg et al. 1982). The ratio of silane to molecular hydrogen is 4.10^{-7} for $r\sim3.10^{15}$cm. The solar Si/H ratio is 4.10^{-5}. The amount of Si locked up in SiH_4 is comparable to that in SiS and larger than that in gaseous SiO (Lafont et al. 1982).

Our measured continuum flux($1.4 \cdot 10^4$ Jy) can be compared with the IRAS 12 μm band value($2.4 \cdot 10^4$). This flux is due to an optically thick dust shell with a radius of 6.10^{14} cm and a temperature of 500 K.

References:

Clegg R.E.S., Hinkle K.H., and Lambert R.L., MNRAS 201, 95 (1982)

Huggins P.J. and Healy A.P., Astrophys.J. 304, 418 (1986)

Lafont S., Lucas R., and Omont A., Astr.Astrophys. 106, 201 (1982)

Rothermel H.,Käufl H.U.,Yu Y.,Astr.Astrophys.126,387(1983)

Schrey, U., MPE Report 198 (1986)

THE CIRCUMSTELLAR ENVIRONMENT OF L2 PUPPIS

Antonio Mario Magalhaes, Instituto Astronomico e Geofisico, Universidade de Sao Paulo, Caixa Postal 30.627, Sao Paulo 01051, BRAZIL and George Vincent Coyne, S.J., Specola Vaticana, V-00120 CITTA' DEL VATICANO.

ABSTRACT. Polarimetric observations of the red semi-regular variable L2 Puppis, obtained over a period of several years, confirm significant variations across the CaI 4226 line and several TiO bands. Together with published infra-red data, the observations point to a fundamental symmetry plane to which all variations with wavelength and time are related. The data are also consistent with a combination of photospheric effects, including a non-uniform distribution of calcium across the stellar disk, and scattering from grains in a cloud in which there is a systematic variation in grain size with distance to the star. Grain growth and dissipation, as evidenced by the observations, occur on a time scale of several years, in contrast to the optical variability time scale of months.

1. OBSERVATIONS.

Previous data collected with three distinct (Minipol, Vatpol and IAG) polarimeters are presented in Magalhaes et al. (1986). Additional data obtained in 1986 are given in Fig. 1. The most noteworthy features of the observations are: (a) for all states of the continuum polarization, the polarization at the CaI 4226 line is larger than the interpolated continuum and at a different position angle; (b) across the TiO bands the polarization may decrease or increase with or without a position angle variation; (c) the general behaviour of the continuum polarization from short to long optical wavelenths presents a definite trend with time, in the sense that a change to less and less pronouced slopes is indicated.

2. INTERPRETATION. The observed secular changes in the wavelength dependence of the polarization in L2 Pup are consistent with grain growth. The overall spectral dependence of the position angle may also be explained, since it would be expected from a systematic variation in grain size and geometrical/optical symmetry plane with distance from the photosphere. As grains form, grow, and dissipate into the cloud, they would cause the spectral dependence of the polarization to be less

555

I. Appenzeller and C. Jordan (eds.), Circumstellar Matter, 555–556.

steep, with an overall position angle reflecting the plane of symmetry indicated by the infrared and near infrared measurements. Our observations indicate that these processes take place on a time scale of several years in contrast with the optical variability time scale of 141 days.

The polarization in the TiO bands depends upon the variation of the ratio of absorption to scattering with optical depth at those wavelengths and upon the relative extent of the gas and dust and the amount to which they are mixed (Coyne and Magalhaes 1979).

The CaI 4226 data require a non-uniform, asymmetric calcium opacity, such as large spots, over the photosphere and a distinct geometry between such distribution and the scattering cloud. Note that this conclusion does not depend on any particular polarization model. For instance, light leaving the stellar surface at this wavelength may already be polarized due to Harrington's (1969) mechanism and/or to resonance scattering. In any case, the symmetry plane of the projected stellar disc in CaI 4226 must be different from that of the dust cloud.

Acknowledgements. The authors wish to thank Mr. W. Velloso for obtaining some of the 1986 data.

REFERENCES

Coyne, G.V. and Magalhaes, A.M. 1979, Astron. J. 84, 1200.

Dyck, H.M., Forbes, F.F., and Shawl, S. 1971, Astron. J. 76, 901.

Harrington, J.P. 1969, Astrophys. Lett. 3, 165.

Magalhaes, A.M., Coyne, G.V., Codina-Landaberry, S.J., and Gneiding, C. 1986, Astron. Astrophys. 154, 1.

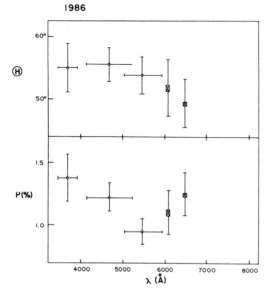

Fig. 1 Linear polarization and position angle for 1986.

TEMPERATURE DISTRIBUTIONS IN CIRCUMSTELLAR DUST SHELLS

Th. Henning and J. Gürtler
University Observatory
Schillergäßchen 2
Jena 6900
German Democratic Republic

1. ANALYTICAL TEMPERATURE DISTRIBUTIONS IN CIRCUMSTELLAR SHELLS

The knowledge of the temperature distribution in circumstellar dust shells (CDS) is needed for both the calculation of the emergent IR spectrum and the modelling of physical processes, e.g., dust formation.

For an optically thin CDS (i.e. re-emission neglected) we have derived analytical expressions (power laws) of the radial dependence of grain temperature $T_d(r)$ from the equation of energy balance, assuming three different types of dust opacity (for the procedure see Henning et al. 1983).

(1) Absorption efficiency $Q_\lambda(a) \sim \lambda^{-n}$

$$T_d(r) = \left(\frac{r}{r_\star}\right)^{-2/(4+n)} \left[\int_0^\infty \pi F^\star \lambda^{-n}\, d\lambda\right]^{1/(4+n)} \frac{hc}{k}\, 8\pi hc^2\, \Gamma(n+1)\, \zeta(n+4)^{-1/(4+n)} \tag{1}$$

(2) The star is presumed to radiate primarily in the UV, where for silicate dust $Q_{UV}(a) \approx 1$. The PLANCK means are represented by power laws.

(a) Absorption efficiency based on observations of thin IR emission of supergiant CDS (Henning et al. 1983); $Q_{10\,\mu m}(a) = 0.1$

$$T_d(r) = \begin{cases} 0.255\ (r_\star^2\, T_\star^4)^{0.335}\ r^{-0.671} & (1500\ K > T_d > 250\ K) \\ 8.939\ (r_\star^2\, T_\star^4)^{0.162}\ r^{-0.324} & (250\ K > T_d > 30\ K) \end{cases} \tag{2}$$

(b) Astronomical silicate (Draine and Lee 1984); grain radius a=0.1 μm

$$T_d(r) = \begin{cases} 0.523\ (r_\star^2\, T_\star^4)^{0.300}\ r^{-0.600} & (1500\ K > T_d > 275\ K) \\ 6.477\ (r_\star^2\, T_\star^4)^{0.179}\ r^{-0.358} & (275\ K > T_d > 30\ K) \end{cases} \tag{3}$$

Symbols used: $F^\star$, $r_\star$, $T_\star$ - flux, radius, and temperature of the star; $\Gamma(x)$ and $\zeta(s)$ Gamma and RIEMANN Zeta functions; h, c, and k - usual meaning.

Schwartz et al. (1983) obtained in the outer regions of S 140 IRS

I. Appenzeller and C. Jordan (eds.), Circumstellar Matter, 557–558.

$T_d(r) \sim r^{-0.3}$ in good agreement with Equations (2) and (3).

For optically thick CDS we developed a rapid-converging procedure for approximating $T_d(r)$. It starts from a linear combination of the upper (only geometrical dilution; no re-emisison) and lower limits (geometrical and optical dilutions; no re-emission) of the temperature. The coefficients are determined by satisfying iteratively the energy balance equations at the inner and outer boundaries of the CDS (Henning 1983). The resulting $T_d(r)$ could be fitted satisfactorily by a power law with an exponent between -0.5 and -0.7, depending on the model parameters.

2. ANALYSIS OF THE SPECTRA OF BECKLIN-NEUGEBAUER (BN) OBJECTS

For BN objects (catalogue - Henning et al. 1984; analysis - Gürtler et al. 1985) we found an inverse correlation between the strength of the silicate absorption feature and the colour temperature in the NIR, extending the analogous relation for Miras and OH/IR stars to higher optical depths (Fig. 1). There is a poor correlation between the optical depths of the silicate and of the ice bands (Fig. 2).

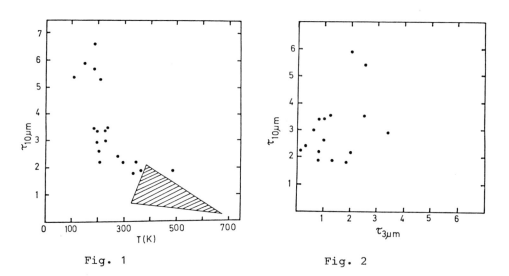

Fig. 1   Fig. 2

REFERENCES

Draine, B.T. and Lee, H.M.: 1984, Astrophys.J. **285**, 89.
Gürtler, J., Henning, Th., Dorschner, J., and Friedemann, C.: 1985, Astron.Nachr. **306**, 311.
Henning, Th.: 1983, PhD Thesis, University of Jena.
Henning, Th., Friedemann, C., Gürtler, J., and Dorschner, J.: 1984, Astron.Nachr. **305**, 67.
Henning, Th., Gürtler, J., and Dorschner, J.: 1983, Astrophys.Space Sci. **94**, 333.
Schwartz, P.R. et al.: 1983, Astrophys.J. **271**, 625.

EXPERIMENTAL STUDIES ON SIMULATED CIRCUMSTELLAR GRAINS

Joseph A. Nuth
Chemical Dynamics Inc.
Severn, Maryland   21144   USA

Bertram Donn
NASA/Goddard Space Flight Center
Greenbelt, Maryland   20771   USA

ABSTRACT. Over the last few years a better understanding of the
chemistry inherent in the refractory nucleation process has emerged
from experimental studies performed in several laboratories. These
studies shed light not only upon the factors which might control the
onset of grain formation, but also on the spectral characteristics of
freshly condensed grains.

The presence of several minor, broad, diagnostic features due to
amorphous silicates over the spectral range from about 8 to 25 microns
should be detectable in certain highly reddened circumstellar regions
such as OH 26.5 + 0.6. We have studied the spectral changes induced
in amorphous silicate smokes as a function of annealing in vacuo
(Figure 1) and have measured the rate of these changes as a function
of temperature (Nuth and Donn, 1984, J. Geophys. Res. (Red), 89,
B657). As a consequence of these studies we can make predictions
concerning the degree of crystallinity of grains ejected from
circumstellar regions if the time-temperature history of the material
can be estimated. We have also measured the rate at which the
infrared spectrum of initially amorphous magnesium silicate grains
changes as a function of hydrous alteration processes in liquid water
(Nuth, Nelson and Donn, 1986, Proc. 17th Lun. Plan. Sci. Conf., J.
Geophys Res. (Red), submitted). Such changes could be induced by the
condensation of a water layer over or into the refractory amorphous
core (cf. Jura and Morris, 1985, Ap. J., 292, 487) in the cooler
regions of the outflow. Hydration produces several diagnostic changes
which should be observable in moderately thick dust shells if such
processing occurs (Nuth, Donn and Nelson, 1986, Ap. J. (Lett).,
submitted). Our laboratory studies have indicated the need for higher
resolution infrared observations taken in a "continuous mode" rather
than "point-by-point" if we wish to learn details of the composition
and degree of crystallinity of circumstellar grains. Laboratory
synthesis of multicomponent silicate smokes which incorporate varying
proportions of the elements Na, K, Mg, Ca, Fe, Si, Al, Ti, O, H and Cl

I. Appenzeller and C. Jordan (eds.), Circumstellar Matter, 559–560.

is currently in progress. The structure and properties of these
multicomponent systems will be determined as a function of the bulk
compositon, thermal history and degree of hydrous alteration. In this
way, data will be available to reliably model the physical properties
of grains in a particular circumstellar environment based upon a
knowledge of the gas composition and the likely temperature/pressure
history of the average outflowing material.

Previous studies of the nucleation of SiO-H_2 (Nuth and Donn, 1982, J.
Chem. Phys., 77, 2639) and Mg-SiO-H_2 (Nuth and Donn, 1983, J. Chem.
Phys., 78, 1618) systems indicate that nucleation at temperatures in
excess of 950K is controlled by the formation of pure $(SiO)_x$ clusters
despite the fact that the Mg concentration is several orders of
magnitude higher than that of SiO. (This is shown in Figure 2 where
at high temperatures the Mg-SiO-H_2 system nucleates at the same SiO
partial pressure as did the SiO-H_2 system.) A multi-element cluster
beam apparatus is under construction which will allow us to measure
the relative stabilities of various refractory, pre-condensation
molecular clusters as a function of the temperature, pressure and
composition of the ambient gas. This system will be used to determine
the molecular pathway from refractory atoms and molecules to solid
particles so that a realistic kinetic model for the formation of
circumstellar grains can be formulated. We will also carry out a
series of experiments aboard NASA's KC-135 Reduced Gravity Research
Aircraft which are designed to measure the conditions under which a
number of refractory materials nucleate from the vapor. These
experiments will be free from the uncertainties associated with the
possible presence of gravity driven convective instabilities which
could have effected our previous measurements of the partial pressure
of the refractory vapor at the time of nucleation. These experiments
may also yield data from which we can derive the "sticking
coefficient" for the coagulation of very small refractory grains
colliding at very low relative velocities.

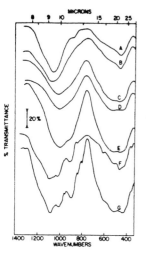

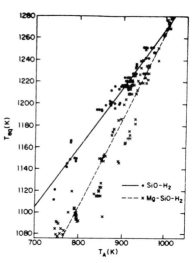

Fig.1 (left) Infrared
spectra of MgSiO smoke
annealed at 1000K in
vacuo for 0,1,2,4,8,
16.5 and 30 hours.

Fig.2 (right) Plot of
the temperature at
which SiO would be in
equilibrium at the
observed pressure vs.
the ambient tempera-
ture of the system
for SiO-H_2 and Mg-SiO
-H_2 nucleation expts.

FUTURE PLANS

NASA PLANS RELEVANT TO THE STUDY OF CIRCUMSTELLAR MATTER

Robert E. Stencel
Center for Astrophysics and Space Astronomy
University of Colorado
Boulder, CO 80309-0391 USA

ABSTRACT. The Astrophysics program of the National Aeronautics and Space Administration of the United States emphasizes use of vehicles to obtain above-the-atmosphere observational advantages, including expanded electromagnetic frequency access, enhanced sensitivity resulting from reduced or eliminated atmospheric absorption of light and image smearing. Space technology provides a superior means for astrophysical inquiry, particularly in the case of circumstellar material. Much of the flight program is undergoing intensive review following the Space Shuttle disaster of January 1986.

1 PROGRAM STRATEGY AND STATUS

As part of a national and international effort, experiments useful for the study of circumstellar matter often must have additional application to a wide range of important current research topics in astronomy, in order to help justify the cost of development. The approach being taken by the NASA Astrophysics Program involves a succession from initial modest experiments, building toward the more ambitious, all underpinned with theory and supporting work in related physics. The theoretical scaffolding that underlies our interpretation of the cosmos must be relied upon as the basis for choice among several attractive options. Also, without the input of physicists and chemists, the interpretation of atomic and molecular spectra, relativistic phenomena and other processes would be impossible.

1.1 THEORY AND LABORATORY ASTROPHYSICS

In addition to ongoing support of individual theoretical investigations, NASA solicitied proposals in 1984 for the first Astrophysical Theory program, intended to support small teams of scientists working on significant modern problems. These first round selections in a kind of "mini-Sonderforschungsbereiche" included studies of UV and X-ray spectral processes, solar and stellar oscillations, physics in the early Universe and of compact objects, star formation and high energy shock processes in astrophysics.

In the realm of laboratory astrophysics, NASA continues to support a variety of lab and theoretical efforts at Universities and its centers for the quantitative spectroscopy of

I. Appenzeller and C. Jordan (eds.), Circumstellar Matter, 563–567.

atoms, ions, molecules and solid state materials relevant to astrophysics. Examples include the atomic data center at the National Bureau of Standards (Wiese and Martin), ultraviolet cross sections and transition probabilities (Parkinson and Smith at CfA, Lee at SDSU), infrared molecular and PAH spectroscopy (Huntress and Poynter at JPL; Allamandola at Ames) and microgravity interstellar particle simulations (Nuth and Donn at Goddard).

1.2 SUB-ORBITAL EXPERIMENTS

The initial experiments in a new area (e.g. extreme ultraviolet astronomy) have been modest, low cost efforts to sample existing celestial sources with state of the art detector systems. These experiments include telescopes and instruments mounted on sounding rockets, under high altitude balloons, or, on the Kuiper Airborne Observatory (KAO) – a telescope-equipped aircraft specializing in infrared work. The KAO has recently returned from a successful Halley observing expedition in Australia where H_2O of cometary origin has been detected for the first time.

The SPARTAN program has begun to move this class of efforts into the Space Shuttle era. SPARTAN is a semi-autonomous satellite, capapble of being carried into orbit on the Shuttle, deployed for a few days as a temporary free-flying experiment, then retrieved and returned. The scientific experiment is pre- programmed and the data captured on tape. SPARTAN, like the sub- orbital efforts, offers a low cost way to pursue timely, well focused experiments, as well as hands-on experience for instrument developers and students. In this way, new technology can be tested and verified in space prior to major investments in larger efforts.

Whereas a dozen SPARTANs were selected for development after the first successful flight of an Xray experiment in June 1985 (Cruddace et al. 1986), the curtailment of the Space Shuttle launches for an indefinite period has forced re-evaluation of these efforts. Many are being reconsidered for sounding rocket interim experiments until Shuttle launches resume.

1.3 MODERATE COST EXPERIMENTS

An important step in this logical succession is to inventory the celestial information that the latest technology can provide. Such sky surveys have been accomplished at several wavelengths so far, most notably in the infrared with IRAS and in the X-ray regime with UHURU. These surveys provide the basis for future work, both in terms of focused, modest scale efforts and for facility class instruments which perform detailed studies. This class of experiments includes the Explorer program, which is a series of space science efforts dating back to the beginning of the space age. Recent Explorers include the International Ultraviolet Explorer (IUE), the Infrared Astronomy Satellite (IRAS), the Solar Mesopheric Explorer (SME), the Dynamics Explorers (DE) and others.

Explorers presently under development include the Cosmic Background Explorer (COBE) and the Extreme Ultraviolet Explorer (EUVE). COBE was being readied for launch in 1987, but is now being refitted for a Delta launch at great cost to the Explorer queue. EUVE is on a slowed development track. NASA had solicited new concepts for the next generation of Explorers and received nearly 50 responses in July 1986, but the fate of those proposals hangs in the balance as the impacts of the Shuttle reprogramming affect the entire agency (see below).

A separate class of comparable cost efforts called Spacelab experiments sadly may never see fruition beyond the flights of Spacelab 1 and 2 because of the Shuttle hiatus. An important payload called Astro was being readied to observe comet Halley with a trio of ultraviolet imaging, spectroscopy and polarimetry telescopes of the meter class. This instrument was also programmed to perform a number of important new astronomical investigations over a several reflight effort, but its fate is presently as dark as the warehouse near Cape Kennedy in which it is being stored.

1.4 MAJOR FACILITIES

The success of the Explorer program and the impressive capabilities of aerospace engineering offered impetus to scientists and NASA officials to consider large experiments which could serve as observatories in space. The Hubble Space Telescope (HST), the approval of which was intricately connected with the approval for the Shuttle program itself, set the tone for a series of proposed "Great Observatories". These include the Advanced X-ray Astrophysics Facility (AXAF), the Space InfraRed Telescope Facility (SIRTF), the Gamma Ray Observatory (GRO) and the Large Deployable Reflector (LDR). The approximate expected capabilities of these instruments are detailed in Table 1. While the HST and the GRO have both been under development and are nearing launch readiness, the others remain unstarted high priorities to the US space astronomy community (see the recent Report of the Astronomy Survey Committee of the National Academy of Sciences). Under present economic conditions, a choice may ultimately have to be made between timely execution of the Great Observatories program and any of the new Explorer concept proposals.

The circumstellar material experiments possible with the Hubble Space Telescope are a good example of the type of science that would be possible with the Great Observatories. For example, the High Resolution Spectrograph (HRS) will be able to examine cool star spectra with enough signal-to-noise and spectral resolution to determine whether the chromospheric motions suggested in several IUE studies are real. These include "downflows" in stars with transition regions (Ayres et al. 1983; Brown et al. 1984), and non-monotonic outflows in red supergiant star winds (Carpenter 1984). To fully assess the energetics and origins of circumstellar matter, a fuller understanding of the underlying stellar atmosphere is necessary. The HRS offers this possibility, particularly given the preparation of scientific questions made possible with IUE studies.

A second important class of experiments will be done with the imaging experiments on HST, including the Wide Field/Planetary Camera and the Faint Object Camera. In either case, the image scale and selection of narrow band filters plus occulting modes should be sufficient to permit unique studies of extended material around nearby or other stars with angular extent in excess of roughly 50 milliarcseconds.

Finally, the capability of the Faint Object Spectrograph to perform polarimetry as well as spectroscopy offers an entre into the important study of inhomogeneities and the

Table 1: Approximate expected capabilities of proposed experiments

	Wavelength or Energy	Angular Resolution	Spectral Resolution	Sensitivity
GRO	0.1-1000MeV	0.1 deg.	6-250	10 x CosB
AXAF	0.1-10 keV	0.5 arcsec	10^3	100 x HEAO-B
HST	1200-8000A	0.1 arcsec	10^5	V = 25
SIRTF	3-700 micron	1-18 arcsec	10^3	1000 x IRAS

role of magnetic fields in circumstellar environments. The impact of the instruments on HST to all phases of stellar astronomy concerned with circumstellar matter promises to be immense.

2   COMMENTS ON THE CHALLENGER DISASTER

Whereas a detailed discussion of the cause of the Space Shuttle accident of 1986 January is beyond the scope of this review, I feel that given the international impact of the event to space astronomy programs, IAU members and other readers deserve whatever additonal insight I can provide based on my recent four years of experience at NASA Headquarters. An excellent and critical review of the origins of the Shuttle program was published by Logsdon (1986) and this is to be recommended as a basis for understanding the present circumstances. The Shuttle program labored under several pressures: (1) the post Apollo era search by NASA for a major mission to justify its existence, plus increasing federal budget pressures, which lead to the agreement to fly a "commercially viable" Shuttle; (2) government-wide increasing reliance on contract labor rather than in-house engineering expertise, which distanced managment and the workers, and, (3) that NASA commanded a declining percentage of US aerospace business, and could no longer influence pricing during high inflation periods.

These factors lead to a configuration of circumstances in 1986 January which contributed to the launch outside of safe limits despite objections from knowledgeable engineers. The Shuttle launch frequency had been growing from every few months toward an ambitious monthly rate. This included opening a second launch pad at Kennedy in time for the planned 1986 May twin launches of Ulysses and Galileo interplanetary probe payloads on Shuttles, as well as opening a new launch facility at Vandenberg in California for Department of Defense use, also in 1986 (March). All of this growth in capability relied heavily upon a central set of crews at Kennedy which already were being worked seven days a week, 24 hours per day. Compounding this was the pressure to meet cometary observation windows. The warning implied by the increasing frequency of launch delays for STS 61-C and 51-L was unfortunately not heeded. Much of the recent pressure to reduce costs while increasing launch frequency came from a Directive from President Reagan in August 1984, ordering NASA to develop a "fully operational and cost effective shuttle program by 1 August 1988" (Goodwin, 1986).

While we can hope that history records this period as a pause to reformulate a successful space exploration strategy, the consequences of the present situation are not yet clear. Whereas the cost to space astronomy and other space sciences will be large in terms of maintaining launch readiness of developed instruments like HST at the expense of starting any new initiatives, there may also be several benefits. First, the previously planned frequency of Shuttle launches would have aggravated a growing problem of low earth orbit debris buildup, which already was manifest in the Shuttle glow problem that harmed certain astronomical observation attempts from Shuttle (e.g. Faust on Spacelab 2). Second, a reassessment of the balance between manned and unmanned launch vehicles is underway. This balance influences the kinds of space astronomy that are possible as a function of payload capacity, pricing, flight frequency and opportunity. Third, before embarking on a major developmental effort such as the proposed Space Station, a larger consideration of the committment for the project may be in order. As Logsdon (1986) points out "Decisions to make capital improvements in major facilities require more than [just] an initial approval. To be effective, they must be accompanied by a political committment to provide the resources required over the lifetime of the program on a timely basis. Further, it makes little sense to invest in a capability intended to enable a wide range of scientific and technological activities if adequate support for those activities is not also provided."

We are fortunate as astronomers that our fellow citizens possess so much interest in exploration of the cosmos, that our proposals for scientific instruments receive national and international support. It is encumbent upon us to channel this support into productive capabilities that enrich the intellectual lives of all thinking persons.

REFERENCES

Ayres, T., Stencel, R., Linsky, J., Simon, T., Jordan, C., Brown, A. and Engvold, O. 1983, *Ap. J.*, **274**, 801.
Brown, A., Jordan, C., Stencel, R., Linsky, J. and Ayres, T. 1984, *Ap. J.*, **283**, 731.
Carpenter, K. 1984, *Ap. J.*, **285**, 181.
Cruddace, R. 1986, preprint.
Goodwin, I. 1986, *Physics Today*, **38**, 41.
Logsdon, J. M. 1986, *Science*, **232**, 1099.

CONCLUDING REMARKS

CIRCUMSTELLAR MATTER, WITH PARTICULAR REFERENCE TO JETS AND MOLECULAR FLOWS

F. D. Kahn
Department of Astronomy
University of Manchester
Manchester M13 9PL
England

ABSTRACT. Circumstellar matter exists in many forms, such as winds from early and from late type stars, in infra-red sources and OH and H_2O masers, in X-ray sources and jets, and molecular flows and Herbig-Haro objects. There is considerable diversity in the nature of the starlike bodies that underlie these phenomena. Here an attempt is made to trace the connection between some of the observations related to pre-main sequence stars and their surroundings, with emphasis on jets and molecular flows.

I.   INTRODUCTION

The astrophysical significance of diffuse matter in space is now well-recognised. It has not always been so; the real change began with the development of radio-astronomical techniques, notably the use of the 21 cm line for the observation of atomic hydrogen in interstellar clouds. Since then many other parts of the spectrum have been opened up, and our understanding has improved with each new band of wavelengths. One only has to consider how the ultra-violet observations made with COPERNICUS entirely changed our views of interstellar physics, and its greatest contribution was the discovery of mass loss from early type stars as far as this Symposium is concerned. Developments in other parts of the spectrum have been equally worthwhile, and so has the application of new techniques; molecular spectroscopy at microwave frequencies has been used to map the structure of circumstellar regions, infra-red observations have probed deep into the dust clouds around newly formed stars, radio interferometers have revealed in great detail the very fine structure that exists in maser sources around both young and evolved stars, and X-ray satellites, notably the EINSTEIN Observatory, have detected significant emission from the vicinity of a wide variety of stars.

Our Organising Committee were very wise to decide that the time had come for a major meeting to review the whole subject of circumstellar matter. Their reward, a harvest of 175 talks and poster papers, testifies to the widespread interest and activity in this branch of astrophysics.

I. Appenzeller and C. Jordan (eds.), Circumstellar Matter, 571–580.

II. THE FORMATION OF MASSIVE STARS

To a large extent the proceedings of this meeting have continued the discussions that took place in Tokyo last November, at the highly successful
IAU Symposium 115, on Star Forming Regions. It is now agreed that star
formation takes place preferentially in giant molecular clouds, and that
certainly in the case of the more luminous stars it occurs mainly near
the potential troughs in the gravitational field caused by the spiral arms
of the Galaxy. The stars themselves condense from diffuse gas following
the collapse of a molecular cloud under its own gravitation. The process
is only moderately efficient at best: it is estimated that typically only
a few per cent of the mass of such a cloud is converted into stars and
that the remainder stays diffuse or is returned to the diffuse state.
Inevitably then the space around newly formed stars is full of circumstellar matter, which manifests itself in quite distinct ways depending
on the mass and luminosity of the newly formed star. Its Kelvin-Helmholtz
time t_{KH} is an important physical parameter, and so is the collapse time
t_{coll} of the mother molecular cloud. Massive stars evolve rapidly towards
the main sequence, so that their t_{KH} is less than t_{coll}. The star inside
will reach its equilibrium state while there is still a copious inflow of
material outside it. This accretion can continue only if the star is
surrounded by a cocoon which converts its optical and ultraviolet emission into infra-red radiation. The direct stellar radiation cannot be
allowed to reach the matter in the collapsing cloud at large: if it were
to do so the repulsion of the gas-dust mixture by radiation pressure would
far exceed the inward attraction by gravitation, and would soon halt the
accretion. No bright star can form in the Galaxy unless it is enclosed
in a circumstellar cocoon: the infra-red radiation which is created by
the cocoon then pumps the OH and H_2O maser sources that often occur nearby. Eventually the star becomes luminous enough to drive away the cocoon,
and the remaining circumstellar material associated with its creation,
and can then be observed in the optical and the UV.

III. X-RAYS AND JETS FROM PROTOSTARS OF LOW MASS

Stars of smaller mass have a lower ratio of luminosity to mass, and also
lower surface temperatures. Radiation repulsion of the accreting gas
and dust around such stars is unimportant; on the other hand the Kelvin-
Helmholtz time is now much longer, so that the mass falls onto the stellar surface much faster than the newly formed star can settle into its
equilibrium state via radiative processes alone, and so the dynamics is
significantly different. The problem with the massive stars was to discover how the accretion process can continue despite the intense repulsion
by the radiation field. By contrast a less massive star cannot emit thermal radiation fast enough to rid itself of the gravitational energy that
is released by the newly accreted material as it settles into its equilibrium configuration. It is reasonable to conjecture that the structure
of the star remains unstable throughout this phase, and that the resulting
mass motion leads to the acceleration of relativistic electrons.
      The EINSTEIN observatory has in fact been used to find many pre-main

sequence stars with a copious production of X-rays. Typically these sources emit photons with energies in the few keV range; the shock-heated gas
that has just fallen on the star cannot radiate at such high frequencies.
The X-rays must instead be produced by relativistic electrons via the synchrotron process, and so a significant magnetic field must exist in the
star. Such a conclusion fits well with other ideas on the process of
star formation: the presence of magnetic fields in and around protostars
has long been conjectured, for without them it would be difficult for the
infalling material to dispose of its angular momentum.

In all these examples the circumstellar material has come from outside the star. Even the relativistic electrons which emit the X-rays have
gained their energy in regions where the ambient density is comparatively
low, that is well above the photosphere.

Pre-main sequence stars of low mass are also sources of fast jets
which have speeds of 200 to 300 km s^{-1} and are detectable by their Hα
emission. Interesting consequences follow from the speculation that a
physical connection exists between the jets and the X-ray emission.

Let the jet be collimated and the electrons be accelerated in a region with typical linear dimensions R. Let $\dot{M}_J$ be the mass-loss rate into
the jet and V the flow speed. The jet luminosity is then $L \equiv \frac{1}{2}\dot{M}_J V^2$, and
the momentum flux

$$\Pi_* = 2L/VR^2 \tag{1}$$

can be used to scale the magnetic field strength B by the relation

$$B^2/8\pi \equiv \beta\Pi_* , \tag{2}$$

with $\beta < 1$. Some of the mechanical energy of the jet will be tapped by
its interaction with the surrounding medium. A relativistic electron can
then gain energy from the mass motion at the rate

$$W_r \approx \frac{\gamma mc^3}{\ell} \left(\frac{V}{c}\right)^\alpha ; \tag{3}$$

in this formula ℓ is the typical distance between locations where successive reflections take place in the path of the relativistic electron, V
is the typical velocity difference between the gas at these places, and
γ is the Lorentz factor. In conventional Fermi acceleration reflections
take place on eddies with uncorrelated motions, and α equals 2. If the
acceleration takes place in a shock then every collision by a particle is
head on, and the index α equals unity.

The X-ray photons from pre-main sequence stars typically have energy
χ of about 5 keV, which corresponds to an angular frequency $\omega_x \sim 10^{19}$ s^{-1}.
By the usual relations for synchrotron radiation

$$\omega_x \sim \gamma^2 e B/mc \tag{4}$$

and the typical electron loses radiant energy at the rate

$$W_- \sim e^2\omega_x^2/\gamma^2 c . \tag{5}$$

Balancing the gains and losses, as given in equations (3) and (5), leads to the estimate

$$\ell \sim \frac{\gamma^3 mc^4}{\omega_x^2 e^2} \left(\frac{V}{c}\right)^{\alpha} \quad . \tag{6}$$

The orbital radius of a relativistic electron in the magnetic field is typically

$$r_o \sim \frac{\gamma mc^2}{eB} \equiv \frac{\gamma^3 c}{\omega_x} \tag{7}$$

and must be small in comparison with ℓ, so that

$$\omega_x < \frac{mc^3}{e^2} \left(\frac{V}{c}\right)^{\alpha} \quad . \tag{8}$$

The appropriate numerical values here are $mc^3/e^2 = 1.1 \times 10^{23}$ s^{-1}, $\omega_x \sim 10^{19}$ s^{-1}, and $V/c \sim 10^{-3}$, and so relation (8) can be satisfied if $\alpha = 1$, but not if $\alpha = 2$. The classical Fermi process cannot accelerate the electrons fast enough to make them emit the X-rays, but acceleration in shocks could do so. The thickness of the layer within which the acceleration takes place is typically of order ℓ, with $\alpha = 1$. The X-rays must also be generated in this layer, and it follows that the fraction of the available volume involved is of order ℓ/R, so that the total emitting volume is typically ℓR^2.

A constraint on this process is that the pressure due to the fast electrons should not exceed the magnetic pressure, or

$$\frac{1}{3}\gamma nmc^2 < \frac{B^2}{8\pi} = \frac{2\beta L}{VR^2}$$

so that

$$n < \frac{6\beta L}{VR^2 \gamma mc^2} \quad , \tag{9}$$

and the total number of radiating electrons is limited by

$$N = n\ell R^2 < \frac{6\beta L \ell}{\gamma Vmc^2} \quad . \tag{10}$$

The total X-ray luminosity is given by

$$L_x = \frac{Ne^2 \omega_x^2}{\gamma^2 c}$$

and its ratio to the jet luminosity must satisfy

$$\frac{L_x}{L} < \frac{6\beta e^2 \omega_x^2 \ell}{\gamma^3 m c^3 V} .$$

(11)

The path length ℓ is given by relation (6) and on substitution it follows that

$$\frac{L_x}{L} < 6\beta .$$

(12)

From observation it turns out that L_x is typically 10^{30} erg s^{-1} and the jet luminosity L of order 10^{33} erg s^{-1}. The inequality (12) is easily satisfied unless β is very small. Physically this result implies that the hypothetical shocks need to cover only a small fraction of the cross-section of the jet and even so a sufficient number of electrons can be accelerated to yield the required energy output at X-ray frequencies.

A favourite theory has it that jets are formed when a stellar wind is collimated by a circumstellar accretion disk. Another, less popular, view is that the jet originates within the star itself, but the EINSTEIN observations seem to tally better with this mechanism. The X-ray luminosity is observed to fluctuate on a timescale of minutes or hours: suppose then that the jet really is collimated within the star. The flow speed must be larger just above the photosphere than at infinity, since the fluid still has to climb out of the stellar gravitational field. A jet moving at 500 km s^{-1} will advance 150 000 km in five minutes, a reasonable scale for the size of the emitting region on the star itself, but much too small for any feature that can realistically be associated with an accretion disk.

IV.   A JET PUSHING OUT INTO CIRCUMSTELLAR SPACE

A jet is collimated on or close to a star and then has to force it way through the molecular gas in circumstellar space. It is natural to seek a connection between the presence of a jet and the large scale bipolar flows commonly observed near pre-main-sequence stars. Accordingly let the molecular gas be initially at rest, with a density distribution given by

$$\rho_o = \mu/r^2 .$$

(13)

As usual dust will be associated with the gas and so the mixture will absorb and scatter radiation. The typical value of the opacity is taken to be $\kappa_o = 200$ cm^{2}gm^{-1}, at visual wavelengths. The optical depth from far outside to within distance r_o of the star depends on the orientation of the line of sight, and is of order $\kappa_o\mu/r_o$. When μ is too large the optical depth is excessive; since fast jets are observed on typical length scales of 10^{17} cm it follows that μ should not be larger than 5 x 10^{14} gm cm^{-1}. The total mass of molecular hydrogen out to one parsec is thus limited to be less than some 10 $M_\odot$. Observed flow speeds in the molecular gas are typically of order 10 km s^{-1}, so that the energy content is of order 10^{46} erg (and proportionately less if the flow does not extend that far).

For comparison the mass loss rate into a jet is typically 10^{-7} $M_\odot$ per year, so that with a speed of 300 km s^{-1} the luminosity for a jet becomes 3×10^{33} erg s^{-1}.

During its early motion the jet is pictured as carving out a conical space with semi-vertical angle ε, say 0.1 radians, within which the gas flows freely. At the far end it passes through a shock, at distance r from the centre. The density in the jet immediately upstream of the shock is

$$\rho \;=\; \frac{\dot{M}_J}{\varepsilon^2 \pi r^2 V_J} \tag{14}$$

where V_J is jet speed, and the newly shocked gas has a cooling time

$$t_c \;=\; \frac{0.02 V_J^3}{q\rho} \;, \tag{15}$$

where q is the usual cooling parameter. When r is sufficiently small the newly shocked gas stays hot very briefly and occupies only a thin layer just behind the shock. The condition is that

$$\frac{V_J}{4} t_c \;\ll\; \varepsilon r \quad ; \tag{16}$$

with numerical values inserted the right hand side becomes 7×10^{16} cm.

The layer of shocked gas itself drives another shock into the molecular gas at a speed w which is determined by balancing the momentum flow in the jet against the ram pressure in the gas outside; formally

$$\frac{\mu}{r^2} w^2 \;=\; \frac{\dot{M}_J V_J}{\pi \varepsilon^2 r^2}$$

or

$$w \;=\; \frac{1}{\varepsilon} \left(\frac{\dot{M}_J V_J}{\pi \mu} \right)^{\frac{1}{2}} . \tag{17}$$

The speed of advance of the end of the jet is found to be 35 km s^{-1}, for representative numerical values, and with μ set equal to 5×10^{14} gm cm^{-1}. A distance of 7×10^{16} cm is reached after 2×10^{10} s, or 700 years. The energy carried by the jet is substantially converted into radiation just behind the shock, but is unlikely to be observed directly since the effect of circumstellar extinction is large at optical and UV wavelengths for a line of sight that comes so close to the star, and because this phase in the evolution of the jet lasts only a short time.

V.   FOCUSSING THE JET BY EXTERNAL PRESSURE

The very early phase in the growth of the jet ends when the characteristic cooling time of the shocked gas becomes too long. A bubble of hot gas then develops, the jet enters it at radial distance r_0. The pressure $P (\equiv \Pi_0 \eta)$ in the bubble scales in terms of the ram pressure

$$\Pi_0 = \frac{\dot{M}_J V_J}{\pi \varpi_0^2} \qquad (18)$$

at distance r_0, and here $\varpi_0 \equiv \varepsilon r_0$. The sudden exposure to external pressure drives shocks into the body of the jet, with interesting dynamical consequences, but here the first question is how long the newly shocked gas stays hot. The adiabatic parameter κ can once again be expressed in terms of the pressure P just behind the shock and the density

$$\rho = \frac{4\dot{M}_J}{\pi \varpi_0^2 V_J} \qquad (19)$$

there, and is given by

$$\kappa^{3/2} = \frac{\eta^{3/2} \pi}{32} \frac{V_J^4 \varpi_0^2}{\dot{M}_J} \quad . \qquad (20)$$

The shocked gas cools after a time

$$t_c = \frac{\kappa^{3/2}}{q} = \frac{\eta^{3/2} \pi}{32} \frac{V_J^4 \varpi_0^2}{q \dot{M}_J} \qquad (21)$$

The external pressure deflects the flow through an angle of order $\sqrt{\eta}$ and so focusses it towards a constriction at a distance of order $\varpi_0 / (\sqrt{\eta} - \varepsilon)$ from the inlet. The gas there still retains its heat from the first shock if

$$\frac{\varpi_0}{V_J} < (\sqrt{\eta} - \varepsilon) t_c$$

or

$$\frac{(\sqrt{\eta} - \varepsilon) \eta^{3/2} \pi}{32} \frac{V_J^5 \varpi_0}{q \dot{M}_J} > 1 \quad . \qquad (22)$$

With the standard values assumed here the limits on η are as follows, for three different distances r_0

r_o (cm)	3×10^{16}	10^{17}	3×10^{17}
η_c	0.22	0.13	0.075
Δz_c (cm)	8×10^{16}	4×10^{16}	1.7×10^{17}
T_c (K)	2.7×10^5	1.6×10^5	9×10^4

In this table Δz_c denotes the distance between the inlet and the constriction, when $\eta = \eta_c$ and T_c denotes the post-shock temperature of the gas. If η exceeds η_c the gas is still hot when it reaches the constriction. Another shock then deflects the flow outward again. If $\eta_c > \eta > \varepsilon^2$ the shocked gas can cool before it reaches the axis. It will then shock a second time and will again cool off, so that a more compact jet reforms. It is tempting to identify these pockets of cooling gas with Herbig-Haro objects. If the flow remains steady then the resulting HH objects will be stationary, even though the gas is streaming through them at high speed. But observation shows that some HH objects have very large proper motions. They too can be fitted in if this description is altered slightly.   The second shock occurs after the gas flow has converged on the axis. Such flows are generally unstable. It is to be expected that the gas is not evenly heated after passing through the second shock, so that some parts of it will cool better than others. A parcel of gas that is heated too well will then disperse again, and another that is not heated enough will cool efficiently, and as it is swept away down the axis it will be observed as an HH object with a high proper motion.

VI.   JETS AND MOLECULAR FLOWS

As time goes on the jet creates an ever-growing bubble, and the kinetic energy that it carries is thermalised after shocking on a working surface at the far end. The expanding volume of hot gas drives a shock into the surrounding molecular gas, which picks up about a quarter of the energy supplied. Radiative heat loss becomes progressively less significant, and eventually three-quarters of the energy from the jet is retained in the hot gas. The temperature in the bubble is consequently given by

$$\frac{kT}{m} = \frac{1}{4} V_J^2 , \qquad (23)$$

the adiabatic constant of the gas is

$$\kappa = \left(\frac{kT}{m}\right)^{5/3} P^{-2/3} = 0.10 \, V_J^{10/3} \, P^{-2/3}$$

and the cooling time is

$$t_c = \frac{\kappa^{3/2}}{q} = \frac{0.031 V_J^5}{Pq} .$$ (24)

The pressure is defined in terms of the ram pressure in the jet at the point where it enters the bubble (see relation (18)) and thus

$$t_c = \frac{\pi}{32} \frac{\varepsilon^2}{\eta} \frac{V_J^4 r_o^2}{q \dot{M}_J} .$$ (25)

Later in the evolution t_c becomes large compared with the dynamical time t, and the mass of hot gas is

$$M_h = \dot{M}_J t .$$ (26)

The density of the hot gas is

$$\rho_h = \frac{P_h}{kT/m} = \frac{4\eta}{\pi\varepsilon^2} \frac{\dot{M}_J}{r_o^2 V_J} ,$$ (27)

and the bubble occupies a volume

$$V = \frac{\pi\varepsilon^2}{4\eta} r_o^2 V_J t$$ (28)

whose radius is

$$r_h = 0.57 \left(\frac{\varepsilon^2}{\eta}\right)^{1/3} r_o^{2/3} V_J^{1/3} t^{1/3}$$ (29)

if its shape is idealised to be spherical. The centre of the bubble is at distance $r_o + r_h$ from the star, but as time goes on the ratio of r_o to r_h becomes small so that r_o can be neglected. The density in the pre-existing molecular cloud is μ/r_h^2 at distance r_h, and the mass of molecular gas displaced by the bubble is

$$M_m = 2\pi\mu r_h = 3.60\mu \left(\frac{\varepsilon^2}{\eta}\right)^{1/3} r_o^{2/3} V_J^{1/3} t^{1/3} .$$ (30)

The molecular gas picks up a quarter of the energy carried by the jet in time t and so

$$\pi\mu r_h w^2 = \frac{1}{8} \dot{M}_J V_J^2 t$$ (31)

or

$$w = 0.26 \frac{\dot{M}_J^{1/2} V_J^{5/6} t^{1/3}}{r_o^{1/3} \mu^{1/2}}$$ (32)

A typical bubble has a radius of 1.5×10^{18} cm, say, and the associated molecular gas moves at 10 km s^{-1}; with typical numerical values inserted in relation (31) the age of such a structure turns out to be 100 000 years, and the parameter

$$\frac{\varepsilon^2}{\eta} = 0.0090 \frac{\dot{M}_J^3 V_J^3 r_h^6}{\mu^3 w^6 r_o^6} , \tag{33}$$

roughly 9 at that time. The mass of swept-up molecular gas is 2.4 $M_\odot$, the pressure in the bubble and in the shell that surrounds it is 7.9 $\times$ 10^{-11} dyne cm^{-2}, and the hot gas has a characteristic cooling time of 800 000 years, much longer than the age of the bubble. The numerical values seem entirely reasonable, but the primitive description that has been given here does not do justice to the observations in one important respect. Flows of molecular gas are observed to be distinctly bipolar, with mean velocities that point outward from the star in each lobe. The present treatment fails to reproduce this effect because of the oversimplification that was introduced by the assumption that the bubbles are spherical. In reality a bubble will expand more easily into the direction away from the star where the ambient gas has a lower density. Bubble and shell both acquire an elongated shape, and the molecular gas reaches a higher speed outside than inside, so that the mean flow will be outwards. Even so the simple argument that has been used here does establish the physical connection existing between the jets and the surrounding molecular gas. Clearly this whole range of problems deserves a decent dynamical treatment. For some of us this is the next item on the agenda.

VII. CONCLUSIONS

A glance at the list of contents shows that the talks presented at this Symposium covered far more astrophysics than just the events associated with the pre-main-sequence stars. In particular apologies are due to the speakers who presented such beautiful results on stellar coronae and mass loss, on masers and infra-red studies, and on the chemistry of circumstellar matter. Given enough time and inspiration it might be possible to bring all the topics of this meeting together into one summary talk. Perhaps such a synthesis will be achieved by the concluding speaker at the next IAU Symposium on Circumstellar Matter. For the present it only remains for us to thank the Organising Committee for letting us take part in such a feast of talks containing so many new and fascinating results.

Y

Z

SUBJECT INDEX

(Based mainly on paper abstracts)